Endoscopic Laser Surgery Handbook

Science and Practice of Surgery

Consulting Editors

John F. Burke
Benedict Professor of Surgery
Harvard Medical School
Massachusetts General Hospital
Boston, Massachusetts

Peter J. Morris
Nuffield Professor of Surgery
University of Oxford
John Radcliffe Hospital
Oxford, England

1. Handbook of Cryosurgery, *edited by Richard J. Ablin*
2. Prostatic Cancer, *edited by Richard J. Ablin*
3. Transient Ischemic Attacks, *edited by Charles Warlow and Peter J. Morris*
4. Hyperalimentation: A Guide for Clinicians, *edited by Mitchell V. Kaminski, Jr.*
5. Colorectal Cancer: Current Concepts in Diagnosis and Treatment, *edited by Glenn Steele, Jr., and Robert T. Osteen*
6. Reoperative Surgery of the Abdomen, *edited by Donald E. Fry*
7. Handbook of Neurotological Diagnosis, *edited by John W. House and Alec Fitzgerald O'Connor*
8. Hepatic and Biliary Cancer, *edited by Harold J. Wanebo*
9. Perioperative Assessment in Vascular Surgery, *edited by D. Preston Flanigan*
10. Endoscopic Laser Surgery Handbook, *edited by Stanley M. Shapshay*
11. Antibiotic Therapy in Head and Neck Surgery, *edited by Jonas T. Johnson*
12. Surgical Management of Morbid Obesity, *edited by Ward O. Griffen, Jr. and Kenneth J. Printen*
13. Visceral Vascular Surgery, *edited by Alfred V. Persson and Paul A. Skudder, Jr.*
14. Reoperative Vascular Surgery, *edited by Hugh H. Trout, III, Joseph M. Giordano, and Ralph G. DePalma*
15. Trauma, Sepsis, Shock: The Physiological Basis of Treatment, *edited by George H. A. Clowes, Jr.*

Additional Volumes in Preparation

Endoscopic Laser Surgery Handbook

edited by

Stanley M. Shapshay
Lahey Clinic Medical Center
Burlington, and
Boston University School of Medicine
Boston, Massachusetts

MARCEL DEKKER, INC. New York and Basel

Library of Congress Cataloging-in-Publication Data

Endoscopic laser surgery handbook.

(Science and practice of surgery ; 10)
Includes bibliographies and index.
1. Lasers in surgery. 2. Endoscope and endoscopy.
I. Shapshay, Stanley M. II. Series. [DNLM: 1. Endoscopy.
2. Laser Surgery. W1 SC679 v.10 / WO 500 E56]
RD73.L3E53 1987 617'.05 87-6832
ISBN 0-8247-7711-5

MARCEL DEKKER, INC.
270 Madison Avenue, New York, New York 10016

Current printing (last digit):
10 9 8 7 6 5 4 3 2 1

PRINTED IN THE UNITED STATES OF AMERICA

Foreword

This book, by experts in their fields, is concerned primarily with operative endoscopy utilizing lasers. The operating microscope has long been an integral part of surgical procedures in laryngology and some aspects of gynecology. Some physicians turned to lasers to improve the results of existing methods of treatment, others used lasers because traditional operative methods through endoscopic channels were inadequate to help their patients, and still others chose to transform open surgery into endoscopic surgery. Endoscopic removal or devitalization of tissue can be extremely demanding of the surgeon, but the benefits to patients of such minimally invasive surgery are great.

The energy emitted by lasers can be focused sharply at the distal end of narrow endoscopic tubes or transmitted through thin, flexible fibers. Lasers have not decreased the demands on the endoscopist; on the contrary, additional knowledge and skills are imperative. Improved medical care and major alleviation of human suffering are the compensations.

With the progressive availability of practical laser equipment in many wavelengths and a broad range of temporal characteristics, the perfecting of fibers and endoscopes, the development of new endoscopic diagnostic means, and the emergence of new ideas (touched on in several chapters of this book), the future for operative endoscopy with lasers is increasingly promising. I am sure this book will stimulate more endoscopists to participate in this exciting field. It is a great pleasure for me to write a short historical introduction to this book.

The discovery by Maiman in 1960 of the first laser light source, the ruby laser, caused great excitement in the scientific world, including the medical community. Ophthalmologists realized that this extraordinary light source might be useful in ophthalmic surgery, and by 1961 experiments were already under way. By 1964, hundreds of patients had been treated successfully. Of course, ophthalmologists already knew that light could be used to create therapeutic lesions in the retina due to the pioneering work of Meyer-Schwickerat in the early 1950s. In the following years, with the help of the Zeiss Company, he developed a retinal photocoagulator based on a high-power xenon arc light source. This photocoagulator was in worldwide use in the early 1960s. Xenon arc and ruby laser were later replaced by ion lasers. Today, the practice of modern ophthalmology without the benefit of laser technology is inconceivable; ophthalmologists are the largest medical users of lasers, the applications of which are still expanding.

Surgeons in other fields became interested in lasers, and soon numerous experimenters began to study the effect of laser radiation on various organs and on tumors grown in or implanted into animals. In these experiments, the ruby laser was used first and then the neodymium-in-glass laser, which had been discovered in 1961. The radiation of these lasers is poorly absorbed by nonpigmented biological tissue. To obtain appreciable effects on sizable tumors, the energy per pulse and the power of these lasers, which operated only in the pulsed mode, had to be increased to high values. High-energy, high-power pulses led to veritable tissue explosions that propelled viable cells into the surrounding area. This important finding, and the less-than-promising surgical results obtained, brought this line of research to an end (1).

Two lasers of great importance for use in surgery, the argon ion laser and the carbon dioxide (CO_2) laser, both operating in the continuous mode, were discovered in 1964. The first was used primarily in the laboratory, owing to its structural complexity. By 1969, however, basic experimental work with this laser in ophthalmic surgery had been accomplished, and soon argon ion lasers were universally used in this field. The power output of the CO_2 laser was at first measured in milliwatts. By 1965, with the addition of nitrogen and helium gases to carbon dioxide, the continuous-wave power output had reached the then unprecedented value of 50 W/m. My colleagues and I, in the research laboratories of the American Optical Corporation, had a CO_2 laser operating on a laboratory workbench. Early in 1966, Yahr, a senior surgical resident at Montefiore Hospital in New York, started surgical experiments with this laser and rapidly recognized its potential surgical utility. In 1966, Yahr and Strully (2) wrote: "Using a nitrogen carbon-dioxide laser . . . clean dry skin incisions which heal as well or better than scalpel controls were fashioned. Bone, liver, kidney, and lung can be cleanly divided . . . . The future medical applications of laser energy are many . . . . One may consider some of the following uses . . . . Many brain tumors, and some congenital defects, block the outflow of the cerebrospinal fluid . . . . A flexible bundle carrying both sight and laser could be introduced into one of the cavities . . . and a simple decompression . . . operation performed as an initial step in the patient's treatment . . . . Laser beams delivered through flexible bundles could be employed to eliminate or fracture stones blocking important biologic duct systems . . . . It appears probable that the nitrogen carbon-dioxide laser will . . . make rapid skin incisions, divide certain organs without blood loss, and . . . anastomose vessels as an adjunct to standard techniques." That was two decades ago. Yahr's and Strully's visions of the many potential uses of lasers in surgery appear uncannily prophetic. What impressed us most at the time was that a "light scalpel" had been found and that surgery with this scalpel seemed "bloodless."

The findings and the enthusiasm of Yahr and Strully led to the development of the first CO_2 laser system for surgery. It became available to surgical experimenters early in 1967. It was a self-contained, movable system requiring connections to a standard electrical power outlet and the water main for cooling the laser tube. An articulated arm with seven mirrors located in rotary joints conveyed the laser beam to a focusing handpiece, freely movable in space (3). Soon numerous surgical investigations were started with this type of system. These studies were all centered on applications where blood loss per se was a major problem: gastric ulcers, debridement of burns, and surgery of the liver, kidney, and brain. Much as the alchemists' search for the philosophical stone contributed immensely to chemistry, so these researchers revealed many important characteristics of the interaction of CO_2 laser beams with biological tissue. With the focused beam of this laser, the surgeon could do everything that could be done with a scalpel and more, such as vaporizing tissue volume, making incisions in tissues without manipulating them, and removing by vaporization thin tissue layers with minimal damage to the underlying tissue. Wound healing was studied, and the often benign nature of the surgical wound was established. However, when the transition was made from small laboratory animals to larger ones or to humans, the hoped-for dramatic hemostatic properties did not materialize. Hemostasis with the CO_2 laser beam was "relative," limited essentially to capillary bleeding. These studies continued well into the 1970s.

In 1967, Jako used the focused CO_2 laser beam to produce a discrete lesion on the vocal cord of a cadaver larynx. This use appeared so promising that he urged my group to develop a device that would permit him to expose the canine vocal cord in vivo to the laser beam. The device developed by Bredemeier in our laboratory was an approximately 6″ × 6″ × 6″ box attached to the distal end of the articulated beam-manipulating arm. The laser beam was deflected by 90° in this box and focused at the distal end of a metal tube 30 cm long with a 1-cm diameter. From the opposite side of the box to which the tube was attached, the effects of the focused laser beam could be observed with ×2 or ×3 magnification. Means were provided to align the axis of the beam to be collinear with that of the tube (4). After performing extensive surgical experiments on canine vocal cords with this device, Jako became convinced the focused CO_2 laser beam was a superior modality for endoscopic laryngeal microsurgery. Bleeding was absent or minimal when operating on the vocal cords and on most tissues of the oral cavity.

For clinical applications, the CO_2 laser was combined with the operating microscope (5). The laser beam was projected through the laryngoscope and focused on the vocal cords or some other region of the oral cavity. The focal point was positioned with precision within the field of view of the operating microscope by means of the fine motions of a mechanical micromanipulator (Bredemeier HC: Stereo Laser Endoscope, U.S. Patent 3,796,220, 1972; issued 1974). Clinical use of this instrumentation was undertaken in 1971 by Strong and Jako at Boston

University School of Medicine. With continued work, the advantages of this modality became apparent. Particularly dramatic was the improvement in treatment of juvenile papillomatosis. Soon the work at Boston University was confirmed by Andrews at the University of Illinois Medical School and by French at the Eye, Ear, Nose, and Throat Hospital in New Orleans.

In 1973, two of Strong's patients presented with peristomal papillomas and two with papillomas extending into the right main bronchus. The endoscopic attachment developed 5 years earlier for the first in vivo experiments on canine larynges, and never intended for clinical work, was brought into use and fitted with a standard ventilating bronchoscope 30 cm long with a 5-mm diameter in our laboratory at American Optical Corporation (Wallace RA, Pejchar J: Endoscopic Surgical Laser System, U.S. Patent 3,906,953, 1974; issued 1975). This instrumentation was successfully used to vaporize the lesions. Strong and colleagues (6) noted in a 1973 paper: "This application of the laser may prove to be useful to other lesions of the trachea and bronchi, such as adenomas, or for palliation of carcinoma causing obstruction at the carina." The obvious importance of this unique application of lasers led rapidly to the development of a more refined instrument for bronchial surgery, using the CO_2 laser (7).

By 1975, close to 1000 endoscopic laryngeal and bronchial laser operations had been performed at the three centers mentioned earlier. Strong (8) reviewed this combined experience in an address to a plenary meeting of the American Academy of Ophthalmology and Otolaryngology in September 1975 and concluded: "The advantages of laser surgery in otolaryngology are so significant that it can be recommended for continued usage. Many applications of this unique surgical instrument have been identified but others need to be explored in the future."

The favorable clinical results, which started to accumulate in laryngeal microsurgery and bronchial surgery, helped to clarify the role of the CO_2 laser in surgery—precision removal of tissue under the visual control of an operating microscope and tissue removal in remote body locations accessible through rigid endoscopic tubes. The application of the CO_2 laser to treatment of cervical and vaginal intraepithelial neoplasia, today the widest surgical application of this laser, was directly inspired by work in laryngeal microsurgery. Indeed, the first such operations were performed by Bellina and French in 1974 at Eye, Ear, Nose, and Throat Hospital in New Orleans (9), and by Cibley and Strong in 1975 at the University Hospital of Boston University School of Medicine.

In the early 1970s, the argon ion and the neodymium:yttrium-aluminum-garnet (Nd:YAG) lasers had become available commercially as practical devices. It was well known to physicists that, in principle, materials existed that were capable of being formed into flexible fibers that could transmit the radiation emitted by these lasers practically without attenuation. In practice, this was a major materials research project. It was undertaken by Nath, a physicist, and his colleagues at the Technical University in Munich, Germany. Nath was associated with Kiefhaber at the University of Munich, who was searching for improved means to treat gastric bleeding. This group was aware of the work in endoscopic coagulation of gastric bleeding with laser radiation undertaken by Goodale in 1969 at the University of Minnesota. Goodale used the CO_2 laser system with a rigid gastroscopic tube fitted to the same attachment first used by Jako in surgery of canine vocal cords. At the distal end of the endoscopic tube a retractable 45° mirror projected an unfocused laser beam at 90° to the axis of the gastroscope. The promising results of this work were published by Goodale's group in 1970 (10). At that time, however, rigid gastroscopes were no longer being used clinically, and flexible fibers to transmit the radiation of the CO_2 laser did not exist. Nath developed the flexible fibers capable of transmitting the high-power radiation of argon ion and Nd:YAG lasers. In 1973, Nath and his co-workers (11) demonstrated reliable transmission of up to 10 W of ion laser radiation through a 2.3 m fiber that could be threaded into the biopsy channel of an Olympus gastroscope (American Cystoscopic Makers, Inc., Stamford, Connecticut 06904). Experimental work on treating gastrointestinal tract bleeding with ion laser radiation was soon started by several groups in Germany and in the United States. Selected patients were treated, the first one by Dwyer in the United States.

Kiefhaber in 1975 compared the hemostatic properties of Nd:YAG and ion lasers in the gastrointestinal tract. The Nd:YAG laser system he used had been developed for medical applications by Messerschmitt-Bolkow-Blohm Corporation in Munich, Germany. He was able to obtain hemostasis of the most severe gastrointestinal tract bleeding with high-power Nd:YAG laser radiation. By the end of 1976 he had been able to control 131 of 140 bleeding episodes in 106 patients presenting for emergency endoscopy. He and his associates (12) published these results in 1977, and the technique was rapidly introduced into many centers.

Through Kiefhaber's work it became clear that the radiation of the Nd:YAG laser had hemostatic properties that were far superior to those of the CO_2 and ion lasers because the radiation of the Nd:YAG laser penetrates several millimeters deep into human tissue. Thus, it was also apparent that radiation from this laser could be used to coagulate sizable tissue volumes endoscopically. Soon this property of the Nd:YAG laser was utilized by other researchers, notably Brunetaud in France, Bown in England, and Dwyer and Fleischer in the United States, for the endoscopic coagulation of benign and malignant lesions in the entire gastrointestinal tract.

Endoscopic treatment of bronchial lesions with the Nd:YAG laser followed. At the meeting of the Société Français de la Tuberculose in December 1978, Toty (13) of the Foch Hospital near Paris reported the promising clinical results he had obtained in a small number of patients by endoscopic treatment of bronchial stenosis using radiation from a high-power Nd:YAG laser transmitted through a flexible fiber. He ended this early report by expressing the hope that his results would be as consistently favorable as those obtained with the CO_2 laser in the glottic region by Freche at the same hospital. [Freche (14) had introduced laryngeal microsurgery with the CO_2 laser into Europe.] Toty's hopes were fully justified, as shown by his subsequent work and that of other pioneers in this field: Dumon, Shapshay, and Unger.

Thus, the decade of the 1970s, which had opened with only a handful of investigators searching for nonophthalmologic medical applications of lasers despite disillusionment and failures, closed with legions of patients who had benefited from improved or unique forms of treatment based on use of lasers.

In the 1980s, more researchers became involved, and the indications were extended and the techniques refined. New laser devices, novel instrumentation, and new ideas, such as endoscopic photodynamic therapy and laser angioplasty, may lead to additional dramatic improvements in medical care by the end of this decade. Even the dream of bloodless surgery may become reality. As the medical historian Castiglione (15) so beautifully stated, "the past is never past, but continues and is very active in every form and every manifestation of the present."

Thomas G. Polanyi, Ph.D.
Chief Scientist
Hospital Division
Cooper LaserSonics, Inc.
Marlboro, Massachusetts

Adjunct Professor of
Otolaryngology
Boston University
School of Medicine
Boston, Massachusetts

REFERENCES

1. Ketcham AS, Hoye RC, and Riggle GC: A surgeon's appraisal of the laser. *Surg. Clin. North Am. 47*:1249-1263, 1967.
2. Yahr WZ and Strully KT: Blood vessel anastomosis by laser and other biomedical applications. *J. Assoc. Adv. Med. Instr. 1*:28-31, 1966.
3. Polanyi TG and Bredemeier HC: CO_2 Laser zur experimentellen Chirurgie. *Laser u. Angewandte Strahlentechnik 4*:58-60, 1969.
4. Polanyi TG, Bredemeier HC, and Davis TW, Jr: A CO_2 laser for surgical research. *Med. Biol. Eng. 8*:541-548, 1970.
5. Jako GJ: Laser surgery of the vocal cords: An experimental study with carbon dioxide lasers on dogs. *Laryngoscope 82*:2204-2216 (1972).
6. Strong MS, Jako GJ, Polanyi TG, and Wallace RA: Laser surgery in the aerodigestive tract. *Am. J. Surg. 126*:529-533, 1973.
7. Strong MS, Vaughan CW, Polanyi TG, and Wallace R: Bronchoscopic carbon dioxide laser surgery. *Ann. Otol. Rhinol. Laryngol. 83*:769-776, 1974.

8. Strong MS, Jako GJ, Vaughan CW, et al.: Use of the CO_2 laser in otolaryngology: A progress report. *Trans. Am. Acad. Ophthalmol. Otolaryngol. 82*: 595-602, 1976.
9. Bellina JH: Gynecology and the laser. *Contemp. Obstet. Gynecol. 4*:24-34, 1974.
10. Goodale RL, Okada A, Gonzales R, Borner JW, Edlich RF, and Wangensteen OH: Rapid endoscopic control of bleeding gastric erosions by laser radiation. *Arch. Surg. 101*:211-214, 1970.
11. Nath G, Gorisch W, Kreitmar A, et al.: Transmission of a powerful argon laser beam through a fiberoptic flexible gastroscope for operative gastroscopy. *Endoscopy 5*:213-215, 1973.
12. Kiefhaber P, Nath G, and Moritz K: Endoscopic control of massive gastrointestinal hemorrhage by irradiation with a high-power neodymium YAG laser. *Prog. Surg. 15*:140-155, 1977.
13. Toty L, Personne C, Hertzog P, et al.: Utilisation d'un faisceau laser (YAG) à conducteur souple, pour le traitement endoscopique de certaines lésions trachéo-bronchique. *Rev. Fr. Mal. Respir. 7*:57-60, 1979.
14. Freche C: Etat actuel et limite de la chirurgie au laser CO_2 et perspectives d'avenir. *J. Fr. Otorhinolaryngol. 33*:221-225, 1984.
15. Castiglione A: *A History of Medicine.* Aronson, New York, 1973.

Preface

My interest in the application of lasers in medicine and, in particular, in otolaryngology began during my residency (1972-1975) at the Boston University Medical Center under the direction of my mentor, M. Stuart Strong. This was an exciting time for me because the CO_2 laser, introduced by Jako and Strong, was rapidly changing the practice of laryngology for both adults and children. Although not initially well accepted by the medical establishment, this precision laser has become the instrument of choice for the management of many laryngeal and some tracheobronchial disorders. Improvements in laser technology and delivery systems, as well as the discovery of new indications for laser applications, further advanced the use of lasers in many specialties. Because of the generous support of the Eleanor Naylor Dana Charitable Trust with the establishment of the Laser Research Laboratory at the Lahey Clinic Medical Center, my work has continued, particularly in endoscopic laser applications with medical fiberoptics.

Endoscopic therapeutics has become an important concept in otolaryngology, gynecology, gastroenterology, pulmonary medicine, and, recently, cardiovascular medicine and surgery. The endoscopic application of laser technology is a natural addition to diagnostic fiberoptic instruments that are now a mainstay in many specialties.

When Marcel Dekker, Inc., asked me to edit a textbook on lasers in medicine, my suggestion was to limit this endeavor to endoscopic applications with lasers. The extensive chapter on laser physics and soft-tissue applications by Fisher begins the text in a technical vein. It is a comprehensive, clear, definitive reference and an excellent source of information. Subsequent chapters, written by knowledgeable leaders in their respective specialties, are clinical presentations. Of particular interest is Isner's chapter on the cardiovascular applications of laser technology; this has become a major research effort worldwide. New concepts of cancer therapy, such as photodynamic therapy, are included because of the laser's endoscopic application. A discussion of potential complications and how to start a laser center are also included.

I wish to thank all of the contributors to this volume for their hard work in the preparation of their excellent chapters, representing the state of the art in the application of lasers in medicine. Special thanks go to Suzanne Setzer and Kaye Newman in our laser laboratory. Eileen Rahamim, our department secretary, was extremely helpful in organizing the project and in contacting the contributors. The cooperation of all other support services at the Lahey Clinic, especially the photographic department, is always appreciated. Most important, much appreciation is extended to my wife Ruth and to my children Sandy and Mara for their continuing support of all my efforts.

It has been an honor and a pleasure for me to edit this textbook on the endoscopic applications of lasers in medicine. I hope the reader will find this book useful, practical, and most of all, stimulating.

Stanley M. Shapshay

Contents

Foreword (*Thomas G. Polanyi, Ph.D.*) iii
Preface xi
Contributors xv
Introduction (*Stanley M. Shapshay*) xvii

1 Basic Laser Physics and Interaction of Laser Light with Soft Tissue 1
John C. Fisher

2 Operating Characteristics of Surgical Lasers and Delivery Systems 131
Terry A. Fuller

3 Laser Surgery of the Oral Cavity and Oropharynx 151
R. Kim Davis

4 Laser Treatment of the Nasal Cavities and Nasopharynx 163
James L. Parkin

5 Laser Surgery in the Larynx 171
Geza J. Jako

6 Applications of the Neodymium:Yttrium-Aluminum-Garnet Laser in Bronchology 189
Jean-Francois Dumon

7 Tracheobronchial Tree: Carbon Dioxide Laser 209
Robert H. Ossoff

8 Bronchoscopic Phototherapy with Hematoporphyrin Derivative and Argon Dye Laser 219
Denis A. Cortese

9 Laser Therapy for the Upper Gastrointestinal Tract 233
David Fleischer

10 Laser Therapy for the Lower Gastrointestinal Tract 255
Stephen N. Joffe

11 Lasers in Hepatobiliary and Pancreatic Surgery 281
Stephen N. Joffe and M. Y. Sankar

12 Laser Surgery for the Cervix and Vagina 301
V. C. Wright

13 Endoscopic Laser Surgery in Gynecology 325
David S. McLaughlin

14 Clinical Applications of Neodymium:Yttrium-Aluminum-Garnet Laser Surgery in Urology 355
John A. Libertino and John F. Cyr

15 Laser Surgery in Cardiovascular Disease 369
Jeffrey M. Isner and Richard H. Clarke

16 Orthopedic Surgery 397
Terry L. Whipple

17 Photodynamic Therapy 423
Thomas J. Dougherty

18 Laser Surgery in the Pediatric Airway 453
Gerald B. Healy

19 Complications of Laser Surgery 463
Marvin P. Fried and Louis Burke

20 Establishment of a Laser Center 475
John F. Beamis, Jr., and Nancy Rizzo

Index 483

Contributors

John F. Beamis, Jr., M.D. Lahey Clinic Medical Center, Burlington, Massachusetts

Louis Burke, M.D. Harvard Medical School, Boston, Massachusetts

Richard H. Clarke, Ph.D. Boston University and Tufts Medical School, Boston, Massachusetts

Denis A. Cortese, M.D. Mayo Clinic and Mayo Foundation and the Mayo Medical School, Rochester, Minnesota

John F. Cyr, B.S., P.A. Wakefield, Massachusetts

R. Kim Davis, M.D. University of Utah School of Medicine, Salt Lake City, Utah

Thomas J. Dougherty, Ph.D. Roswell Park Memorial Institute, Buffalo, New York

Jean-Francois Dumon, M.D., F.C.C.P. Salvator Hospital, Marseilles, France

John C. Fisher, Sc.D. St. Barnabas Medical Center, Livingston, New Jersey, St. Luke's Hospital, Milwaukee, Wisconsin, and University of Cincinnati, Cincinnati, Ohio

David Fleischer, M.D. Georgetown University Hospital, Washington, D.C.

Marvin P. Fried, M.D. Harvard Medical School, Boston, Massachusetts

Terry A. Fuller, Ph.D. Fuller Research Corporation, Vernon Hills, and Northwestern University, Evanston, Illinois

Gerald B. Healy, M.D. The Children's Hospital, Boston, Massachusetts

Jeffrey M. Isner, M.D. Tufts Medical School and New England Medical Center, Boston, Massachusetts

Geza J. Jako, M.D. Boston University School of Medicine, Boston, and Melrose Hospital, Melrose, Massachusetts

Stephen N. Joffe, M.D. University of Cincinnati Medical Center, Cincinnati, Ohio

John A. Libertino, M.D. Lahey Clinic Medical Center, Burlington, Massachusetts

David S. McLaughlin, M.D. Humana Women's Hospital, Indianapolis Fertility Center, Inc., Indianapolis, Indiana

Robert H. Ossoff, D.M.D., M.D. Vanderbilt University Medical School, Nashville, Tennessee

James L. Parkin, M.D. University of Utah School of Medicine, Salt Lake City, Utah

Nancy Rizzo, M.S. Lahey Clinic Medical Center, Burlington, Massachusetts

M. Y. Sankar, M.D. University of Cincinnati Medical Center, Cincinnati, Ohio

Terry L. Whipple, M.D. Orthopedic Research of Virginia, Medical College of Virginia, University of Virginia School of Medicine, and Humana Hospital-St. Lukes Center for Orthopedics, Richmond, Virginia

V. C. Wright, M.D. The University of Western Ontario, London, Ontario, Canada

Introduction

The routine application of laser technology in medicine and surgery has gradually become an accepted therapeutic modality over the past 15 years. The application of various lasers to different specialities has stimulated research into newer laser wavelengths and delivery systems. The *Endoscopic Laser Surgery Handbook* is specifically directed toward endoscopic application because in my opinion this is one of the most important applications of laser technology in medicine. The use of lasers in ophthalmology, although of great importance, has been excluded from discussion because this application requires separate handling and is not considered to be of an endoscopic nature. Endoscopy is strictly defined as examination of the interior of a canal or hollow viscus. Laser applications in dermatology and plastic surgery are dealt with in other texts.

I. THERAPEUTIC ENDOSCOPY

The term therapeutic endoscopy, a buzzword of the 1980s, has taken on special importance for medical specialists in particular. With the development of flexible fiberoptics for endoscopy in gastroenterology and pulmonary medicine, more sophisticated endoscopy could be performed on an outpatient basis with the patient under local anesthesia and with less morbidity. For example, bronchoscopy carried out with the rigid instrument was the domain of bronchoscopists, such as otolaryngologists trained in the Chevalier Jackson School of Endoscopy (Temple University School of Medicine, Philadelphia) and thoracic surgeons. Similarly, endoscopy of the esophagus was usually performed by surgeons using rigid esophagoscopes, usually without any telescopic magnification. With the introduction of the flexible fiberoptic endoscopes, procedure-oriented internists began to flourish. Endoscopic application of electrocautery and cryosurgery probes became commonplace. In addition, flexible forceps for use in specialized diagnostic procedures and for removal of foreign bodies were developed for endoscopic use. Fiberoptic capabilities for laser technology have recently extended this diagnostic capability into the field of therapeutic applications. In gastroenterology and gastrointestinal surgery, a virtual renaissance of interest has occurred in applying therapy through endoscopes. Examples of this innovative approach are the use of both argon and neodymium:yttrium-aluminum-garnet (Nd:YAG) lasers through gastroscopes and colonoscopes for removal of tissue (e.g., benign polyps), for the termination of gastrointestinal bleeding, and for photoablation of vascular lesions such as telangiectasia and angiodysplasia. Reports of the use of the Nd:YAG laser for palliation of malignant biliary obstruction have appeared in which the laser was introduced through a percutaneous catheter tract or intraoperatively through a choledochoscope. Palliation of obstructing gastrointestinal malignant disease through an endoscopic approach has met with great success when used by experienced endoscopists trained in laser techniques. When laser procedures are successful, more-dangerous external palliative approaches have been avoided. In pulmonary medicine and other specialties in which bronchoscopy is used (thoracic surgery and otolaryngology), both the carbon dioxide (CO_2) and the Nd:YAG lasers have been used through rigid (CO_2 and Nd:YAG) and flexible (Nd:YAG) bronchoscopes for palliation of obstructing tracheobronchial malignant disease, relief of benign stenosis, removal of benign neoplasms, and other selected indications. Over the past 10 years, the techniques of laser bronchoscopy have been refined, and successful results have been duplicated by many investigators throughout the world.

Recent examples of therapeutic endoscopy utilizing laser technology have occurred in urology, orthopedic surgery, and vascular medicine and surgery. Cystoscopy and transurethral resection performed with rigid instrumentation have become routine procedures for urologists but only recently has the laser been added to the endoscopic equipment. Pioneering work by Hofstetter and co-workers (1) treating early tumors in the bladder with Nd:YAG laser photocoagulation has been duplicated by other investigators. The major unanswered questions are: is this relatively expensive technology superior to conventional transurethral techniques utilizing electrocautery; and will this approach obviate the necessity for performing more aggressive external procedures, such as cystectomy? Careful pathologic evaluation of specimens taken from the bladder after laser application and follow-up data will be necessary to answer these questions.

Specialists in otolaryngology and gynecology have led the way in applying laser technology in endoscopic procedures. The CO_2 laser was used through a microlaryngeal approach by Strong and Jako (2) as early as 1972. Great precision was made possible when the operating microscope was wed to the CO_2 laser. Starting with benign abnormalities, such as tracheobronchial papillomatosis, ordinary polyps, and nodules, and progressing to early malignant disease, microlaryngeal laser techniques have revolutionized the field of laryngology. The most exciting evidence comes from a number of investigators who have proved the usefulness and safety of endoscopic laser excision of early T1 cancer of the vocal cord. Cancer control rates at three-year follow-up study are equal to those of radiotherapy with much less overall morbidity. Application of the CO_2 laser in the oral cavity and oropharynx has also been shown to be advantageous when compared with conventional techniques, i.e., cryosurgery and electrocautery, because of more precise soft-tissue interaction, leading to less postoperative edema and pain and to uncomplicated wound healing. Gynecologists have similarly applied the CO_2 laser with microscopic precision through colposcopes for treatment of cervical cancer and dysplasia in an early stage (3). Although gynecologists continue to debate whether the laser is superior to other techniques, such as cryosurgery and electrocautery, much information is accumulating to substantiate the effectiveness of laser endoscopy both in cervical and vaginal diseases.

Arthroscopic and intravascular applications of laser technology are areas of exciting investigation in therapeutic endoscopy. The state-of-the-art technology in these two areas is presented in this text and holds great promise for the future. Therapeutic endoscopy of various vascular systems (cerebral, coronary, and peripherovascular) awaits newer developments in fiberoptic delivery systems and perhaps more precise laser wavelengths. At the present time, removal of arterial plaque is possible, but perforation of the vessel has been a problem. The potential of endoscopic manipulation of vascular disease through a percutaneous approach, obviating the greater morbidity associated with open procedures, is an exciting new area in medicine, attracting our best research efforts.

Two chapters are included on the endoscopic application of the interaction of injectable dyes and dye-laser technology in the field of bronchology. This subject, an exciting area of investigation, was pioneered by Dr. Thomas Dougherty, who presents an elegant concept of selective death of cancer cells through a photochemical interaction. This concept applies to a number of different medical and surgical specialties but is most impressive, in my opinion, in the area of endoscopy. The endoscopic application of photodynamic therapy in the treatment of cancers of the lung and bladder is leading the areas of investigation. The major shortcoming thus far has been phototoxic side effects to the patient and problems with dosimetry, i.e., getting enough light to the photosensitized cancer.

Although many applications of lasers in medicine and surgery employing lasers through external delivery systems are excellent, some are computerized and robotic in nature, and they are the subject of several other textbooks. The combination of lasers and sophisticated fiberoptic endoscopic delivery systems is an ideal marriage for these technologies. Endoscopy, in general, obviates the morbidity usually associated with an external approach. Without an external wound to heal, endoscopic therapy permits quicker recovery of the patient, less pain, less disturbance to normal body functions, and, most important to health care planners and to the economy, a shorter hospital stay. Cost-effectiveness is best fulfilled by the concept of therapeutic endoscopy. The laser has a unique role to play in this form of therapy, as it is useful in diagnosis and for photocoagulation, thermal ablation, and photostimulation of injected dyes.

II. WHY USE A LASER?

A general basic criterion for utilization of laser technology is whether the laser provides an advantage over conventional modalities. If electrocautery, cryosurgical probes, or other more conventional, modalities can be used with appreciably less expense, the laser is not needed. The investigator must prove that the laser has a decided advantage over other techniques if it is to be accepted. Potential advantages provided by the laser to the endoscopist are: standard endoscopes can be used for certain laser wavelengths; the surgeon has a choice of multiple wavelengths with particular absorption by various types of tissues; unlike conventional electrocautery, the laser has a more defined interaction with soft tissue; contact with tissue is unnecessary and therefore the endoscopist's vision is unimpaired; the operator can vaporize or ablate tissue or coagulate or incise by varying the power and time exposure of the application; tissue edema and postoperative pain are lessened; and destruction of surrounding tissue is decreased, thus reducing the risk of infectious complications.

These advantages must be weighed against higher cost to the patient and to the health facility and the requirements of the complex maintenance of an evolving technology. The special training of laser endoscopists must also be taken into consideration.

III. TRAINING THE LASER ENDOSCOPIST

The utilization of laser technology through an endoscopic approach requires a skillful endoscopist. Therapeutic endoscopy, particularly when dealing with pathologic anatomy, can be a challenging procedure. The clinician must be an experienced endoscopist in his or her specialty before learning to use laser technology. Special courses and training, which should include basic laser physics and soft tissue application as well as safety features, should be part of the minimal requirements. These laser endoscopic courses should include a two-day training program and a minimum of four hours of hands-on experience. Ideally, the prospective laser endoscopist should spend time working with experienced laser endoscopists applying the laser in the clinical setting. In addition, the new endoscopists should be supervised closely, particularly during the first several procedures. The establishment of a laser surgery unit in the endoscopist's hospital and a laser committee is extremely important and is the subject of a separate chapter. The committee or task force on laser safety is concerned with accrediting personnel, maintenance of laser equipment, safety, and continuing education.

IV. PRESENT PROBLEMS WITH EXISTING LASER TECHNOLOGY

Problems with dependability and excessive or frequent downtime seem to be improving. Articulating arms, presently necessary for CO_2 lasers until fibers are available, are often jarred and put out of alignment. Miniaturization and greater portability are needed to facilitate use by multiple clinical departments in different physical locations. Improved delivery systems with focusing handpieces and different lens adapters for free fibers would increase overall utilization. To borrow terminology from the computer field, laser technology should be "user-friendly" so that trained assistants can easily operate the equipment. Finally, in this field of rapidly evolving and changing technology, expense must be minimized because today's state-of-the-art laser might be suitable for the antique room in the science museum tomorrow.

V. FUTURE APPLICATIONS

Advances in the medical application of lasers depend to a great extent on a close working relationship between clinicians and laser engineers. The clinician should be able to communicate problem areas encountered in endoscopic therapy to the engineer. The need for a change in delivery systems or laser wavelengths or both should stimulate avenues of research and development. All too often, the laser industry creates a technology and then asks the clinician to apply it in some fashion. Clinical input should be incorporated into the developmental process before the product is completed. The present machines, CO_2, argon, Nd:YAG, ultraviolet, and dye lasers, have certain advantages and disadvantages. For example, the CO_2 laser has the most precise soft-tissue interaction but has poor hemostatic properties (microcirculation-size vessels). The Nd:YAG laser, on the other hand, has excellent photocoagulation properties for hemostasis but lacks the precision and cutting qualities of the CO_2 laser. Combining or blending wavelengths might solve this problem as would the discovery of a new wavelength that would combine both properties. In addition, at the present time the CO_2 laser cannot be transmitted through an efficient fiberoptic system, limiting its usefulness to a more clumsy articulating arm-mirror system. A tunable laser system, such as the free-electron type, might well be the answer to this problem and awaits future development. Ideally, a single-fiberoptic flexible-catheter system that could handle all the available laser wavelengths would be a great advantage as well. Special laser endoscopes small enough to carry fibers into vessels the size of coronary arteries would enable the endoscopist to control visually the access of laser radiation.

In oncologic endoscopy, development of more tumor-specific and less phototoxic dyes would permit precise laser photoactivation for cancer cell death. The development of more efficient, penetrant wavelengths with absorption spectra for the injected dyes would greatly facilitate a more precise dosimetry, so crucial for the effectiveness of the system. The addition of computer technology to this photoactivation process would also assist the clinician. The creation of prospective randomized protocols utilizing standard modalities of cancer therapies, such as radiation and chemotherapy, would help establish the effectiveness of this treatment. Careful scientific scrutiny with randomized protocols using dyes or dye-laser therapy is clearly needed to determine whether photoradiation therapy has a significant place in cancer therapy.

Another potential future endoscopic approach is the endoscopic "welding" of tissues. The ability of lasers to weld tissues together, such as with a milliwatt CO_2 laser, may have important applications not only for the obvious microanastomosis of blood vessels, nerves, and reproductive tubes but also for the approximation of tissue grafts, such as grafting of tympanic membrane and placement of mucosal grafts in such hard-to-reach places as the pharynx and trachea.

Laser technology may play an important role in diagnostic medicine when combined with computer programming and video technology. Laser stimulation of injected dyes and spectrophotometry of tissue for matching of wavelengths are promising areas of future research. With the advent of microchip types of cameras, photographic documentation is possible by placing the camera on the distal end of the endoscope.

In summary, combining newer imaging approaches, such as magnetic resonance imaging, computer programming, and advanced fiberoptic transmissions of laser wavelengths, promises exciting possibilities. Just as robotics revolutionized many industrial plants, advanced technology should aid the clinician in delivering better care to patients. Laser technology will not replace many types of conventional therapies but offers certain definite advantages in many areas of medicine and surgery. The laser is much more than a unidimensional generic technology, such as cryotherapy, because it offers a myriad of different wavelengths, each with its unique soft-tissue interaction.

The purpose of this *Endoscopic Laser Surgery Handbook* is to present proved and future endoscopic applications of lasers in medicine. A thorough understanding of basic laser physics and soft tissue interaction is obviously the framework for this understanding, and an extensive presentation of this technology is given in Chapter 1. The use of lasers in medicine is an evolving technology, and this textbook provides the necessary background and state-of-the-art coverage.

Stanley M. Shapshay

REFERENCES

1. Hofstetter A, Frank F: *The Neodymium-YAG Laser in Urology*. Hoffmann-LaRoche Inc., Basel, Switzerland, 1980.
2. Strong MS and Jako GJ: Laser surgery in the larynx: Early clinical experience with continuous carbon dioxide laser. *Ann. Otol. Rhinol. Laryngol.* *81*:791-798, 1972.
3. Baggish MS: High-power–density carbon dioxide laser therapy for early cervical neoplasia. *Am. J. Obstet. Gynecol.* *136*:117-125, 1980.

1

Basic Laser Physics and Interaction of Laser Light with Soft Tissue

JOHN C. FISHER
St. Barnabas Medical Center, Livingston, New Jersey, St. Luke's Hospital, Milwaukee, Wisconsin, and University of Cincinnati, Cincinnati, Ohio

I. INTRODUCTION

The term *laser* is an acronym composed of the first letters of the words *l*ight *a*mplification by *s*timulated *e*mission of *r*adiation. Physically speaking, *radiation* is the most basic of these terms. The others are concerned with the mechanism by which lasers produce radiation. We shall discuss these terms in some detail in the material that follows. Most of the discussion will be nonmathematical, resorting to mathematics only when it is important to a basic understanding of laser physics. The statements made are presented without rigorous proof, not to burden the medical reader with a clutter of physical details. By its very nature, an exposition such as this can be neither exhaustive nor encyclopedic. We shall concern ourselves in this chapter with the practical details of only those types of laser most commonly used in surgery today, namely the argon (Ar) ion, the carbon dioxide (CO_2), and the neodymium (Nd) yttrium-aluminum-garnet (YAG). However, comprehension of the principles of these lasers will readily enable the reader to grasp the basics of all other lasers currently used in medical applications.

Sections I, II, and III of this chapter deal with the basic physics and technology of lasers, with specific reference to the Ar, CO_2, and Nd:YAG. Section IV is concerned with the interaction of laser radiation with living tissue, a subject that the practitioner must understand to use lasers safely and effectively in treating patients. In the title of this chapter, the word *light* is used to describe the output of lasers. Strictly speaking, light is only that portion of the electromagnetic spectrum lying in the range of wavelengths that our eyes can perceive. However for our purposes in the discussion throughout this chapter, the term *light* will be used interchangeably with the term *radiation* to describe any radiation of wavelength shorter than 0.1 mm but longer than 10 nm (1×10^{-8} m). In the range of wavelengths longer than our arbitrary span of light we find microwaves and radio waves; in the span of shorter wavelengths we have x-rays, gamma rays, and cosmic rays. Our rationale for this arbitrary classification is that today the range of wavelengths produced by all devices technically defined as lasers is that which we have used for our definition of light.

II. THE NATURE OF RADIATION

In the most general sense, *radiation* can be defined as the transmission of energy from one point in space to another without the need for a material medium between the points. Radiant energy can also be transmitted through an intervening material that fills the space between the two points. Light, as we have defined it, is a special case of *electromagnetic* radiation, which does not require a material medium between a source and a receptor. It can flow through empty space, as rays of the sun do for most of their 90 million-mile journey to the earth. It can also pass through the gases of the earth's atmosphere, the windows of our buildings, and the lenses of our cameras. The transmission of electromagnetic radiation involves no transport of *mass*. (See Bibliography Ref. 1 for definitions of energy and mass.)

Radiation can also be *mechanical*: e.g., the transmission of *vibrations* through a building from a heavy vehicle on a nearby street, or the passage of *sound* through the air from the lips of a speaker to the ears of a listener. The ultrasonic waves of special high-frequency generators used in medical diagnosis, and the waves that are created by the fall of a pebble into a still pond or the sweep of wind across an ocean's surface are other forms of mechanical radiation. However, unlike electromagnetic radiation, these types do require the presence of a material medium that has both mass and elasticity, although vibration does not involve *mass convection* of the medium. It is simply a transfer of energy from one particle of the medium to the next by harmonic oscillation of each particle about its average position, in which it is bound by the elastic forces holding the material together.

Finally, radiation can be *particulate,* i.e., a stream of material particles such as electrons, protons, neutrons, or other subatomic fragments. This type of radiation needs no material medium but often can pass through gases, liquids, and solids. Unlike the transmission of energy by electromagnetic and mechanical radiation, particulate radiation requires the transfer of mass from the source to the receptor, because its energy is the *kinetic* energy of the particles.

Because electromagnetic radiation is what lasers produce, we shall henceforth be concerned with only this type, and for simplicity we shall simply refer to it as radiation.

To understand how laser radiation, or light, differs from that of all other sources, such as the sun, incandescent lamps, flash tubes, etc., it is first necessary to know the physical nature of radiation. Unfortunately, not all of the observed phenomena associated with radiation can be explained by a single, unified theory. To account for all the effects, we require two distinct views of radiation: (1) that it is the propagation of *waves* of electric and magnetic fields through space; and (2) that it is a stream of massless *particles* of energy, called *photons.* The phenomena of reflection, refraction, diffraction, interference, and polarization of light can be explained only by the wave theory. However, the facts of emission and absorption of light by atoms cannot be accounted for by the wave concept; for these we must resort to the particulate theory. This apparent duality in the nature of light has resisted all attempts of the greatest scientific minds of history to resolve it into a single, omniscient theory, from the ancient Greek Eratosthenes (who first measured the earth's circumference by simultaneous observations of the inclination of the rays of the midday sun at two separate points in the Egyptian desert) to Albert Einstein (who first predicted the possibility of laser light). The duality is presently unavoidable. We shall use one or the other view to explain laser phenomena as best suits the effect under consideration.

A. The Wave Theory of Radiation

In the wave theory view of electromagnetic radiation, energy is propagated through space, either free or containing a material medium, in lines that are straight as long as the properties of the medium are constant at all points. Along these *rays* of propagation, *vectors* of electric field intensity (denoted by the symbol **E**) project in all radial directions perpendicular to the ray, and for each such **E** vector there is a corresponding vector of magnetic field intensity (denoted by **H**) whose direction is perpendicular to both the ray and the **E** vector. A vector is a physical entity having direction as well as magnitude. If an observer were to look down the ray in the direction of propagation, the direction of the magnetic vector at any point would be displaced 90° in a clockwise sense from that of its corresponding electric vector.

When viewed in any plane that contains the ray direction, or axis of propagation, at any instant fixed in time, the **E** vectors exhibit a variation in magnitude (length) with distance along the ray, such that a continuous curve connecting the tips of all the **E** vectors in that plane has a *sinusoidal* shape symmetrically disposed about the axis of propagation. Thus each positive crest of this E wave is as far to one side of the ray as the next negative crest is to the opposite side. In another plane containing the axis but rotated 90° from the plane of the E wave, the tips of all the **H** vectors lie on a sinusoidal curve symmetrically disposed about the ray and having its intersections with the ray at the same points as those of the E wave. These waves are illustrated schematically in Fig. 1, in

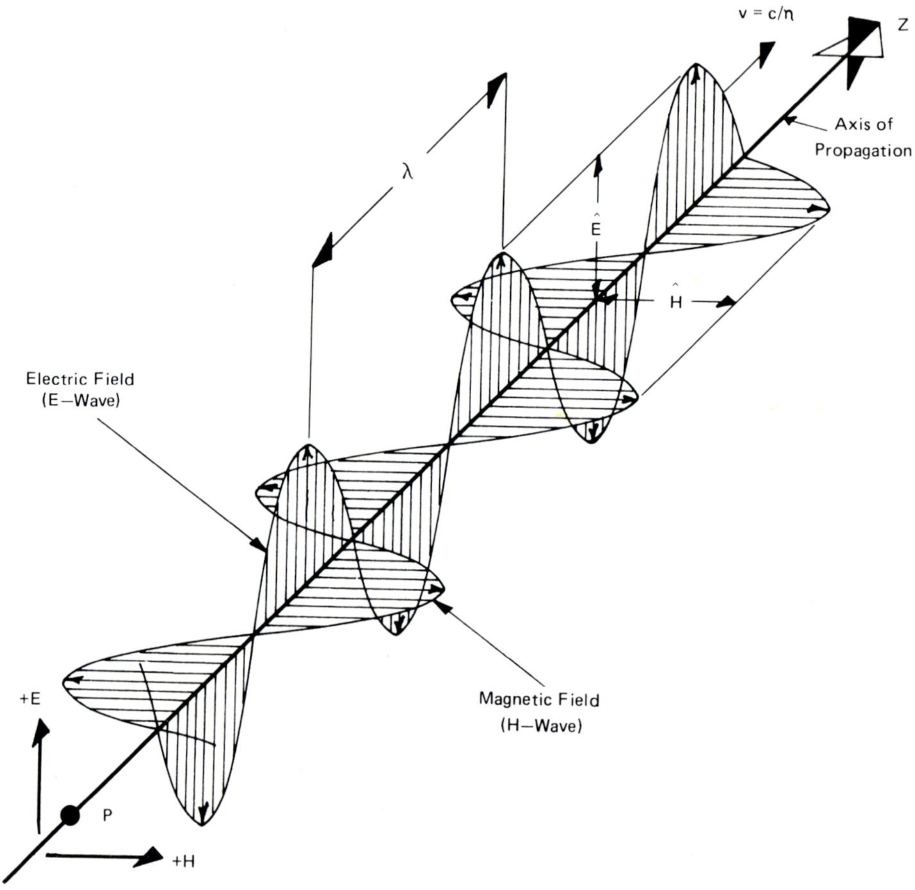

Figure 1 A plane-polarized ray of light traveling through a homogeneous medium of refractive index η at a constant speed $v = c/\eta$, where c is the speed of propagation in free space = 2.998×10^8 m/s. Light is here depicted as orthogonal traveling waves of electric and magnetic fields.

which only one set of perpendicular planes and waves is shown, for the sake of simplicity and ease of visualization. The ray of radiation of Fig. 1 is *plane polarized,* because the **E** vectors and **H** vectors exist only in one plane each. Unpolarized light would have electric and magnetic vectors in all planes containing the axis of propagation, projecting radially outward like the bristles of a long bottle-brush.

At the forward tip of the ray, the wave front is traveling at a constant speed (in a homogeneous medium). The magnitude of the speed of propagation is highest in free space, having a value usually denoted by the letter c:

$$c = 2.998 \times 10^8 \text{ m/s}$$

If this ray of light passes through any homogeneous medium other than free space, its speed of propagation is reduced:

$$v = c/\eta \ (DISTANCE)/(TIME) \qquad [1]$$

where

η = the *index of refraction* of the medium (*NUMERIC*); $\eta \geqslant 1$

Because the speed of propagation decreases with increasing index of refraction, a ray of light obliquely striking the interface between two different media will be changed in its direction of propagation so that its angle of incidence (off the normal) is larger in the medium of lower index and some of its intensity will be reflected away from the incident side of the interface at an angle equal to the angle of incidence. These effects depend upon the magnitude of the angle of incidence and the relative values of the refractive indices. They are utilized in lenses to focus beams of light and in optical fibers to produce repeated internal reflections of light rays that enter the fiber at angles off the axis less than the *acceptance* angle. Such fibers must have a core material that has a higher refractive index than the cladding. As long as the curvature of the fiber does not cause the internally reflected rays to graze the core-cladding interface at angles greater than the acceptance angle, *total* internal reflection will occur. This phenomenon is discussed in more detail later.

In Fig. 1, a stationary observer located at point P would perceive a sinusoidal fluctuation with *time* of both the electric and magnetic field intensities as the respective space waves of **E** and **H** pass by the observation point at the speed $v = c/\eta$. On the other hand, an observer moving with these waves through space at speed v could look up or down the ray's direction and perceive a *spatial* sinusoidal variation of **E** and **H** with distance from his location on the wave train. If the frequency of the temporal variation seen by the stationary observer is f and the distance between two successive wave crests seen by the moving observer is λ, which is the *wavelength*, then the combined variations of **E** and **H** are described by the simple *traveling-wave equations*:

$$E = \hat{E} \sin \frac{2\pi}{\lambda} (z - vt), (POTENTIAL)/(LENGTH) \quad [2a]$$

and

$$H = \hat{H} \sin \frac{2\pi}{\lambda} (z - vt), (CURRENT)/(LENGTH) \quad [2b]$$

where z = distance along the axis of propagation, measured from the viewpoint of the moving observer at a positive-going axis crossing of either the E wave or the H wave;

t = time measured by the stationary observer from the instant at which any positive-going zero-cross passes his viewpoint;

$\hat{E}$ and $\hat{H}$ = the instantaneous values of electric and magnetic field intensities, respectively, measurable by a stationary observer at various points separated from P by distances z.

In the meter-kilogram-second (*mks*) system of units, which is universally used nowadays in practical calculations, the electric field intensity E is expressed in volts per meter, and the magnetic field intensity H in amperes per meter. The symbols in Eqs. [2a] and [2b] denote crest, or maximal values, The frequency of variation perceived by the stationary observer is related to the wavelength seen by the moving observer:

$$f = v/\lambda, (CYCLES)/(TIME) \quad [3]$$

Because the speed of propagation is lower in material media than in free space, either f or λ in Eq. [3] must be correspondingly reduced. The frequency of the oscillator producing the wave is constant, irrespective of the medium through which the wave passes after leaving the oscillator, and so the wavelength is reduced in proportion to the speed.

1. Electric and Magnetic Fields

a. Electric Fields. An electric field exists wherever electric charges of opposite signs are separated. It is essentially a volume of space between separated charges within which a force will be exerted upon any unit charge placed into that space. It is usually depicted by *lines of force* beginning on the positive electric charge and terminating on the negative electric charge. These lines of force represent the paths that would be followed by a unit positive charge placed into the field at various points and allowed to move under the influence of the electric force that the unit charge experiences as a result of being in the field. The density of these lines of force, i.e., the number that pass through any

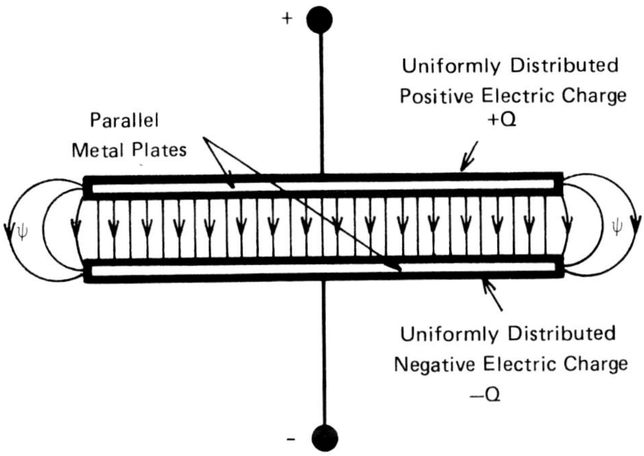

Figure 2 The electric field between two flat metal plates having uniform distribution of electric charge on each plate.

small element of area normal to the lines, is everywhere proportional to the electric field intensity E. The geometric configuration of the lines in three-dimensional space is dependent upon the distribution of the separated charges within that space. Figure 2 shows the lines of the electric field between two large, flat metal plates having equal magnitudes of electric charge distributed uniformly over their surfaces, with the positive on one plate and the negative on the other. In this simple case, the density of the lines of force, expressed as (*LINES*)/(*AREA*), is uniform. It is called the *electric displacement density* and is usually denoted by the symbol D, where

$$D \equiv \frac{d\psi}{dA} = k_0 K E \qquad [4]$$

and we have ψ = the *electric displacement*
k_0 = the *absolute dielectric constant,* or *permittivity,* of free space (a primary physical constant) = 1.11265×10^{-10}
K = the *relative dielectric constant* of the medium in which the electric field exists (*NUMERIC*)

Because the total number of field lines, or *displacement,* is proportional *only to the total electric charge* on either plate, and the field is inherently associated with an electric charge, the physical dimension of D is taken to be (*CHARGE*)/(*AREA*). The *mks* unit of electric charge is the *coulomb* (C).

The *electric potential* of the field between separated charges, such as those of Fig. 2, is defined as the *work* (energy) required to pull or push a unit positive charge from infinitely far away to some point within the field. It is often denoted by the symbol V. The potential is the line integral of E multiplied by distance along the path followed by the unit positive charge in arriving at a point P from very far away:

$$V_P = \int_{\infty}^{P} E\,ds \qquad [5]$$

where V_P = the electric potential at point P $(ENERGY)/(CHARGE)$
s = distance measured along the path by which the unit positive charge arrives at P, $(LENGTH)$

As long as there are no dissipative (frictionlike) forces present, the potential at point P is independent of the path followed by the approaching charge. In most practical situations, we are concerned with *differences* of potential between two points rather than with the potential at one point. Potential difference, in nondissipative fields, depends upon only relative positions and not upon the path of transit of a charge between these positions. Such situations often exist within and between atoms.

If electric charges are free to move within an electric field, there results an electric *current,* usually denoted by the symbol I (or i). The physical dimensions of I are $(CHARGE)/(TIME)$. If the field exists within a good electrical conductor, such as a metal wire, whose ends are connected to a perpetual source of charge separation, such as a battery, a steady current will flow through this electric *circuit.* If the source of charge separation maintains a potential difference of constant polarity, the current is said to be *direct*; if the polarity of the potential difference reverses in cyclic fashion, the current is said to be *alternating.* Alternating currents are most often sinusoidal in their temporal variation.

It should be noted that, in their normal state, atoms, which are the ultimate sources of all electric charges, are neutral, i.e., their positive and negative charges are equal in magnitude. Thus in the macroscopic world, large aggregations of matter usually are not subject to electric fields unless acted upon by external agents that create separation of their atomic charges.

b. Magnetic Fields. Every closed circuit carrying an electric current, either direct or alternating, produces a *magnetic* field. Magnetic fields are depicted customarily as lines of force which, unlike the lines of an electric field, do not terminate on charges, but form *closed loops* that are linked with the current circuit like bracelets on a keyring. Figure 3 depicts a circular loop of wire carrying an electric current and producing an interlinked magnetic field. Each line of force has, at every point of its length, the direction that would be assumed by

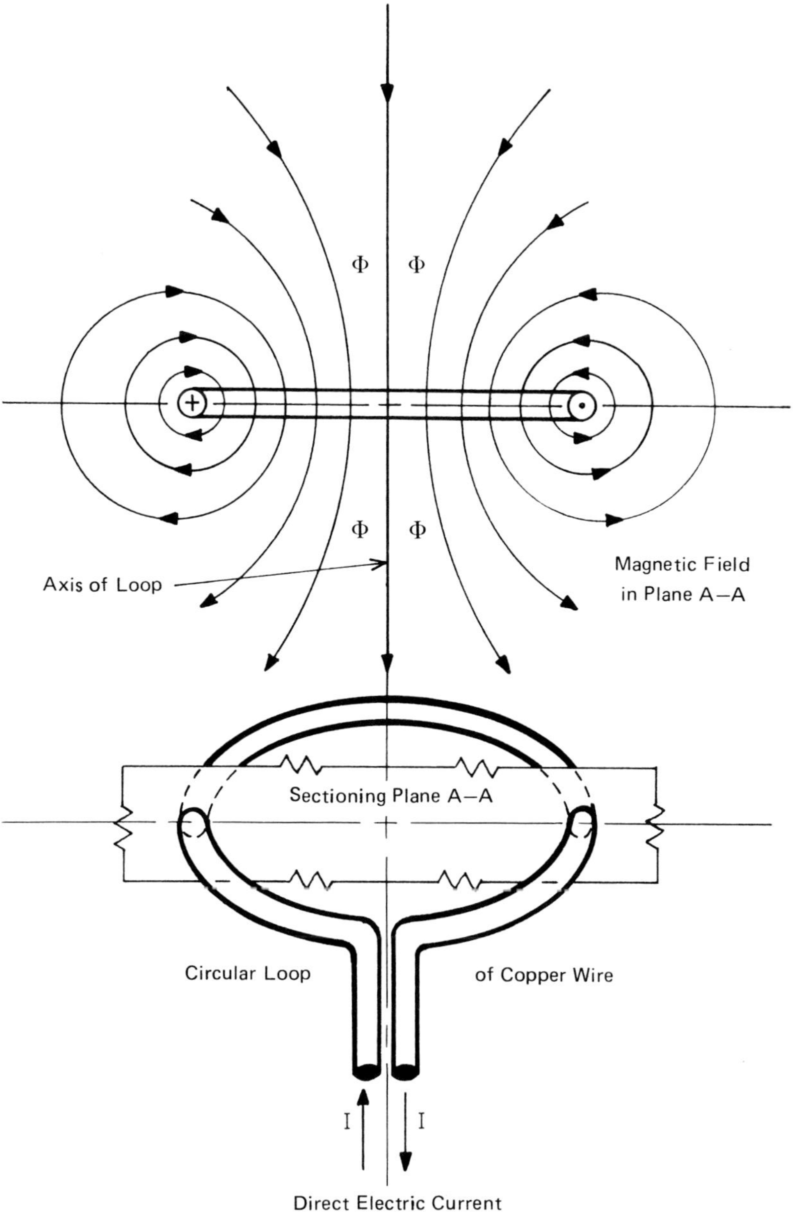

Figure 3 Magnetic field produced by a plane loop of copper wire carrying a direct electric current. Sectioning plane A-A passes through the axis of the loop. All lines of magnetic flux are closed plane curves linked with the loop of copper wire.

the needle of a small pocket compass situated at that point (i.e., the axis of the compass needle would be tangent to the line of force). In the physics of the late nineteenth and early twentieth centuries, it was thought that there might be unit magnetic poles (analogous to unit electric charges) like those at the ends of a permanent magnet. However, if one attempts to cut off the pole at either end of a permanent magnet, the severed piece exhibits both north and south poles, as does the remainder of the original magnet. All attempts by scientists to demonstrate the existence of unit magnetic poles have failed, and today the accepted view is that magnetic fields, even those of permanent magnets, are produced by electric currents. Because of the indivisibility of the poles of a permanent magnet, a small compass needle will experience no net force impelling it along a line of the magnetic field in either direction (i.e., the force on the north pole is equal and opposite to that on the south pole if the axis of the needle is tangent to the line of force). The lines of force of a magnetic field collectively constitute a *magnetic flux,* usually designated by the symbol Φ. The number of such lines per unit of area perpendicular to their direction is called the *magnetic flux density*, customarily designated by the symbol B. The magnetic field intensity H is related to the flux density B by a simple equation like Eq. [4]:

$$B = \mu_0 \mu H \qquad [6]$$

where μ_0 = the *absolute permeability of free space* (a primary physical constant) $= 1 \times 10^{-7}$
μ = the relative permeability of the material medium (if one is present) in which the magnetic field exists (*NUMERIC*).

Magnetization, either in free space or in a material medium of any kind, is produced by electric currents, which can be electric charges moving through a metal wire, as in electric motors; or through air, as in discharges of lightning; or in orbits around an atomic nucleus, as in all chemical elements, e.g., the iron, cobalt, and nickel of permanent magnets.

In Fig. 3, if distance along a line of magnetic flux is denoted by s, and the product of the flux density B at each point and the element of path length ds at which B is measured, or $B \times ds$, is integrated around the entire length of the line, this line integral is related to the current in the loop of wire by the *circuital law of the magnetic field*:

$$\oint B\, ds = 4\pi \mu_0 \mu I \qquad [7]$$

where I = the total electric current flowing in the cross-section of the wire circuit.

Note that, if the wire circuit is made up of several concentric turns laid close together (i.e., a coil), then I in Eq. [7] is the product of the number of turns and the current in each turn. If we differentiate both sides of Eq. [7] with respect to s and rearrange the factors, we obtain

$$B = 4 \pi \mu_0 \mu \frac{dI}{ds} \qquad [8]$$

where the differential element of current dI is that part of the total I that is needed to magnetize the element of length ds of the closed line of force. The current in the circuit may be thought of as the magnetizing force which produces the magnetic flux density B; and the quantity $4\pi I$ in Eq. [7] is customarily called the *magnetomotive force* of the magnetic field, by analogy with the quantity V_P in Eq. [5], which is often called the *electromotive force* of the electric circuit.

A comparison of Eq. [6] with Eq. [8] shows that we may write

$$H = 4 \pi \frac{dI}{ds} \qquad [9]$$

The physical dimensions of H can thus be seen as *(CURRENT)/(LENGTH)*.

The units of the various physical entities discussed in the previous paragraphs are, in the *mks* system of measurement, as shown in Table 1. The *mks* system has two versions, the *rationalized* and the *unrationalized.* In the rationalized *mks* system, the numerical value of the permittivity of free space is that of the unrationalized system divided by 4π:

$$k_{r0} = k_0/4\pi = 8.854 \times 10^{-12}$$

and the numerical value of the permeability of free space is that of the unrationalized system multiplied by 4π:

$$\mu_{r0} = \mu_0 \times 4\pi = 1.2566 \times 10^{-6}$$

Thus the difference between the two versions is simply in the values of the primary constants and basic parameters of electric and magnetic fields, which are adjusted by a factor of $(4\pi)^{\pm 1}$ in the rationalized, so that the displacement of an isolated electric charge, $\psi = 4\pi Q$ in the unrationalized system, becomes simply $\psi = Q$ in the rationalized. The formula relating magnetic flux density produced by a current-carrying curcuit to the current flowing in that circuit, i.e., Eq. [7] can be written simply as $\oint B\, ds = \mu_{r0}\, \mu I$.

Table 1 The Dimensions of Physical Entities in Terms of the Primary Physical Dimensions (*LENGTH*), (*MASS*), (*TIME*), and (*ELECTRIC CHARGE*)

Symbol(s)	Entity	Physical dimensions
$d, \lambda, r, R, w, x, z$	Length	$(LENGTH) \equiv (L)$
m	Mass	$(MASS) \equiv (M)$
t, τ	Time	$(TIME) \equiv (T)$
Q	Electric charge	$(CHARGE) \equiv (Q)$
F	Force	$(M)\ (L)/(T)^2$
E, e	Energy, work	$(M)\ (L)^2/(T)^2$
P, P	Power	$(M)\ (L)^2/(T)^3$
v, c	Velocity, speed	$(L)/(T)$
a	Acceleration	$(L)/(T)^2$
θ	Angle	$(NUMERIC)$
f	Frequency	$1/(T) = (T)^{-1}$
p	Power density	$(M)/(T)^3$
T	Temperature	Average kinetic energy per molecule of an ideal gas, $(M)(L)^2/(T)^2 \equiv (\mathbf{T})$
c_t	Thermal conductivity	$(M)\ (L)/(T)^3\ (\mathbf{T})$
a, A	Attenuation coefficient	$(L)^{-1}$
g	Gain coefficient	$(L)^{-1}$
α	Absorption coefficient	$(L)^{-1}$
σ	Scattering coefficient	$(L)^{-1}$
ψ	Electric displacement	(Q)
D	Displacement density	$(Q)/(L)^2$
I, i	Electric current	$(Q)/(T)$
E	Electric field intensity	$(L)\ (M)/(T)^2(Q)$
V	Electromotive force, potential difference	$(L)^2(M)/(T)^2(Q)$
R	Electric resistance	$(L)^2(M)/(T)\ (Q)^2$
C	Electric capacitance	$(Q)^2(T)^2/(L)^2(M)$
Φ	Magnetic flux	$(L)^2(M)/(T)\ (Q)$
B	Magnetic flux density	$(M)/(T)\ (Q)$
$4\pi I$	Magnetomotive force	$(Q)/(T)$

(continued)

Table 1 (continued)

Symbol(s)	Entity	Physical dimensions
H	Magnetic field intensity	$(Q)/(L)\,(T)$
L	Magnetic inductance	$(L)^2(M)/(Q)^2$
k_0	Absolute permittivity of free space	$(T)^2(Q)^2/(L)^3(M)$
μ_0	Absolute permeability of free space	$(L)\,(M)/(Q)^2$
μ	Relative permeability	$(NUMERIC)$
K	Relative dielectric constant	$(NUMERIC)$
mv	Linear momentum	$(M)\,(L)/(T)$
mvr, J, L, S	Angular momentum	$(M)\,(L)^2/(T)$
η	Index of refraction	$(NUMERIC)$
Z	Atomic number	$(NUMERIC)$
n	Principal quantum number	$(NUMERIC)$
l	Orbital quantum number	$(NUMERIC)$
m_l	Orientation quantum number	$(NUMERIC)$
m_s	Spin quantum number	$(NUMERIC)$
h	Planck's constant	$(M)\,(L)^2/(T)$
κ	Boltzmann's constant	$(M)\,(L)^2/(T)^2(\mathrm{T})$
ϵ	Base of natural logarithms	$(NUMERIC)$
π	Ratio of circumference to diameter of a circle	$(NUMERIC)$
N, ξ	Integer number	$(NUMERIC)$

However, irrespective of which version of the *mks* system is used, a very fundamental relationship exists between the entities k_0 (or k_{r0}) and μ_0 (or μ_{r0}):

$$k_0\,\mu_0 = 1/c^2 \qquad [10]$$

The discovery that an electric current produces a magnetic field was made by the Danish physicist Hans Christian Oersted (1777-1851) in April of 1820. In the year 1831, Michael Faraday (1791-1867), the renowned English physicist and chemist, discovered the complementary effect, namely, that a *changing* magnetic field will induce an electric field around every closed path that is linked with this magnetic field. The intensity of the induced electric field is related to the time-rate of change of the magnetic flux:

$$\oint E\, ds = -d\Phi/dt \qquad [11]$$

where $d\Phi$ = the incremental change in the magnetic flux Φ which occurs in a time-increment dt; $d\Phi$ is expressed in *webers* and dt in *seconds*

ds = the increment of length of the closed path, along which E is measured

$\oint E\, ds$ = the electromotive force (EMF) induced around the complete path of length S by the changing magnetic flux linked with this path; the EMF is expressed in *volts.*

The minus sign in Eq. [11] indicates that the electromotive force is always in such a direction as to produce an electric current (if the closed path were replaced by a ring of metal wire) whose magnetic field would be in a direction opposite to the change $d\Phi$. It should be noted in passing that the electric field E given by Eq. [11] cannot be depicted by lines of force terminating on electric charges, as in the case of *electrostatic* fields described earlier, because an electric field can be induced by a changing magnetic field in free space, where there are no atoms to provide charges that can be separated. In the case of Eq. [11], the lines of the electric field are *closed loops* linked with the magnetic field lines.

c. Electromagnetic Waves. In the year 1864, it fell to the analytical brilliance of the English physicist James Clerk Maxwell (1831-1879) to combine the earlier observations of Oersted and Faraday into one comprehensive theory of all the known phenomena of electricity and magnetism. In his classic paper (1) he set forth the now-famous Maxwell's equations (see the Appendix at the end of this chapter), four relationships (when expressed in modern vector notation) from which he derived his traveling wave equations to describe the propagation of energy through space without the need for a material medium for such transmission. Basic to Maxwell's formulation of all electromagnetic phenomena was his postulate that a changing electric field is equivalent to an electric current of a kind not requiring a conducting medium such as a metal wire. He named these *displacement currents,* which we shall denote by the symbol I_d:

$$I_d = \frac{1}{4\pi}\frac{d\psi}{dt} \qquad [12]$$

where $d\psi$ = the incremental change in electric displacement ψ ($\psi = 4\pi Q$, in electrostatic fields), (*CHARGE*)

dt = the incremental time during which $d\psi$ occurs

An important consequence of Maxwell's derivation of his traveling wave equations was the relationship among the primary electric and magnetic constants of free space, μ_0 and k_0, and the speed of light in free space, c, which we have presented as Eq. [10]. The startling aspect of Maxwell's theoretical description of radiation as traveling waves of electric and magnetic fields in empty space, which we have presented in Fig. 1 and Eqs. [2a] and [2b], was that it predicted a phenomenon that was not experimentally demonstrated until 23 years later, when Heinrich Rudolf Hertz (1857-1894) performed his dramatic experiment. Hertz connected a large, flat metal plate to each of two spheres forming a spark gap, with the plates apart yet lying in the same plane. When the spark gap was flashed over by an electric arc produced by a high-voltage induction coil, the current in the arc alternated at very high frequency (about 10^8 Hz), so that a plane-polarized E wave was radiated in a plane perpendicular to both plates at their vertical midline, with a wavelength of about 3 m. A second spark gap like the first, but connected to a circular loop of metal wire such as to provide a conducting path between the spheres of the second gap, was placed a few meters away. When the plane of the loop of wire was parallel to the plane of the polarized E wave radiated by the first spark gap, the second gap would flash over because of the electromotive force induced in it by the traveling E wave. However, when the plane of the loop was placed perpendicular to the plane of the E wave, no spark would occur at the second gap. Hertz was able to reflect these waves with large metal mirrors, and to refract them through a "prism" made of pitch. By moving the second spark gap along the axis of propagation of the wave from the first gap, he was able to determine the wavelength by noting the distance between the nodes of the E wave (i.e., those points at which E was zero and no spark occurred at the second gap). From a theoretical calculation of the frequency of oscillation, based upon the inductance and capacitance of the first circuit, together with the measured wavelength, Hertz was able to calculate the speed of propagation of his waves. His calculated value agreed well with that computed from Eq. [10] by using known values for μ_0 and k_0. Thus, eight years after Maxwell's death, his bold theoretical prediction was experimentally verified with remarkable accuracy! Figure 4 shows schematically the details of Hertz' experiments.

At great distances from circuits carrying conduction currents, the only magnetic fields of importance are those due to displacement currents. When the frequency of alternation is sufficiently high (100,000 Hz or more), the magnetic fields resulting from the displacement currents can be of a magnitude to propagate enough power for detection by receivers at very long distances away. The displacement currents, which are actually changing electric fields, cause changing magnetic fields, which in turn cause changing electric fields, and so energy is radiated away from the conduction circuit (e.g., a radio antenna) in the form of electromagnetic waves. A final important consequence of Maxwell's equations is the ability to calculate the *radiant power density,* which we shall designate as p_0 in free space:

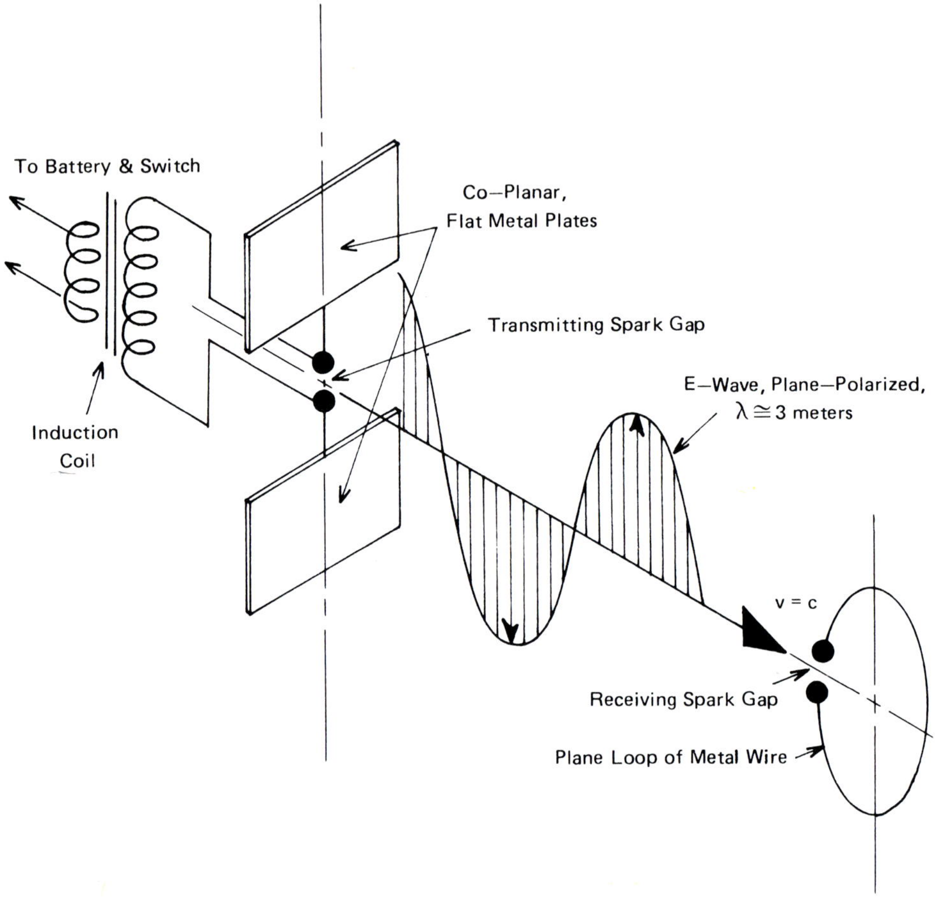

Figure 4 Schematic diagram of apparatus used by Heinrich Hertz to demonstrate the propagation of electromagnetic waves through free space from one circuit to another.

$$p_0 = \widehat{E}\,\widehat{H}/8\,\pi = k_0\, c\,\widehat{E}^2/8\,\pi, \ (POWER)/(AREA) \qquad [13]$$

where p_0 = the *time-averaged* value of power density, or the energy per unit time transmitted through a unit area normal to the axis of propagation.

As we shall see later, the power density of a laser beam is the single most important operating variable for surgical and therapeutic purposes.

2. The Electromagnetic Spectrum

In the maxwellian theory, radiation can be categorized either by its wavelength or by its frequency. When either of these variables is plotted as a distance along one axis of a two-dimensional graph, we have what is called a *spectrum.* The spectrum of all known radiation is vast, ranging from cosmic rays at the short end of the wavelength scale to radio waves and very low frequency waves at the long end:

$$1 \times 10^{-15} \leqslant \lambda \leqslant 1 \times 10^{7} \text{ m} \qquad [14]$$

This whole spectrum is illustrated schematically in Fig. 5, in which wavelength is plotted at one side and frequency of oscillation at the other. From this figure it can be seen that Homo sapiens and all the lower animals are virtually color-blind, because our eyes respond to only an infinitesimal fraction, known as the *visible spectrum,* ranging from 400×10^{-9} m to 700×10^{-9} m, which constitutes but one part in 30 trillion of the whole gamut of electromagnetic radiation (when measured by wavelength)!

Although the physicists of the nineteenth century lacked the means of producing light waves from electromagnetic devices, the great success of Maxwell's theory in predicting the existence of traveling waves, the refraction and reflection of light, and the interference phenomena left no doubt in the minds of scientists that the wave theory of light was correct. However, early in the twentieth century, the maxwellian view of radiation began to falter in attempts to explain the emission of radiation from hot bodies.

B. The Particle Theory of Radiation

The conspicuous failure of the wave theory to explain the empirically determined spectral distribution of radiant power density from a black body as a function of temperature led the German physicist Max Planck (1858-1947) to introduce a new hypothesis. He showed that the maxwellian theory could satisfactorily account for black-body radiation only if the energy carried by an electromagnetic wave were not continuously subdivisible, but instead were made up of *finite increments,* now called *quanta,* whose value was wavelength-dependent. The energy of such a quantum is directly proportional to the frequency of the wave and inversely proportional to its wavelength:

$$e_p = h\,f = h\,c/\lambda \qquad [15]$$

where e_p = the value of a quantum of energy
h = Planck's constant = 6.626×10^{-34} joule sec
f = the frequency of the wave in hertz, or cycles/second.

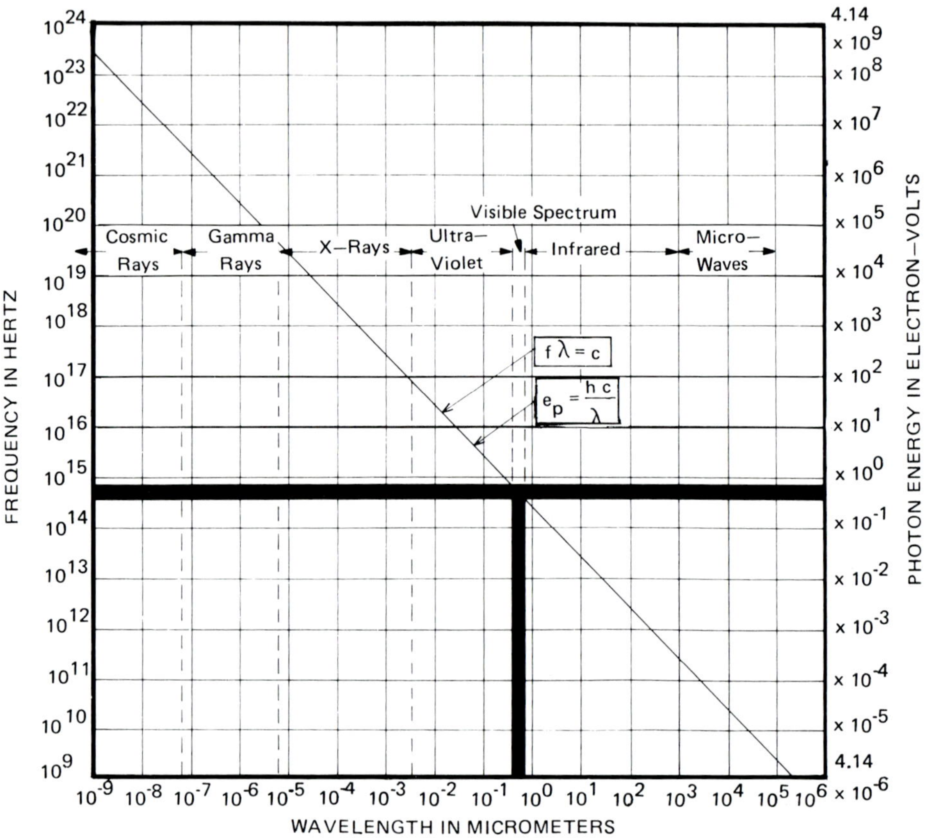

Figure 5 The electromagnetic spectrum from 1 fm (1×10^{-15} m) to 1 m of wavelength. The black bands show the very limited extent of the visible spectrum. To the right of the vertical line marked 10^3 μm lie all the wavelengths produced by electric circuit devices (klystrons, magnetrons, and radio oscillators). To the left of this line lie all the wavelengths radiated by molecular, atomic, and subatomic sources.

In that same year, 1905, another German-born scientist, Albert Einstein (1879-1955), then unrecognized for his theoretical genius, proposed that electromagnetic radiation was explainable as a *stream of particles,* which he named *photons.* He was led to this conclusion by his attempt to explain the *photoelectric effect,* first discovered by Sir Joseph John Thomson (1856-1940) in England. Thomson, who discovered the electron, showed that electrons could be ejected from a metal surface by irradiation with ultraviolet light (10 nm $\leqslant \lambda \leqslant$ 400 nm). In Einstein's view, a photon was a particle of radiant energy having the value given by Eq. [15]. Furthermore, it had no mass, although it was later demonstrated that photons have the equivalent of momentum, (*MASS*) $\times$ (*VELOCITY*), and always traveled at the speed of light in free space, $c = 2.998 \times 10^8$ m/sec. Although this characteristic of photons appears to be inconsistent with the wave-related phenomenon of refraction, in which a material medium, because its index of refraction is greater than unity, reduces the speed of light waves traveling through it, it can be reconciled by the following observations. On a microscopic scale, where dimensions are measured in units of the diameter of an atom, material substances are actually empty space partially occupied by atoms and molecules. In the empty spaces between atomic nuclei, a photon travels at the speed c. When it is absorbed or scattered by an atom, the finite time required for absorption and reemission introduces a delay into its journey, as well as a change of direction. The aggregate delay caused by encounters with large numbers of atoms increases the total time of travel through the medium, but does not proportionally increase the total distance traversed. Hence the *average* speed of a photon in a material medium is reduced. In the wave theory of light, this reduction is described quantitatively by the index of refraction, η.

Figure 6 shows schematically a ray of light depicted as a linear stream of photons. As it later became clear, the photonic theory of radiation was able to predict and accurately describe the process by which laser radiation is created. In the year 1917, Einstein published his now-famous paper (2) on the quantum theory of radiation, in which he suggested the possibility of *stimulated emission,* the essential phenomenon of laser action.

Henceforth in this chapter we shall speak of photons and waves interchangeably, perhaps causing some bewilderment to the reader. The duality of light as wave and particle is unavoidable, but need not be confusing if we think of a photon as being equivalent to a *wavetrain of finite length* in space, which we shall call a *wavelet.* The wavelength of this wavelet is specified in Eq. [15]. For atomic sources of radiation, the duration of a wavelet is relatively short, ranging from about 1×10^{-8} sec to about 1×10^{-2} sec. It is equal to the time required for an atom to make a downward transition between a higher and a lower level of energy. The length of a wavelet in its direction of travel is, therefore, between 3 m and 3 million m, and the number of complete wavelengths it contains is between 3×10^4 and 3×10^{14} for the whole spectrum we have defined as light. If we think of radiation as composed of waves, it will appear to be continuous when there is a large enough number of atoms radiating at random instants so that their wavelets overlap in both time and distance along any possible ray direction.

On the other hand, if we think of light as streams of photons, any one atom may shoot out, with maximum external excitation, between 30 and 30 million

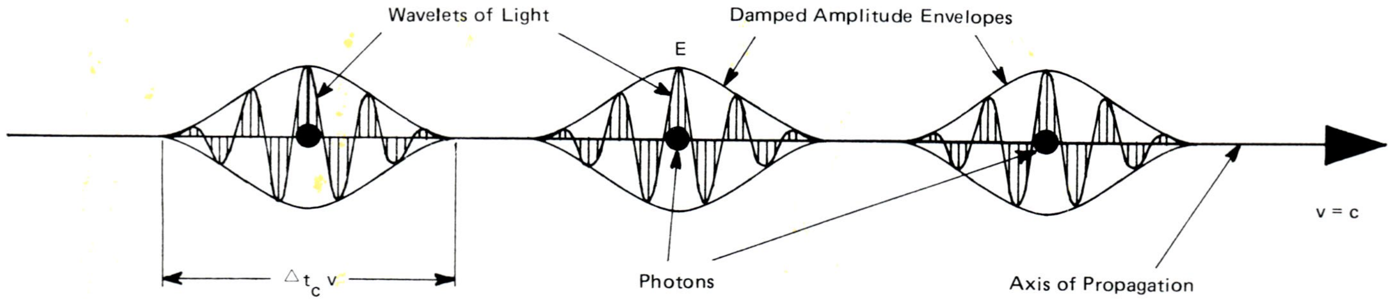

Figure 6 Light depicted as a stream of particles of energy, or photons, traveling in a straight line within a homogeneous medium. The equivalent wavelets of light corresponding to the photons are also shown. Note that the number of cycles of the E waves shown here is small, for simplicity of illustration. Actually, the number of cycles per wavelet would be very large (e.g., at $\lambda = 10.6\ \mu m$ and $\Delta t_c = 1 \times 10^{-8}$ sec, the number of cycles per wavelet would be 2.8×10^5.

photons/sec in any random direction within a solid angle of 4π steradians (a complete sphere of unit diameter). The minimum possible spacing between successive photons along any possible ray direction lies between 2 m and 2 million m; the maximum spacing can be vastly greater because the spontaneous emission of photons is equally probable in all directions. This radiation will appear continuous only when there is a large enough number of radiating atoms within a given space so that any comparable volume of space nearby is densely filled with photons.

1. Emission and Absorption of Energy by Atoms and Molecules

For wavelengths shorter than 0.1 mm, atoms and molecules are more efficient generators of radiation than the electric circuits used to produce radio waves and the resonant chamber devices that generate microwaves. To comprehend the fundamental principles of emission and absorption of radiation by atoms and molecules, it is prerequisite to have a basic understanding of the structure of atoms, which exist in more than 100 species, or chemical elements, and are the building blocks of molecules. The modern view of atomic structure is an outgrowth of a theory of the structure of the hydrogen atom (the simplest among all elements) proposed in the year 1913 by the Danish physicist, Niels Bohr (3) (1885-1962). In the following sections, the essentials of Bohr's theory are presented, together with the subsequent modifications thereof by the German physicist A. Sommerfeld (4) in the year 1916. Although the medical practitioner reading this chapter may feel that the quantum theory of atomic structure is too abstruse to be relevant to the uses of lasers in surgery, it is so fundamental to an understanding of the generation of laser light and its interaction with living tissue that no exposition of the physics and biophysics of lasers can be complete without it.

a. The Quantum Theory of Atomic Structure. Every atom consists of three basic particles, which for our purposes will be considered as indivisible, despite the ever-proliferating list of subparticles that modern atomic physicists keep adding to their collection to explain the ongoing experimental discoveries. Our three basic particles are *protons, neutrons,* and *electrons*. The atomic nucleus, which is the center of an atom and contains nearly all of its mass, is made up of at least one proton (as in the hydrogen atom) and one or more neutrons (as in all elements heavier than hydrogen). The proton is a particle whose mass is 1.6726×10^{-27} kg; it carries a positive electric charge of 1.6022×10^{-19} C. The neutron is electrically neutral (uncharged), and has a mass of 1.6749×10^{-27} kg. The *atomic number,* which we shall designate by the symbol Z, is equal to the number of protons in the nucleus, and to the total number of electrons, in a neutral atom. Moving around the nucleus, in orbits that may be either circular or elliptical, are the electrons. Each electron has a negative electric charge of 1.6022×10^{-19} C and a mass of 9.1096×10^{-31} kg. The electrons are held in their orbits around the nucleus by the forces of electrostatic attraction between themselves and the positively charged protons in the nucleus, in much the same way that the planets are held in their orbits around the sun by the force of gravity. Atoms that have the same atomic number, Z, but different numbers of neutrons in the nucleus are called *isotopes.* Isotopes of an element have the same chemical and physical properties except for mass.

As each electron moves in its orbit, it has a radius r_e from the nucleus and a velocity of magnitude v_e, which is directed tangentially to the orbit. By virtue of its velocity it has *kinetic energy,* and by virtue of the attractive electrostatic force between itself and the nucleus it has *potential energy.* The force between the negative electric charge of the electron and the clumped positive charges of the protons on the nucleus can be calculated from the *inverse-square law,* first formulated by the French physicist Charles Augustin Coulomb (1736-1806), which states that the force between any two electric charges is directly proportional to their product, and inversely proportional to the square of the distance between them:

$$F_e = Q_1\, Q_2 / k_0\, r_e^2 \qquad [16]$$

where Q_1 and Q_2 are the magnitudes of the electric charges. If F is positive in Eq. [16], it is a force of *repulsion,* and vice versa. For the orbiting electron the charge is $Q_e = -1.6022 \times 10^{-19}$ C, and for the atomic nucleus the total protonic charge is $ZQ_e = 1.6022 \times 10^{-19}\, Z$. Hence Eq. [16] becomes

$$F_e = Z\, Q_e^2 / k_0\, r_e^2 \qquad [17]$$

Because the electric field intensity at the electron caused by the clumped protonic charges on the nucleus, is $E_e = Z\, Q_e / k_0\, r_e^2$, we could write Eq. [17] as

$$F_e = E_e\, Q_e \qquad [18]$$

which shows that the force on an electric charge is equal to the product of the charge and the intensity of the electric field in which it is located. Thus we see that the physical dimensions of electric field intensity, which were given as (*POTENTIAL*)/(*LENGTH*) in Eq. [2a], can also be expressed as (*FORCE*)/(*CHARGE*). These dimensional expressions are consistent, because electric potential is expressible as (*WORK*)/(*CHARGE*) or (*FORCE*) × (*LENGTH*)/(*CHARGE*).

The orbiting electron has another physical property known as *angular momentum,* which is the product of its tangential speed, its mass, and its radius from the nucleus. As it travels its orbit around the nucleus, it is constantly changing the direction of its velocity vector. This means that it is experiencing an acceleration toward the center of its path, i.e., a *centripetal* acceleration whose magnitude is given by the ratio of the square of its tangential speed to its radius from the nucleus:

$$\text{Centripetal acceleration} = v_e^2/r_e \qquad [19]$$

Bohr's theory, which dealt originally with the atom of hydrogen (the simplest of all elements), postulated that the possible orbits of the single electron ($Z = 1$) were *circles* whose radii could have only certain discrete values r_e, such that the angular momentum of the electron could have only the series of values given by:

$$\text{Electronic angular momentum} = m_e\, v_e\, r_e = n\,h/2\,\pi \qquad [20]$$

where m_e = the mass of the electron = 9.1096×10^{-31} kg
n = the *principal quantum number* (*NUMERIC*)
= 1, 2, 3, 4, 5, 6, . . .
h = Planck's constant = 6.626×10^{-34} joule sec.

This postulate is the foundation of the modern quantum theory of atomic structure. The radius r_e may be calculated by noting that the electrostatic force of attraction holding the electron in orbit must be exactly equal to the inertial force corresponding to its centripetal acceleration toward the nucleus:

$$r_e = n^2 \frac{h^2\, k_0}{4\,\pi^2 m_e\, Q_e^2\, Z} = 5.2914 \times 10^{-11}\, n^2 \text{ m} \qquad [21]$$

where k_0 = the permittivity of free space
Q_e = the charge of the electron = -1.6022×10^{-19} C.

In a similar fashion, the orbital speed of the electron can be derived:

$$v_e = \frac{2\,\pi\, Q_e^2\, Z}{n\, h\, k_0} = 2.1877 \times 10^6\, n^{-1} \text{ m/sec} \qquad [22]$$

From Eqs. [21] and [22] it can be seen that the orbital speed becomes smaller as the orbital radius grows larger, and vice versa.

If an electron jumps from one orbit to another, its total energy (kinetic plus potential) must change by an amount

$$\Delta E_{ij} = \frac{2\,\pi^2\, m_e\, Q_e^4\, Z^2}{h^2\, k_0^2}\,(n_i^{-2} - n_j^{-2}) \qquad [23]$$

where $n_i < n_j$, and n_i and n_j are the principal quantum numbers of the orbits involved in the transition. Equation [23] shows that the energy of an electron increases as it moves to a larger orbit, and vice versa.

The great success of Bohr's theory of the hydrogen atom was that the change of energy predicted by Eq. [24], when substituted into Eq. [15], gives with high accuracy the wavelength of every previously observed spectral line of hydrogen. However, all the other chemical elements have more complex atoms than the simple proton-electron combination of hydrogen. To explain the structure of these, Bohr proposed that the circular orbits permitted by his theory for hydrogen are the only possible ones for all other atoms, but that each orbit may contain more than one electron. To aid in comprehension of several electrons orbiting about the nucleus at the same radius, Bohr introduced the concept that each possible orbit was a *spherical shell* of the radius prescribed by Eq. [21], with the several electrons of each shell moving in circular orbits that are meridians (great circles) of the sphere. He postulated that the maximum possible number of electrons in each shell is

$$(N_e)_{max} = 2\,n^2 \qquad [24]$$

where n is the principal quantum number of the orbit, as before. According to Eq. [24], the maximum number of electrons in each shell follows the sequence $(N_e)_{max} = 2, 8, 18, 32, 50, 72, \ldots$, corresponding to the integer values of the principal quantum number n. However, some of the heavier elements do not have all of their shells filled according to this rule. For example, mercury has only 18 electrons in the fifth shell and only two in the sixth.

To explain this and other peculiarities of the structure of heavier atoms, the German physicist A. Sommerfeld (4) extended Bohr's theory of quantized circular orbits to include quantized elliptical orbits as well. An ellipse is a closed plane curve having the property that every point on the curve has a constant sum of its distances from two fixed points called the *foci.* A circle is actually a special case of an ellipse whose foci coincide at its center. The planets of our solar system orbit the sun in elliptical paths having one common focus at the sun. The characteristic dimensions of an ellipse are its *major* and *minor axes.* The major axis is the length of the ellipse from apex to apex through the foci, and the minor axis is its width on a line perpendicular to the major axis at its center. Corresponding to each of Bohr's principal quantum numbers n, Sommerfeld's theory stipulates that there are n possible orbits, each of which has the atomic nucleus as one focus. Of these n orbits, $n - 1$ are elliptical and one is circular. The elliptical orbits all have the same length of major axis, which is equal to the diameter of the circular orbit, as given by $2r_e$ from Eq. [21]. The minor axes of these $n - 1$ ellipses have lengths of

$$2r_e,\ 2r_e \times 2/n,\ 2r_e \times 3/n, \ldots\ 2r_e\,(n - 1)/n.$$

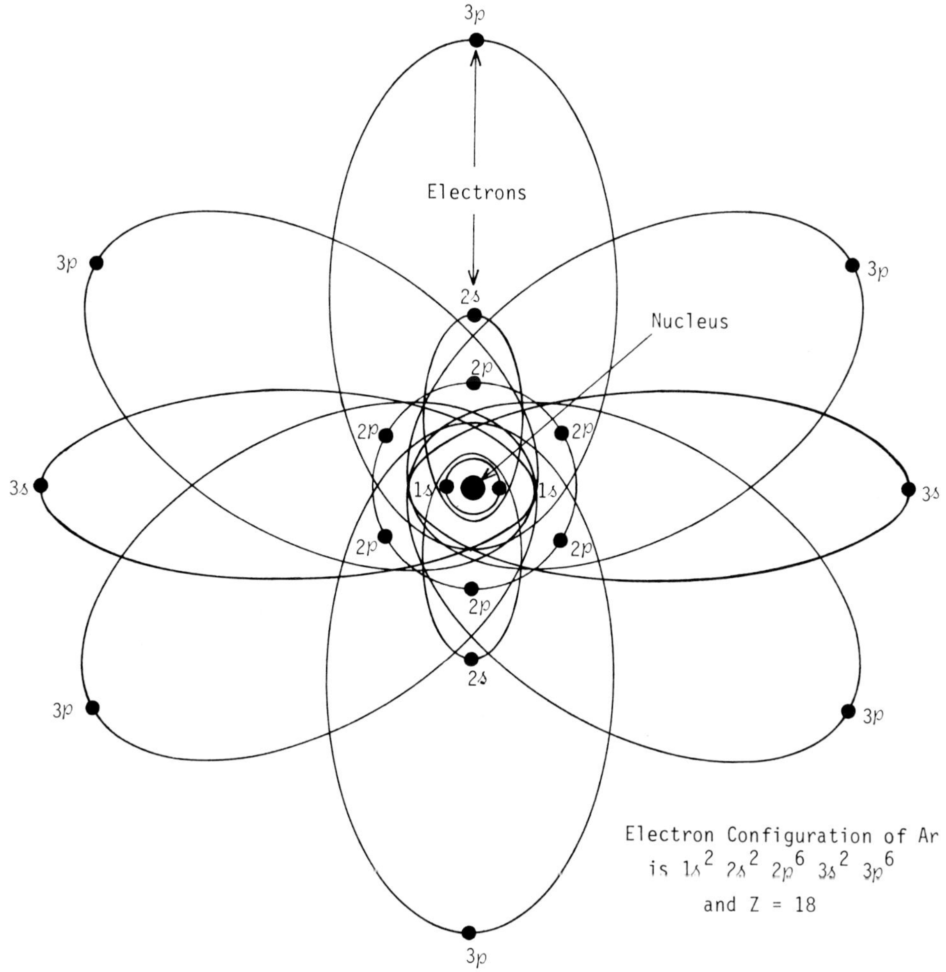

Figure 7 Schematic representation of the electronic orbits of an argon atom. Because all of the existing shells are filled with electrons, argon is chemically inert in its ground state. In excited states, it can combine with other elements, notably fluorine, to form the excimer (excited dimer) argon fluoride, which produces laser radiation at a wavelength of 193 nm.

An electron in any of these orbits having the same principal quantum number has the same average energy as that given by Bohr's theory for the corresponding value of *n*. Figure 7 shows schematically the nucleus, electrons, and orbits for the argon atom.

In current practice, an arbitrary system of designation is used to categorize all these possible orbits. An *orbital quantum number, l,* is assigned to each of the possible orbits corresponding to every integer value of *n,* such that

$$l = 0, 1, 2, 3, \ldots, (n - 1)$$

The lowest value of l is given to the narrowest ellipse, and the highest value designates the circular orbit of this group. It is customary now to assign to each value of l a letter, according to the arbitrary code

$$\begin{array}{rcccccccccl} l = & 0, & 1, & 2, & 3, & 4, & 5, & 6, & 7, & \ldots \\ \text{letter} = & s, & p, & d, & f, & g, & h, & i, & j, & \ldots \end{array}$$

For example, in this code the circular orbit for $n = 4$ is designated as *4f,* and the associated elliptical orbits are *4d, 4p,* and *4s,* with that corresponding to $l = 0$ being the narrowest. Each of the possible orbital *shapes* is called a *subshell,* and it is filled with electrons, or *closed,* when it contains a maximum number of electrons equal to $2(2l + 1)$. All of these facts are summarized in Table 2.

The outermost subshells of the various elements are usually not filled, having less than the maximum number of electrons permitted by the rule $2(2l + 1)$. These outermost electrons are called *valence electrons*; they are responsible for the chemical properties of the atom. Those elements whose outer shells are closed (filled) are more or less chemically inert, not readily forming compounds (molecules) with other elements. The electrons in all filled (closed) shells and subshells are disposed in space with spherical symmetry about the nucleus, but valence electrons moving in unfilled orbits have an unbalanced spatial distribution thereabout.

The Bohr-Sommerfeld theory of atomic structure is still regarded as fundamentally correct, and it provides the best means by which people other than atomic physicists can mentally visualize atoms. However, many phenomena empirically observed since 1913 are not easily explicable by the basic theory, and so it has been modified and expanded over the succeeding years. In 1924 the French physicist Louis de Broglie (5) suggested that moving particles, in particular electrons, can produce effects like those of electromagnetic waves, and are operationally equivalent to photons or wavelets. His *wave equation* gives the wavelength corresponding to a moving particle as

$$\lambda = h/m\,v \qquad [25]$$

where h is Planck's constant, m is the mass of the particle and v is its speed. At high speeds, such as those attained by electrons in electron microscopes, the

Table 2 Designations of the Possible Orbits of Electrons and the Maximum Number of Electrons Per Shell

Shell number, n	Subshell number, l						
	0	1	2	3	4	5	6
1	$1s^2$						
2	$2s^2$	$2p^6$					
3	$3s^2$	$3p^6$	$3d^{10}$				
4	$4s^2$	$4p^6$	$4d^{10}$	$4f^{14}$			
5	$5s^2$	$5p^6$	$5d^{10}$	$5f^{14}$	$5g^{18}$		
6	$6s^2$	$6p^6$	$6d^{10}$	$6f^{14}$	$6g^{18}$	$6h^{22}$	
7	$7s^2$	$7p^6$	$7d^{10}$	$7f^{14}$	$7g^{18}$	$7h^{22}$	$7i^{26}$

NOTE: The superscript numeral in each box indicates the maximum number of electrons that can occupy that particular subshell. If there is more than one electron in a subshell, that subshell will consist of n_e distinct orbits, with one electron in each orbit. If a subshell is elliptical and has n_e electrons, it will consist of n_e *separate* elliptical orbits, each containing one electron; when this subshell is filled, these elliptical orbits will be symmetrically disposed in space about the nucleus. If a subshell is circular and has n_e electrons, the n_e separate circular orbits may be combined into a single circle on which the electrons are equally spaced if the shell is filled. To determine the maximum possible number of electrons per shell, simply add the superscript numerals in any horizontal row of the table, including all of the boxes that have characters in them.

wavelength can be very short (a few nanometers), thus enabling the imaging of objects too small to be seen by visible light through optical microscopes. The wave property of electrons accounts for the *bremstrahlung* emitted by the electrons moving in circular paths within a cyclotron accelerator, and for the radiation emitted from the laterally accelerated electron beam in a *free-electron laser.*

In the year 1926, the Austrian physicist Erwin Schrödinger (6) (1887-1961) extended de Broglie's concept to formulate his theory of *wave mechanics,* in which the electrons of Bohr's atomic model are replaced by three-dimensional standing waves called *spherical harmonics.* From this concept, Schrödinger derived his now-famous *wave equation* (expressed below in coordinates x, y, and z):

$$\frac{1}{2m}\left(\frac{h}{2\pi}\right)^2\left(\frac{\partial^2}{\partial x^2}+\frac{\partial^2}{\partial y^2}+\frac{\partial^2}{\partial z^2}\right)\Psi - V\Psi = \frac{h}{\sqrt{-1}}\frac{\partial\Psi}{\partial t} \qquad [26]$$

where the symbols $\partial^2/\partial x^2$, etc., are second partial derivatives, with respect to the cartesian coordinates of the variable Ψ, which is the *wave function* describing the particle (electron) in terms of its velocity, energy, position, etc. As before, h is Planck's constant, V is the potential of the electron in the field of the nucleus, m is the mass, and t is time.

Then, in 1928, the English physicist Paul Adrien Maurice Dirac (7) included in the Schrödinger wave equation the effects of the angular momentum of an electron due to its *spin* about its own axis, which can have only the values $\pm\frac{1}{2}(h/2\pi)$. This essentially completed the theory of wave mechanics as it is accepted today. Although the wave mechanics theory provides us with no readily interpretable physical picture of an atom, like the one we can infer from the Bohr-Sommerfeld theory, it has been very useful in explaining and predicting the permissible energy levels of atoms.

In the theory of wave mechanics, each electron within an atom is completely characterized by four quantum numbers, which are tabulated in Table 3.

The vector of orbital angular momentum (OAM) of an electron is constrained to have only $2(l + 1)$ discrete directions of orientation relative to an external electric field, rather than being continuously variable from 0 to 4π steradians. The maximum permissible magnitude of the component of the OAM vector along the direction of any external electric field is the value given by Eq. [20], namely $n(h/2\pi)$. As in the Bohr-Sommerfeld theory, all filled shells have electrons arranged in symmetrically disposed pairs, so that their vectors of momentum cancel one another in an undisturbed atom.

In atoms with a relatively simple structure, such as those of hydrogen, helium, lithium, beryllium, etc., the primary quantum number is the principal determiner of the energy of any electron. When an atom contains many electrons, the electrons in all filled shells may be disregarded for computation of *changes* in the energy of the whole atom, because their vectors of total angular momentum add to zero. Changes of total atomic energy are determined only by the electrons orbiting in unfilled shells. As in the Bohr-Sommerfeld theory, an electron is designated as *s, p, d, f, g, h, i, j, k,* etc., according to its value of l, as shown in Table 2. The German physicist Wolfgang Pauli (1900-1958) added his *exclusion principle* to the theory in the decade of the 1920s. This states that *no two electrons within an atom can have the same set of quantum numbers n, l,* m_l, and m_s.

These four quantum numbers characterizing an electron in orbit determine the *state* of this electron, and to each state there corresponds a specific value of electron energy. The complete electronic configuration of an atom, as described by the set of all existing quartets of values of the quantum numbers, is called its *atomic state.* When the individual electronic energies are summed for all electrons within the atom, a value of total *internal energy* is determined for that atom, *leaving*

Table 3 Wave-Mechanical Quantum Numbers

Symbol	Name and description	Possible values
n	Principal quantum number; describes the Bohr shells for orbiting electrons	1, 2, 3, 4, 5, 6, ...
l	Orbital quantum number; describes the orbital angular momentum, whose vector has a magnitude of $(h/2\pi)\sqrt{l(l+1)}$	0, 1, 2, 3, 4, 5, ...
m_l	Orientation quantum number; describes the direction of the OAM vector relative to an external electric field E	$-l$, $(-l+1)$, . . . , 0, . . . $(l-1)$, l
m_s	Spin quantum number; describes the vector of spin angular momentum as parallel or opposite to the vector of OAM	+½, −½

out of consideration any constant kinetic energy that the atom might have by virtue of a steady velocity, e.g., its thermal motion between collisions with others in a gas or liquid. The permissible numerical values of internal energy, atomic or electronic, are termed *levels.* It is possible for certain states to have the same numerical values of energy, and these levels are called *degenerate.* In certain texts, the discussion of energy levels does not always clearly distinguish between atomic energy and electronic energy. This is not usually critical, because we are frequently interested in changes of energy, which are often determined for the whole atom by the behavior of just one valence electron.

An atom exists in its lowest possible energy level when all of its electrons are in their smallest permitted orbits and/or have those quartets of quantum numbers corresponding to minimum energy. This is the normal situation for most atoms at finite temperatures, in the absence of any outside influences. This minimum possible atomic energy is called the *ground level* and the lowest possible energy for each electron is given the same designation, although the numerical value of the total atomic energy is not equal to the value of energy for any one electron. The corresponding state, whether of the whole atom or of just one electron, is called the *ground state.* All higher levels and their respective states are said to be *excited.* Creation of excited states usually involves the transition of one or more electrons from existing to larger orbits, although the reversal of the vector of spin angular momentum for just one electron can cause a small change in level.

For most of the species of atoms and ions of interest in lasers, the total angular momentum of an atom can be determined by first adding the vectors of orbital angular momentum for all the electrons, then adding the vectors of spin angular momentum of all these electrons, and finally adding the vector sum of all the orbital momenta to the vector sum of all the spin momenta. Because a vector is a physical quantity having both a magnitude and a direction, the addition of two vectors must be done by drawing in the common plane of these vectors a parallelogram, two adjacent sides of which are represented to scale by the vectors; the diagonal of this parallelogram, which is intermediate in direction between the two vectors being added, is their vector sum. The diagonal connecting the tips of the added vectors represents their difference (with appropriate sign). Therefore the vector of total momentum will, in general, differ in both magnitude and direction from those of any of the component vectors. The process of vector addition described above is represented mathematically by

$$\mathbf{J} = \mathbf{L} + \mathbf{S} \qquad [27]$$

where **J** is the vector sum of all angular momenta, **L** the vector sum of all orbital angular momenta, and **S** the vector sum of all spin angular momenta.

In undisturbed atoms L (the magnitude of the vector) is zero for all orbital shells that are closed, as is S. However in any *n*th shell having an odd number of electrons, $S_n = \pm\frac{1}{2}(h/2\pi)$, and $L_n = \pm(h/2\pi)$. The modern quantum theory of atomic structure shows that, for each fixed numerical value of S, there are (2S + 1) possible different spin configurations for a whole atom. It is customary to measure S and L in units of the basic quantum of angular momentum, $(h/2\pi)$. Therefore the numerical values of S are 0, 1/2, 2/2, 3/2, 4/2, . . . , and those of L are 0, 1, 2, 3, 4, ... Therefore the values of (2S + 1) are 1, 2, 3, 4, ... The corresponding states are called *singlets, doublets, triplets,* etc. The transitions of atomic energy between spin states that are singlets yield radiation of only one wavelength, transitions between doublets give two wavelengths, transitions between triplets produce three wavelengths, and so on.

The reader may have seen papers dealing with photodynamic therapy, in which a parenterally injected photosensitizer undergoes photochemical change by weak visible irradiation in malignant tissue, releasing atoms or molecules that destroy the cancer cells. Hematoporphyrin derivative, the agent most widely used in current photodynamic therapy, produces singlet oxygen, in a highly energetic state. The ground state for oxygen is a triplet.

The trio of values J, L, and S determines the energy level of the atom as a whole. This level is characterized by the symbol

$$^{(2S+1)}X_J$$

where the letter X stands for one of the letters $S, P, D, F, H, I, \ldots$, depending upon the magnitude of the total orbital angular momentum, L, as follows:

$$\mathrm{L} = 0,\ 1,\ 2,\ 3,\ 4,\ 5,\ 6,\ \ldots$$

$$X = S,\ P,\ D,\ F,\ G, H, I,\ \ldots$$

The superscript (2S + 1) before the letter denoted by X characterizes the level as singlet, doublet, triplet, etc. In typical energy level diagrams like that for Nd^{3+} in YAG (yttrium-aluminum-garnet), shown in Fig. 8, the important levels are designated in this way.

b. Absorption and Emission of Radiation by Atoms. To drive an atom from its ground state to any other state having higher total energy, there must be an input of energy from an external source. If this input has exactly the magnitude corresponding to the difference in level between the two states, it will cause an electron in an outer shell to change its quartet of quantum numbers (Table 3) to a new set of values associated with the higher level. This process usually involves a jump of the electron to a larger orbit. When this electron later returns to its original level, energy must be emitted by the atom. Such relaxative emission can be accomplished by one of two basic means: *mechanical transfer* or *radiative transfer.* Mechanical transfer may occur by *collision* of the atom with another of lower energy (as in gases or liquids), or by *excitation of vibrations* (as in solids). Radiative transfer is the emission of a photon (a wavelet), whose wavelength is always just equal to

$$\lambda = v/f = v\ h/\Delta E_{ji} = v\ h/(E_j - E_i) \qquad [28]$$

where v is the speed of light in the medium surrounding the atom, E_j denotes the upper level of energy, and E_i is the lower level of energy. The symbol E is not to be confused with the symbol E of earlier equations, which denotes electric field intensity. At this point it is appropriate to note that the *frequency* of the emitted radiation is independent of the surrounding medium, and is *always* equal to $f = \Delta E_{ji}/h$. As the wavelet propagates through surrounding material of refractive index η, its velocity is reduced, with a corresponding decrease in the wavelength λ. Thus the *photonic energy* ΔE_{ji} is always proportional to the frequency of the wavelet.

This photonic energy may be expressed in several different units, three of which are in common use:

1. Electron volts (eV)
2. Reciprocal centimeters (cm^{-1})
3. Joules (J) (1 J = 1 W $\times$ 1 sec)

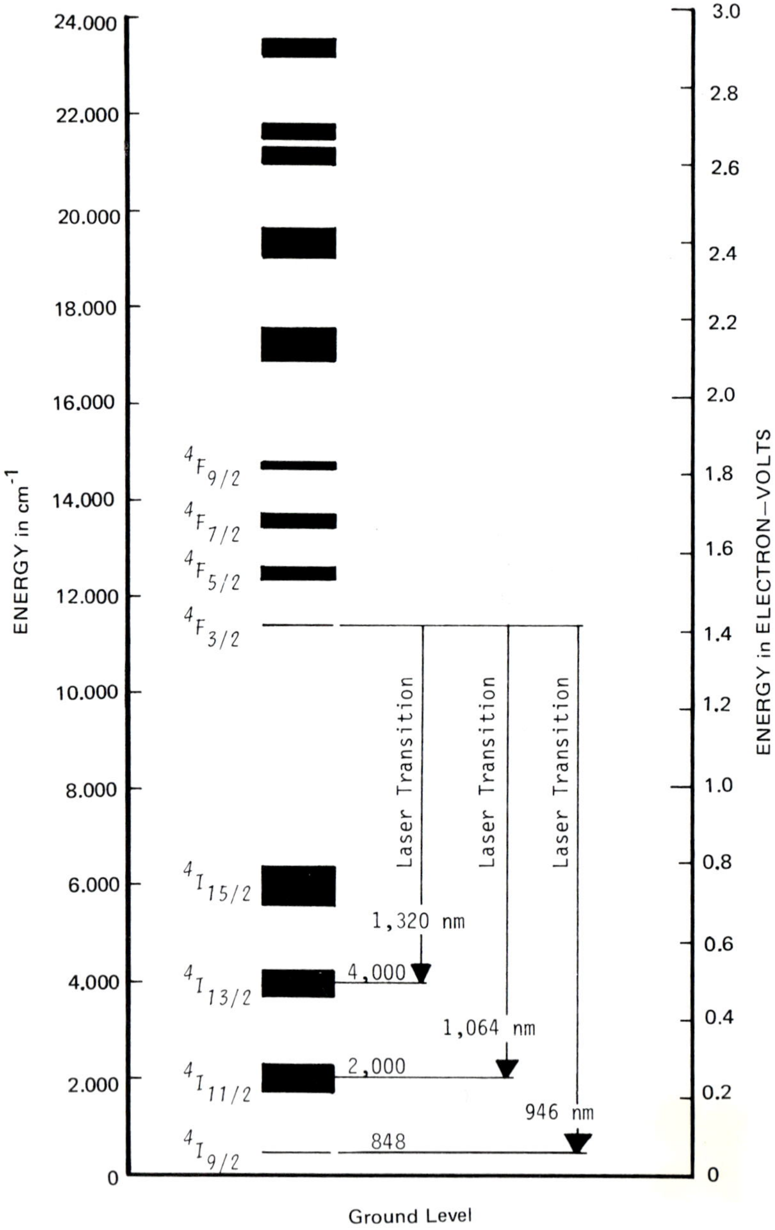

Figure 8 Energy level diagram for neodymium ions (Nd^{3+}) in a matrix of crystalline YAG (yttrium-aluminum-garnet). Note that the quantum efficiencies of these three laser transitions are 64.9, 82.5, and 92.6%, respectively.

One electron volt is the energy required to move an electron in an electric field through a potential difference of 1 V; it is equal in magnitude to 1.6022×10^{-19} J, because energy is the product of electric charge and potential difference. One reciprocal centimeter, which is a measure of photonic energy because the energy of a photon is inversely proportional to its corresponding wavelength, is equivalent to 1.9865×10^{-23} J. Finally, 1 eV is equivalent to 8065 cm^{-1}.

The downward transitions of electrons between upper and lower levels are governed by certain selection rules that determine which transitions are possible in given circumstances, which can be made by emission of radiation, and which can be made only by mechanical exchange between an excited atom and another of lower level. At each level above ground, there is a probable *lifetime* (or relaxation time) before the electron descends spontaneously to a lower level by emission of energy, ranging from about 1×10^{-8} sec to several seconds. The lifetime of a level or state is the reciprocal of the probability that any one atom will undergo a transition from that condition to another within a long period. Those levels or states that have relatively long lifetimes are said to be *metastable.* An atom can *quickly* descend from a metastable level or state to a lower one only through exchange of energy by mechanical means.

If the excitation of an atom from ground to a higher level occurs by absorption of radiation whose photonic energy just equals the difference between the upper level and ground, or between an excited level and some higher level, the process is called *resonant absorption.* This phenomenon is clearly observable only in gases, and is somewhat blurred in liquids and solids for reasons to be explained later. If such absorption is followed by *radiative* downward level transitions of magnitude ΔE_{ji} less than ΔE_{gj} absorbed, the emitted radiation will always be of greater wavelength than the exciting radiation because wavelength is inversely proportional to photonic energy. This secondary radiation is called *fluorescence.* If it persists after the exciting radiation has ceased, it is termed *phosphorescence.* Fluorescence plays a major role in the operation of many practical lasers. It is a special case of a more general phenomenon known as *spontaneous emission,* which is the name given to any unstimulated downward transition in level of an excited atom that occurs with emission of radiation. *The light from all natural sources, and most manmade sources, is produced by spontaneous emission from atoms, ions, or molecules,* except for some reported, but unproved, laser emissions from other planets. Ions are atoms in which one or more electrons are removed (positive ions) or added (negative ions), so that the negative charges of the electrons do not exactly balance the positive charges of the protons in the nucleus. For spontaneous emission to occur, there must be a preceding excitation by some external source(s) of energy in the form of heat, light, changing electric fields, collision with energetic particles, or chemical reaction.

c. Emission and Absorption of Radiation by Molecules. The energy levels of molecules, which are aggregations of two or more atoms of the same or different elements bound together by electric forces, are those of the constituent atoms plus those associated with internal *vibrations* of atoms relative to one another, with internal *angular oscillations* of groups of atoms relative to other parts of the molecule, and with *rotations* of the molecule as a whole. At the power densities of radiation occurring in most natural situations, most of the changes of level involved in absorption and emission of light by molecules are those associated with these mechanical phenomena. In general, such transitions are of smaller magnitude than electronic transitions, and so the wavelengths of the exciting or spontaneously emitted radiation are longer than those of the light involved in changes of electronic energy. The same types of quantum rules as those we have previously described for atomic states govern changes of molecular state. However, as molecules become more complex (i.e., organic molecules or inorganic polymers), the number of permissible states and levels can be enormous. Thus in the fluorescent dyes used in lasers, a large spectrum of wavelengths is emitted, and there can be a wide absorption spectrum having many broad peaks. The most important of all molecular lasers for surgery is the carbon dioxide laser, whose output wavelengths lie in the midinfrared portion of the spectrum of electromagnetic radiation, near 10 μm (10×10^{-6} m).

d. Energy Levels in Gases, Liquids, and Solids. The chief characteristic that distinguishes solids, liquids, and gases from one another is *density,* or (*MASS*)/(*VOLUME*). For any given element, density is proportional to the number of atoms per unit volume, which increases as the element changes its *physical* state (not to be confused with its atomic-electron state) from gas, to liquid, to solid. This change occurs by virtue of a decrease in temperature or an increase in pressure or both. In appearance, the transition from gas to liquid is marked by the onset of *visibility* because gases are normally invisible, but liquids can reflect light. The transition from liquid to solid is characterized by the ability of solids to sustain *tensile* forces, whereas the same elements in liquid or gaseous form can sustain only *shearing* and *compressive* forces. For our present purposes, we may think of a gas→liquid→solid transition as a process in which the interatomic distance decreases from very large multiples of the atomic diameter down to just a few. From Eq. [21] for the radius of an electron's orbit, we may estimate a typical atomic diameter by doubling the value of r_e for $n = 6$: $2r_e = 3.81 \times 10^{-9}$ m. Thus about 4 nm may be considered the low limit of interatomic spacing in solids, and billions of nanometers as the upper limit in gases at the fringes of the earth's atmosphere. One important consequence of the reduction of the distance between adjacent atoms in a gas→liquid→solid transition is the *broadening* of the permitted energy levels from sharp lines or sets of lines into *bands* of energy. This effect is due to the influence of the electric fields of adjacent atoms upon one another. As would be expected, the broadening of energy levels begins at larger interatomic distances for the upper levels (outermost electrons) and results in wider bands at these upper levels for any given interatomic spacing, as shown in Fig. 9. Furthermore, mechanical exchange of internal

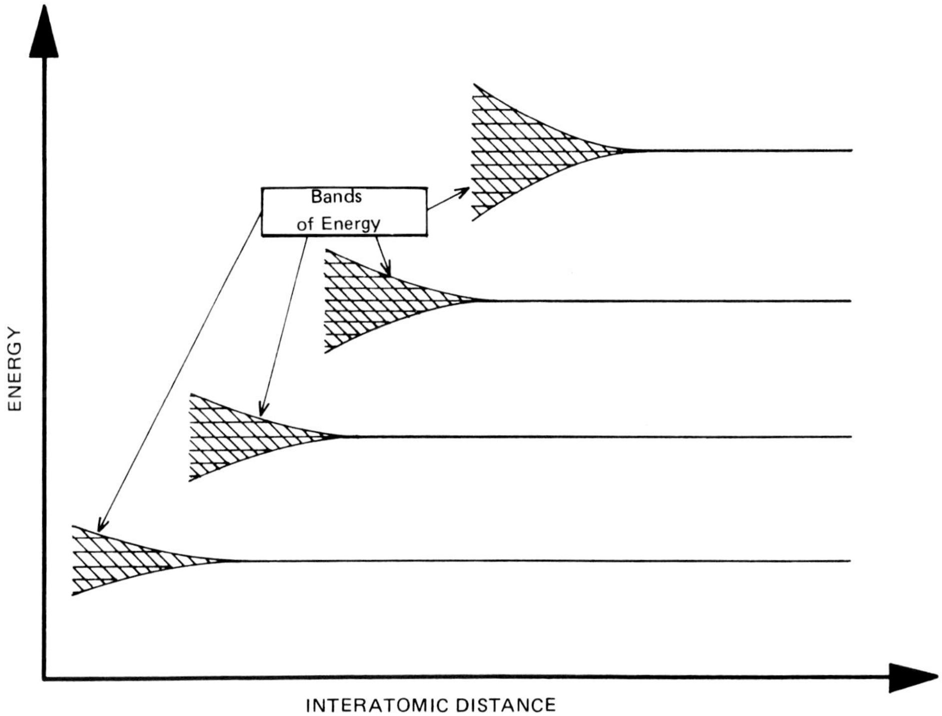

Figure 9 Broadening of discrete energy levels into bands in solids, because of the interaction of the electric fields of adjacent atoms. The energy levels of outer-shell electrons are affected first as the interatomic distance is reduced in a gas→liquid→solid transition.

energy from atom to atom occurs in solids by *vibration,* or *phonon* transfer, rather than by the *collisions* that occur because of thermal motion in gases and liquids. A phonon is a quantum of vibrational energy, somewhat analogous to a photon as a unit of radiant energy. Only in a gas are the discrete levels of quantum theory clearly discernible by the sharpness of the observed spectral lines of emitted or absorbed light.

e. Effects of Temperature on the Distribution of Atoms and Molecules Among Possible Energy Levels. In the following discussion, wherever the word *atom* appears, it can be considered interchangeable with the word *molecule.* Under conditions of *thermal equilibrium* (i.e., constancy of temperature in time and space), any homogeneous collection of atoms in a given space will be preponderantly in lower levels unless excited by some external agency. If we neglect degenerate levels, the distribution of the total atomic population, numbering N_0, among all possible levels, is given by Boltzmann's equation, which describes the probability of conversion of thermal energy to internal atomic energy.

$$N_j/N_i = \epsilon^{-(E_j - E_i)/\kappa T}, \text{ (NUMERIC)} \quad [29]$$

where N_j = the number of atoms in any upper level j
N_i = the number of atoms in any lower level i
ϵ = the base of the natural logarithms = 2.71828 . . .
E_j = the energy of the upper level, in joules
E_i = the energy of the lower level, in joules
κ = Boltzmann's constant = 1.3805×10^{-23} J/°K
T = the absolute temperature of the atoms, in °K

Equation [29] says, in essence, that the ratio N_j/N_i must always be less than unity, because the exponent of ϵ is always a negative number. In other words, there will always be fewer atoms in any upper energy level than there are in any lower level. If the number of atoms at the ground level (which we shall arbitrarily set equal to zero energy) is designated as N_g, then the ratio of N/N_g, where N is the number at any higher level E, varies with E as shown in Fig. 10. The graph in this figure is shown as continuous, for ease of comprehension, although the actual values of energy permitted by the quantum rules are only discrete levels. From the figure it can be seen that the population of atoms in any energy level above ground will be only a fraction of that in the ground level, and that this fraction becomes smaller as the energy of any upper level, E, increases. At absolute zero temperature, only the ground level would be populated, and at infinite temperature all energy levels would be equally populated. This situation is not favorable to the phenomenon known as *stimulated emission,* which is the basis of laser action.

2. Stimulated Emission, Population Inversion, and Pumping

In his fundamental paper (2) published in 1917, Albert Einstein predicted the possibility of stimulated emission, although no one at that time possessed the physical means to produce it. If an atom (or a molecule) in an excited state from which there is a possible radiative transition to a lower level should be struck by a photon (a wavelet) whose energy is exactly equal to the energy change for this downward transition, the excited atom will drop to the lower level and emit a second photon (wavelet) that has exactly the same energy, frequency, wavelength, temporal phase, spatial phase, polarization, and direction of propagation as the stimulating one. The first wavelet is not absorbed in this process, but propagates onward in perfect

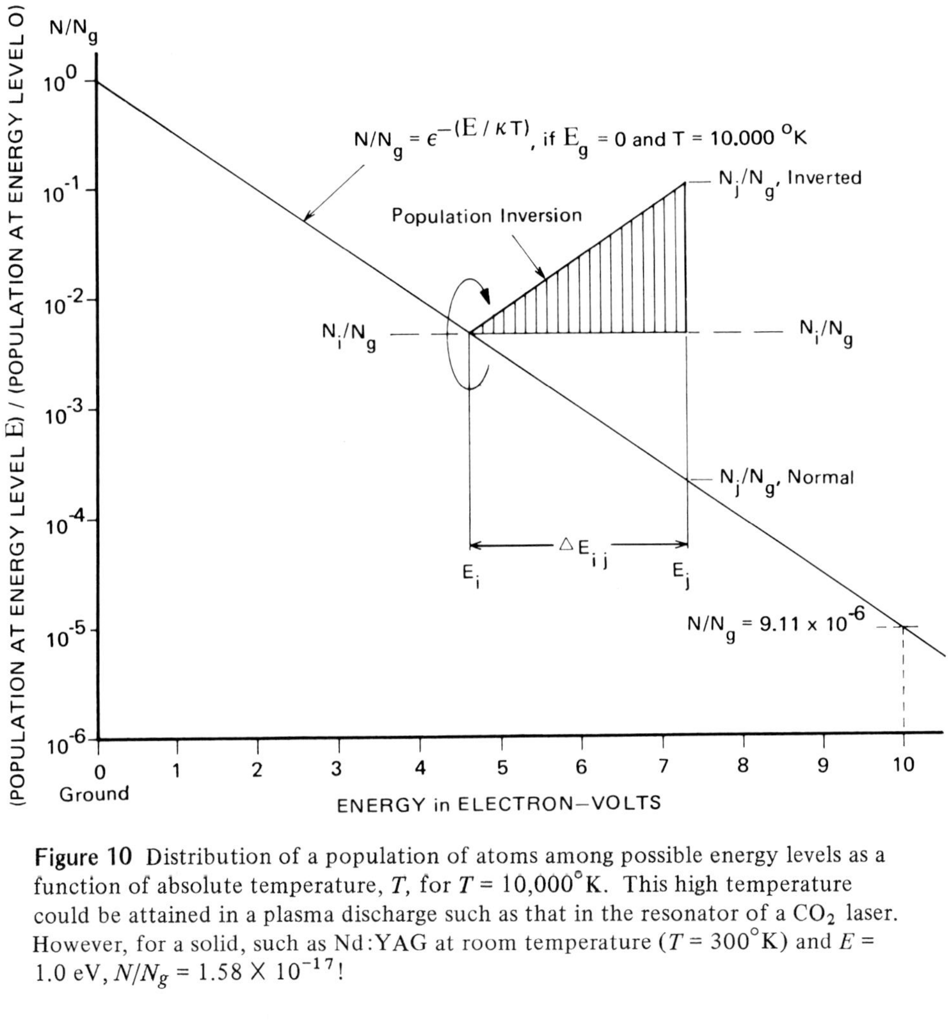

Figure 10 Distribution of a population of atoms among possible energy levels as a function of absolute temperature, T, for $T = 10{,}000^\circ$K. This high temperature could be attained in a plasma discharge such as that in the resonator of a CO_2 laser. However, for a solid, such as Nd:YAG at room temperature ($T = 300^\circ$K) and E = 1.0 eV, $N/N_g = 1.58 \times 10^{-17}$!

synchronism with the second. Thus an *amplification* has occurred. Although this stimulated emission can occur naturally in any collection of excited atoms, it will not be sustained under normal conditions, but will be sporadic. This is so because there are so many more atoms in low energy states than in highly excited ones that spontaneously emitted wavelets have a much higher probability of being absorbed into unexcited or low-level excited atoms than of striking high-level excited atoms that can be stimulated to emit *coherent* radiation. As wavelets of $\lambda = c\ h/\eta\ \Delta E_{ij}$

pass through a collection of atoms in thermal equilibrium, their intensity is *attenuated* because of resonant absorption and scattering by less-excited atoms of the same element and by dissipative absorption into atoms of other substances (e.g., the walls of the container):

$$p_z = p_o \epsilon^{-az}, (POWER)/(AREA) \qquad [30a]$$

where p_o = the power density of the incident wavelets at the boundary of the space containing the atoms
p_z = the power density of the wavelets at some point along the path of the incident rays inside the atom-containing space
a = the *attenuation coefficient,* $1/(LENGTH)$
z = distance along the path of the wavelets, $(LENGTH)$

Simultaneously, the intensity of these wavelets may be augmented by the process of stimulated emission from atoms already in the corresponding upper excited state:

$$p_z = p_o \epsilon^{+gz}, (POWER)/(AREA) \qquad [30b]$$

where g = *gain coefficient,* $1/(LENGTH)$

It is evident that sustained stimulated emission will occur only when $g \geqslant a$. For this condition to exist, there must be more atoms in the upper excited state than in the lower state of the corresponding downward transition in energy: $N_j > N_i$. Such a condition is called a *population inversion,* because it implies that part of the graph of Fig. 10 must be turned upside down about the horizontal line whose ordinate is N_i/N_g, so that for some energy levels E_j and E_i the ratio N_j/N_i is greater than 1. For this to occur, two factors are needed:

1. A source of external energy that can be injected into the system
2. A metastable level somewhere above ground, from which there is at least one possible radiative downward transition to a level at or above ground

The process of creating a population inversion is called *pumping.*

a. Problems of Pumping: Two-Level Energy Systems. Pumping is not as simple as the name would suggest. It is essentially the addition of energy to a collection of atoms of one species in such a way that at least one upper level of energy becomes more densely populated than some other level lying below it. Because most of the upper energy levels of an atom have relatively short (about 1×10^{-8} sec) lifetimes and, from Eq. [29], will have very small populations at ordinary temperatures, pumping an atomic system between just two energy levels (ground and one upper level) is very difficult. *At least half* the population of the ground level must be pumped to the upper level, from which it rapidly falls back to the ground level in a steady cascade. This is analogous to a water fountain in which the water falls back into the pool as fast as it is sprayed upward by the pump. Let us assume for simplicity that the total time required to pump a single two-level atom to its upper state and allow it to relax back to ground is three times the lifetime of the upper state: $\tau = 3\tau_u$, where τ_u is the lifetime of the upper level. Let us also assume a modest energy difference between ground and the upper level of 1.0 eV. Then the power required from the pumping source, averaged over the whole cycle, is 1.6×10^{-19} J divided by $3\tau_u$; if τ_u is assumed to be 2×10^{-8} sec, then the average pump power per atom is 2.7×10^{-12} W. If the material is a solid, for which the density of atoms is about 5×10^{22}/cc, the total average pump power needed just to attain the start of a population inversion (i.e., to put 50% of the atoms in the upper level) is 6.7×10^{10} W/cc! In this situation, the power from the pump is all lost as spontaneous emission from the atoms returning to the ground level.

b. Three-Level Systems. It is much more effective to use three energy levels of a material, by pumping atoms out of the ground state to some upper level lying far enough above ground so that there is at least one intermediate level that is metastable. If this is the case, then a source of moderate power can pump atoms up out of the ground state at a leisurely rate to the upper level, from which they rapidly relax back to the intermediate level, where they become *trapped* temporarily because of the comparatively long lifetime of that level. A population inversion can then be easily created by a much lower power input from the pumping source than is required in a two-level system, because there is no rapid downward cascade of atoms back to the ground state. However, in a three-level material, the number or atoms in the metastable intermediate level must still be at least 50% of the number in the ground level, because ground is the terminal level of the radiative transition, and no sustained stimulated emission can occur until there are *more* atoms in the intermediate level than in the ground. Very few practical lasers use active materials having three-level energy systems. A notable exception is the ruby laser, the first working laser ever built.

c. Four-Level Systems. Although the pumping power needed for a three-level system is much smaller than for a two-level system, the *total threshold energy* input is still very high, because at least half the atoms normally in the ground state must be pumped up to the metastable level, which, at ordinary temperatures, will be virtually unpopulated. Furthermore, the primary upper level to which the atoms are pumped before relaxing back to the metastable intermediate level must have an energy *higher* than that of the metastable one, and consequently the energy difference between the primary level and the metastable level is *lost* for the purpose of producing stimulated emission. The threshold energy input for stimulated emission can be greatly reduced if a material can be found that has a fourth energy level, lying between the metastable and ground, such that a radiative transition is possible from the metastable level to this fourth level. The advantage of such a system is that the lower (fourth) level of the radiative transition will be nearly unpopulated at ordinary temperatures. Therefore attainment of a population inversion requires pumping only a *small number* of atoms up to the primary level and thence by rapid relaxation to the metastable level, where almost any number of atoms greater than zero will exceed the number in the fourth level. Most practical lasers use four-level materials.

Figure 8 shows the energy level diagram of neodymium ions (Nd^{3+}) distributed in low concentration (0.5-2.0%) in the crystal lattice of the material yttrium-aluminum-garnet, whose chemical symbol is $Y_3Al_5O_{12}$. Pumping of this system is usually done by light energy obtained from a high-intensity lamp using an electric discharge in xenon gas, which produces an intense, near-white light having a broad, continuous spectrum. The pumping energy can excite the neodymium ions to any level lying in the range above the metastable level designated by ${}^4F_{3/2}$, according to the symbol described at the end of Sec. II.B.1.a. The lifetime of this state is about 230 μsec. The photonic energy of the light from the xenon lamp will seldom be of exactly the right magnitude to excite Nd^{3+} ions into one of the upper bands, and the excess will appear as vibrational energy of these ions in the crystal lattice. From one or more of these upper bands, the Nd^{3+} ions relax back to the ${}^4F_{3/2}$ level by nonradiative transitions. From this metastable level, they return by radiative transitions to one of the three intermediate levels or bands designated by ${}^4I_{13/2}$, ${}^4I_{11/2}$, or ${}^4I_{9/2}$, whose respective central energies are 4000 cm^{-1}, 2000 cm^{-1}, and 848 cm^{-1}. These transitions give rise to laser radiation at wavelengths of 1320 nm, 1064 nm, and 946 nm, respectively. Of these, the 1064-nm wavelength is usually emphasized in laser design, to the exclusion of the others.

When the neodymium ion is dispersed at low concentration in a matrix of glass (an *amorphous*, or noncrystalline material), the dominant wavelength is 1060 nm. Because the lifetime of the upper laser level in Nd:YAG is considerably longer than that of the lower, this material is suitable for either pulsed or continuous wave operation. In Nd:glass, however, although the relative lifetimes are favorable, the high absorption losses and comparatively low thermal conductivity of glass limit this material to pulsed operation.

Figure 11 shows the energy level diagram of the carbon dioxide molecule. The upper energy levels of CO_2 are those of the symmetrical stretch, the asymmetrical stretch, and the bending modes of molecular vibration, split into multiplets by the combination of these with rotations of the molecule. Note that the upper laser level of the CO_2 molecule is nearly equal to an excited level of the nitrogen molecule (N_2). Hence nitrogen gas is commonly mixed with carbon dioxide in practical lasers so that the N_2 molecules, which are easily excited by an electric discharge, can excite the CO_2 molecules by collision. Helium gas is also added to the mixture,

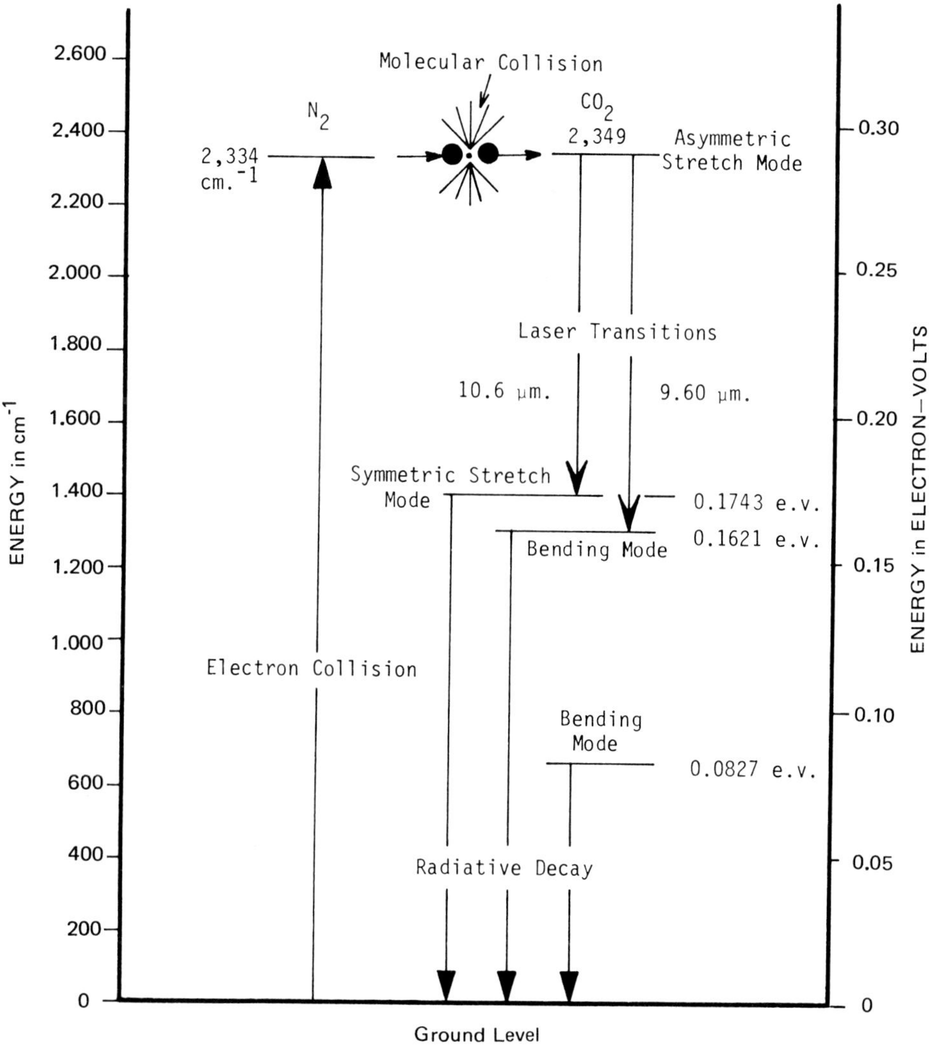

Figure 11 Energy level diagram for nitrogen and carbon dioxide in a CO_2 laser, showing mechanisms of energy exchange and modes of vibration of the CO_2 molecule. Rotational multiplets of energy level are not shown, for simplicity, but they would lie above the corresponding vibrational mode levels.

making up the bulk of the composite gas, to provide cooling (i.e., removal of the energy from the CO_2 molecules in the lower intermediate level 010). Nitrogen is a good energy-exchange medium for pumping the CO_2 molecule because the excited state of the N_2 molecule has a long lifetime, making a population inversion in this species easy to achieve. The commonly used wavelength of 10,600 nm is produced by transition of the CO_2 molecule from the (001) asymmetrical stretch mode to the (100) symmetrical stretch mode. The 9600-nm wavelength is usually suppressed by appropriate design. For simplicity in Fig. 11 the rotation-induced multiplets of energy level are not shown.

The energy level diagram for argon ions is not shown in this chapter because it adds nothing to our basic understanding of energy levels. Suffice it to say that because the laser transitions of argon ions are much higher in level than those of neutral argon atoms, the intrinsic efficiency of the argon ion laser is considerably lower than that of an atomic argon laser, which is, in turn, considerably lower than those of most molecular gas lasers. The major laser transitions of the argon ion laser occur between upper levels in the vicinity of 20 eV and lower levels around 17 eV.

III. GENERATION OF COHERENT RADIATION BY PRACTICAL LASERS

A. Laser Resonators

1. General Principles

The kind of stimulated emission that we have described in the foregoing sections would be neither continuous, nor coherent, nor collimated (i.e., having parallel rays) without the presence of a *resonator* (also often called a *cavity*). The term *resonator* is more physically correct, because a cavity, by definition, is an empty space surrounded by some enclosing material. Only in a gas laser can the resonator be thought of as being empty, and then only partially so. Clearly the sort of cumulative process of generating stimulated emission previously discussed herein would be difficult or impossible to achieve unless some means were provided to ensure that the spontaneously emitted wavelets and their stimulated, coherent companion wavelets travel more or less in one direction, and groups of such wavelets resulting from the spatially random emission of myriads of atoms are somehow coordinated in space and time. Both of these conditions are achieved by forming the active material, or *laser medium,* into a circular-cylindrical shape whose axial length is usually much greater than its diameter, and by locating this cylinder of material between two *mirrors* whose reflective axes are optically collinear with the axis of the cylinder. Many surface geometries are possible for these mirrors, but only a few produce *stable* patterns of reflection through the active material, and of these few only a handful are employed in practical lasers. The combination of a cylinder of active material located coaxially between two mirrors of suitable surface contours and reflectivity is called a *laser resonator.*

a. Stability. In practical lasers, appropriate mechanical means must be provided to hold the cylinder of laser material in the proper position relative to the mirrors, to allow independent alignment of the mirrors, and to permit the introduction of pumping energy. The surface geometries of the mirrors are critical to proper operation. In general terms, all surface geometries will be stable that produce a pattern of reflection such that any ray originating within the material in a direction nearly parallel to the resonator's axis will always remain within a definite volume of space about the same size as, and lying mostly within, the cylinder of laser material, no matter how many reflected transits it makes between the mirrors. Unstable geometries of the mirrors will, sooner or later, reflect a ray out of the cylinder of active material so that it is thereafter lost and will not again pass through the laser medium to cause stimulation of emission from excited atoms. Plane, parallel mirrors were used by Theodore Maiman (8) in 1960 in building the very first working laser, utilizing a cylindrical rod of ruby (aluminum oxide, Al_2O_3, doped with chromium ions, Cr^{3+}) with flat, polished end surfaces silvered for reflection. Although plane mirrors seem to be a logical choice, the practical difficulties of achieving perfect parallelism are such that they are seldom used in working lasers. A very slight deviation of one-mirror plane from being parallel to the other quickly results in escape of the reflected rays from the cylinder of laser material.

For several reasons, the surface geometry most commonly employed is a *spherical sector* having a radius greater than the axial separation of the mirrors. The two mirrors are usually of the same radius and are always aligned so that their centers of curvature lie on the optical axis of the laser cylinder. The criterion for stability of spherical sector mirrors can be stated quite simply: *Either the center of curvature of one of the mirrors, or the mirror itself, but not both, must be located between the other mirror and its center of curvature* (9).

b. Longitudinal Modes. The rays reflected repeatedly back and forth between the mirrors of a stable resonator will always lie within a three-dimensional surface having a cross-section in any plane normal to the axis of the resonator that is geometrically similar to the axially viewed silhouette of the mirrors. For plane, round mirrors this surface would be a simple circular cylinder. For spherical mirrors, it will have the shape of a slender hourglass, whose center will lie at the center of the resonator if the mirrors are of equal radius and concave facing each other (the most common arrangement). This limiting three-dimensional geometric surface is called the *external envelope of rays,* and the space it encloses is called the *ray bundle.* It can be seen intuitively that every reflected ray of laser light lying within this bundle is the axis of propagation of a wavelet whose frequency and wavelength must be constant during each passage between the mirrors' surfaces, and whose wavelength must have a definite relationship to the distance between these surfaces, for the successive reflections to reinforce (coincide), rather than to interfere with one another. The reflected intensity at each mirror ideally should be very nearly 100% to produce maximum amplification within the resonator. However, in a practical laser, some of the radiation must be allowed to emerge for external use, and therefore one of the mirrors is made partially transmitting so that part of the radiation striking its surface will pass through. The other mirror must have a reflectivity greater than 99%. Such a highly reflecting surface can be made of a polished *metal* or of successive thin layers of *dielectric* material whose refractive indices are alternately high and low.

The physical laws of reflection from smooth surfaces require that, at or near normal incidence to the reflecting surface, there must be a shift in the spatial phase of the light wave by one-half a cycle (180°), as well as a reversal of the direction of propagation. The net effect of these two changes is equivalent to folding the plane of the E wave back upon itself along a crease line parallel to the **E** vector at the mirror, and then turning the E wave upside down around the ray axis in this folded-back half plane. If the reflected wave is to reinforce the incident wave, both waves must have common zero points in space and their crests must coincide (i.e., they must be exactly in space phase with one another). This will occur *only if a zero point of each wave occurs at each reflecting surface.* If such is the case, a *standing wave* will exist between the mirrors. For this to be so, the distance between the mirrors along each reflected ray must be *an integer multiple of one-half the wavelength:*

$$Z_m = \xi\lambda/2 \qquad [31]$$

where $\xi = 1, 2, 3, 4, 5, 6, 7, \ldots$ (*NUMERIC*)
Z_m = the distance between the mirrors' surfaces

Because $\lambda = v/f = c/\eta f$, we can express Eq. [31] in terms of frequency and speed of propagation as well as wavelength:

$$f = \xi v/2Z_m = \xi/\Delta t_m \text{ Hz} \qquad [32]$$

where Δt_m = the time required for the wave to make one round-trip between the mirrors' surfaces.

Because the values of frequency given by Eq. [32] are a series of integer multiples of the reciprocal of the round-trip transit time, the difference between successive frequencies satisfying Eq. [32] is simply $1/\Delta t_m$. For a mirror separation of 1 m, the difference between the frequencies of any standing wave and the next is very nearly $\Delta f = 150 \times 10^6$ Hz.

Because the wavelengths of radiation emitted by lasers are many orders of magnitude smaller than the length of a typical resonator, the numerical values of ξ are exceedingly large integers. For example, in a resonator of 1 m in length, at a wavelength of 1000 nm, the value of ξ is 2 million.

The difference between successive wavelengths of standing waves in the resonator can be calculated by differentiating the basic relation among wavelength, frequency, and speed of propagation, $\lambda f = v$:

$$\Delta\lambda = -(\lambda/f)\,\Delta f = -\lambda^2/2Z_m \qquad [33]$$

The minus sign in Eq. [33] merely signifies that an increase in frequency corresponds to a decrease in wavelength, and vice versa. As an example, if the central wavelength of a laser is 1×10^{-6} m (1000 nm) and the resonator is 1 m long, the adjacent wavelengths are displaced by successive increments of only 0.0005 nm from the central wavelength. For a Nd:YAG laser operating at a central wavelength of 1064 nm, the width of the upper laser band of energy and the natural spreading of frequency caused by the finite radiating time of each individual neodymium ion will force the laser to radiate over a bandwidth of about 0.0491 nm centered around the nominal wavelength. The width of this *atomic "line"* is large enough to include about 98 sequential standing wavelengths in the resonator. These wavelengths which both satisfy the standing-wave criterion (Eq. [31]) of the resonator and lie within the width of the atomic laser line are called the *longitudinal modes* of the laser.

The gain of the resonator will be greater than the attenuation (see Eqs. [30a] and [30b]) for only a few of these possible longitudinal modes whose wavelengths are near the central one. If it is desirable to limit the laser output to just one longitudinal mode, a tuning device such as an *etalon,* essentially a small Fabry-Perot interferometer which, by being tilted off the perpendicular to the resonator's axis, selects just one of the many possible wavelengths, can be used to suppress the others by destructive interference.

2. Practical Details of Laser Construction

a. General Considerations for All Lasers. Obviously, the supporting structure that holds the components of a resonator must be rigid, because small deformations of the assembly can cause misalignments of mirrors and etalons that are critically detrimental to the operation of a laser. Furthermore, the structure ideally should be built such that changes of temperature, and their resultant changes of linear dimensions, do not cause critical misalignments. In lasers having long resonators, such as CO_2 and argon ion, with gases contained inside tubes of glass, quartz, or ceramics, the supporting structure may be made of a metal having a very small temperature coefficient of expansion, such as Invar, a special nickel-iron alloy.

The mirrors must be securely mounted in such a way as to allow for precise microadjustment of alignment. This is usually accomplished by placing the mirrors in structurally rigid mounting brackets, each of which has three adjusting screws whose axes are parallel to the optical axis of the laser. The mirror is held in firm contact with these screws at three equidistant points of its outer circumference by springs that eliminate backlash. Careful adjustments of these screws allow the optical axis of each mirror to coincide almost perfectly with the optical axis of the other mirror and with that of the laser cylinder. Mirrors must be replaceable, because they undergo slow deterioration in reflectivity with age.

The laser cylinder must be provided with adequate means for removing the waste heat generated by the pumping process, because both the pumping system and the laser material have losses. In most practical lasers, the overall efficiencies are rather low, ranging from a fraction of 1% to perhaps 25%. Therefore, the cooling system must be able to carry away large amounts of thermal power: in a 100-W Nd:YAG laser, for example, the overall efficiency is typically 2% or less, and so some 4900 W must be removed by the coolant. In this process, the temperatures within the resonator must be kept within the range in which laser operation is most effective and structural deterioration is minimal. Structural temperatures, except for those in ceramic laser tubes, must be kept near or below the boiling point of water, 100°C.

In gas lasers, seals must be provided at the ends of the laser tube to prevent escape of the laser gas or the entry of atmospheric gases. Some lasers, like the higher-powered CO_2 units, use flowing-gas systems, which require a source of premixed gases at controlled inlet pressures, and a vacuum pump to remove the exhausted gas while maintaining the correct pressure within the tube. Some lower-powered gas lasers utilize sealed tubes that are prefilled with gas at the correct pressure. The shelf life and operating life of such tubes are finite, usually in the range of a few months to 1 or 2 years before recharging or replacement is needed.

Pumping systems can be of many types, but those most commonly used in practical lasers with power outputs of up to 1 kw are the optical excitation and electrical discharge systems. Optical excitation by high-intensity lamps containing xenon gas is customarily used in solid lasers like the Nd:YAG, Nd:glass, and ruby. The pumping lamp is usually linear and parallel to the laser rod, with the axes of lamp and rod located at the focal axes of an elliptical reflective enclosure that surrounds both, providing reflection of the pumping light into the laser rod and a duct through which liquid coolant can be forced for removal of heat. The ruby laser built by Maiman used a coaxial helix of flashtube surrounding the ruby rod, but this is inefficient because the light emitted at the outer periphery of the helix is lost. In gas lasers, hollow cylindrical metal electrodes are often used, located at opposite ends of, and concentric with, the laser tube; between these electrodes a direct or alternating electric current will flow once the gas has been ionized by connection of the electrodes to an external high-voltage power supply. Such supplies provide voltages ranging from a few hundred to 50,000 V, and currents from a few milliamperes to hundreds of amperes. Alternatively, sealed tubes may be excited by radiofrequency electric fields applied externally to tubes made of dielectric materials, or internally in metal waveguide tubes.

Delivery of the laser beam from the resonator to the point of application on a target may be accomplished by one or more of three basic schemes.

1. Transmission through a flexible *optical fiber*
2. Reflection through an *articulated arm*
3. *Direct irradiation* from the resonator via lenses, mirrors, and beam-splitters

The first of these is used to the virtual exclusion of all others in lasers producing wavelengths in the visible (400-700 nm) and near-infrared (700-1400 nm) portions of the electromagnetic spectrum, because these wavelengths can be efficiently transmitted through fibers of optical glass or quartz (fused silicon dioxide). The second method is used with most carbon dioxide and other midinfrared (1400-20,000 nm) lasers because there are few materials suitable for transmissive fibers at these wavelengths, and those that have been developed to date are beset with a variety of problems such as toxicity, brittleness, water solubility, and lack of suitable solid cladding materials. Articulated arms, which will be discussed in some detail in later sections, are used also in the very short pulsed Nd:YAG lasers for ophthalmic applications, because these produce such high power densities in their beams that solids and liquids cannot transmit them without destructive effects such as self-trapping and self-focusing. Only mirrors of high reflectivity can deliver these beams by multiple reflections, and it is usually necessary to enlarge the beams for their *transflection* through the mirror sequence so as not to damage the mirrors by heating. The third scheme is used with some ultrashort-pulsed Nd:YAG lasers and also with relatively low-powered visible lasers used in *scanning* or *designating* applications, such as bar-code readers at the checkout counters of supermarkets and therapeutic lasers for analgesia or the stimulation of wound healing.

Finally, lasers for medical applications must be provided with power consoles containing the necessary components for energizing, cooling, and controlling the laser resonator. These consoles most often are mobile, and provide controls for laser-output power, exposure duration, and temporal mode of beam delivery (i.e., pulsed or continuous). The United States Food and Drug Administration (FDA) requires that certain warning labels be affixed to the laser structure and to each port (opening) through which laser rays can emerge; it further mandates that each laser used for medical purposes have an indicator showing the power output in the laser beam, usually called a *power meter.* To prevent accidental irradiation of objects and persons in the vicinity of the laser, a mechanical *shutter* is required to cut off the beam internally. This shutter is usually controlled by a *footswitch.* Certain *safety interlocks* are required by the FDA for all medical laser products to prevent electrocution and accidental irradiation when the enclosure of the laser system is opened for servicing or inspection.

b. Neodymium:Yttrium-Aluminum-Garnet Lasers. Lasers of this type designed to deliver continuous outputs of 100 W or more are often relatively immobile, because they require three-phase electric power for their energization and forced-flow water cooling from a wall faucet. Some newer designs producing lower output power for limited times of delivery can be operated from single-phase electric power lines and with internal liquid cooling that disposes of the waste heat via an air-cooled radiator in the console, much like the cooling systems used in most automobiles. These design problems are consequences of the low (< 2%) overall efficiency of the Nd:YAG laser, which is largely a result of the inefficiency of the xenon lamps used for pumping the YAG crystals. However, in terms of simplicity, durability, and dependability, the Nd:YAG laser has few equals among the whole gamut of modern lasers.

As previously mentioned, the pumping lamp is most commonly placed parallel to the laser rod, with lamp and rod each located at one focal axis of an internally silvered cylindrical housing having an elliptical cross-section normal to the axes of the rod and lamp. For maximization of output power from a single YAG rod, the reflective housing of the resonator may contain two or more lamp tubes located at the focal axes of an equal number of ellipsoidal shells, which are structurally combined so that they all have a common focal axis on which the laser rod is located. By means such as this, continuous wave output power of more than a kilowatt can be achieved. This scheme does not, however, significantly increase the overall efficiency of the laser because the xenon lamps are the major sources of loss in the Nd:YAG laser. With a single pumping lamp, 200 W of output power can be produced by a YAG rod about 2 cm in diameter and 20 cm in length.

Although the mirror surfaces may be formed by grinding, polishing, and silvering the ends of the Nd:YAG rod, it is more convenient in terms of adjustability to have the laser mirrors outside the rod. This arrangement also allows for tuning devices (e.g., etalons or prisms) and for various *switching* devices that can be used to generate short pulses of very high peak power. The rod ends may then protrude through the end plates of the elliptical resonator housing, so that the laser rays do not have to be transmitted through the liquid coolant, which is usually silicone oil or deionized water, both of which have appreciable absorption of radiation at a wavelength of 1064 nm. A liquid-tight seal around each rod end keeps the coolant from leaking out. At the crystal-air interface at each end of the YAG rod in this arrangement there will be reflections of the rays traversing the resonator between the mirrors. These rod end reflections can be reduced to zero for waves polarized with the **E** vector parallel to the plane of incidence, by grinding an oblique-plane face on each end of the YAG crystal, the normal to each face being inclined off the axis of the corresponding external mirror by the *Brewster angle.* The plane of incidence is that plane containing the axis of the YAG rod and lying perpendicular to the plane of the end face of the rod. These plane-polarized waves undergo zero reflection at the end face in either direction of transmission. Because there would

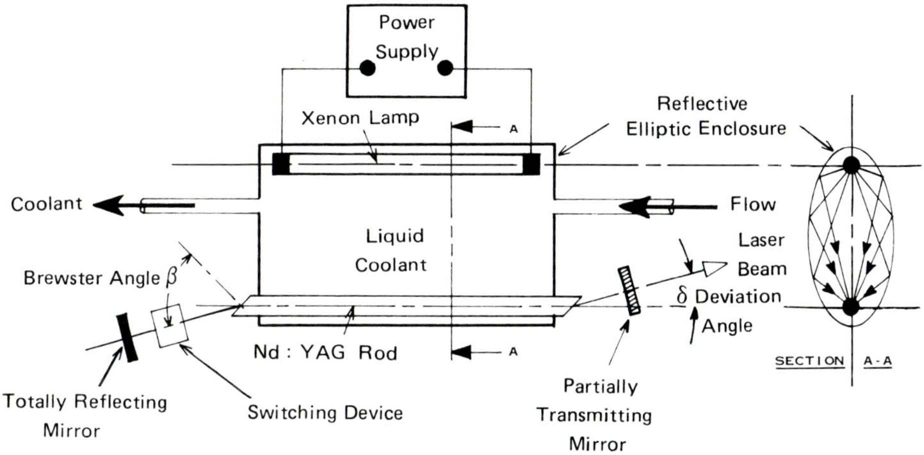

Figure 12 Schematic diagram of the essential details of construction of a typical Nd:YAG laser.

be a reflection of about 4% at each incidence of a laser ray upon an end face if it were perpendicular to the axis of the rod, inclining these end faces eliminates the interference of these small, but numerous, reflections with the development of resonance between the mirrors. For YAG, whose index of refraction is 1.823 at the wavelength of 1064 nm, the magnitude of the Brewster angle is 61.25°. Because of the *refraction* of the polarized waves of laser light as they pass out of the YAG rod into the air, the axis of each mirror must be turned from the axis of the rod by the *angle of deviation* in the YAG crystal caused by refraction.

The foregoing constructional features of a typical Nd:YAG laser are depicted schematically in Fig. 12 for a system using just one pumping lamp. In this figure a switching device is illustrated symbolically by a box interposed in the central line of the laser rays outside the YAG rod. Such a device can be used to create pulses of output power even if the pumping lamp is continuously operating. Of course, the pumping lamp itself can be pulsed by the electric power supply, but such operation does not produce the shortest duration or highest power in the pulses.

At or beyond the output mirror of the resonator of a typical Nd:YAG laser, a set of *lenses* must be employed to reduce the diameter of the laser beam and to focus it on the input end of the quartz fiber that transmits it to the target, perhaps several meters away along the length of the fiber. The convergence angle of this focused beam where it enters the fiber must be less than the acceptance angle of the fiber, and this may range from a total included angle of 15° down to one of about 10°. An antireflective coating on the fiber end helps to minimize reflection loss. Somewhere before or after this lens system, a mechanical shutter, controlled by a solenoid, is interposed in the beam to block it when the laser is not in active use. The laser beam that emerges from the distal end of the fiber will have a divergence angle about equal to that of the convergence of the beam at the proximal end. The fiber must have an optical cladding over its quartz core, and this whole waveguide must be enclosed within a suitable plastic catheter having an annular clearance space between itself and the outside of the cladding through which a stream of nonoxidizing gas can be forced to prevent organic debris from being splattered onto the output end of the fiber and also to cool the target surface during irradiation by the laser beam.

Because the beam of the Nd:YAG laser is invisible to the human eye, an *aiming beam* of visible light, usually from a helium-neon laser delivering 1 or 2 mw of power, must be introduced coaxially with the Nd:YAG beam. This is usually done by means of a *beam-splitter* that reflects the He:Ne beam into the path of the Nd:YAG beam while allowing the Nd:YAG beam to pass through the reflective coating of the beam-splitter with little attenuation. This aiming beam must be preconditioned by appropriate lenses, or by mirrors, or by both, so that it is of a diameter and convergence angle equal to those of the Nd:YAG beam at entry into the fiber. The Ne:Ne beam has a red wavelength of 633 nm.

c. Carbon Dioxide Lasers. The important features of a typical flowing-gas carbon dioxide laser are shown schematically in Fig. 13. The resonator consists of a glass or quartz tube whose length/diameter ratio is typically between 25 and 50. A cooling jacket is formed integrally with the gas tube, surrounding the gas duct concentrically over its entire length. Brewster windows are not always used in CO_2 lasers for surgery, and the mirrors are usually contained in separate, concentric end pieces that are joined to the laser tube by ring-type gas seals of medium-durometer elastomeric material. The whole structure is customarily supported in a metal frame with elastomeric cushions between the metallic parts and the laser tube. A premixed gas at regulated pressure of about 1 kg/cm^2 flows from the supply cylinder into the laser tube at one end, and is pumped out at the other end by a vacuum pump that discharges the exhausted gases into the room. This causes no hazard to operating personnel because the mixture consists of helium, CO_2, and N_2, in proportions approximating 76, 8, and 16%, respectively. The coolant is usually silicone oil or deionized or distilled water, circulated through an internal cooling system by a pump and giving up its heat in a finned convector cooled by forced air from one or more fans. For a given power output, the cooling of a CO_2 laser is much easier than that of any other common laser because of the relatively high overall efficiency of the carbon dioxide system, which runs between 10 and 20% in typical medical units.

The excitation of the gas mixture is most commonly achieved by a direct current flowing through the ionized nitrogen gas. At the customary operating pressure of 30-70 torr in the laser tube, the operating current is about 20 mA, driven

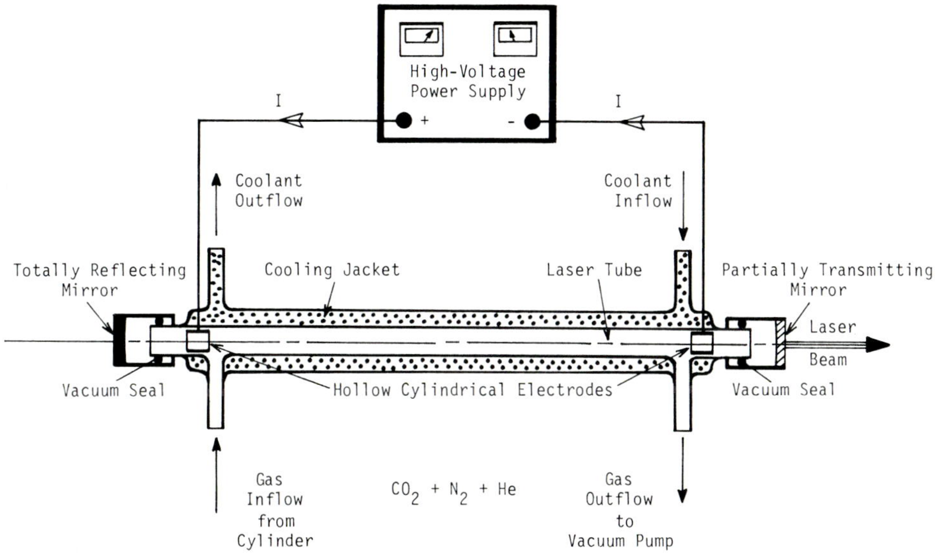

Figure 13 Schematic diagram of the essential details of construction of a carbon dioxide laser. Depicted here is a flowing-gas system. The laser tube and cooling jacket are usually made as one unit, typically of high-temperature glass. The gas is a mixture of carbon dioxide, nitrogen, and helium delivered from a high-pressure cylinder through a reducing valve and regulator of pressure. The vacuum pump maintains the pressure in the tube at a value between 20 and 80 torr.

by a potential difference of 15,000-50,000 V from an external power supply. Control of the output power of the laser is accomplished by varying the gas pressure in the laser tube, the current, or both. In sealed-tube lasers, excitation of the electric discharge within the gas is accomplished by a high-frequency alternating electric field that can be applied through a dielectric wall from outside electrodes or between electrodes inside the resonator. In either case the frequency needed is 30 MHz or higher. Sealed-tube lasers typically produce less than half the specific output of flowing-gas CO_2 systems. The latter yield between 0.5 and 1.0 W/cm of tube length in surgical lasers. Sealed-tube systems gradually decline in performance because of diffusive leakage of the helium through the wall of the tube and chemical breakdown of the CO_2 to form oxides of nitrogen and carbon monoxide under the high temperatures attained in the plasma (ionized gas and electrons) in the tube during operation.

As with the Nd:YAG laser, a helium-neon aiming beam is usually employed to mark the path of the invisible main laser beam (λ = 10,600 nm). A footswitch-controlled, solenoid-actuated mechanical shutter interrupts the beam when the laser is not being fired at a target.

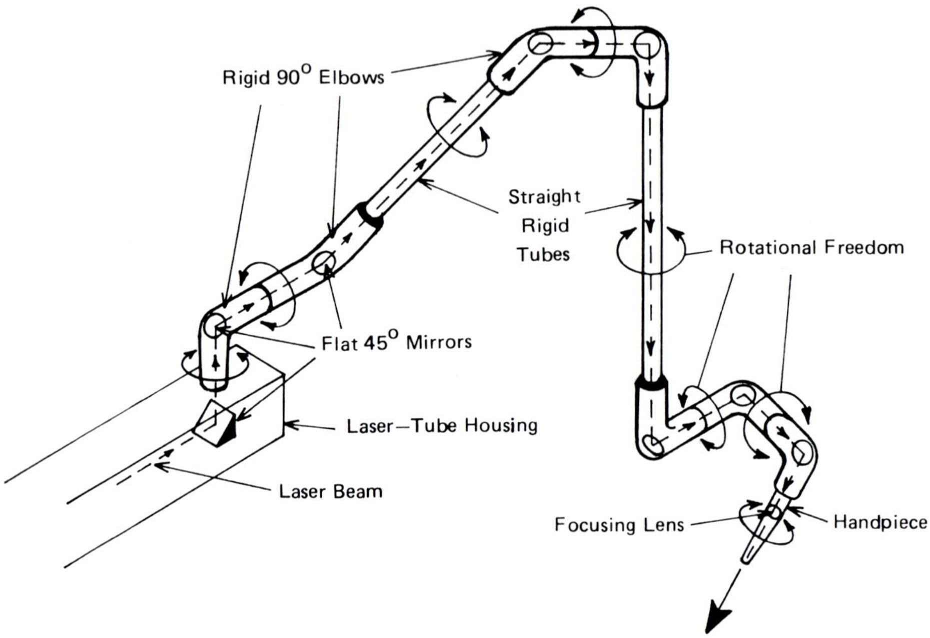

Figure 14 Schematic diagram of a typical articulated arm for a CO_2 laser. The elbows are usually made of aluminum alloy, and the tubular segments may be of aluminum alloy or of graphite fibers in matrix of thermosetting plastic. Not shown here is the counterbalance, which would be attached to the second elbow from the laser and would rotate about the optical axis between the first and second mirrors.

The beam of a typical CO_2 laser is transmitted from the resonator to the target via a multisegmented articulated arm, shown schematically in Fig. 14. This arm consists of rigid 90° elbows, each containing an adjustable mirror set at 45° to the axes of the inlet and outlet stubs of the elbow, with one stub each connected rigidly to a relatively long and straight segment of rigid circular tubing, so that each straight segment of tubing has one elbow firmly attached at each end. The elbows occur in pairs that are closely coupled at those stubs that are not attached to the straight tubular segments. This coupling maintains coaxial alignment of the adjacent elbow stubs, yet allows one stub to rotate relative to the mating stub of the adjacent elbow. When the mirrors are properly aligned, the laser beam is reflected from the center of each mirror coaxially down the bore of the next elbow stub or tube segment to the succeeding mirror, finally emerging at the distal end to be focused and aimed by a hand-held piece or by a *steerable mirror* whose plane can be rotated in space about its center under the guidance of a *control stick* or *joystick* operated by the thumb and forefinger of the user of the laser. The steerable mirror and control stick are always used in conjunction with a binocular operating microscope. In surgery, a well-designed CO_2 laser used in conjunction with a microscope and steering mirror has no equal for precise destruction of tissue among all the other surgical instruments ever devised and presently approved for surgical use.

The articulated arm has two chief disadvantages when compared with a flexible optical fiber: susceptibility to misalignment of the mirrors from impacts against walls and other hard objects, and a somewhat awkward restriction of the surgeon's hand when using a handpiece. However it has two offsetting advantages over optical fibers: it can transmit much higher levels of power in the laser beam, and it preserves the *spatial coherence* of the beam (which is *lost* in fiberoptic transmission). Other factors being equal, a well-designed articulated-arm-and-steering-mirror system allows a *smaller focal spot* to be attained, because it *preserves* the *transverse electromagnetic mode* of the laser beam, whereas an optical fiber, by multiple transposition of the internally reflected rays, tends to flatten and broaden the profile of power density within the beam when the fiber has curves along its length.

In general, only metallic halide salts, zinc sulfide, zinc selenide, chalcogenide glass, and certain intermetallic compounds transmit midinfrared radiation with tolerable losses. Therefore the focusing lenses for CO_2 lasers must be made of rather expensive materials such as zinc selenide or zinc sulfide. Some compounds, such as thallium bromide, thallium iodide, and arsenic trisulfide, are too toxic to risk using in endoscopic systems for surgery and have certain undesirable mechanical properties. Although these have been used in fiber systems to transmit CO_2 laser beams from laser resonators to microscopes or handpieces, they are unsuitable for endoscopic transmission in living patients. The search continues for a material possessing low absorption, good flexibility, high refractive index, long flex life, resistance to body fluids, and fiber-drawing capability, usable at midinfrared wavelengths. So far, no one has commercially produced an optical fiber for midinfrared that is completely suitable for surgery.

d. Argon Ion Lasers. By comparison with the Nd:YAG and CO_2 lasers, the argon ion laser is inefficient and difficult to manufacture. Because the energy levels of argon ions lie above those of argon atoms, multiple collisions of electrons with argon ions in an electric discharge are needed to pump these ions to the upper laser levels. To produce these, high current densities are needed in the laser tube, running well over 100 A/cm^2. The total current through the typical argon ion laser tube is between 5 and 50 A, driven by potential differences ranging from 100 to 500 V. To attain the high current densities, the electric discharge in the ionized argon gas must be confined to a narrow cylindrical volume near the center of the laser tube. This can be accomplished by placing disks of metal having a high melting point and good thermal conductivity at short distances apart along the axis of the tube, with their planes perpendicular to this axis, each disk having a small hole at its center through which the electric current flows. Coaxial magnetic solenoids are often employed to pinch the electric discharge at the axis of the tube. The high temperature in the discharge column and the large thermal power that must be removed by the cooling system cause severe deterioration of the laser tube and the electrodes, if these are placed inside the tube. The optimum gas pressure in an argon ion laser tube is about 0.5 torr, much lower than the pressure common in CO_2 lasers. The tube is best made of a high-temperature ceramic such as beryllium oxide, which withstands the intense heat but is difficult to manufacture. Because the operating life of an argon ion laser tube is limited, many practical lasers energize the tube only when output power is acutally required. The laser tube is prefilled with argon gas and sealed.

Most argon ion lasers are water cooled. The very low overall efficiency, approximately 0.1%, necessitates the removal of an enormous thermal power flow and the input of exceedingly high electric power. The output power from typical argon ion lasers ranges from 1 W or so up to 15 W. The size and weight of these lasers are very high in relation to their output. Because the common output wavelengths of the argon ion laser are 488 nm and 514 nm, blue and green in the visible spectrum, this system needs no separate laser to provide an aiming beam.

Transmission of the beam from the resonator to the target is most always accomplished through an optical glass fiber, although the early ophthalmic argon ion lasers sometimes employed an articulated arm for this purpose. When used with slit lamps and ophthalmoscopes for retinal coagulation and other intraocular procedures, the beam of the argon ion laser is customarily controlled by a steering mirror and joystick.

Because the Nd:YAG laser can be used with a frequency-doubling crystal to produce a wavelength of 532 nm, it is quite possible that the argon laser may someday be replaced in medical applications by frequency-doubled Nd:YAG lasers. As of this writing, one company has been formed in the United States to manufacture such devices for medical uses.

Figures 15 through 17 show photographs of typical Nd:YAG, CO_2, and argon ion lasers currently used in medical and surgical applications.

Table 4 lists the types of laser most commonly used in surgery and therapy, their

---

Figure 15 A typical modern Nd:YAG laser, the Cilas Model YM-101, rated at 100 W continuously, and delivering its beam through a nitrogen-cooled quartz fiber. The footswitch control for the output is on the floor in front of the console. This instrument has a microprocessor for regulation of the output power level and for self-checking of internal functions. The unit is mobile and can be operated either with external water cooling or internal air cooling (*Source*: photograph furnished by CILAS, Inc., of Needham, Massachusetts).

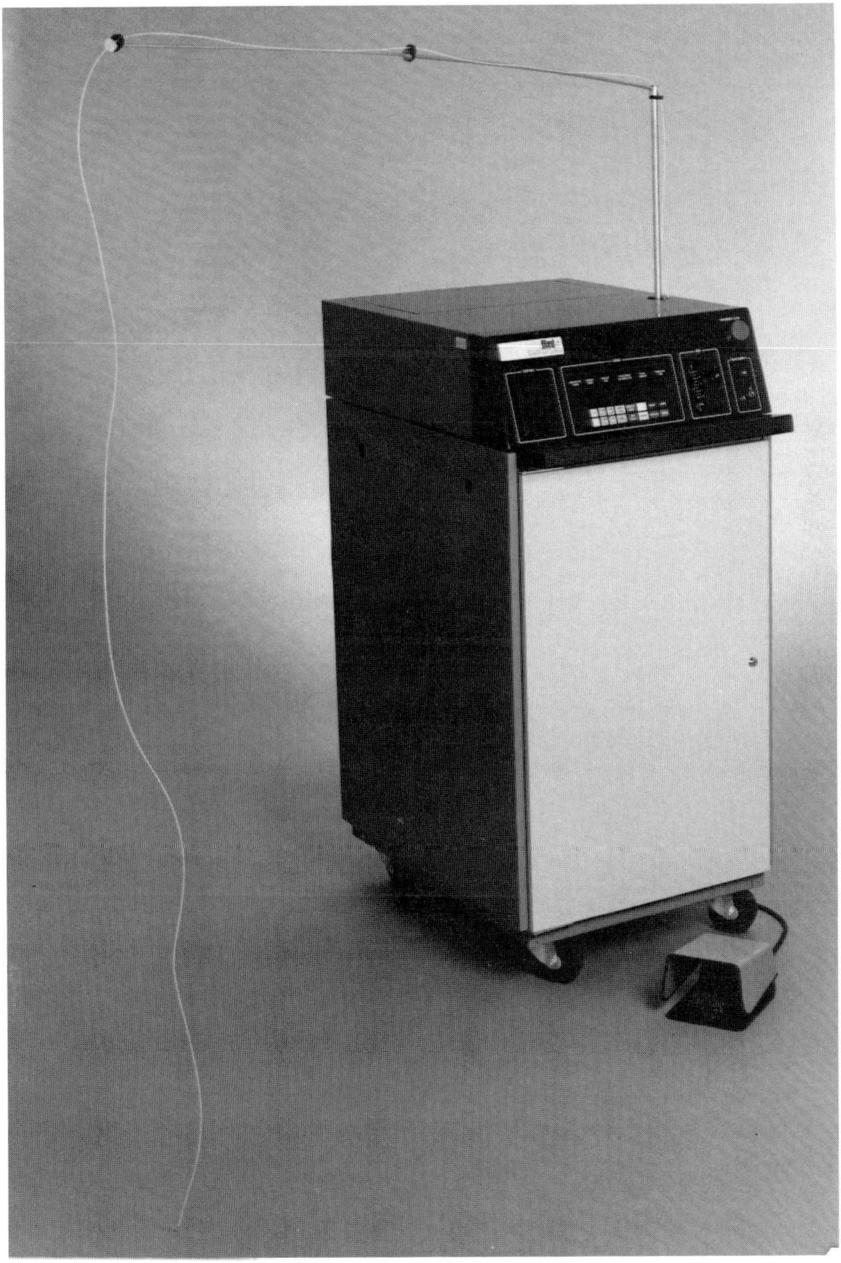

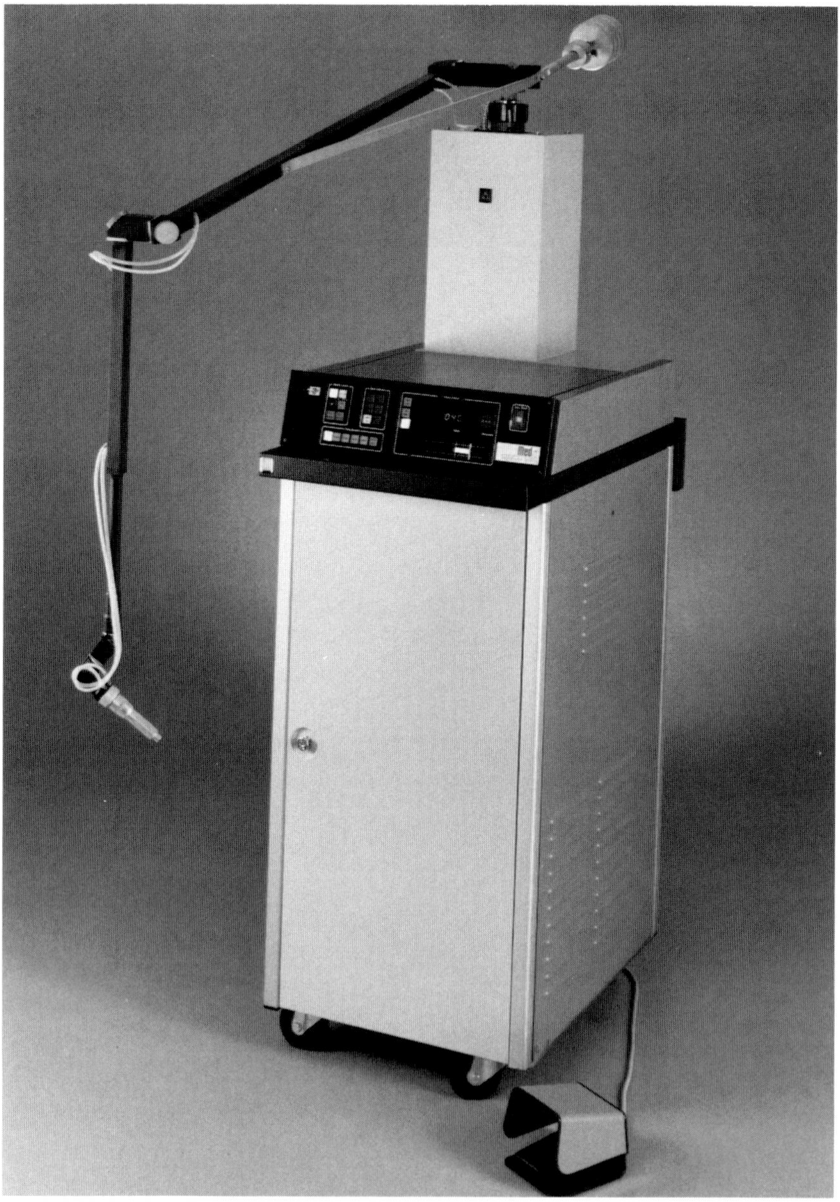

operating characteristics, and their major applications. The reader must remember that hundreds of laser wavelengths are potentially available, many of which have been used experimentally on living tissue; however, at the time of this writing, only those described in the table have achieved broad use in medical work. Long-term trends are always difficult to discern from short-term perspectives, but there now appears to be a growing realization among physicians and surgeons who use lasers that the dominant medical lasers of the future will offer either *continuous adjustability of wavelength* or a *wide selection of discrete wavelengths.* Currently, only the dye laser offers continuous adjustability of wavelength over more than a narrow range. Its major medical use is for the activation of photosensitizers in photodynamic therapy.

B. Temporal Operating Modes of Lasers

The temporal operating mode of a laser is characterized by the variation of its output of radiant power with time. There are two such modes: *pulsed* and *continuous wave.* In the pulsed mode, as the named implies, the output power is delivered in episodes of finite duration, which may be isolated or repetitive. In the continuous wave mode, the output power is delivered for a relatively long period without interruption. Although the distinction may seem trivial, these two modes of operation have profoundly different requirements for the laser itself and physical effects upon the target. Pulses may be classified as *short, medium,* or *long.* Continuous wave output may be divided into *bursts* and *steady firing.* These categories are shown in tabular form in Table 5, in terms of output duration and the means by which this is controlled.

1. Pulsed Operation

Today it is recognized that there are valid reasons for employing pulsed operation of lasers in surgery, despite the criticisms made by Alfred S. Ketcham (10) and his

---

Figure 16 A typical modern carbon dioxide laser for surgery. This unit, the Cilas Model 80S, is capable of delivering 0-80 W of output power through a counterbalanced articulated arm, to either a handpiece (shown here) cooled by flowing nitrogen gas or a microscopic coupler and micromanipulator (not shown here). The footswitch control for the output beam is on the floor in front of the console. The laser resonator is mounted vertically inside the tower atop the console. The unit is fully mobile, having an internal forced-liquid cooling system discharging its heat via a fan-cooled radiator. It provides automatic shutter-timed bursts of output power, rapid repetitive pulses, and continuous wave delivery (*Source*: photograph furnished by CILAS, Inc., of Needham, Massachusetts).

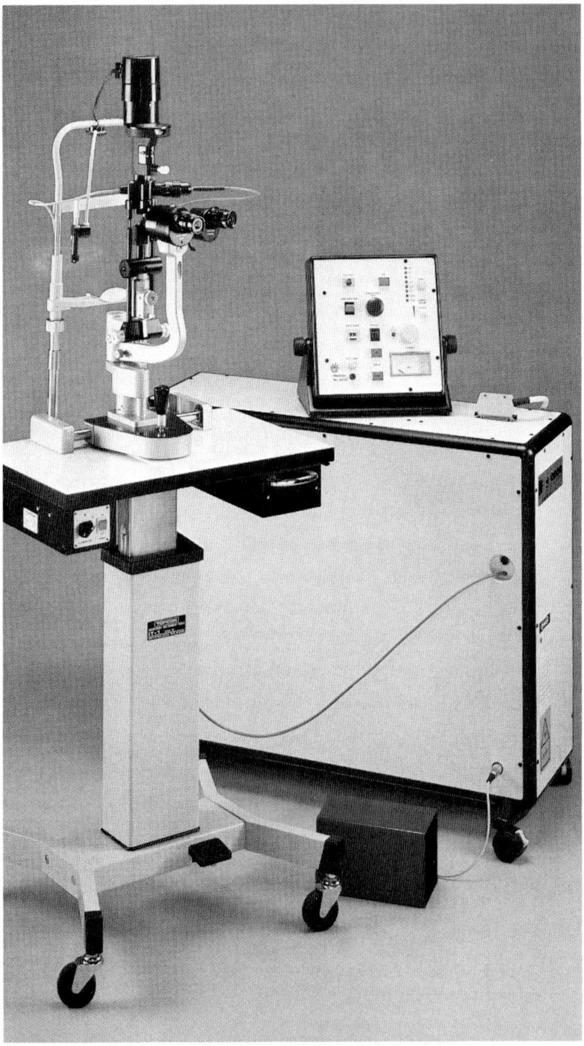

Figure 17 A modern argon ion laser designed for ophthalmic and dermatological use. On the right is the mobile power console, with the control panel on its top surface. On the floor in the center is the footswitch. On the left is the mobile stand holding the slit-lamp, ophthalmoscope, and headrest for the patient. The delivery of the beam is via the flexible optical fiber from the console to the slit-lamp, through a lens system providing focal spot sizes from 50 μm to 2000 μm, and exposure times from 0.02 to 2.0 sec, automatically controlled. It provides 1.0 W at 514.5 nm. The system can be adapted to various slit-lamps, and can be used with manual probes for dermatology (*Source*: photograph courtesy of Meditec of America, Inc., Clearwater, Florida).

Table 4 Important Lasers Used in Medicine and Surgery

Laser type	Major wavelengths (nm)	Timing of radiation	Delivery of beam	Maximum power (W)	Present applications
Argon ion (gas)	488 514	Continuous wave Long-pulsed Medium-pulsed	Fiberoptic cable Handpiece Slit-lamp and microscope Flexible endoscopes Sapphire tips	1.0-20	Photocoagulation in eye Photocoagulation of vascular skin lesions Photocoagulation of ectopic endometriomas Vaporization of pigmented skin lesions Pumping of dye lasers Ablation of intravascular atheromas
Carbon dioxide (gas) in mixtures of helium and nitrogen	10,600	Continuous wave Long-pulsed Medium-pulsed	Articulated arm Handpiece Microscope Micromanipulator	20-150	Precise general surgery Cutting, vaporizing, and drilling of tissues throughout the body Debridement of necrotic and infected lesions
Helium-neon (gas)	633	Continuous wave Long-pulsed Medium-pulsed	Fiberoptic cable Handpiece Articulated arm Micromanipulator	0.001-0.025	Visible designation for aiming infrared lasers Biostimulation for healing wounds, etc. Analgesia for pain in various parts of body

(continued)

Table 4 (continued)

Laser type	Major wavelengths (nm)	Timing of radiation	Delivery of beam	Maximum power (W)	Present applications
Nd:YAG (solid) neodymium ions in crystals of yttrium-aluminum-garnet	1,064 1,320 946	Continuous wave Long-pulsed Medium-pulsed	Fiberoptic cable Handpiece Sapphire tips Flexible and rigid endoscopes	50-150	Voluminous heating of tissues for hemostasis, and coagulative necrosis of obstructive tumors in airway and gastrointestinal tract Incisive and coagulative surgery throughout the body
		Short-pulsed	Articulated arm Micromanipulator	10^{10} and up	Ultraprecise focal destruction of intraocular structures
Semiconductor (solid) junctions between semiconducting metallic substances containing aluminum, antimony, arsenic,	590-5,300	Continuous wave Long-pulsed Medium-pulsed	Fiberoptic cable Handpiece	0.0005-0.0050	Biostimulation for various purposes, such as healing of wounds, relief of pain, regeneration of neural tissue, and neovascularization of ischemic tissues

gallium, indium, and phosphorous, notably					
GaAs	830-920				
and					
AlGaAs	620-900				
Dye (liquid) fluorescent organic dyes (e.g., rhodamines) in liquid solutions (e.g., dimethyl sulfoxide)	360-670	Continuous wave Long-pulsed	Fiberoptic cable Handpiece Interstitial needles Sapphire tips	1.0-10	Excitation of photosensitizers, such as hematoporphyrin derivative in photodynamic therapy of malignant tumors Color-selective treatment of pigmented skin lesions, such as port-wine hemangiomas and nevi
Krypton ion (gas)	476 521 568 647	Continuous wave Long-pulsed Medium-pulsed	Fiberoptic cable Handpiece Slit-lamp and microscope Flexible endoscopes Sapphire tips	1.0-10	Photocoagulation in eye Treatment of pigmented skin lesions Pumping of dye lasers
Ruby (solid) chromium ions in matrix of aluminum oxide	694	Long-pulsed Medium-pulsed	Fiberoptic cable Handpiece Slit-lamp microscope	10^8	Photocoagulation of pigmented lesions in the eye and skin

Table 5 Temporal Operating Modes

Pulsed			Continuous wave	
Short	Medium	Long	Bursts	Steady
1×10^{-12} to 1×10^{-7} sec	1×10^{-7} to 1×10^{-2} sec	1×10^{-2} to 1×0.5 sec	0.5-5 sec	5-∞ sec
Mode-locking Q-switching Cavity-dumping	Q-switching Pump-pulsing	Pump-pulsing Auto shutter timing	Operator control by shutter and footswitch	

associates in the latter part of the 1960s. One of these reasons is the reduction of thermal trauma adjacent to the impact site of the beam of a carbon dioxide surgical laser, which can be accomplished by application of the laser energy in pulses of short duration separated by relatively long intervals of zero intensity during which the nonimpacted tissue has time to cool. This type of operation is usually achieved by pulsing the pumping device at a rapid rate between 10 and 100 pulses/sec, with a pulse duration of 50% or less of the interpulse interval. Although it is designated by the unfortunate and inaccurate name of *superpulsing,* this mode creates pulses of low enough peak power density to avoid the *blast effects* caused by the early ruby and pulsed Nd:glass lasers, which Ketcham et al. so rightly condemned. This form of pulsing is of greatest benefit with the CO_2 laser, but of little or no value with argon ion and Nd:YAG lasers. The Nd:YAG, whose radiant energy *scatters* copiously in tissue, is used surgically to produce *voluminous coagulation* of tissue for necrosis and/or hemostasis, in most applications. Therefore it derives no advantage from superpulsing in such situations. However, a new class of short-pulsed Nd:YAG lasers, first developed in Europe in 1977 by Danielle Aron-Rosa (11) in Paris and Franz Frankhauser (12) in Bern, for the purpose of creating small, precise holes in opacified posterior lens capsules of patients following removal of cataractous natural lenses and replacement by plastic implants, operates by virtue of ultrashort pulses of exceedingly high peak power but *low energy per pulse.* Such lasers require much shorter pulses than can be achieved by pulsing of their pumping sources. The tissue effects of these lasers are discussed in Sec. IV of this chapter. Suffice it here to say that thermal damage and mechanical damage to surrounding tissue are limited by the low value of energy in each pulse so that *macroscopic* blast effects do not occur.

As indicated in Table 5, there are three basic means of creating very short pulses from lasers: *mode-locking, Q-switching,* and *cavity-dumping.* All of these methods involve the use of a switching device *within* the resonator, as opposed to *pump-pulsing,* which involves electronically switching on and off the energy flow to the pumping system. These three methods are discussed in some detail in the following sections, because it is important that the surgical user of short-pulsed lasers understand the fundamental differences in the means by which short pulses are produced.

a. Q-Switching. For any resonant system, such as a laser, there is a *quality factor,* designated by the symbol Q, which is a measure of the *sharpness* of the resonance. As we have seen, the resonator of a laser produces a set of longitudinal modes of oscillation, each of which is energized by the pumping source via the excitation of atoms or molecules, spontaneous emission, and stimulated emission of various frequencies (and wavelengths) of light. Each of the waves of light having one of the permissible frequencies for the laser resonator represents *storage of energy* in the resonator in the form of radiation reflecting back and forth between the mirrors. As these waves traverse the resonator repeatedly between the mirrors, some energy is lost from each wave (consisting of many synchronized wavelets) on each round trip. These losses are caused primarily by less-than-total reflection at each mirror, by absorption into underexcited atoms (or molecules), and by travel in ray directions that do not lie within the ray bundle. Once the standing wave pattern has been established in the resonator, that is, once the population inversion is sufficiently large that sustained stimulated emission occurs, the only important loss of energy within the resonator is that due to imperfect reflections at the mirrors. To simplify the analysis, we shall first assume that the mirrors are both of equal reflectivity of nearly 100%.

If the time-rate of energy loss is designated by the symbol P_l, the frequency of the coherent radiation by f, and the energy stored within the resonator by E_R, then the quality factor of the resonator is given by

$$Q = 2\pi f E_R / P_l, \ (NUMERIC) \qquad [34]$$

If the fraction of the incident radiant energy reflected by each mirror is designated by r, then the total number of traverses made by each wavelet before extinction will be equal to $1/(1 - r)$. The quantum rules require that any one photon or wavelet be extinguished (give up its energy) *all at once.* Hence any individual wavelet will not lose a fraction $(1 - r)$ of its energy at each reflection. However if we think of a *large number* of wavelets making up the continuous coherent radiation being reflected back and forth between the mirrors, $(1 - r)$ of these will be extinguished at every reflection, with a final result that *any one* wavelet will, on average, make a *probable number* $1/(1 - r)$ of traverses between mirrors before extinction. The time required for this number of traverses the total distance traveled, divided by the speed of travel: $t_e = Z_m/v\ (1 - r)$. Therefore the average rate of energy loss represented by the extinction of each coherent wavelet is the energy per wavelet divided by this time to extinction:

$$P_l/N_w = \Delta E_{ji}/t_e = hfv\,(1 - r)/Z_m,\ (POWER) \qquad [35]$$

where N_w = the total number of coherent wavelets within the resonator

and h = Planck's constant

The total energy stored within the resonator is simply the energy of one wavelet multiplied by the number of wavelets, or

$$E_R = N_w\,\Delta E_{ji} = hfN_w \qquad [36]$$

Substitution of E_R from Eq. [36] and P_l from Eq. [35] into Eq. [34] gives us the final expression for the quality factor:

$$Q = 2\pi f\,Z_m/v(1 - r) = 2\pi Z_m/\lambda(1 - r) \qquad [37]$$

Equation [37] shows the critical importance of high reflectivity in the mirrors. As Fig. 13 shows, one of the mirrors must be made partially transmitting so that some of the laser radiation may emerge to do work externally. Typically, this transmissivity will be about 20%, which far exceeds the usual 0.05% loss at the other mirror. Therefore significant loss occurs only *once* on each round trip, rather than twice. Thus the distance between successive lossy reflections becomes $2Z_m$, and r decreases to about 0.80 in Eq. [37], so that we have $Q = 20\pi Z_m/\lambda$ for most practical lasers. For a resonator length of 1 m and a wavelength of 1000 nm, the quality factor is 62.8 million; this is far higher than can be attained by resonant electric circuits, and so the spectral purity of light from a laser is inherently much greater than that of radiation generated by radio transmitters or microwave devices, such as magnetrons and klystrons.

Obviously, anything that interferes with the resonance will reduce the value of Q, and when it falls below a certain value laser action will cease, even if an adequate population inversion is maintained. If, for example, the ray path to one or both mirrors is blocked by an opaque object or deviated by deliberate misalignment of a mirror, the coherent radiation within the resonator will cease. Meanwhile, if the pumping source is still energized, the population inversion will continue to increase to some high equilibrium value. Thereafter, any sudden restoration of the quality factor to its proper value will result in a very short pulse of coherent radiation having a very high peak intensity. In lasers having a favorable (large) upper-level lifetime/lower-level lifetime ratio, this pulsing process can be rapidly repeated. The Nd:YAG laser is capable of cyclic pulsing at rates up to 50,000/sec, with pulse durations ranging from 10 ns to 500 ns, when operated in this fashion; peak power per pulse ranges from a few kilowatts to several megawatts. Shorter pulses correspond to higher power, and vice versa, other factors being equal. Any process that systematically reduces the *Q* of the resonator below the critical value of oscillation and then suddenly restores it to normal is called *Q-switching.*

A simple method of Q-switching is to rotate or angularly oscillate the totally reflecting mirror. More sophisticated techniques involve the use of electrooptic modulators, called *Kerr cells* and *Pockels cells,* which in effect, rotate the plane of polarization of incoming polarized light so that some or most of it can be transmitted through a subsequent material that propagates polarized light only in a plane perpendicular to that of the light entering the electrooptic cell. Such devices can be used to cause very abrupt switching on of the coherent radiation within the ray bundle outside the laser material, thus allowing a sudden growth of laser amplification within the resonator. Because of their very short response times, electrooptic devices can produce laser pulses lasting only a few nanoseconds and having peak power of 100 MW, more or less.

There are dozens of variations on this scheme that can be employed in practical lasers, but the nature of this exposition does not permit us to explore them here. Q-switching systems are employed in ruby lasers, which inherently are unsuitable for continuous wave operation because of the large threshold energy required for pumping the chromium ions to bands lying above the upper laser level (about 10,000 cm^{-1}) and because it is a three-level system needing 50% of its atoms in the upper laser level before stimulated emission can occur. Q-switching is possible with any type of laser using a conventional resonator, as long as the ray bundle is accessible to a suitable switching device.

b. Cavity-Dumping. Cavity-dumping is a system in which the *Q* of the resonator is left at a high value well above the threshold for laser action, but the coherent radiation within the resonator is suddenly allowed to emerge to the outside, usually by means of an acoustooptic device that diffracts the laser rays at an angle that is a function of the electric driving signal applied to the device. For example, by changing the direction of the coherent rays striking one mirror, these rays can be made to impinge on either a totally reflecting (typically 99% or higher) part of the mirror or on a partially transmitting region (about 80% reflecting) that lets part of the internal laser beam out of the resonator. With such schemes, the repetition rate of pulsing can be extended into the range of millions per second, and the average power output is about equal to the steady power that could be obtained by continuous wave operation, while the peak power per pulse may be as high as 100 times the continuous wave value.

c. Mode-Locking. Mode-locking is a technique in which the switching device (Fig. 12) is a cell containing an organic dye, which may be in a liquid solution or uniformly dispersed in a plastic film. This dye has the characteristic property of strongly absorbing most of the light falling on it within a certain range of wavelengths (which is chosen to match those of the laser) until the intensity of the light exceeds a certain threshold value, above which the dye becomes transparent. Such saturable absorbing dyes are usually called *bleachable* dyes. Rigorous mathematical analysis of the operation of a bleachable dye cell in a laser resonator is complicated and beyond the scope of this exposition. However, their operation can be easily understood by a qualitative analysis. To operate effectively, the dye cell must be located just in front of one of the mirrors, and it must revert to its opaque condition within a few picoseconds (1 ps = 1×10^{-12} sec) after the incident light intensity first reaches the threshold level, that is, the dye *automatically* reverts to opacity within a few picoseconds after it first becomes transparent. This has the effect of limiting the time duration of each pulse of light that passes through it and thus cutting off the leading and trailing *edges* that the pulse shows when viewed as a function of *time* (as, for example, on a fast oscilloscope). When the dye is opaque, and the laser is first switched on, the process of spontaneous emission from the Nd^{3+} ions is random and sporadic, occurring in brief pulses. Each pulse consists of a wavelet traveling by chance in some direction. When this direction is more-or-less parallel to the axis of the YAG rod, one of the pulses of radiation will travel toward the dye cell, gaining intensity as it moves, until it is strong enough to saturate the dye and make the cell transparent. If the transparent time of the dye is 30 ps, the length of this traveling pulse will be about 9 mm, clipped off at its leading and trailing edges in space while making two passes through the cell, one toward the adjacent mirror and the other back from the mirror after reflection. The cell maintains opacity while this pulse travels back across the resonator, gaining amplitude by successive encounters with excited ions of neodymium. It completes two full transits of the resonator before returning to the dye cell. On its next passage through the dye cell, the pulse has grown in amplitude and in axial length because of repeated increments from stimulated emission. This passage, the second complete traverse of the dye, again clips the leading and trailing edges of the pulse, limiting it to 30 ps and an axial length of about 9 mm (3×10^8 m/sec $\times 30 \times 10^{-12}$ sec). The growing pulse makes between 10 and 100 round trips between the mirrors, depleting the population inversion and inhibiting the formation of other, competing spikes that are not synchronized with it. The laser emits a train of such pulses through its partially transmitting mirror, one for each time the traveling internal pulse strikes this mirror. The time interval between pulses, therefore, is

$$\Delta t_m = 2Z_m/v \qquad [38]$$

At the beginning of the growth process of the traveling pulse, its peak intensity grows rapidly. As the population inversion is depleted by repeated transits of the pulse through the YAG rod, the pulse height reaches a maximum and then wanes until it is no longer sufficient to bleach the dye. Typically, this takes about 20 round trips through the resonator. The amplitude envelope of the pulses has the

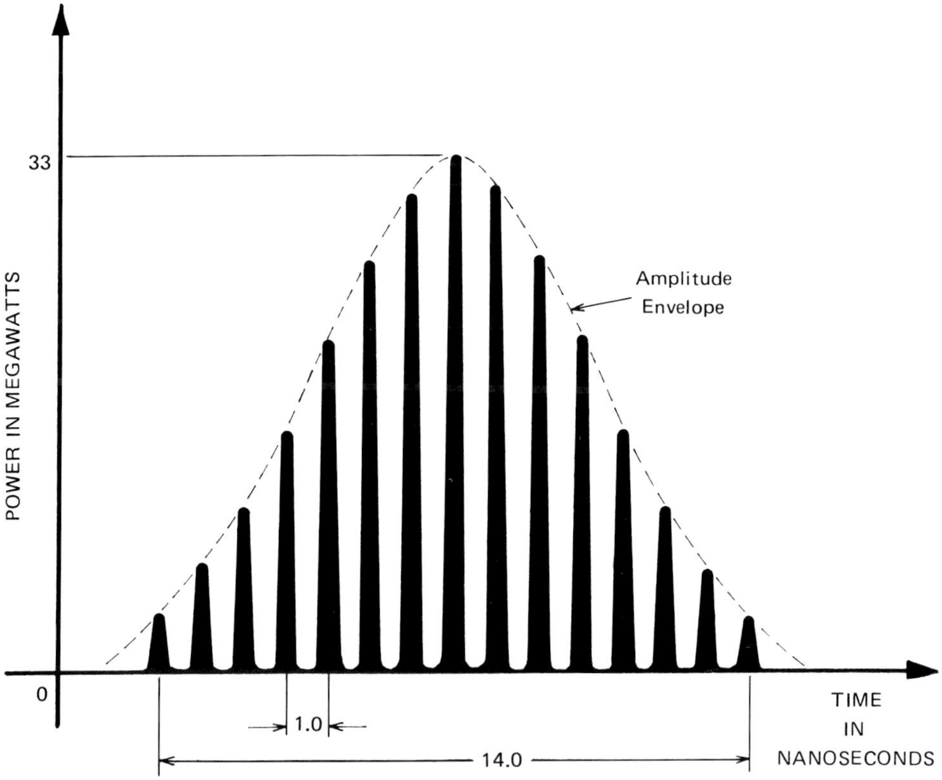

Figure 18 Train of pulses of output power from a mode-locked laser having $Z_m = 0.15$ m, $\Delta\tau_m = 30$ ps, and 1.0 mJ in the highest pulse. Note that the duration of each pulse has been greatly exaggerated here for clarity of viewing.

shape of a bell, as shown in Fig. 18. The energy per pulse is small, typically about one millijoule (1×10^{-3} J), and this would give a peak power (in the highest pulse) of about 33 million W.

If this process were analyzed in terms of the standing waves which represent the possible longitudinal modes of the resonator, that is, those frequencies or wavelengths of radiation that fall within the width of the ionic "line" of Nd^{3+} centered about 1064 nm, it would be seen that the dye cell has the net effect of forcing all of these standing waves in the resonator to oscillate *in time phase with each other* (they are, of course, already constrained to be nearly in space phase because of the fixed spacing between the mirrors). Thus a complete synchronization of all the standing waves in the resonator is achieved, with a resulting maximization of the stimulated emission. It is this synchronization which gives rise to the name *mode-locking*.

If we refer to Eq. [32], which gives the possible frequencies of the standing waves in the resonator as an integer multiple of a base frequency $f_1 = 1/\Delta t_m$, and remember that the values of the integer ξ are exceedingly large numbers, we see that the frequencies of the longitudinal modes of a typical Nd:YAG laser are very close together, and therefore so are their wavelengths. For example, if the resonator is 1 m long, $f_1 = 150$ million Hz (cps). If we approximate the Nd:YAG wavelength as 1000 nm, then the central frequency of the Nd:YAG band is 300 million $\times\ 10^6$ Hz. Therefore the value of the integer multiple corresponding to the center of the band is $\xi = 2$ million. If we assume that the band is wide enough to hold 100 such frequencies, and we express each of these in megahertz, then the frequencies of these longitudinal modes are f = 299,992,650; 299,992,800; 299,992,950; . . . 299,999,850; 300,000,000; 300,000,150; . . . 300,007,050; 300,007,200; 300,007,350 MHz. The ratio of the highest of these frequencies to the lowest is only 1.000049! In other words, they are all virtually identical, and can be thought of as being produced by atomic oscillators (Nd^{3+} ions) whose time phases are drifting randomly about a central frequency of 300 million MHz. The effect of mode-locking is to synchronize all these 100 oscillators and force them to radiate just at the central frequency, by stopping their random drifting of frequency (or phase).

Thus mode-locking can achieve much higher peak power per pulse by adding the outputs of many atomic oscillators. If the number of longitudinal modes is N_m, then the peak power per mode-locked pulse will be approximately $N_m^{\ 2}$ times the continuous wave power of any one of these oscillators. Furthermore, the spectral purity of the laser output in each pulse is greatly improved: the bandwidth of frequencies contained in each pulse is approximately

$$\Delta f_m = \tau_m^{\ -1} \qquad [39]$$

where τ_m = the total duration of each pulse.

If the transparent time of the dye cell is chosen optimally, it will be approximately

$$\tau_m = \Delta t_m / N_m \qquad [40]$$

From the foregoing considerations, it can be seen that a laser material having a relatively large atomic bandwidth (i.e., a large number of longitudinal modes) will give the shortest pulses and the highest peak power. The material Nd:YAG is well suited to mode-locked operation because its natural atomic bandwidth is about 13,000 MHz: for $Z_m = 1.0\ m$ there are $N_m \simeq 87$ longitudinal modes. Neodymium-glass is even better, with a bandwidth of 1.5 million MHz and $N_m \simeq 10{,}000$ modes. However the thermal conductivity of glass is much lower than

than that of YAG and its absorption at 1060 nm is considerably greater, so that it cannot withstand such high internal power densities as can YAG.

2. Continuous Wave Operation

In general, any laser can be operated in continuous wave fashion if its active material is a four-level energy system having an upper-laser level whose lifetime is five to ten times as long as that of the lower laser level. Under these conditions, the major limitations on output power are those imposed by the pumping system (available power and efficiency) and by the cooling system needed to remove the wasted thermal power of the pumping system and the laser material. In surgical applications, the duration of the output from the laser can be chosen by the surgeon via the footswitch and mechanical shutter. When the laser beam is applied in *bursts,* the lower limit of exposure duration is set by the neuromuscular reflex time of the surgeon, which is about 0.25 sec in the quickest individuals and around 0.5 sec for most people.

a. Superpulsing. This is a quasi-continuous wave mode of temporal operation in which the output power of the laser is switched rapidly on and off in a regular sequence of pulses whose duration is 50% or less of the repetition period. It is achieved by electronically pulsing the pumping source, and its purpose is to allow the tissue surrounding the impact zone of a CO_2 laser beam to cool between applications of radiant energy. It is offered by some commercial manufacturers of CO_2 surgical lasers. However, the superpulse systems now available have two disadvantages: (1) even if the pulses of power to the pumping source have a perfectly rectangular time shape, the corresponding pulses of *radiant* power show a steeply rising wave front followed by a relatively long, slowly decaying exponential tail; and (2) the duty factor (ratio of on-time to repetition period) of the pulses of radiation is too great, at minimum, to allow the adjacent tissue to maintain a near-normal temperature. Because the cooling time constant of most soft tissue is many seconds, while the heating time constant is much shorter (typically a fraction of a second) and *decreases* with rising radiant power, the maximum duty factor that will permit a safe temperature in tissue around the impact spot is 10% or less, well below the minimum available on most commercial lasers for surgery. This problem is compounded by the long, trailing tails of the typical superpulses, which continue to add heat to the tissue when it should be cooling. One manufacturer offers a superpulse system in which the output radiation of the CO_2 laser is continuous, but *chopped* into a regular sequence of pulses having a near-perfect rectangular time shape, by means of a rotating wheel having sharp-edged slots that let the beam pass through when they are aligned with the laser aperture. Between these slots, the wheel has opaque sectors that either reflect or absorb the beam. The chopper wheel eliminates the problem of amplitude instability in the pulses of radiant power produced by some of the electronically pulsed systems, in which the peak power of the typical pulse will be 10-20 times the average power of the same laser when not repetitively pulsed. However, it is achieved, superpulsing necesitates a reduction in the time-averaged radiant power delivered to the surgical target; if the average power is kept constant at normal continuous wave levels by dint of increasing the peak power of the pulses, the benefit of superpulsing is largely lost.

b. Automatic Scanning of the Laser Beam. This system, offered by at least one manufacturer of CO_2 surgical lasers, allows the operator to preselect (within certain limits) the size and shape of a surface area to be irradiated by the laser beam, simply by manually tracing the outline of the target area with the He:Ne aiming beam. The CO_2 beam will then sweep horizontally and/or vertically across *only* the *traced* area, at constant linear speed of sweep, with no unexposed regions between the lines of this *sweep raster.* The sweep raster is generated by a microcomputer that controls the deflection system of the laser beam. The rate of sweep is usually selectable by the operator over a limited range. This operational mode is extremely useful in any surgical application in which an area of tissue must be ablated to a specified depth within a defined periphery (e.g., a superficial hemangioma of the skin) with minimum thermal trauma to underlying tissue. As with superpulsing, automatic scanning reduces the time-averaged radiant power delivered to each and every element of the irradiated area, if the continuous wave power of the laser beam is constant.

Automatic scanning, if done with high enough sweep rates, can enable the beam of a CO_2 laser to vaporize just one cell layer of tissue per raster, with no generation of smoke and no carbonization of adjacent tissue. However, the systems presently available do not provide sweep speeds fast enough to attain this precise control, being limited to a maximum of about 15 cm/sec, whereas the ultimate resolution in depth control achievable with this technique requires rates of more than 500 cm/sec. To attain such sweep speeds in a system that also allows delineation of an arbitrary boundary necessitates the use of high-frequency, taut-band galvanometers on which the deflection mirrors for the laser beam are mounted. In the mid-1970s Cavitron Corporation's CO_2 Surgical Laser Group developed a prototype deflection system using a single steering mirror that was angularly oscillated in random fashion about two mutually perpendicular axes at high speeds. With this device, it was possible to ablate the skin of an apple with no damage to the soft tissue underneath.

c. Future Requirements. As more surgeons learn to use lasers and more hospitals purchase them, there will be an increasing demand from the rank and file of physicians for lasers with more sophisticated control systems for their temporal operation. The author, in company with many of the pioneers in this field, is aware of operations that could be performed most advantageously or solely by lasers, if the necessary operational features were available. Unfortunately, most of the manufacturers of lasers for surgery and therapy have little or no understanding of the real needs of the operating room, and they bedazzle the prospective purchaser with touchbutton controls, flashing lights, and digital displays that merely add complexity without providing commensurate utility. Such gadgetry delights the souls of laser engineers, but it does little to assist the surgeon.

C. Characteristics of Radiation from Real Lasers

1. General Properties of Laser Light

The three important properties of laser light are *coherence, collimation,* and *monochromaticity.* No real laser produces light having these characteristics absolutely, but lasers provide the closest approximation to such light yet devised. Coherence has already been discussed in the preceding sections, but it has two aspects: *spatial coherence* and *temporal coherence.* These will be further discussed later. Collimation is the near-*parallelism* of the rays emerging from the laser resonator. Only with flat mirrors or spherical mirrors of very large radius is collimation maximized. It is always degraded by the *diffraction* of light from the laser beam because of *Huygens' principle,* which states, among other things, that *each point of an advancing wave front acts like a point-source of new waves,* thus spreading the power of the beam over a larger cross-section than it would otherwise have. Monochromaticity has been discussed at length in previous sections, and we have seen that it is never perfect, even in mode-locked lasers. However, the *spectral purity* of common lasers is far greater than can be achieved with noncoherent sources of light. Spectral purity, which is defined as the ratio of central wavelength to output bandwidth, is a numerical parameter, which we shall designate by the symbol Λ. It is given by the relationship

$$\Lambda \equiv \lambda/\Delta\lambda = f/\Delta f \; (NUMERIC) \qquad [41]$$

The value of A is generally higher for gaseous lasers than for liquid or solid lasers. Table 6 shows the values of f, Δf, λ, $\Delta\lambda$, and Λ for six common types of laser. Note that the highest purity is found in the red light of the helium-neon laser.

a. Coherence of Laser Light. Let us first consider temporal coherence. Figure 19 shows several trains of light waves, which we shall imagine to be portions of *infinitely long* wavetrains, moving on parallel rays in one direction in one plane, but displaced laterally from one another. For simplicity, only the E waves are depicted. If the oscillators producing these waves run for an infinitely long time at exactly the same frequency (by virtue of some ideal synchronizing scheme that we shall not discuss here), each wave has perfect temporal coherence with itself. This means that a stationary observer standing at a point P on one particular ray will

Table 6 Spectral Purity of Common Lasers

Laser types	f(gHz)	Δf (gHz)	λ (nm)	$\Delta\lambda$ (nm)	Λ[a]
Carbon dioxide	28,280	0.100	10,600	0.03748	282,820
Helium-neon	473,800	1.500	632.8	0.002004	315,770
Argon ion	614,300	5.000	488.0	0.003972	122,860
Nd:YAG	281,770	13.000	1,064	0.04909	21,674
Ruby	431,800	30.000	694.3	0.04824	14,393
Nd:glass	282,830	1,500	1,060	5.622	188.55

[a] $\Lambda \equiv f/\Delta f \equiv \lambda/\Delta\lambda$.

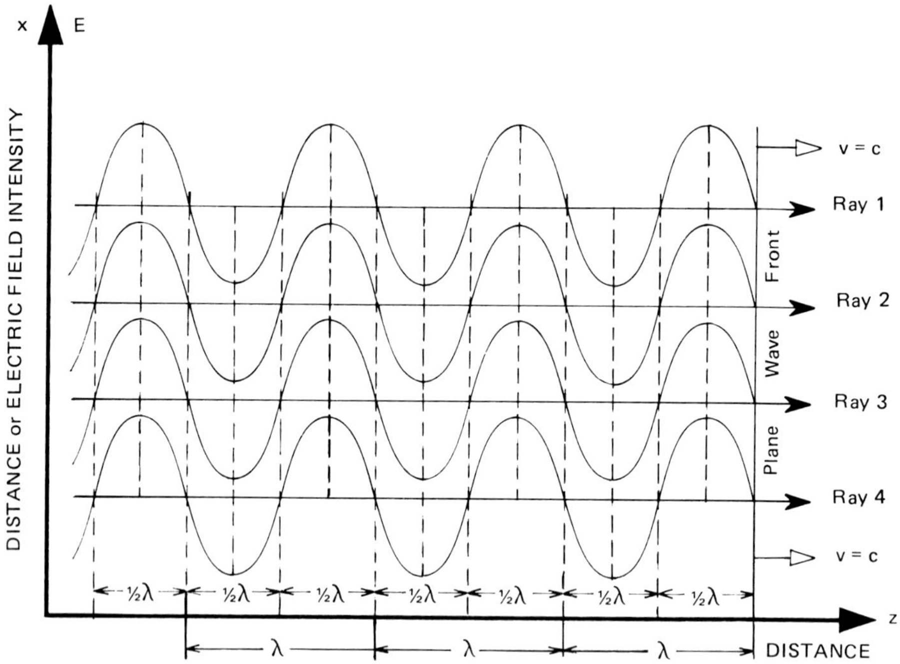

Figure 19 Plane polarized light waves from a collimated, coherent source, traveling through free space along parallel rays. Spatial coherence is evident from the coincidence of crests, valleys, and zeros of the waves in the z-coordinate. Temporal coherence is evident from the constancy of the wavelength λ, because the frequency of the source is constant and the speed of propagation is constant.

perceive a sinusoidally varying E field as a function of time measured from some arbitrary starting instant at which $E_p = 0$. Because of the constancy of the frequency of oscillation and the invariant speed of light (we shall assume free space here), the observer knows that the time wave of E field that he measures at P will be *exactly in time phase* with those that he could measure at a series of discrete points, fixed in space along the same ray that are separated in distance from one another and from P by λ, 2λ, 3λ, 4λ, 5λ, ... That is, at all fixed points along the ray that are distant from P by integer multiples of the constant wavelength, the observer knows that the instantaneous value of E will be zero *simultaneously with the zero value at point P.* In a like manner, the positive and negative maxima of the time wave observed at these fixed points will occur simultaneously with those at P. In other words, *the instantaneous value of* E_p *will be identical to that at any and every point distant from P along this ray by an integer multiple of the wavelength, at all instants of time.*

Next, let us consider spatial coherence. In the hypothetical case just discussed, the rays of light are all exactly parallel in direction. Because of the constancy of the wavelength, frequency, and speed of propagation, the stationary observer at point P knows that, *along a straight line perpendicular to the ray direction at P, the instantaneous value of E on every other wave will be equal to that at P for all instants of time.* Furthermore, if the observer were moving forward with the speed of light on the wave whose ray passes through *P, a straight line normal to his ray at his location on the moving wave would intersect all the other waves in this plane at points where the value of E is constant and identical to that at the intersection of this perpendicular with his ray.* Any *smooth* line drawn through points of equal E value at the leading edges of all coherent waves is called the *wave front* of the beam. It is not necessary that the wave front be a straight line (as in this case of waves whose rays are parallel and whose **E** vectors all lie in one plane), or that the wave front be a plane surface (as in the case of a three-dimensional beam having perfect collimation). If the rays diverge or converge, then the wave front will be a *smooth plane curve or curved surface* that is everywhere perpendicular to the rays of the beam. A smooth curve or surface is one that is always convex or concave to the same side and never reverses its curvature (i.e., never undulates).

Any fixed locus occupied by the wave front at an earlier instant of time is called a *phase front.* Spatial coherence requires that three conditions be fulfilled.

1. Along any one phase front all waves must have the same value of E at all instants of time.
2. Except in the vicinity of a *waist* (i.e., a region of relative minimum beam diameter), any phase front of this beam must be parallel to all other phase fronts.
3. All waves that are polarized must be polarized identically.

For the record, it should be noted here that polarization, in the most general sense, means that *the amplitudes* (i.e., *maximum values*) *of the* **E** *vectors radiating outward from the axis* (*ray*) *of a light wave are not all equal when viewed in any plane perpendicular to the ray.*

However, in the real world, no oscillator runs forever, and if it did there would be some measurable drift of its frequency over long periods. Therefore no observations by the observer fixed at point P can ascertain that the frequency he observes in a wavetrain of *finite duration* is, in fact, constant over any time interval longer than his period of observation. Similarly, a wavetrain of *finite length* (i.e., a wavelet) cannot be said by the observer to have a truly constant wavelength. The actual oscillators in a lasing medium are atoms, ions, or molecules, each of which oscillates for a limited period, called the *coherence time,* which is equal to the duration of the corresponding radiative transition in energy. This transition time is longest for an atom that is isolated from all external influences except radiation from distant sources. Such a condition does not exist in real lasers but is approximated in gas lasers operating at low pressures. For an isolated atom of any species, the transition time is about 1×10^{-8} sec, which corresponds to a wavelet having an axial length (*coherence length*) of about 3 m. The proximity of atoms in a liquid or solid causes the effective coherence time to decrease, with a corresponding increase in the spectral width of the emitted radiation.

b. Monochromaticity of Laser Light. Each transition of an atom, ion, or molecule from an upper to a lower laser level causes the emission of a wavelet (photon), whose E wave starts at zero amplitude (crest value), rises to a maximum, and then falls to zero again at the end of the transition. For radiation resulting from a single atom undergoing a single downward transition in energy, where the energy change ΔE_{ji} has sharply defined upper and lower levels, the amplitude of the emitted wavelet changes with time according to a *gaussian* envelope, symmetrical about the time center of the wavelet. Such a curve has a bell-like shape; it will be discussed later in some detail. The wavelet is therefore said to be *amplitude modulated.* By a mathematical technique known as *Fourier analysis,* an amplitude modulated wave of frequency f_0 and finite time duration Δt_c can be demonstrated to be physically equivalent to a set of wavetrains of *infinite length* with a *continuous distribution* (i.e., a spectrum) of frequencies symmetrically disposed about the frequency f_0. The shape of the curve of amplitude versus frequency for this continuous distribution will have a maximum amplitude at the center (where $f = f_0$), and a frequency width at the *half-intensity level* (i.e., half-*power density,* that is proportional to E^2) denoted by Δf_a. For a gaussian amplitude modulation, the width of this so-called *atomic line* (really a spectrum, not a true line) at the half-intensity points is given by

$$\Delta f_a = 1/\Delta t_c \qquad [42]$$

Not all of the frequencies within this spectral curve will be amplified by the resonator because only the longitudinal modes defined by Eq. [25] satisfy the wavelength criterion for reinforcement among the forward and backward wavetrains in the resonator. Furthermore, the gain curve of the resonator is highest at the center of the atomic line and falls below the threshold for sustained stimulated emission at frequencies lying near the outer edges of the atomic line. Nevertheless, Eq. [42] gives a good indication of the spectral purity.

Other processes within the laser material also act to broaden the frequency span of the atomic line. The most important of these for a solid-state laser are the banding of energy levels and the vibratory motions of the emitting atoms in the crystal lattice of the host material. In gases, the thermal motions of the molecules or atoms cause a *Doppler shift* of frequency proportional to the speed of the emitter. Also in gases, sudden collisions of the atoms or molecules with one another can interrupt the radiative process by an abrupt transfer of energy to the impacted element. This has the effect of shortening the coherence time and thus increasing the width of the line in terms of frequency. Doppler broadening has a gaussian line shape, but collision broadening has a so-called *lorentzian* shape. In general, spectral purity improves with increasing radiative lifetime (coherence time).

The width of the atomic line in terms of wavelength is easily calculated from Eq. [33]: $\Delta\lambda = \lambda/f\Delta f = \lambda^2/c\Delta f$, where the minus sign has been omitted because here we are interested only in the magnitude of $\Delta\lambda$. For a typical gas laser, such as the CO_2, $\Delta\lambda$ is a small fraction of a nanometer.

c. Collimation of Laser Light. Even if a beam were to emerge from a laser with perfect collimation, this would be degraded by diffraction at the wave front and by scattering in any material medium through which the beam passes. Only in lasers used for surveying and alignment is extreme collimation of critical importance.

2. Transverse Electromagnetic Modes of a Laser Resonator

a. Possible Transverse Electromagnetic Modes. The acronym TEM is used to designate the transverse electromagnetic mode(s) of a laser's resonator. The TEM is described by the *distribution of power density* over any cross-section of a laser beam normal to its axis of propagation. The acronym is customarily written with two subscript numerals following it: TEM_{mn}, where m and n are small integers. The most desirable transverse distribution of power density, for several reasons, is the so-called TEM_{00}, or gaussian, or *fundamental mode.* The subscripts 00 in this case indicate that the power-density distribution is high in the center of the beam and falls gradually to very low values at the edges, so that a short exposure of a photographic film to the beam would show a black spot at the center, gradually varying through declining shades of gray to transparent at the outer rim of the beam. The significance of the subscripts 00 is that there are no striations of trans-

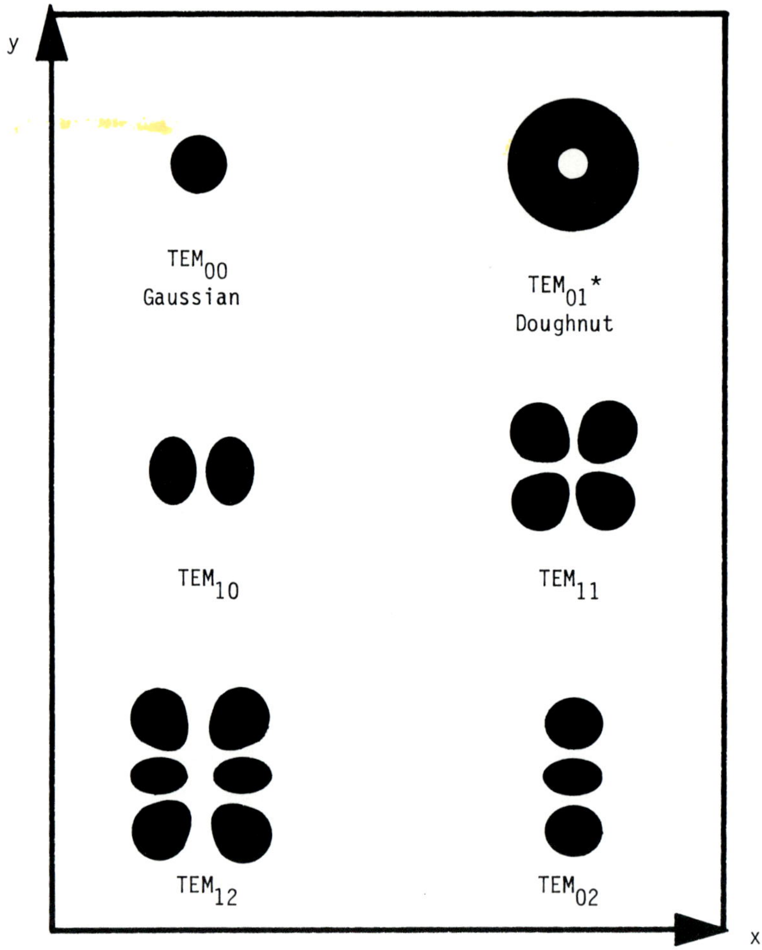

Figure 20 Low-order transverse electromagnetic modes of a laser, as they would be seen by exposing a photographic film to the laser beam. The subscript numerals denote the number of bands of zero intensity in the x and y directions, the first being x and the second y. The patterns are somewhat idealized here; they would be more fuzzy at the edges in a real laser.

parent film (zero intensity) within the dark spot, in either the vertical or horizontal directions. Modes of higher order will have m and n different from zero. These are shown in Fig. 20 for a few of the more common modes. The figure is self-explanatory.

It is possible for a laser to operate in several TEM_{mn} simultaneously, a condition that is known as *multimode operation.* In such cases the beam patterns of Figure 20 become rather complex, and can resemble Rorschach ink blots, indescribable by any simple two-subscript notation. Although fundamental mode operation is usually most desirable, multimode operation produces higher output power, if the inverted population is large enough to feed all modes simultaneously and other factors are equal. Most real lasers, especially those whose laser medium is configured in a cylinder of a length not much greater than its diameter, will not operate in the fundamental mode exclusively. Furthermore, as changes of temperature cause small changes in linear dimensions of the resonator's structure, varying misalignments of the mirrors may occur unless there are compensating measures. The small shifts of the mirror orientations can change the TEM of the laser beam rather dramatically. Plane mirrors are quite susceptible to unwanted multimode operation. Nonuniform curvature and/or reflectivity across the surface of one or both mirrors can cause distortions of the TEM.

To achieve the best approximation to the fundamental mode, the following factors should receive careful attention:

1. The mirrors should be clean, uniform, and properly aligned.
2. The length/diameter ratio of the cylindrical laser medium should be as large as possible in view of other design constraints.
3. The structure of the resonator should be operated at uniform temperature, or be compensated for temperature changes, or both.
4. The ray bundle should be limited in its diameter at its *waist* by the use of an iris or aperture placed coaxially into the resonator, so that the hourglass boundary surface of the ray bundle is relatively long and slender. The waist is the smallest diameter.
5. The pumping power should be introduced into the laser medium with as uniform a volume density (watts/cubic centimeter) as possible.
6. The distribution of the lasing elements should be as uniform as possible throughout the volume of the laser material.
7. The laser medium should be at a uniform temperature in both space and time to avoid variations of population inversion. This requires careful attention to the design of the cooling system.
8. The laser medium should be homogeneous in both the axial and radial directions. This places severe requirements on the growing of crystals for lasers, such as the ruby and Nd:YAG.

b. The Gaussian, or TEM_{00}, Mode. From here onward we will confine our attention to the fundamental, or TEM_{00}, mode. Most well-designed medical lasers deliver beams having at least a fair approximation to this mode. The TEM is of greatest importance in CO_2 lasers, which are used for precise cutting and vaporization of tissue. The high absorption coefficient and negligible scattering of this laser's beam in tissue limit the zone of surgical irradiation (for stationary-beam target exposures) to a very shallow volume having a diameter equal to the effective diameter of the beam. For all the other types of laser used in surgery the scattering effect is important, especially for the Nd:YAG. This subject is discussed at length in Sec. II of this chapter. It is sufficient for now merely to state that scattering within living tissue causes a drastic change in the transverse distribution of power density of the beam *in the tissue.* Therefore the exact power density profile of the beam outside of the tissue is not of critical importance for a Nd:YAG laser. Most significantly, moreover, *the surgeon has no means of determining the histological power density of a Nd:YAG laser's beam during surgery,* even if the incident power density is accurately known. For this reason it is justifiable for us to confine our attention here just to the fundamental transverse mode. An understanding of the TEM_{00} mode will enable the reader to extrapolate the histological effects to beams of higher-order modes.

Figure 21 schematically depicts a laser resonator having, for simplicity, the mirrors directly at the ends of the medium, which can be a gas, liquid, or solid. The mirrors depicted are spherical and have radii of curvature that are much greater than their separation Z_m. If these radii are equal, there will be a *waist* in the envelope of rays at the center of the resonator. The rays emerging through the partially transmitting mirror at the right side will be diverging slightly because of the curvature of the mirrors and the consequent hourglass shape of the ray envelope, unless the output mirror has a lens formed on its outer surface to collimate the rays externally. To the right of the output mirror, a focusing lens is shown. If the focal length of this lens is F and the aperture (i.e., the diameter of the beam at the lens) is D, then the minimum possible diameter of the beam at the focal plane, for a gaussian TEM, is given by

$$d_e = 2w_0 = 4F\lambda/\pi D \qquad [43]$$

where d_e = the *effective* diameter of the beam.

The diameter d_e is the smallest effective diameter that can be achieved by a *perfect* lens focusing a gaussian beam. In actual practice, this minimum can rarely be achieved because of lens imperfections and deviations of the TEM from gaussian.

Note in Fig. 21 that any waist of the beam (e.g., the section of minimum diameter at the focal plane) is a cross-section where the phase front of the beam is plane. At all sections other than a waist (there is another waist within the resonator) the phase front of a diverging or converging laser beam of gaussian profile is a sector of a sphere. Because the diameter of the beam at the focal point can

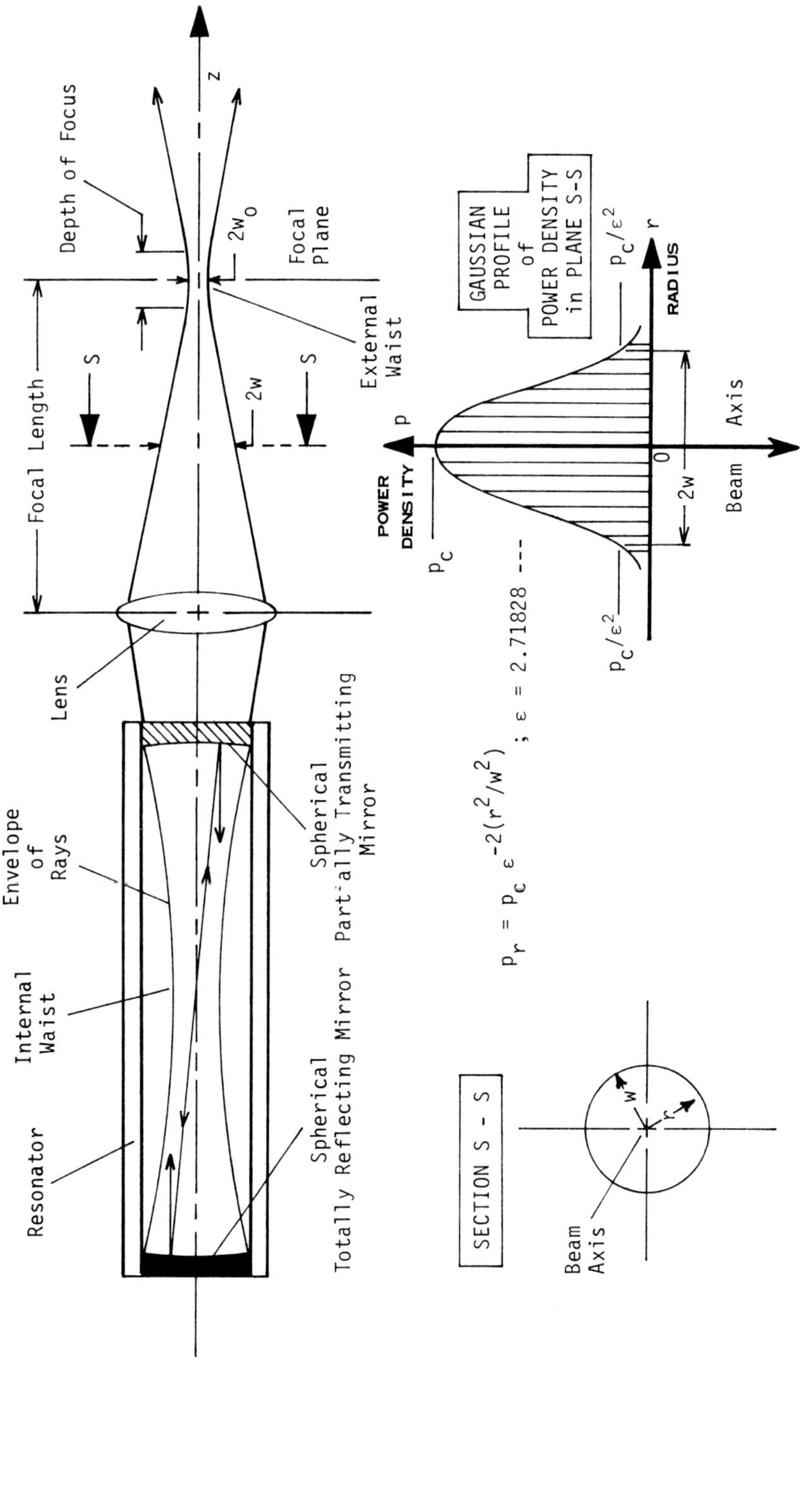

Figure 21 Schematic diagram of a laser resonator with spherical-section mirrors and an external focusing lens. Note that section S-S is shown enlarged for clarity of viewing. A gaussian profile of power density is assumed.

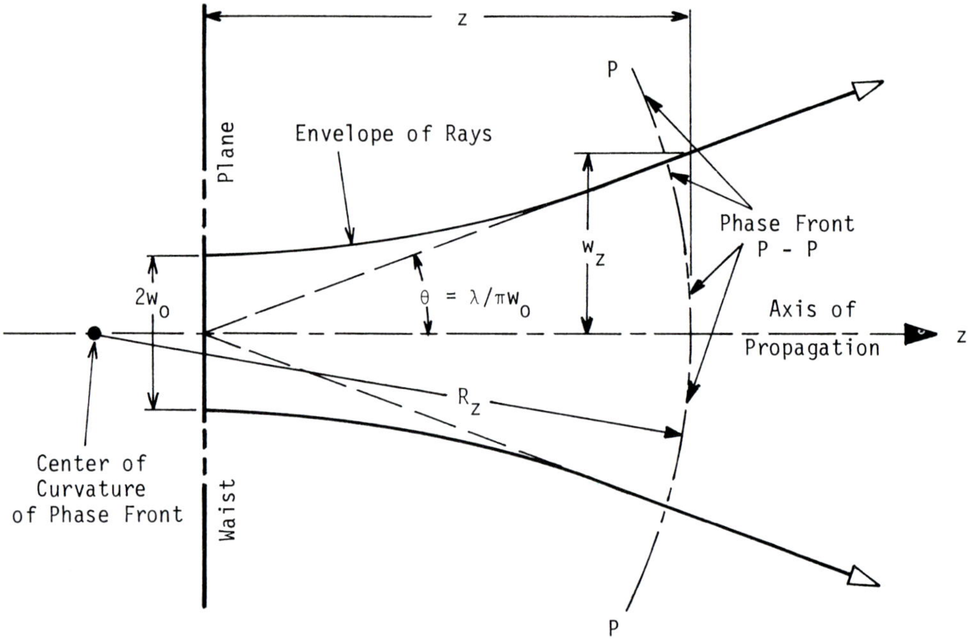

Figure 22 Divergence of a gaussian laser beam from an external waist (focal plane). Note that the center of curvature of the phase front does not coincide with the center of divergence of the rays distant from the waist.

never be zero, the ray envelope at the waist is curved hyperbolically within a short axial span centered about the focal plane, known as the *depth of focus.* The depth of focus is greatest for small angles of convergence θ, and vice versa. Conversely, the minimum focal spot diameter is achieved with large angles of convergence, which implies a low ratio of focal length to lens aperture (i.e., the so-called *f* number of the lens). Only at axial distances greater than the depth of focus does the ray envelope become conical again, and at all such sections the angle of convergence or divergence is constant. Because of the foregoing phenomena, the center of curvature of the phase front is displaced axially from the focal plane (or waist) of the beam, as shown in Fig. 22.

For a detailed analysis, which is beyond the scope of this chapter, the reader should refer to the definitive paper of Kogelnik and Li (B4). The following formulas for the radius of curvature of the phase front and the change of effective diameter with axial distance from the waist of a gaussian beam are presented here without proof. The effective radius of the beam at any cross-sectional distant z from the waist is given by

$$w_z = w_0 \, [1 + (\lambda z/\pi w_0{}^2)^2]^{1/2} \qquad [44]$$

The radius of curvature of the phase front is given by

$$R_z = z \, [1 + (\pi w_0{}^2/\lambda z)^2] \qquad [45]$$

The half-angle of the beam's divergence is given by

$$\theta = \lambda/\pi w_0, \text{ radians } (1 \text{ rad} = 57.3^\circ) \qquad [46]$$

The power density profile in any cross-section S-S in Fig. 21 has the characteristic bell-shaped gaussian curve when power density at any point within that section is plotted versus the radial distance of that point from the axis of the beam. The mathematical description of this gaussian curve is

$$p = p_0(w_0/w_z)^2 \, \epsilon^{-2(r^2/w_z{}^2)} \qquad [47]$$

Because this profile is the same when measured across any diameter of the beam in section S-S, it would have the shape of a three-dimensional bell if it could be seen by an observer. Figure 23 shows a perspective view of a gaussian laser beam impinging on a plane surface, with one quadrant of the beam cut away to reveal the radial profile of power density.

It should be noted that power density in Eq. [47] falls to zero *asymptotically*, so that some radiation exists even at very great distances from the axis. For this reason, it is impossible to define a single value of diameter that encompasses all the radiation of the beam. For the sake of convenience, we define the effective diameter of the beam as that for which the point power density has decreased to $1/\epsilon^2$ of its central value (on the axis), p_c:

$$d_e \equiv 2w \qquad [48]$$

One useful property to remember about a gaussian laser beam is that 86.3% of the total power of the beam is transmitted within a coaxial circle whose diameter is the effective diameter. We can arbitrarily use this fact to define an effective diameter for nongaussian beams. Table 7 shows the fraction of total power transmitted through coaxial circles in any section S-S of a gaussian beam, and the ratio of average power density across such circles to that at the axis, for various values of radius expressed as a fraction of the effective radius. From this table it can be seen that more than 98% of the total power is transmitted through a coaxial circle of 1.5 times the effective diameter.

Because the *average* power density of the beam within any diameter is simply the total power flowing through a coaxial circle of that diameter, divided by the

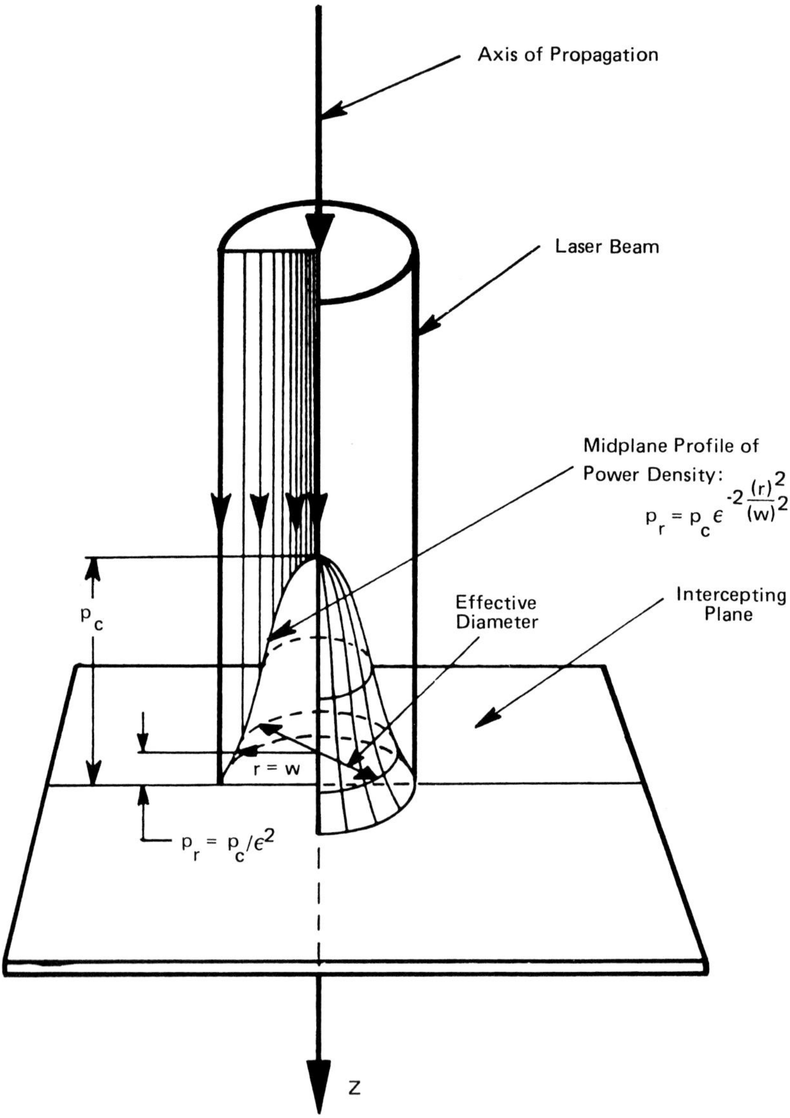

Figure 23 Gaussian laser beam impinging on a plane target surface. One quadrant is cut away to show the midplane profile of power density in the beam (*Source*: from Fisher JC: *Lasers Surg. Med. 2*:301-315, 1983. Reproduced with permission).

Table 7 Variation of Power and Power Density of a Gaussian Laser Beam with Radius of a Circle Coaxial with the Beam

r/w	P_r/P_0	p_{ar}/p_c
0.05	0.004976	0.995190
0.10	0.019752	0.987596
0.15	0.043896	0.975464
0.20	0.076700	0.952570
0.25	0.117224	0.937790
0.30	0.164344	0.913020
0.35	0.216796	0.884879
0.40	0.273236	0.853861
0.45	0.332296	0.820482
0.50	0.392636	0.785271
0.55	0.452996	0.748752
0.60	0.512236	0.711438
0.65	0.569364	0.673802
0.70	0.623556	0.636280
0.75	0.674176	0.599267
0.80	0.720768	0.563099
0.85	0.763048	0.528060
0.90	0.800896	0.494379
0.95	0.834332	0.462233
1.00	0.863492	0.431745
1.05	0.888600	0.402992
1.10	0.909956	0.376014
1.15	0.927904	0.350814
1.20	0.942804	0.331608
1.25	0.955032	0.305610
1.30	0.964952	0.285488
1.35	0.972904	0.266914
1.40	0.979208	0.249797
1.45	0.984148	0.234042
1.50	0.987976	0.219550
	A	*B*
Ratio of radius of circle to effective radius	Fraction of total beam power passing through a coaxial circle of radius *r*	Ratio of average power density over circle of radius *r* to that at the axis of the laser beam

r = the radius of the coaxial circle; w = the effective radius of the beam; P_r = power flowing through the circle; P_0 = total power of the laser beam; p_{ar} = average power density over the circle of radius r; p_c = power density at the center of the laser beam.

Source: from Fisher JC: *Lasers Med. Surg.* 2:301-315, 1983. Reproduced with permission.

area of that circle, or πr^2, we compute the average power density of the *beam* as the total beam power divided by πw^2 and multiplied by 0.8635 (the fraction of total power flowing within the effective diameter). This gives

$$p_{aw} = 0.8635\, P_0/\pi\, w^2 = 0.27486\, P_0/w^2 \qquad [49]$$

If we express Eq. [49] in terms of the effective diameter, $2w$, it becomes

$$p_{aw} = 1.0994\, P_0/(2w)^2 \qquad [50]$$

where P_0 in both Eq. [49] and Eq. [50] denotes the *total* power transmitted by the beam. If, for practical calculations, we express P_0 in watts, and $2w$ in millimeters, then we must multiply the numerator of the right-hand side of Eq. [41] by 100 to obtain p_{aw} in watts per square centimeter:

$$p_{aw} = 110\,(P_0 \text{ in watts})/(2w \text{ in millimeters})^2 \text{ W/cm}^2 \qquad [51]$$

For simplicity in remembering the formula, the factor of 110 in Eq. [51] can be approximated as simply 100. This is accurate enough for most calculations, in view of the difficulty of accurately measuring $2w$ and the uncertainties about the exact TEM in most real lasers.

D. Lasers of Other Types Having Possible Medical Uses

In the foregoing sections we have given a basic physical description of the important phenomena of radiation and lasers, as well as a practical review of the construction and operation of lasers commonly used in surgery and therapy. The presentation, of necessity, has not been encyclopedic. For more detailed study, the reader can refer to one or more of the works listed in the References and Bibliography. These will lead, in turn, to still more sources of information. However, it is useful here to discuss briefly three types of lasers that are not presently employed in the treatment of human diseases, but that may become widely used for this purpose within the next decade. Despite skepticism that now exists regarding lasers in general for medical applications, and especially for these three types of lasers, the physician who is newly learning the principles and applications of lasers to his or her practice would be well advised to keep an open mind on future developments.

1. The Diode Laser

The diode laser is essentially a junction of a wafer of p-type semiconductor with a wafer of n-type semiconductor at a plane interface. When an electric field is applied across the interface, a current flows, and this gives rise to rather monochromatic radiation emerging from the outer periphery of the wafers and directed within a narrow angle centered on the interface plane. At a certain threshold current, the radiant intensity increases sharply and it becomes highly directional and monochromatic. Although the quantum efficiency of diode lasers is higher than that of all other types of lasers, diode lasers are commonly operated intermittently at low-power output because of the high power density of heat generated at the junction. The spectral purity of diode laser radiation is usually much lower than that of the coherent radiation from other lasers. Presently, the most important applications of diode lasers are in communications, where signals are transmitted through long fiber-optic cables by modulation of the light intensity produced by a diode laser and passed through the fiber with very low losses owing to absorption. Most of the diode lasers now commercially available produce wavelengths in the near-infrared portion of the spectrum (700-1400 nm) or in the far-infrared range (1400-100,000 nm). The wavelength can be varied in such lasers by pressure applied normal to the interface plane or by subjecting the junction to a magnetic field.

The only current medical applications for diode lasers are in the relief of pain or inflammation and in the stimulation of wound healing. Although these uses still lack adequate clinical documentation to convince the United States FDA, and many American physicians, of their safety and validity, there is a growing body of anecdotal evidence from Europe and the Far East to suggest that some biochemical effects of importance are produced by low-intensity radiation in the near-infrared range of wavelengths. Many veterinarians and animal trainers in North America are using low-powered diode laser systems to irradiate the leg muscles, ligaments, and tendons of racehorses for the palliation of inflammation and pain. The major evidence of these people to substantiate their claims of effectiveness is in the form of improved racing performance in horses so treated, as compared with that of untreated animals, or for the same animals before and after treatment. Although such information is not of the kind that some veterinarians consider clinically acceptable, one can say with assurance that horses are not susceptible to placebo effects by suggestion. If a formerly lame horse runs well after irradiation, the improvement in track performance is a primary datum of observation that can be statistically correlated with the application or withholding of radiant energy to the inflamed parts. One such study (13) has been made in Canada.

2. The Excimer Laser

The term *excimer* refers to a metastable molecule formed by the transient combination of an atom of a noble gas (helium, neon, xenon, argon, krypton, and radon) with an atom of a halogen (fluorine, chlorine, bromine, or iodine) while each atom is in an excited state. After a downward transition in energy giving rise to emission of radiation, the transitory molecule decomposes into its constituent atoms. Because there is no ground state for the excimer as a molecule, excimer lasers are easy to pump. Their quantum efficiency is good, and overall efficiencies are in the order of 1-2%. However, because the lifetime of the excimer molecule is only a few nanoseconds, such lasers cannot be operated in a continuous wave mode. Power output of up to 25 W, time averaged, is possible with repetitive pulsing. The major wavelengths produced are 193 nm, 248 nm, 308 nm, and 351 nm, all of which lie in the ultraviolet range of the spectrum. Excimer lasers employ high-speed electric discharge pumping systems and sealed-off tubes containing gas mixtures at absolute pressures of 2.5-3.0 atm. These mixtures typically consist of two parts of halogen, 20 parts of noble gas, and 978 parts of helium. Because of the extreme chemical activity of the halogen atoms, the tube, electrodes, mirrors, and seals deteriorate rapidly, and the life of a tube in an excimer laser is limited to a maximum number of pulses, of approximately a few million.

Excimer lasers have been used experimentally in the removal of atheromatous plaque from the interior of autopsied cadaver arteries (14), and have demonstrated the ability to cut this plaque with extreme precision, partly because of the smaller focal spots afforded by these short wavelengths and partly because the action is largely photochemical rather than photothermal. This *photochemolysis* is the result of the high photonic energy of ultraviolet light (3.10 eV or higher), which breaks the interatomic bonds in organic molecules and converts them to atomic debris. The power density required was about 2.5 W/cm^2, far below the level at which vaporization of the plaque could occur. The precision of tissue destruction that is possible with excimer lasers is two orders of magnitude greater than that which can be achieved with the carbon dioxide laser. This holds out the hope that surgical removal of tumors on critical structures such as the optic nerve may be possible without damage to fragile normal tissue nearby. However, ultraviolet radiation can also cause ionization of atoms in irradiated tissue, and therefore has a carcinogenic effect even at low-power densities. Much careful study will be needed to determine whether the potential carcinogenicity of ultraviolet lasers outweighs the potential benefits to be derived from the ability of these devices to cut tissue nonthermally with a resolution of a few micrometers. Because quartz optical fibers have 90% transmission at 200 nm in lengths of 1 or 2 m, ultraviolet laser rays can be used for recanalization of obstructed arteries. The results of the few human studies, to date, done with argon ion and Nd:YAG lasers have been disappointing in terms of the postoperative duration of patency. Choy, Fournial, Stertzer et al. have employed the argon ion laser in Toulouse, France (15) for recanalization of human coronary arteries, and Geschwind et al. (16), in Paris, have used the Nd:YAG for opening obstructed femoral, popliteal, and tibeal arteries in humans. The overall rate of patency in these patients at 1 year postoperatively has been less than 20%. Although the causes of reformation of atheromas in these laser-treated arteries are not understood, it may be that thermal trauma to the intima and media of the arteries is a contributing factor. If so, the use of a nonthermal technique may offer the possibility of improved long-term patency.

3. The Free Electron Laser

The free electron laser is essentially an evacuated chamber into which a collimated beam of electrons is injected at high velocity from a linear accelerator. When the electrons in the beam are accelerated laterally by means of transverse magnetic fields having an alternating spatial distribution of intensity, they give off electromagnetic radiation whose wavelength can be varied by changing the spatial frequency and/or peak intensity of the transverse fields, as well as by altering the velocity of the electrons entering the chamber. Theoretically, this allows the generation of wavelengths ranging from x-rays to microwaves. However the size, cost, and complexity of such systems are formidable obstacles to their widespread adoption in medical applications. At present, the free electron laser remains a laboratory curiosity that intrigues the imaginations of visionaries in the medical and scientific communities. Given the phenomenal acceptance of large, expensive devices like computer-assisted tomographic (CAT) scanners and nuclear magnetic resonance (NMR) scanners, though, the mere cost and size of free electron lasers may not deter their wide acceptance, provided that a sufficient range of surgical or therapeutic procedures can be done only by their use, or with significant advantages over other lasers or other modalities. The unique advantages of beams of light as tools for cutting, vaporizing, coagulating, or photochemically altering tissue have been demonstrated beyond a shadow of doubt. As the skeptical, complacent generation of physicians and surgeons who were trained before the advent of lasers retires and is replaced by young graduates of medical schools and hospitals that teach the new arts of photosurgery and phototherapy, the laser modality will finally attain its recognition as the most important advance in surgery in the twentieth century, ranking with the invention of anesthesia by Morton and Wells in 1846 and the discoveries of Pasteur and Lister in antisepsis later in the nineteenth century. The free electron laser might well become a near-universal surgical instrument in the twenty-first.

IV. INTERACTION OF LASER LIGHT WITH LIVING TISSUE

A. Relative Importance of the Special Properties of Laser Light in Surgery and Therapy

In essence, a laser is simply a generator of a unique form of light having three special properties: coherence, collimation, and monochromaticity. Surprising as it may seem to those unfamiliar with biophysics and biochemistry, these properties are not all of equal importance for surgery and therapy. We discuss them in detail in the following sections.

1. Coherence

Coherence of laser light is not essential to the surgical uses of lasers producing focal power densities of less than 1×10^6 W/cm^2. Certain applications of lasers, such as interferometry and holography, rely upon the spatial and temporal coherence of laser radiation. It appears quite likely that holography, the recording of three-dimensional images of solid objects, may have significant applications in the study of cell biology within the next few years. However, for cutting, vaporizing, and coagulating of tissue, coherence is not critical except insofar as it contributes to the attainment of small focal spots and high-power densities. For any effects that depend upon *heating* of tissue, which is the result of *nonresonant absorption* of laser light by the atoms and molecules, neither spatial nor temporal coherence is essential. For *photochemical effects,* which do involve resonant absorption and can occur at low power densities throughout the entire ultraviolet, visible, and near-infrared portions of the spectrum, but are most pronounced in the ultraviolet, coherence may significantly facilitate whatever reaction is occurring, especially with large, complex organic molecules or long-chain polymers.

A little thought will reveal that the rays of laser light that are transmitted through optical fibers (e.g., argon ion, dye, and Nd:YAG) lose their spatial coherence even if the fiber is configured in a straight line because rays that pass through the fiber in any direction other than parallel to the axis of the core will experience multiple reflections from the interface between the core and the cladding, thus traversing a number of straight-line segments between reflections, whose total length is always greater than that of the fiber measured along its axis. This effect is exacerbated by flexing of the fiber, which is necessary if it is to be of any use in surgery. Only a perfectly collimated beam passing through a perfectly straight fiber will retain its spatial coherence. In fibers that are of *graded* index of refraction, that is, the index of refraction decreases continuously with radial distance from the axis of the fiber, this effect may be reduced. Such fibers often have a refractive index that decreases parabolically with radial distance from the axis, and these transmit nonparallel rays in approximately sinusoidal paths through the core. However, transmission of a laser beam through any fiber that is not perfectly straight will degrade the spatial coherence of the light waves. Nevertheless, it is a common observation that the beams of visible and near-infrared lasers coagulate tissue just as effectively at the distal end of a fiber as at the proximal, provided that the average power density is the same at both ends.

2. Collimation

Of the three special characteristics of laser light, this is the most important, for it allows a system of lenses or mirrors to capture *all* of the light emerging from the source (the laser) and focus it upon the target. While it is true that certain light source configurations, such as a long, straight lamp of small diameter at the focal line of a parabolic reflector, can deliver virtually all of their output to a target, no source except a laser can do this with such a high degree of concentration of its radiant power *prior to focusing*. In the 1960s, a helium-neon laser (λ = 694 nm) was directed at the surface of the moon; the diameter of the beam on the lunar surface was about 2.5 miles at a distance of 239,000 miles from the laser, corresponding to a divergence angle of 1.0×10^{-5} radian, or about 57×10^{-5} degrees!

3. Monochromaticity

Monochromaticity is the second most important property of laser light for biomedical purposes. However the degree of spectral purity afforded by any laser (e.g., $\lambda/\Delta\lambda$ = 1060/5.6 = 189 for the Nd:glass laser, relatively low on this scale of value) is far higher than necessary for resonant activation of any biochemical system. This fact is evident from Fig. 24, which depicts an in vitro absorption spectrum of hematoporphyrin derivative (HPD), the photosensitizer used in photodynamic therapy. The half-power width of the broadest peak of this spectrum is about 50 nm (the blue peak, located at 408 nm), whereas the spectral width of the output from the Nd:glass laser is only 5.6 nm, or 11% of the linewidth of HPD. Although the Nd:glass laser is not used in photodynamic therapy, we have used it as a basis for comparison here because it has one of the lowest values of spectral purity among common lasers. Thus we can see that the helium-neon laser, with its principal output at 633 nm and an output linewidth of 0.002 nm, which could be used to excite HPD at its lowest in vivo peak of 628 nm if the laser could deliver several watts, would represent a waste of spectral purity in such an application. Essentially the same comment is true for the dye laser, most commonly used to excite HPD in photodynamic therapy. The extreme spectral purity of most lasers, which is an inherent result of the generation of coherent radiation, is a technical overkill in biomedical applications, comparable with using a large mainframe computer to do simple arithmetic.

The broad absorption band of HPD around any of its major peaks is evidence of an underlying photochemical resonance that is not sharply tuned in terms of wavelength, but instead responds to many wavelengths lying within a range centered on the most effective one. The breadth of such resonances is characteristic of large, complex organic molecules, which, by virtue of the very intricacy of their molecular structure, have wide bands of possible energy levels corresponding to the permitted energies of the electrons in myriads of constituent atoms, combined with the multiple energy levels corresponding to rotation of the molecule as a whole and to numerous degrees of vibrational freedom within the structure of the molecule. Even in simple inorganic compounds such as carbon dioxide, the spectral absorption peaks are considerably broader than the linewidths

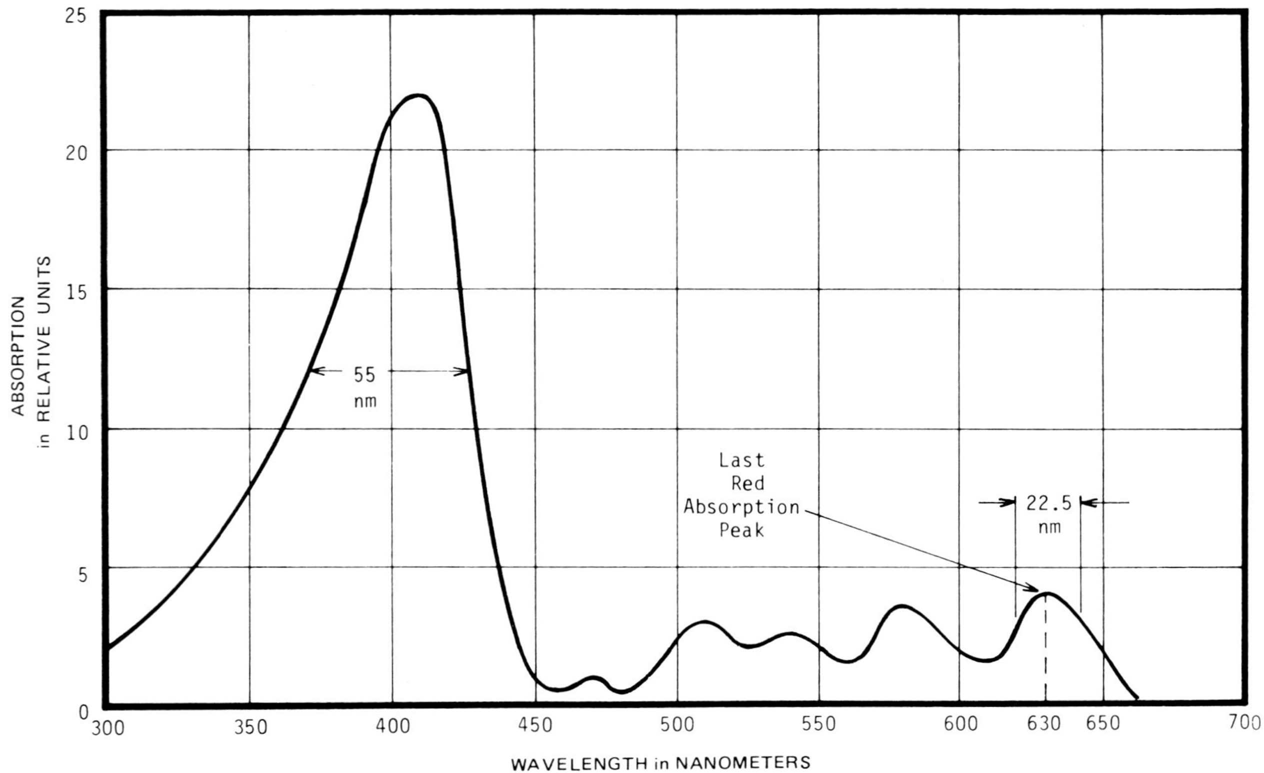

Figure 24 The absorption spectrum of hematoporphyrin derivative (HPD) in human serum, plotted by the author from the data of Thomas Dougherty and associates at Roswell Park Memorial Institute, Buffalo, New York. Note the width of the several peaks. The last red peak at 630 nm is usually used in photodynamic therapy because this wavelength affords good penetration of overlying tissue by the stimulating light.

of most lasers. As we saw in our discussion of resonant absorption in Sec. II. B.1.b, the photoexcitation of laser substances that have three-level or four-level energy-exchange systems can often be achieved by *broadbanded,* noncoherent sources of light. This is so because there are multiple permitted energies lying above the metastable upper laser level, from which the pumped laser material can quickly relax into the upper laser level and wait for stimulated emission to occur.

The great advantage of lasers as light sources for biological uses is the *spatial and spectral concentration of their radiant power* so that it is *neither widely spread over a range of wavelengths nor emitted in a large solid angle* like the light of noncoherent sources such as incandescent bodies. If a hypothetical, noncoherent source of light were available to produce a 100-W beam of radiation 25 nm wide spectrally and 10° wide spatially, it would be quite satisfactory for medical and surgical applications, provided its central wavelength could be tuned or chosen over a range from 400 nm to 10,000 nm. Unfortunately, such a source does not now exist.

B. General Effects of Light on Tissue

Because even primitive organisms exhibit mind-boggling complexity of chemical, physical, and biological structure, a description of the interaction of radiation with living tissue suitable for the guidance of the practicing physician who uses lasers must, of necessity, be an approximation to the actual facts, a complete recitation of which would require an encyclopedic treatise longer than this entire book. However we can provide here a basic, general account of the important principles and phenomena, so that the user of lasers who desires later to conduct an exhaustive study has a solid foundation upon which to build better understanding.

For our present purposes, it is useful to regard the human body as having five levels of structure: cells, tissues, components, organs, and organ systems. None of these parts of structure can be regarded as homogeneous, let alone isotropic. However, it is instructive for us to idealize cells and tissues to some extent. Let us think of cells as being sacs having very thin, impermeable, elastic membranes containing a homogeneous, isotropic liquid whose physical properties are weighted averages of the actual properties of the inhomogeneously distributed cellular constituents (genes, chromosomes, nucleoli, protoplasm, and organelles). Let us think of tissues as being quasi-homogeneous aggregations of our ideal cells and/or connective substances (collagen, elastin, reticulin, keratin, and ground substance) with histological properties that are weighted averages of the physical properties of our idealized cells and the connective substances. Thus we view tissues, in essence, as being composed of a homogeneous, isotropic liquid that is subdivided by thin, impervious, elastic barriers into many small compartments, interspersed with fibers, bonding materials, and other liquids within the interstices. The significant difference between this model of soft cellular tissue and a gross volume of protoplasm is that our model permits *no bulk convection* of the liquid. As we shall see later, this has an important bearing on the thermodynamic behavior of irradiated tissue.

When real tissue consists chiefly of noncellular elements, such as the connective tissue of the dermis or the vitreous gel of the eye, there may be a substantial fraction of water that strongly affects the volume-averaged physical properties of the tissue. This liquid is often bound within a matrix, such as the hyaluronic acid of the eye, and so is incapable of convection. Except for liquid passing through the gastrointestinal tract, urine in the bladder, blood in large vessels, and the cerebrospinal fluid, most of the water in the body is nonconvective. However, real living tissue is discretely interspersed with blood vessels and lymphatic ducts, which serve as sources or sinks of heat to maintain the average histological temperature nearly constant despite flow of heat to or from the environment. This network of liquid conduits, ranging in size from micrometers to centimeters, plays a vital role in limiting the spatial extent of *thermal* damage when tissue is exposed to laser light.

When an incident laser ray strikes the surface of a mass of tissue, four basic phenomena occur:

1. *Reflection* from the first surface
2. *Transmission* into or through the tissue
3. *Scattering* within the tissue, including *diffraction*
4. *Absorption* by the tissue between scattering points

All of these effects are dependent in magnitude upon the wavelength of the light and the physical properties of the tissue.

Quantitatively, the relationship among these four biooptic phenomena can be described in terms of the radiant power density measured along the line of any incident ray of light, from outside the first surface to the region beyond the mass of tissue under consideration. This power density will be greatest just outside of the first surface, and will progressively change in magnitude along the ray direction with distance into the tissue measured from this surface. We shall assume that the first surface is exposed to air, as it usually will be in laser surgery, although in certain cases the laser beam may be transmitted to the first surface through a liquid such as normal saline or a gas such as carbon dioxide. In general, the *forward* power density will diminish with distance into the tissue along the ray direction (which we shall take as *perpendicular* to the surface). However, at some wavelengths, notably around 1000 nm, it may actually increase over the first part of the ray's path into the tissue, because of forward scattering. The interrelation among the four basic biooptic phenomena is given by

$$p_0 = p_r + p_s + p_a + p_t \quad (POWER)/(AREA) \qquad [52]$$

where p_0 is the forward power density in the ray direction just outside the first surface.

p_r is the backward power density along the ray, caused by reflection and/or backscattering from the surface and the tissue just beyond.

p_s is the cumulative loss of forward power density between the surface and the interior point on the ray at which the net forward power density is measured, caused by scattering and diffraction.

p_a is the cumulative loss of forward power density between the surface and this interior point because of absorption by the tissue.

p_t is the net forward power density on the ray at this interior point.

The suitability of a particular type of laser for a given surgical or therapeutic application is determined largely by the magnitudes of p_s and p_a, and their ratio, as functions of depth into the tissue.

1. Reflection

What appears to be reflection from the first surface, if viewed with the eye or seen by a reflection spectrophotometer, may actually include a substantial fraction of radiation scattered in the backward ray direction, especially in the near-infrared part of the spectrum, in which some of the laser light penetrates well into the tissue and is then scattered back out again into the air. Furthermore, the *reflectance* (i.e., the fraction of the incident power density that is reflected from the first surface) is strongly dependent upon the angle of incidence of the incoming ray, varying from nearly 100% at or near 90° from the normal (grazing incidence) to a minimum value at normal incidence, for any wavelength. This fact is of great importance for lasers used endoscopically within the lumina of hollow organs, where the objective is to irradiate an obstructive mass of tissue to cause coagulation or vaporization without damaging the wall. If the laser rays are delivered so that they are nearly or perfectly parallel to the wall, its reflectance will be very high. Hence it is important to limit the angle of divergence of a laser beam emerging from the distal end of an endoscopic fiber and to be careful not to direct the fiber's axis perpendicular to the wall (except where the objective is to coagulate the wall itself, as with bleeding gastric ulcers).

For normal incidence, the reflectance of human skin has been measured by several investigators, but little is known about the reflectance of internal organs. Figure 25 shows the spectral variation of epidermal reflectance for fair-complexioned and dark-complexioned individuals, plotted by the author from data of Jacquez and Kuppenheim (17) for visible (400-700 nm) and near-infrared (700-1400 nm) portions of the spectrum, and from the data of Hardy et al. (18) for the ultraviolet (200-400 nm) and infrared (1400-50,000 nm). Figure 26 shows

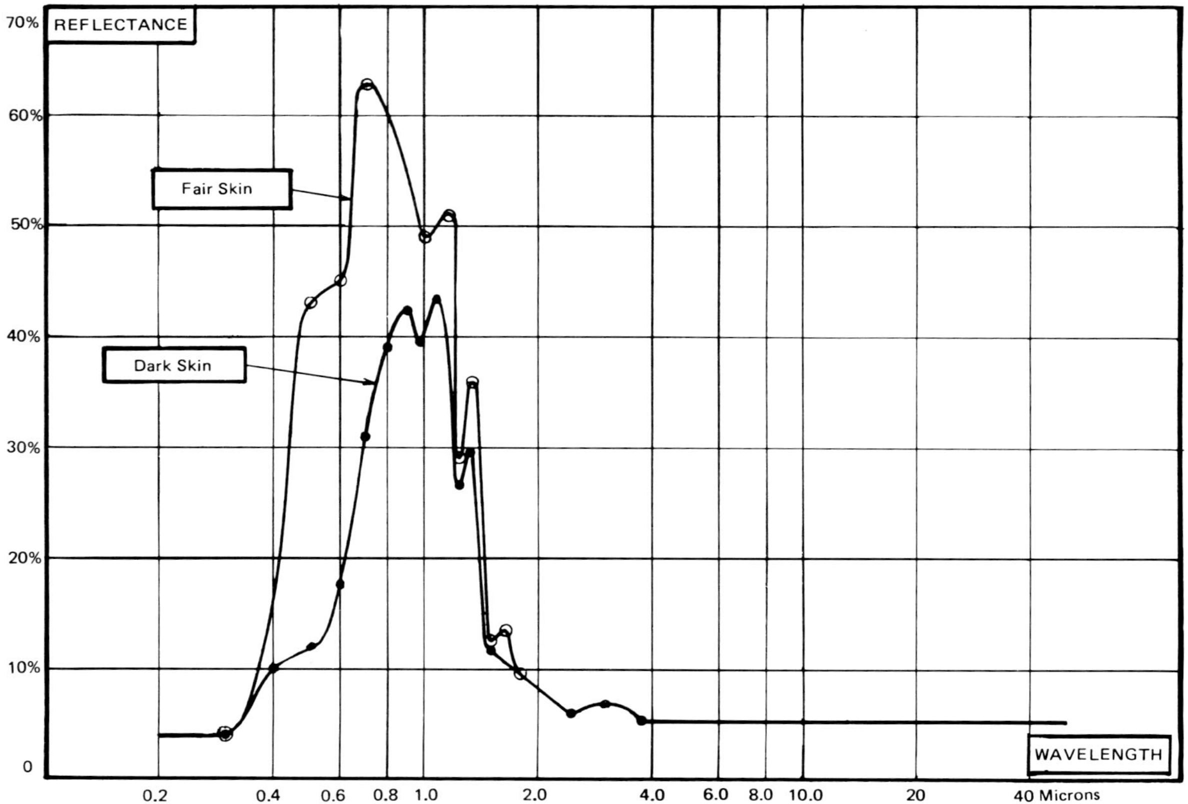

Figure 25 Spectral variation of epidermal reflectance for fair-complexioned and dark-complexioned individuals, as plotted by the author from the data of Jacquez and Kuppenheim (17) for the visible and near-infrared, and from the data of Hardy et al. (18) for the ultraviolet and far-infrared regions of the spectrum.

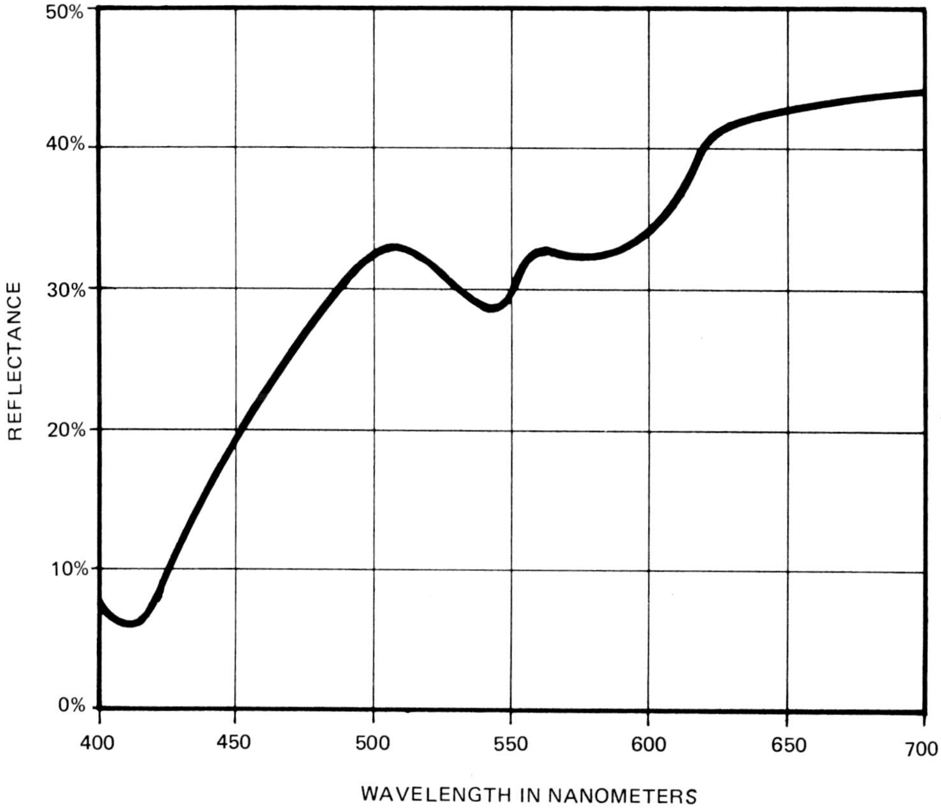

Figure 26 Spectral variation of reflectance of human blood in vivo, as determined by T. Ohshiro using a direct-reading reflection spectrophotometer, and plotted by the author from Ohshiro's data (19).

spectral variation of the reflectance of liquid blood, as plotted by the author from the data of Ohshiro (19) for the visible range, at normal incidence. Liquid blood often covers the surfaces of targets for laser beams within the body, and the surgeon is well advised to remove it by suction or swabbing before irradiating his target because it has a high reflectance throughout most of the visible range. From Fig. 25, it is probably safe to assume that most soft human tissue will strongly reflect light in the visible and near-infrared regions of the spectrum, but poorly in the ultraviolet and infrared. Most surgeons are not aware of the high reflectance of soft tissue for laser wavelengths that can be transmitted efficiently through optical fibers suitable for endoscopy. For example, about 50% of the light in the beam of an Nd:YAG laser can be reflected from skin at normal incidence. While this does not pose a serious hazard to medical personnel because of the diffused nature of the reflected light, it does represent a serious waste of a large part of the available radiant power, whose capital cost currently is more than 1000 dollars/W for surgical lasers.

2. Absorption

Absorption of radiation takes place at the level of atoms and molecules, whose energies are thereby increased. Radiant energy can be absorbed also by ions or by electrons. The details of this process were discussed at some length in preceding sections, particularly II.B.1 and II.B.2. In photochemistry, *radicals* are capable of absorbing radiation. A radical is a submolecular fragment consisting of one or more species of atoms bonded together as a unit that can pass through a chemical reaction unchanged in its internal bonding.

In the most general terms, absorption is a process of conversion of radiant energy at one or more incident wavelengths into other forms of energy as follows:

1. Excited levels of atoms, ions, molecules, and radicals
2. Chemical energy, i.e., the quasi-permanent *difference* between the possible levels of a whole molecule and the sums of the corresponding levels of all its atomic constituents or smaller molecular constituents, notably the ground levels of these entities, before and after a chemical reaction
3. Reradiated energy at other wavelengths, usually longer than that or those of the incident radiation
4. Thermal energy, i.e., the random kinetic energy of molecules or molecular constituents in liquids and gases, and the random or ordered vibrations of atoms and ions in solids, always associated with temperatures above absolute zero
5. Kinetic energy of electrons stripped away from their parent atoms by radiation of high photonic energy ($e_p \geqslant 3.9$ eV or 6.2×10^{-19} J, or $\lambda \geqslant 318$ nm); or high power density ($p \geqslant 1 \times 10^{10}$ W/cm^2) equivalent to multiphoton absorption

In the preceding list, the conversion of radiant energy to the forms 1, 2, 3, 4, and 5 is never totally reversible, but some forms have higher reversibility than others. Forms 1 and 2 can have as much as 40% reversibility. Form 3 has much lower reversibility, because reradiated energy is most always of longer wavelength than the excitation, and these longer wavelengths trend ultimately toward conversions into thermal energy, or heat. Form 4 is largely nonreversible because heat normally flows from regions of higher temperature to regions of lower temperature, thereby decreasing the probability of thermally producing excited states, according to Eq. [29]. Form 5 is only partially reversible, because some of the kinetic energy of a free electron is lost as heat during inelastic collisions with atoms and molecules.

If the radiant power density of the incident light is high enough, the temperature of the absorbing units (ions, atoms, molecules, radicals, etc.) will rise temporarily, and this can cause conversion of some of the excess heat to an increase in the *gross mechanical energy* of the population, which is the product of the *pressure* and *volume* of this population. This product is dimensionally equivalent to $(FORCE) \times (LENGTH)^3/(LENGTH)^2$, or $(FORCE) \times (LENGTH)$, which has the dimensions of *work*. The gross mechanical energy of the population in situations of inequilibrium between the population and the environment represents the *potential* energy available to do macroscopic work on the surroundings. If the thermal energy of the population suddenly increases, either the temperature, the pressure, the volume, or any combination of these, may increase. In particular, if the volume increases, the population can do work on the environment, as when the hot gases in the cylinders of an automobile engine push the pistons. Alternatively, if the increase of volume is nearly unlimited (as in an explosion), the population may experience a sudden increase of kinetic energy in nonrandom directions.

In the absorption of laser light by living tissue, forms 2 and 4 are the predominant effects at power densities below 1×10^6 W/cm^2 and wavelengths longer than 400 nm, with almost all of the radiant energy being converted into heat. At wavelengths from 100 to 400 nm, chemical energy (form 2) and kinetic energy of ejected electrons (form 5) are likely to be the dominant forms.

Figure 27 schematically depicts the energy balance of a numerically and spatially finite population of molecules and submolecules. These submolecules are electrons, ions, atoms, radicals, and smaller molecules, as well as excimers. The molecules may be those normally found in living tissue, including enzymes (organic catalysts), plus photosensitizers. A photosensitizer is a molecule capable of absorbing light at certain wavelengths and transferring this energy to other molecules that are not themselves capable of radiant absorption at these wavelengths.

In Fig. 27 it is assumed that the population (shown inside the box of dot-dashed lines) is in thermal, chemical, mechanical, and photonic equilibrium with its environment (everything outside the box) and with itself. The flows of energy among the various forms it can take are shown as dashed lines, with arrowheads indicating the direction of flow. If, at some instant $t = 0$, the population is suddenly irradiated by light of one wavelength that was not present before, or at a higher intensity than before, this equilibrium will be disturbed. If the new radiant energy is delivered in a period much shorter than the time needed for this population to reach equilibrium again, then the various flows of energy will be altered in directions shown by the solid arrows. The added radiant energy quick-

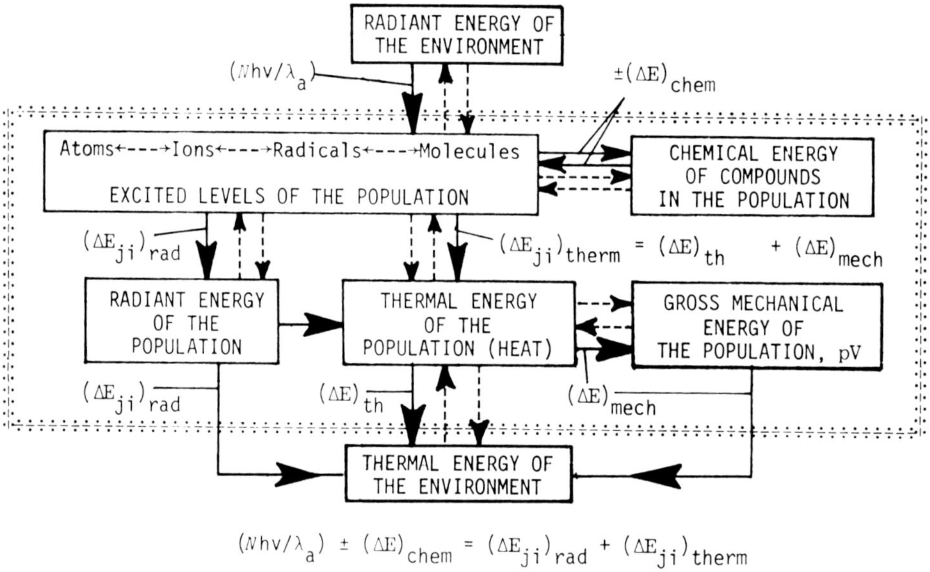

Figure 27 Schematic diagram of the energy balance between a population of molecules and submolecules, initially in equilibrium with the environment, which is upset by the sudden absorption of radiant energy from the environment. The dashed arrows show the flows of energy among the various possible forms in equilibrium, and the solid arrows show the flows resulting from the absorption of radiant energy from the surroundings. Ultimately, equilibrium is restored when the external radiation stops.

ly increases the sum of all excited levels, which returns rapidly to its former value, with emission of heat and reradiation:

$$(Nh\nu/\lambda_a) \pm (\Delta E)_{chem} = (\Delta E_{ji})_{rad} + (\Delta E_{ji})_{therm} \qquad [53]$$

where N is the number of new photons absorbed, h is Planck's constant, ν is the speed of light, and λ_a is the wavelength of the new incident radiation. In Eq. [53], the left-hand side, the net change in the total internal energy of the population, can be either increased or decreased by the change of chemical energy resulting from the new radiation. For example, in the hydrogen-fluoride laser, the chemical energy liberated by the reaction of molecular hydrogen and carbon tetrafluoride causes a population inversion among the resulting molecules of hydrogen fluoride.

The sum of the thermal energy of the population and its gross mechanical energy is known in thermodynamic terminology as the *enthalpy*. In all the lasers currently used for surgery, changes in the enthalpy of the tissue caused by irradiation are responsible for the major effects of the beam, namely *coagulation* and *vaporization*. Coagulation is a process of altering the molecular structure of organic compounds, notably proteins, by raising their temperature to values at which molecular breakdown occurs. This is the result of changes in the internal energy levels of the molecular constituents, but these level changes are thermally excited, for most constituents, rather than photonically, at wavelengths greater than 400 nm. Vaporization is a process requiring the *change of physical state* of at least part of the population of the tissue, usually water molecules, from liquid to vapor. For water, this requires the addition of 2522 J/g to take it from liquid at 37°C (normal body temperature) to steam at 100°C, of which 2257 J/g is needed to provide the *latent heat of vaporization* necessary for the liquid-vapor transition. Once the histological water is converted to steam, the gross mechanical energy of this vapor is available to do work on the remaining solid elements of the population, rupturing cell membranes, breaking fibers, and ejecting particles out of the tissue. This destructive process is in sharp contrast to that of ultraviolet laser light, which fractures chemical bonds directly by alteration of the balance of levels of the population, or by ionization of the molecular constituents, or by both, with little or no change in temperature.

Along the direction of any incident ray of light, extended into the tissue without deviation (because we shall consider scattering and diffraction separately), the power density will ultimately diminish in monotonic fashion with distance beyond the first surface. If this distance is denoted by z, the power density remaining at any point for further transmission is given by

$$p_t = (p_0 - p_r)\, \epsilon^{-Az} \qquad [54]$$

where A is the *attenuation coefficient* of the tissue and ϵ is the base of the natural logarithms ($\epsilon = 2.71828\ldots$). The value of A is dependent upon both the wavelength of the light and the properties of the tissue, which we shall again assume to be homogeneous and isotropic, an assumption justified by the fact that in macroscopic volumes of tissue Eq. [54] gives a fair approximation of the real situation if the value of A is chosen to represent the weighted average of the properties of the many diverse histological components. The attenuation coefficient is itself the sum of two parts:

$$A = \alpha + \sigma\ (LENGTH)^{-1} \qquad [55]$$

where α is the *absorption coefficient*, $(LENGTH)^{-1}$
σ is the *scattering coefficient*, $(LENGTH)^{-1}$

As we shall see later, the ratio σ/α primarily determines the suitability of a particular wavelength for a given surgical purpose in the spectral range of wavelengths longer than 400 nm. Equation [54] has been attributed to both the French scientist Pierre Bouguer (1698-1758) and the German physicist Johann Lambert (1728-1777). It says, in essence that the transmitted power density in the ray direction will lose a constant fraction of itself in each increment of path length of fixed value Δz.

Accurate values of α, σ, and A for tissue of various types in human and animal bodies have not been determined. However, it is usually assumed that the values of these coefficients for water are valid for living tissue because water is the major constituent of all soft tissue. This is a crude approximation that is useful only for a general understanding of the phenomena of absorption and scattering as functions of wavelength. Much experimental work remains to be done in measuring these values in vivo. Fortunately, there has been a substantial amount of investigation by various experimenters in measuring values of α as spectral functions for such biological substances as water, melanin, and hemoglobin. The data of various studies have been plotted by the author in Fig. 28 for these substances.

3. Scattering and Diffraction

For our purposes, scattering and diffraction will be regarded as one lumped phenomenon in which the direction of a photon or wavelet is changed, but its frequency remains the same. This is not strictly true, because certain kinds of scattering, such as that discovered by the Indian physicist C. V. Raman (1888-1971) in the latter part of the 1920s, involve a change of the frequency. Our combined phenomenon, as observed in living tissue, is actually a composite of these fundamental optical effects:

1. *Diffuse reflection* from solid particles that are much larger than the wavelength
2. *Diffraction* of light rays at wavefronts
3. *Refraction* of light rays as they pass obliquely from one substance to another in the microscopically inhomogeneous tissue
4. *True scattering* from molecules, radicals, atoms, and ions

---

Figure 28 Absorption coefficient of hemoglobin and melanin for wavelengths in the visible and near-infrared regions of the spectrum. Melanin in light brown skin would have an absorption curve about one order of magnitude lower on the vertical scale. Curves plotted by the author from the data of Franz Hillenkamp [38].

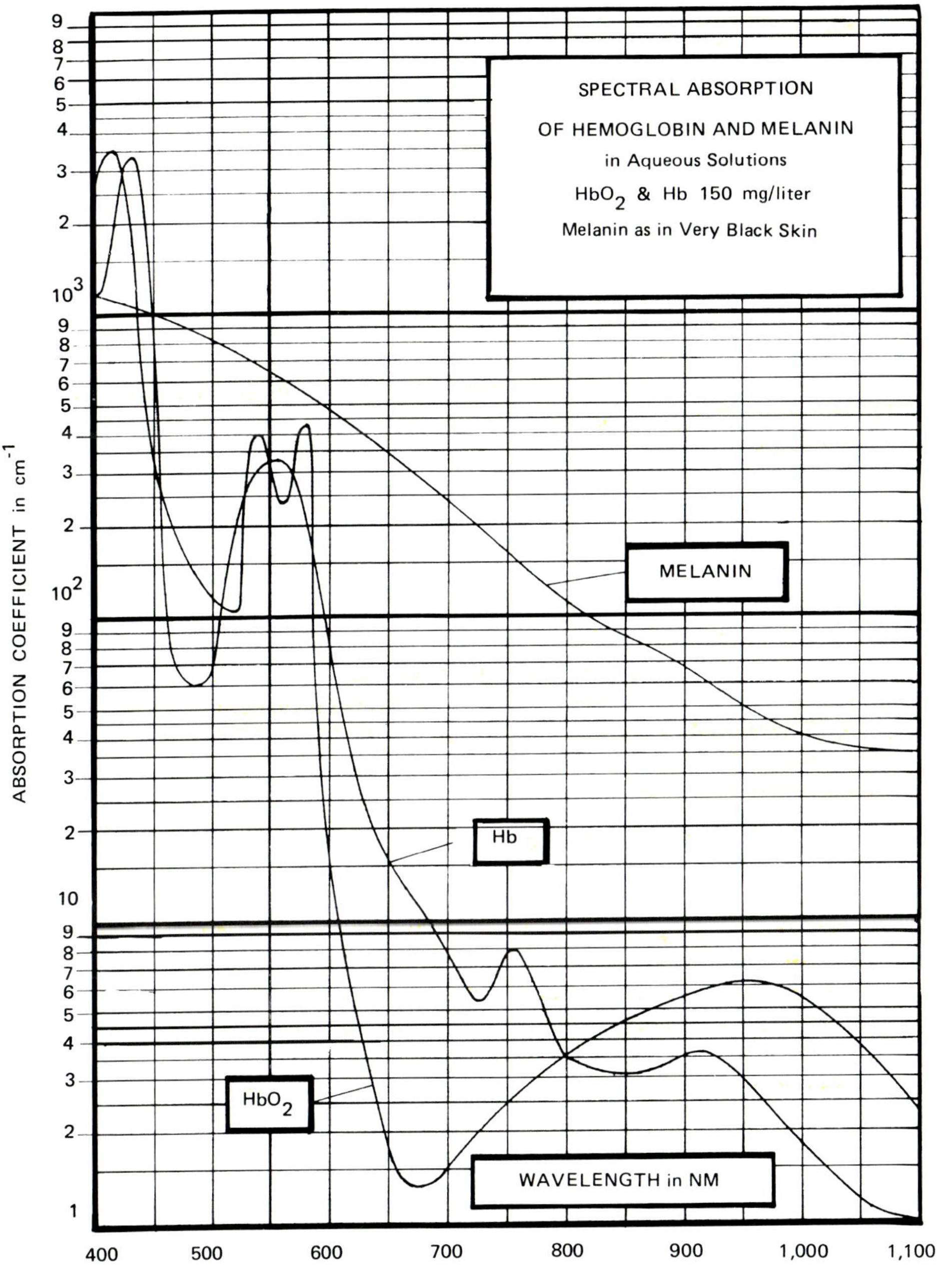
SPECTRAL ABSORPTION
OF HEMOGLOBIN AND MELANIN
in Aqueous Solutions
HbO2 & Hb 150 mg/liter
Melanin as in Very Black Skin
MELANIN
Hb
HbO2
WAVELENGTH in NM
ABSORPTION COEFFICIENT in cm-1
400
500
600
700
800
900
1,000
1,100
1
10
10^2
10^3

Diffuse reflection was first formulated by Lambert, who described the reflected intensity of light from a rough surface normal to the incident ray as varying with the cosine of the angle between the viewing direction and the incident ray, if viewing distance is fixed. If viewing distance varies, the observed intensity at any fixed angle is inversely proportional to the square of the viewing distance. Diffuse reflection occurs because of myriads of irregularities on the surface, which range in size from larger than the wavelength to near equality with the wavelength, each oriented randomly with respect to the others and to the direction of the incident ray. Because diffuse reflection of the true lambertian variety falls to zero at viewing angles near 90°, it produces strong *backscattering* from any one particle. However, when large numbers of randomly oriented particles lie close together, their net effect is to reduce the variation of observed intensity with viewing angle.

Diffraction is the radiation of energy in directions away from the line of propagation of an advancing wave of light because of *Huygens' principle,* which states, among other things, that every point of an advancing wavefront (even a plane one) acts like a point source of new waves. Because of diffraction from the ray bundle of a laser beam, even if it is perfectly collimated at its emergence from the output port of the laser, it will produce some measurable power density in any direction not parallel to the axis of the beam, and this diffracted power density progressively reduces the forward (axial) power density of the beam with increasing distance from the output port. This is why the effective diameter of even a perfect gaussian beam at the focal point of a perfect lens is said to be *diffraction limited.*

Refraction of light rays in passing from one medium into another of different index of refraction results in a change of the direction of each ray at the point where it crosses the interface between the two media, unless the rays strike the interface perpendicularly. When any wavefront, plane or curved, passes through the junctional surface into a medium of higher index, its speed of propagation is reduced. If we consider any two rays lying in a plane normal to the interface but traveling into the second medium obliquely, the points P_1 and P_2 where these rays intersect the wavefront will not reach the interface simultaneously. During the time interval between the arrival of the nearer point P_1 and the subsequent arrival of P_2 at the interface plane, P_1 will travel a shorter distance into the second medium than P_2 travels in the first. Thus all points of the wavefront within the second medium are retarded relative to those still outside. The net result is to rotate the wavefront in the second medium toward a direction more nearly parallel to the interface and every ray toward a direction normal to the interface. Cell membranes, organelles, nuclei, and protoplasm all have differing indices of refraction, as do lymph, blood, and other biological materials. The cumulative effect of refraction from many interfaces among these constituents of tissue is to scatter the rays of a laser beam in all directions.

True scattering is an atomic or molecular phenomenon involving the *temporary absorption* of wavelets (photons) by some members of a population whose levels at the moment of wavelet arrival are such that all or part of the incoming photonic energy can be subsequently reradiated, rather than being emitted by collision or vibration, i.e., as heat. If we think only of resonant absorption, that is, the excitation of an atom or molecule to a permissible higher level such that the difference between the initial level and this higher level is exactly equal to the energy of the incoming wavelet, we see that the units of the population that are excited in this way can subsequently emit all of the absorbed radiant energy as light of exactly the same wavelength. Unlike stimulated emission however, this spontaneous emission can be emitted in any direction, not just that of the incoming wavelet. Therefore the net result is a change in direction without a loss of energy. The probability of these events is dependent upon the atomic or molecular species in the population, the temperature of the population, and the wavelength of the incoming radiation.

If we relate the amount of power density diverted from its original direction in any material to the size of the scattering elements, we find that its dependence upon wavelength is strongly affected by the ratio of the wavelength to the linear dimensions of these elements. If the scattering elements are much smaller than the wavelength, we have what is known as *Rayleigh* scattering, named after the English mathematician and physicist, who was born John William Strutt (1842-1919), and was knighted Lord Rayleigh for his scientific achievements. At the other extreme, if the mean size of the scattering elements is much larger than the wavelength, we have *Mie* scattering. The Rayleigh scattering coefficient is proportional to λ^{-4}, and its dependence upon viewing angle is given by

$$p_s = p_{sn}\,(1 + \cos^2 \Theta) \qquad [56]$$

where p_{sn} is the scattered power density observed at some fixed distance from a laser beam along a line making an angle of 90° with the forward beam direction. Equation [56] shows that Rayleigh scattering is equally strong in both the forward ($\Theta = 0$) and backward ($\Theta = 180°$) directions, and weakest normal to the beam. The blue color of the sky in daylight is caused by Rayleigh scattering of the blue part of the visible spectrum by the earth's atmosphere. The Mie scattering coefficient is much less wavelength-dependent, varying roughly in proportion to $\lambda^{-1/2}$, and its dependence upon viewing angle is a function of wavelength, but tends to show preponderantly forward scattering.

To estimate the relative importance of these two kinds of scattering, it is useful to have an approximate idea of the range of sizes of scattering elements present in living tissue. We shall restrict our attention to soft tissue, because bone is so strongly absorbing of all laser wavelengths that scattering is not significant there. In very approximate numbers, the size ranges for the important biological scattering agents are as follows:

Element	Range of size (nm)
Atoms and inorganic molecules	10-100
Organic molecules	100-1,000
Organelles	1,000-100,000
Cells	10,000-1 million

From this tabulation we can see that atoms and inorganic molecules cause predominantly Rayleigh scattering for all wavelengths used in medical lasers today. Organic molecules of the large varieties found in living tissue cause a scattering that is intermediate between Rayleigh and Mie for visible and near-infrared lasers, and more like Rayleigh for midinfrared. Organelles cause primarily Mie scattering for visible lasers, intermediate and Mie for near-infrared, and a blend of Rayleigh, intermediate, and Mie for midinfrared.

With reference to the four basic optical effects that we have lumped into what we call scattering, it is evident that the only true kind of scattering, in the strict sense of the term, is the Rayleigh variety. Mie scattering is a composite of diffuse reflection and refraction. Diffraction is present for all wavelengths, but it is most pronounced when the distance between scattering elements is comparable with the wavelength. Therefore it is a function of both the volume density of scattering elements and their size.

In summary, scattering is an exceedingly complex phenomenon about which we know very little in living tissue. It is also a very important factor in the effect of laser light in surgery and therapy. Much study needs to be done.

4. Thermal Conduction and Change of Temperature

When a volume of biological tissue is irradiated by light, the ultimate effect of greatest importance is conversion of the radiant energy into heat, for wavelengths longer than 400 nm. The blood vessels and lymphatic ducts in animal tissue act as convective systems to remove any excess heat created in the tissue, either as a result of physiological disturbances or as a result of heat coming from the environment. However, this excess heat can reach these thermal conduits from the sites of its appearance *only by conduction through the intervening tissue.* Such conduction occurs primarily through the water and its dissolved biochemicals, such as salts and carbohydrates, that are the major constituents of animal bodies, by weight. When the thermal power flowing into, or arising within, living tissue exceeds the rate at which heat can be abducted by conduction to the vascular and lymphatic channels, the temperature of the tissue will rise. When the temperature of the tissue has reached its limiting value, namely, the boiling point of its major constituent (water in soft tissue, or minerals in bone), the rate of thermal abduction reaches its maximum, namely the *characteristic power density of thermal conduction.*

If the radiant power density being converted to heat in the tissue is lower than the characteristic power density of thermal conduction, the temperatures of all points within the tissue will reach equilibrium values above normal body temperature (37°C) but below the minimum boiling point of water (100°C for fresh water at atmospheric pressure). Conversely, if the absorbed power density is much greater than the characteristic power density, the temperature of all points in the irradiated volume will rise rapidly to the boiling point of cytosol, with consequent destruction of cytological and histological architecture by explosively expanding vapor (steam). For histological temperatures at or below 100°C, the time needed to carry away significant quantities of heat is several seconds. If we assume that the temperatures of all points within a given mass of tissue rise quickly to 100°C because of irradiation at high power density, then the volume occupied by this tissue acts as an *isothermal source* (i.e., a source at constant temperature), because its temperature cannot rise until all the water is vaporized. From this source, heat will be conducted away to the vascular heat sink in a quasi-steady-state situation. The power density thus removed is given by Fourier's equation of heat conduction, named for the French mathematician Jean Baptiste Joseph Fourier (1768-1830):

$$p_{ct} = -c_t \frac{dT}{dr}, \ (POWER)/(AREA) \qquad [57]$$

where p_{ct} is the characteristic power density of thermal conduction, c_t is the thermal conductivity *(POWER)/(LENGTH)(TEMPERATURE DIFFERENCE)*, and dT/dr is the rate of change of temperature T with distance r from the boiling zone. The minus (-) sign in Eq. [57] merely signifies that the temperature falls in the direction of heat flow. The average value of c_t for pure water between 37°C and 100°C is 6.92×10^{-3} W/cm-°C. It is not known for cytosol.

Measurements made by Mihashi et al. (20) in the mid-1970s on a mongrel dog's tongue irradiated by a CO_2 laser showed that the temperature gradient near the small (about 2 mm in diameter) boiling crater at the impact spot of the laser beam was about 1500°C/cm. If this value is put into Eq. [57], together with the average thermal conductivity for water between 37°C and 100°C, the value of p_{ct} is 10.4 W/cm^2. This is only an approximate value, but it gives us an indication of the magnitude of absorbed power density in tissue that marks the transition from coagulation to onset of vaporization by a laser beam. Absorbed power densities of radiation in tissue well above this value produce rapid vaporization of histological water, whereas values equal to or lower than it produce only heating of the tissue to temperatures below 100°C.

If the absorbed power density in the tissue exceeds p_{ct} by several orders of magnitude, the vaporization of cytosol will be so rapid that the resulting steam becomes superheated because it cannot expand quickly enough to maintain its pressure at atmospheric value. When this occurs, the pressure within the ball of vapor can rise to very high values in very short times (i.e., microseconds or less), producing a blast, a shock wave of high pressure that propagates spherically outward from the region of absorption. This phenomenon was responsible for the explosions observed by Ketcham et al (10) in the early surgical use of ruby and Nd: glass lasers. Absorbed radiant power density below 1×10^6 W/cm^2 does not cause mechanical effects of sufficient intensity to tear soft tissue into fragments containing still-viable tissue, although the acoustic waves resulting from irradiation at histological power densities between 100,000 and 1 million W/cm^2 can cause cell death by rupture of membranes.

Strictly speaking, the line of reasoning that leads us to the conclusion that the radiant power density absorbed in the tissue must equal p_{ct} if vaporization is to occur is based upon the tacit assumption that the absorption coefficient for the particular tissue and incident wavelength is such that the *extinction length* of the laser beam is much less than its diameter. We define this length as the distance into the tissue along the ray path from the first surface over which 99% of the radiant energy has been attenuated. Note that some authors define it as the distance corresponding to 63% or to 90%. According to our definition, the extinction length is $z_e = 4.6\,A^{-1}$, where A is the attenuation coefficient from Eq. [55]. If the extinction length is much smaller than the effective beam diameter, d_e, all of its radiant power will be converted to heat within a thin disk of tissue whose diameter is that of the beam (d_e), and whose axial thickness is z_e. Because of the narrow edge of this disk, its lateral surface area is very small and thus the heat generated within it must be conducted away through its distal face, whose effective area is equal to that of the beam. Therefore the thermally abducted power density must be equal to the incident radiant power density if thermal equilibrium is maintained. In this situation it is clear that the specific power density of thermal conduction must be less than the incident power density of the beam if sustained boiling of water within the absorption disk is to be maintained. We have here ignored the scattering coefficient, because its value for soft tissues and the wavelengths currently used in surgery is almost always less than 10 cm^{-1}, and typical laser beams for surgical vaporization are 3 mm or less in effective diameter, which requires $A >> 15$ cm^{-1} if z_e is to be $<< d_e$.

In situations where the extinction length is much greater than the effective diameter of the beam, we may approximate the thermal geometry by assuming that the illuminated volume of tissue is a hemisphere of radius $R = z_e$, whose equator lies in the (plane) surface of first incidence. If the interior of this hemisphere equilibrates at the boiling temperature of water and we neglect the loss of heat from the first surface, all of the thermal power must escape from the hemispherical surface whose radius is z_e. The power flowing into the hemisphere from the laser beam is, from Eq. [50], $P_0 = d_e^2\, p_{aw}/1.10$, where p_{aw} is the average power density within the effective diameter. If this power is converted to heat and conducted away through the hemispherical surface at an average power density of p_{ct}, then p_{ct} must be equal to P_0 divided by the surface area of the hemisphere. Therefore we have $p_{ct} = P_0/2\pi z_e^2 = p_{aw}\, d_e^2/2.2\pi z_e^2$, with the final result (neglecting reflectance) that

$$p_{aw} = \frac{146}{(d_e A)^2}\, p_{ct} \qquad [58]$$

Thus, if $A = 10\ \text{cm}^{-1}$ (approximately the value for Nd:YAG rays in unpigmented tissue), and the effective beam diameter is 1.0 mm, the power density in the incident beam must be at least 146 times p_{ct}, or 1460 W/cm^2 to produce sustained boiling of histological water.

Finally, we must remember that all of the foregoing commentary applies to only lasers whose destructive effects in tissue depend upon thermal phenomena. Ultraviolet lasers (i.e., excimer or frequency-quadrupled Nd:YAG) destroy tissue primarily by *photochemolysis*; therefore the destructive thresholds of incident power density are nearly independent of histological temperature and thermal conductivity. The chemolysis of ultraviolet light is somewhat enhanced by elevated temperature of large, complex organic molecules. Furthermore, it does not depend upon absorption in water, which attenuates the power density available for breaking chemical bonds.

The damage to soft tissue resulting from temperatures above 37°C is a function of the duration of such temperatures. Tissue death is not instantaneous unless the histological temperature is close to 100°C. The data of Henriques and Moritz (21,21a) for thermal damage to living tissue show that the threshold temperature, T_c, for necrosis varies with the time of thermal exposure, t, in a monotonically inverse manner. The author has plotted their data for T_c vs. t, and found that it can be approximated to within a tolerably small error by the simple formula

$$T_c = 37 + 31.5\, t^{-0.164}\ ^{\circ}\text{C} \qquad [59]$$

where t is expressed in seconds, and lies between 1.0 sec and 1000 sec. Equation [59] is probably valid to times as short as 0.1 sec, but for thermal exposures ranging from picoseconds to milliseconds (1×10^{-12} to 1×10^{-3} sec) analytical calculations become mathematically complex and of little value to surgeons using lasers in long-pulsed or continuous wave temporal modes of operation. It must be remembered that T_c in Eq. [59] represents the center of a *band* of temperature several degrees wide because of statistical variations among individuals, tissue types, and patients of different ages. It is important to remember that thermal necrosis of tissues and organs is a process *that does not stop at the cessation of elevated temperature.* Because of physiological synergy, cells on the periphery of a coagulated tissue mass, even though not themselves heated to critical temperature, may subsequently die. As in all facets of the application of lasers to medical purposes, the laser practitioner is well advised to use caution when coagulating tissue with a laser beam.

5. Basic Photonic Processes in Living Tissue

The fundamental mechanisms by which laser light interacts with living matter are functions of both wavelength and histological power density. All of them can occur at all wavelengths from 100 to 10,000 nm, but the photochemical effects are most pronounced at the ultraviolet end of the spectrum and the thermal effects at the midinfrared (1400-20,000 nm). These effects can be most easily categorized according to the *ranges of histological power density* in which they are dominant. This stratification is shown in Fig. 29. The remainder of this section consists of definitions and brief descriptions of these basic photonic processes.

a. Photochemolysis and Photochemosynthesis. These are the stimulation of biochemical reactions in either the direction of lysis or the direction of synthesis by light waves at low intensity, typically below 1 W/cm^2. Photosensitizers or photocatalysts may or may not be required. For example, chlorophyll catalyzes photosynthesis in green plants absorbing visible light, but dermatoheliosis of human skin by the ultraviolet light from the sun occurs even in the absence of photosensitizers (e.g., the psoralens). Dwyer, Castro and co-workers (22-24) have demonstrated convincingly that the rays of the Nd:YAG laser can cause lysis of collagen in vitro and in vivo at power densities too low to produce significant thermal effects. At low radiant intensities, photochemolysis and photochemosynthesis are dose (energy per unit area) dependent. This is a highly promising field of study that may be of enormous importance for therapy in the next decade. Outside of the ultraviolet range, photochemical effects are found throughout the visible and into the near-infrared part of the spectrum.

b. Photopyrolysis. Photopyrolysis is the heating of tissues to temperatures below 100°C by direct conversion of laser light into heat in the tissue, to cause coagulation for either thermal necrosis or thermal hemostasis. It depends heavily upon the scattering and absorption coefficients of light in tissue of various types. It is the major effect of the Nd:YAG laser, and a major effect of all visible lasers, for which it is dependent upon the color of the tissue. For the Nd:YAG laser, it is affected by tissue color only when this is very dark or black, especially when free amorphous carbon is present after charring by the laser beam. The dominant effect from 1 to 100 W/cm^2 is photopyrolysis.

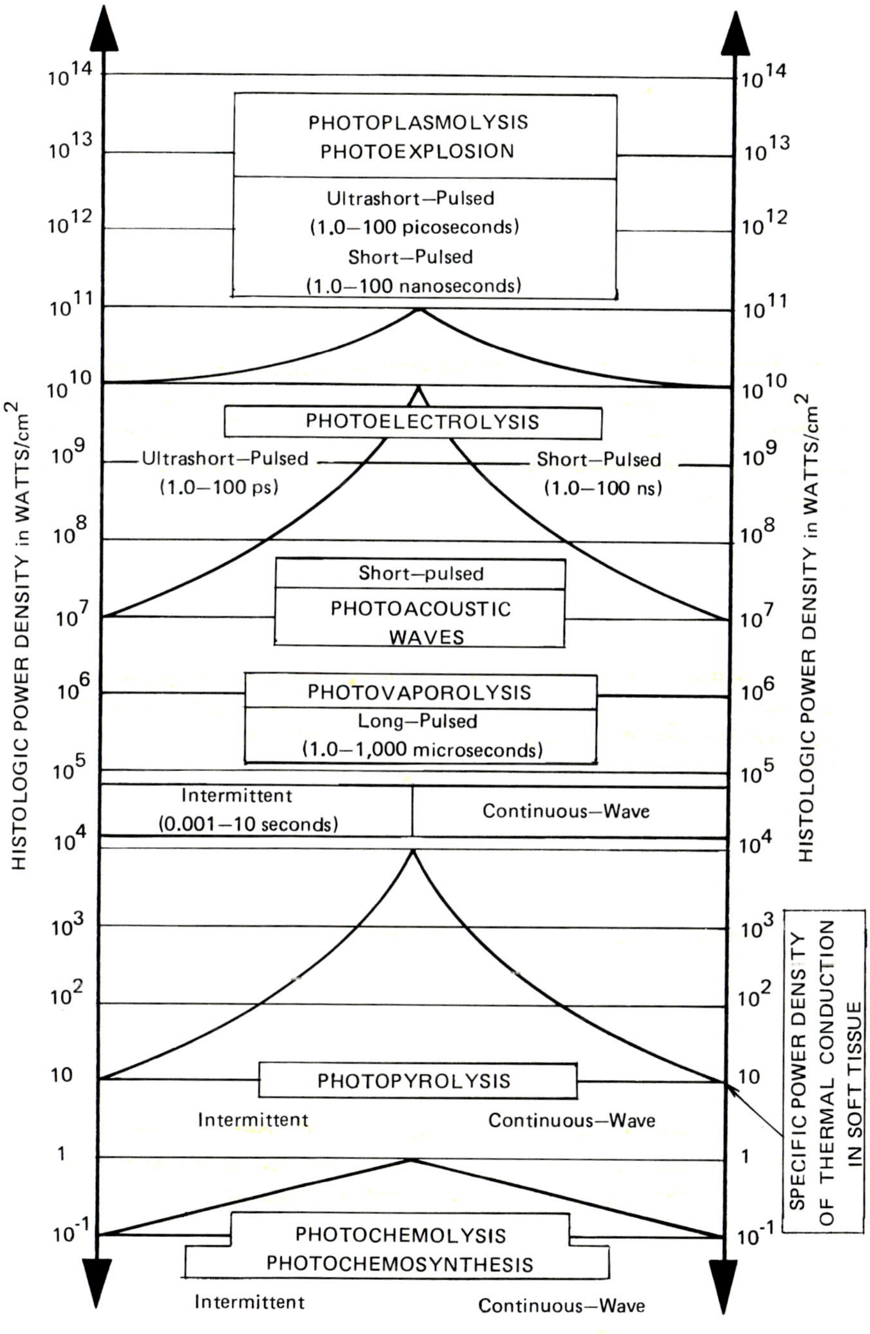

Figure 29 Biologic effects of laser radiation between 100 nm and 12,000 nm as functions of power density in soft tissue.

c. Photovaporolysis. Photovaporolysis is the rapid vaporization of cytological and histological water to form steam at or near atmospheric pressure, by the absorption of laser light and conversion into heat. Cytological and histological architecture is destroyed by *mechanical disruption* caused by expanding steam. It can also occur in bone, but at the much higher temperatures at which osteominerals vaporize. This is the major destructive process of the carbon dioxide and other midinfrared lasers. It is predominant in the range of power densities from 100 to 1 million W/cm^2.

d. Photoacoustic Waves. Pressure waves radiate outward from the point of impact of a laser beam. They can cause cytological and histological disruption by mechanical effects ranging from disruption of cell membranes at the low end of the scale, to tearing of vessel walls and nerve fibers at the high end. These waves occur at power densities from 1000 to 1×10^9 W/cm^2. We define them here as resulting from laser pulses having less than 1.0 J (W-sec) of energy in each pulse. These waves can be detected at power densities as low as 1.0 W/cm^2 by means of sensitive instruments known as *photoacoustic spectrophotometers,* which utilize chopped (cyclically interrupted) beams of laser light and lock-in amplifiers to detect the small cyclic changes of pressure within a tiny sample of tissue in a special cuvette.

e. Photoelectrolysis. Photoelectrolysis is destruction of cell membranes by the strong electric fields of light waves, at intensities of 1×10^7 V/m and higher, corresponding to power densities of 2.66×10^7 W/cm^2 or more. There is no upper limit of power density for photoelectrolysis, but it is the dominant effect up to 1×10^{10} W/cm^2 if the energy per pulse is less than 1.0 J. It cannot be produced by continuous wave lasers presently available for medical purposes.

f. Photoexplosion. Photoexplosion is the production of shock waves and blast effects by the heating of tissue constituents in very short times to form superheated vapors at pressures up to several hundred atmospheres. The extent of the mechanical damage depends upon the energy per pulse because this effect cannot be produced by the continuous wave lasers presently available for medical use. If the energy per pulse is less than 0.1 J, the mechanical damage is minimal. However, if the energy per pulse is 10 J or higher, chunks of viable tissue can be ripped out and hurled away from the impact site of the laser beam. Photoexplosion can be produced by direct absorption of light in tissue without the formation of an electric plasma or as a result of the heat generated in a plasma. It occurs at power densities from 1×10^8 W/cm^2 upward, if the energy per pulse is 1.0 joule or higher.

g. Photoplasmolysis. Photoplasmolysis is the destruction of molecular, cytological, and histological architecture by the ultrastrong electric fields of the light rays from short-pulsed lasers (either Q-switched or mode-locked) with pulse dura-

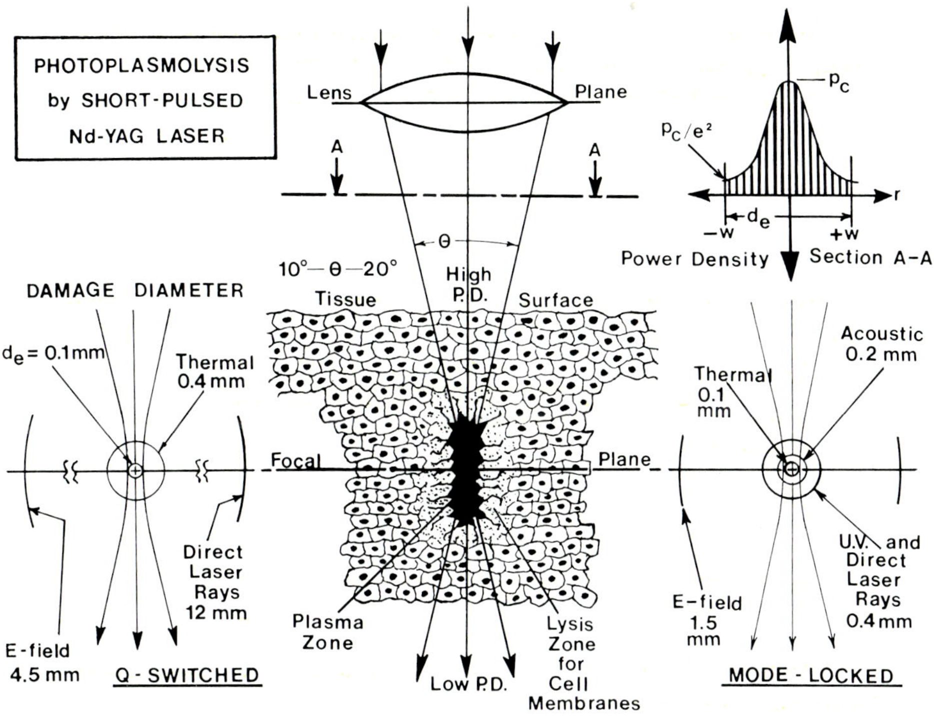

Figure 30 Schematic illustration of the destruction of soft tissue by photoplasmolysis. The mode-locked Nd:YAG laser causes smaller damage zones than the Q-switched, other factors being equal, because the plasma forms earlier in the irradiation period, and shields structures distal to the focal plane by absorption of the laser beam.

tions of picoseconds or nanoseconds. These intense electric fields pull electrons out of their atomic orbits to form a plasma, a hot gas of electrons and positive ions in nearly equal concentrations, whose temperature can be 15,000°C or higher. For this effect to occur, the absorption in tissue must be low. It occurs at power densities from 1×10^{10} W/cm^2 upward. Because cells are totally destroyed by photoplasmolysis, it does not cause carcinogenesis, despite the ionization. The Nd:YAG laser is best suited to produce photoplasmolysis in soft tissue because its coefficient of absorption in lightly pigmented tissue is minimal. Absorption of light by tissue overlying the intended site of destruction reduces the power density at the focal point, where it must be at least 10 billion W/cm^2 to cause photoplasmolysis. Figure 30 illustrates plasmolysis.

C. Specific Effects of Various Lasers on Tissue

The suitability of a particular laser for a given surgical or therapeutic application depends primarily upon the values of α and σ for its wavelength and the tissue of interest, and secondarily upon the radiant power it can deliver in either pulsed or continuous wave operation. The following general observations are pertinent and are based upon the assumption that the laser beam at the first tissue surface is approximately gaussian in its transverse distribution and has an angle of convergence or divergence between 0 and 15°. They are not applicable to laser beams delivered by dielectric waveguides of special surface geometry that are placed in contact with the tissue (see Sec. IV.C.3).

1. Lasers whose extinction length z_e is less than 0.1 mm for all tissues, and whose σ/α ratio is less than 0.001 make good photonic scalpels but poor coagulators. Their power output must be at least 20 W for effective and expeditious incisive or ablative surgery by photovaporolysis. Such wavelengths are all longer than 3000 nm if we exclude the ultraviolet lasers, which currently are not approved by the United States FDA for surgery.
2. Lasers whose extinction length is 5 mm or more, and whose σ/α ratio is larger than 10 make good coagulators but poor scalpels. They must be able to deliver at least 50 W to heat a cubic centimeter or more of soft tissue in a few seconds to 60°C. Such wavelengths are all in the near-infrared (700-1400 nm) region. The Nd:YAG is the best of present lasers for voluminous photopyrolysis in all soft tissues.
3. Visible lasers, for which z_e and σ/α are very dependent upon the color of the irradiated biological materials (see Fig. 28), may be useful for special purposes in specific tissues. They are not suitable as precise cutting instruments but only as coagulators that can cause photopyrolysis *selectively* on the basis of color and physical properties of a biological substance. For example, the argon ion laser, at 514 nm, is very effective in coagulating small ectatic vasculature, such as exists in port-wine marks, because oxyhemoglobin at this wavelength has an absorption coefficient, α, of about 150 cm^{-1}, well above that for nearby fair skin or underlying dermis.

In the following sections we examine the histological effects of those lasers that are presently suitable for endoscopy; either rigid, in which light is transmitted by multiple mirrors and lenses; or flexible, in which light is carried by slender optical fibers. The short-pulsed Nd:YAG lasers used for ophthalmic purposes are excluded here because the extremely high-power densities they produce cause *self-focusing* and *self-trapping* in optical fibers, whose index of refraction increases with power density at the values needed for photoplasmolysis, thus reducing the beam diameter within the fiber until the material from which it is made is destroyed by the ever-increasing power density. The use of sequential lenses and mirrors for such laser beams is also precluded because the attainment of focal power densities above 1×10^{10} W/cm^2 necessitates lenses of large aperture and short focal length, as shown by Eq. [43], which are not feasible within the barrel of a rigid endoscope.

1. Carbon Dioxide Lasers

At a wavelength of 10,600 nm, scattering is negligible and α for water is about 770 cm^{-1}, so that z_e is 0.06 mm. For virtually all other materials, except those that strongly reflect (polished metals) or efficiently transmit (intermetallic compounds and metallic halides), α is about 1000 cm^{-1}. A stationary, quasi-collimated CO_2 laser beam at normal incidence upon a tissue surface and having a gaussian distribution of power density that is equal to or greater than 100 W/cm^2 everywhere within some coaxial circle, will ultimately bore a straight hole through the mass of tissue. The transverse cross-section of this final hole will be circular, and the longitudinal cross-section in any plane containing the axis of the beam will be a *conical frustum.* The included angle of this cone and its convergence, or divergence, in the forward direction of the beam depend upon the focal length of the last focusing lens, the diameter of the beam entering that lens, and the location of the focal plane relative to the first surface of the tissue. The proximal diameter of the hole at the first tissue surface will be at least as large as the *boiling diameter* of the beam at this surface (i.e., the diameter of a coaxial circle at which the point power density is just high enough to cause reliable boiling of the intra- and intercellular water, or about 100 W/cm^2). If the focal plane is at the first surface, the cone of radiation will diverge toward the bottom of the hole, with an included angle that is larger for shorter focal lengths, if the geometry of the beam proximal to the lens is constant.

During the *boring process,* the longitudinal cross-section of the hole *between the shoulder* (i.e., the proximal boiling diameter) and the *distal apex* will have a gaussian profile if the central depth is not more than 1.5 times the diameter at the shoulder. This whole *boiling front,* where the laser's rays strike the unvaporized tissue, is at or somewhat above the normal boiling temperature of water (100°C), depending upon the pressure of the steam on the boiling surface if the power density is below 1×10^6 W/cm^2. From the boiling front, steam and the solid residues of ruptured cells are ejected upward (proximally) at speeds between 1 and 10 m/sec (20) into the free air above the hole. There, separated from any protective water, they are heated to incandescence by the laser beam, because α for these residues is about 1000 cm^{-1}. These hot particles burn in the presence of atmospheric oxygen, creating a dense, malodorous smoke. The smoke leads incognizant observers to the erroneous conclusion that *burning* is the basic process of tissue destruction. The generation of smoke can be stopped by flooding the impact zone with a nonoxidizing gas, such as carbon dioxide, but *vaporiza-*

tion at the boiling front goes on. The smoke plume contains *chemical carcinogens* (25), as does most smoke, but *no viable cells* (26,27), although some cells retrieved from the smoke may *appear morphologically intact.*

Along the conical surface of the laser hole proximal to the shoulder, there is virtually no absorption of radiant energy because the rays are parallel to this surface. The temperatures of points in the tissue lateral to this part of the hole rise quickly to peak values, as the boiling front goes by distally, and then decline slowly for up to 20 sec after frontal passage. At the first surface, the power density in the fringe of the gaussian beam becomes equal to p_{ct} and from there outward will cause only *coagulation and charring without vaporization.* This produces a raised rim around the hole, like that of a volcanic crater.

In holes whose depth is less than about five times their diameter, the time rate (speed) at which the boiling front bores deeper into the tissue, for incident power densities above 100 W/cm^2, is given approximately by

$$dz_c/dt = \frac{0.96\ (P_0 \text{ in watts})}{(d_e \text{ in mm})^2} \text{ mm/sec} \qquad [60a]$$

where the constant 0.96 has the dimensions of cubic millimeters per joule and takes account of the heat needed to warm and vaporize 1 mm^3 of tissue, as well as of the gaussian distribution of power density and of the 5% reflection from the boiling surface. Observed values of dz_c/dt will be somewhat lower than those calculated using Eq. [60a], because the actual *heat of ablation* for proteinaceous soft tissue is higher than the value of 2.522 J/mm^3 used in deriving the value of 0.96. Thus a 100-W CO_2 laser beam of 1 mm in effective diameter would bore deeper at a calculated rate of 96 mm/sec.

At any off-axis point of the boiling front between the shoulder and the apex, the time rate of increase of depth is proportional to the incident power density at that point:

$$dz_\rho/dt = 0.00377\ p_\rho \text{ mm/sec} \qquad [60b]$$

where ρ is the radial distance measured normally from the axis to the point. Equation [60b] is valid for the central (axial) point also, if p_ρ is set equal to p_c. Therefore it is evident that the boring speed will be greatest on the axis of the hole and least at the wall. The final wall of the hole is that concentric surface at which the incident power density is just high enough to cause sustained vaporization (100 W/cm^2). From the foregoing discussion, we see that the boiling surface will approximate a three-dimensional replica of the gaussian power density profile of the beam, which is *progressively elongated* because the boring speed is highest at the apex and declines toward the shoulder. This can readily be seen by firing sequential shots of a CO_2 laser beam at the edge of a polished plexiglas block at constant power and increasing exposure time, and noting the depths of successive holes.

The diameter of the hole at the proximal end is not equal to the effective diameter of the beam unless the power density at that diameter is just high enough to sustain vaporization of the tissue. In general, *the proximal diameter of the hole will increase as the total power of the beam increases,* even if the effective diameter of the beam is constant. We must remember that the effective diameter is a mathematical concept, namely, that diameter for which the point power density of the beam falls to p_c/ϵ^2. The numerical value of power density at diameter d_e is always 13.53% of that at the center of the beam and therefore will be dependent upon the total power P_0, which determines p_c if the profile of the beam is gaussian. However, the power density needed to cause reliable vaporization of tissue is a function of only the tissue type and its initial temperature; it is independent of the characteristics of the laser beam.

As the center of the hole at its distal apex goes progressively deeper, the angle of incidence (measured from the perpendicular) of the incoming rays on the wall of the crater off the axis becomes steadily greater, because the rays (assumed to be quasi collimated) are striking a surface of increasing slope. The off-axis rays approach a condition of *grazing incidence* with increasing time and/or depth of the hole. This obliquity of incidence increases the reflection and decreases the absorption of the peripheral rays at their points of first contact with the wall of the crater. This wall is not a smooth surface but rather a highly irregular one consisting of a layer of boiling water (in soft tissue) whose microscopic contours are always changing because of the turbulence of the boiling process. The reflected rays will be deflected randomly toward the apex (bottom) of the hole, finally undergoing total absorption at the wall near the apex. Only near the bottom of the hole do these rays strike the surface approximately perpendicular to it. The net result of all this is to *increase the boring speed at points near the apex beyond the value given by Eqs.* [60a] *and* [60b] when the central depth is greater than perhaps 1.5 times the proximal diameter of the hole.

If the apex of the hole passes completely through a mass of tissue and out into open space on the other side, the peripheral rays of the laser beam will gradually bore deeper, ultimately making a hole whose geometric shape is a replica of the ray bundle within which the power density is everywhere equal to, or greater than, the value required for reliable vaporization (about 100 W/cm^2 in soft tissue). If the focal plane of the bêam is at the first tissue surface, this ultimate hole will be a distally diverging cone whose included angle depends upon the focal length of the lens and the effective diameter of the beam (assumed collimated on the proximal side of the lens) where it enters the lens. If the focal plane lies between the first surface and the last surface of the tissue, the hole will have a waist between the two surfaces. If the focal plane is at the last surface, the hole will be a converging cone with maximum diameter at the first surface.

It is intuitively evident that points lying in the tissue just outside the ultimate periphery of the hole will be subjected to rising temperatures caused by *conduction* of heat laterally from the isothermal wall (the boiling surface). During the *passage of the shoulder clear through the tissue,* points lying closest in the z coordinate to the first surface will be exposed for the shortest times and those nearest the last surface exposed for the longest times. Biological structures that pass partially or totally through the wall of the final hole will be cut by the beam. In particular, blood vessels may be transected. Whether or not such vessels bleed depends upon the temperatures to which they are exposed before transection and the duration of exposure.

If the apex of the beam cuts a blood vessel in such a way that the axis of the beam passes through the center of the vessel's cross-section, there will be minimal time for the conducted heat to coagulate the vessel before the beam punctures the wall. Conversely, if the vessel lies off the axis of the beam with its wall just barely protruding into the ultimate lumen of the hole, the vessel will be thermally exposed for a longer time before transection, if the z coordinate is the same in both cases. Therefore on-axis vessels are less likely to be fully coagulated than are off-axis vessels of the same diameter at the same depth below the first surface. Hemostasis by thermal coagulation caused by heat from the laser beam must be effected before puncture of the vessel's wall by the beam, because once the liquid blood flows copiously into the path of the beam, it totally absorbs the radiant energy in a depth of 0.1 mm and carries the resultant heat away by mass convection.

Hemostasis effected solely by the drilling beam of a CO_2 laser will be greatest for a vessel lying anywhere within the path of this beam if the power density at the *center* is about 100 W/cm^2, which means that the value at the effective radius is only 13 W/cm^2, just a little higher than p_{ct}. Under this condition, the boring speed at the axis is only 0.38 mm/sec. The tissue lateral and distal to the boiling wall of the crater is then in a quasi-steady-state thermal situation, for which the data of Mihashi et al. (20) suggest that the curve of temperature versus perpendicular distance from the boiling wall is describable by the relation

$$T_r = T_b + (100 - T_b)\, \epsilon^{-1.5r}\, ^\circ C \qquad [61]$$

where r is the distance in millimeters, T_b is the normal body temperature in degrees Celsius, and T_r is the temperature in degrees Celsius at distance r.

If we imagine a vessel whose axis intersects that of the beam normally at a point below the advancing apex of the crater, this vessel will be coagulated completely before puncture by the apex, if the temperature at its distal wall is equal to or higher than the critical temperature T_c from Eq. [59] corresponding to the time needed by the apex to advance from the first surface to the distal wall of the vessel. For an apical boring speed of 0.38 mm (p_c = 100 W/cm^2), simultaneous solution by trial and error of Eqs. [59] and [61] gives r = 0.49 mm.

This suggests that the maximum diameter of a blood vessel that can be coagulated by a *boring* CO_2 laser beam that punctures the vessel centrally is about 0.5 mm. For higher power densities, and correspondingly higher boring speeds, this diameter will decrease. Blood vessels lying off the axis of the boring beam and close to the first surface can be coagulated in diameters up to 0.76 mm, if the central boring speed is 0.38 mm/sec. *In general, the higher the central power density of a boring beam, the poorer will be the hemostasis of vessels centrally punctured or laterally grazed by it.*

If the beam of the laser is swept laterally (i.e., in a direction perpendicular to its axis) at constant linear speed across the tissue, it will cut a *furrow* whose profile in a cross-sectional plane normal to the sweep direction is essentially the same as that for the same beam when stationary. The central depth of this groove is

$$z_c = \frac{0.060\ (P_0 \text{ in watts})}{(v_s \text{ in cm/sec})\,(d_e \text{ in mm})}\ \text{mm} \qquad [62]$$

where v_s is the speed of sweep, and the constant 0.060 has the dimensions of cm × mm^2/J. Thus we see that rapid sweeps cut shallow grooves and vice versa. Measured values of z_c will be somewhat lower than those predicted by Eq. [62], because the true heat of ablation is higher than the 2.522 J/mm^3 used in deriving the equation. The effect of reflection is included in Eq. [62], for grooves whose central depth is less than about five beam diameters. For grooves cut by beams moving at lower speeds, the cross-section of the furrow in a plane passing through the beam's axis and normal to the sweep line will be elongated progressively as the sweep speed diminishes, in much the same way as for a stationary beam, so that the actual depth at the center will be greater than that predicted by Eq. [62]. The duration of thermal exposure to heat conducted outward from the boiling wall of the groove depends upon the location of the exposed point relative to the path of the sweeping beam, with those points lying directly on the midplane of this path having the shortest exposures and those lying laterally the longest. The maximum diameter of a blood vessel that can be coagulated solely by the heat conducted from a sweeping laser beam is probably somewhere between 1.0 and 2.0 mm, even for very low speeds of sweep. Because thermal conduction in soft tissue can transport only about 10 W/cm^2, the intraluminal convection of blood in larger vessels can carry heat away faster than it arrives at the adventitia when thermal conduction is the sole mechanism of heating the vessel. As sweep speed increases to very high values (e.g., in the range of 100 cm/sec), the depth

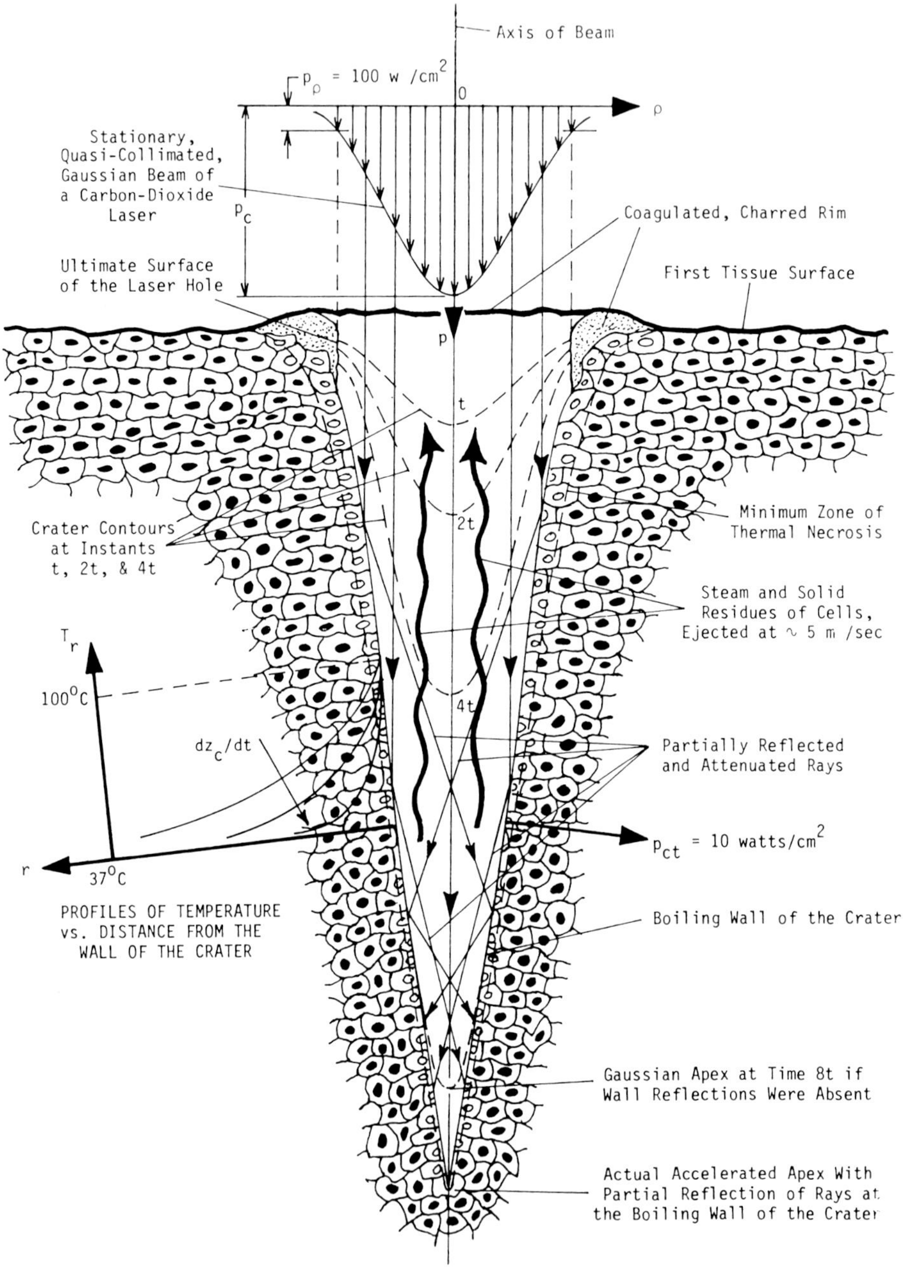

Figure 31 Schematic cross-section of the hole bored by a stationary gaussian beam of a carbon dioxide laser in soft tissue, showing the elongation of the boiling surface at various times t, $2t$, $4t$, and $8t$ after the onset of irradiation. If the beam were swept across the tissue surface at constant speed, the apical depth of the groove would be constant and inversely proportional to the speed of sweep.

of furrow decreases unless higher power is employed. At such high speeds, the hemostasis of transected vessels will be poor or nonexistent. However the use of automatically programmed sweeps at very high speeds can ablate tissue with negligible thermal injury to adjacent regions.

The foregoing observations are depicted schematically in Fig. 31, which is valid for ultraviolet lasers as well, with the exceptions noted in the next section.

2. Excimer and Frequency-Quadrupled Nd:YAG Lasers

The beam of an ultraviolet laser is capable of incising or focally destroying tissue with extreme precision, perhaps two orders of magnitude more precisely than that of a CO_2 laser. The reasons are as follows: first, it can be focused to a spot of effective diameter about 1% as large as that for a CO_2 beam, other things being equal; and second, it produces very little thermal damage to adjacent tissue when used at low power densities. Because ultraviolet rays at surgical power densities produce almost no heat in tissue, they are incapable of coagulation. Ultraviolet light *disintegrates* complex organic molecules (and synthetic polymers, too), at relatively low power densities, typically in the range of 1.0-10 W/cm^2. Equations of the form of Eqs. [60a], [60b], and [62] are valid for ultraviolet laser beams, if the constants are adjusted to reflect the appropriate values of the *specific energy of photochemical disintegration* in joules per cubic millimeter. This entity is a function of the molecular species, and has not, to this author's knowledge, been measured for physiological substances. At a wavelength of 193 nm (the argon-fluoride excimer), the values of absorption and scattering coefficients in water are about $\alpha = 0.15\ cm^{-1}$ and $\sigma = 4.0\ cm^{-1}$, where σ is calculated as for Rayleigh scattering by inverse fourth-power extrapolation from the value for water at 514 nm (the argon ion laser). At power densities of 10 W/cm^2 or lower, *thermal* effects of ultraviolet laser beams are not important in soft cellular tissue because of the low value of α in water (0.15 cm^{-1}). However, in densely concentrated organic molecules, the value of α can be very high. According to the findings of Linsker et al. (14), the extinction length is less than 3.5×10^{-4} mm in fatty arterial plaque at $\lambda = 193$ nm. This implies that the corresponding value of α is at least 131,000 cm^{-1}. Ultraviolet lasers were not found by Linsker and his associates to be effective destroyers of calcium deposits or bone at the exposures and power densities they used, which probably reflects a high-bonding energy for these minerals. However the results of Grundfest and his group (28) indicate that calcium deposits can be effectively lysed by ultraviolet lasers *if the total absorbed radiant dosage exceeds a certain threshold value.* Furthermore, Isner et al. (29) have succeeded in cutting calcified aortic valves with excimer laser rays at several different wavelengths.

In papers presented by Isner and associates (30-34) at the 6th Congress of the International Society for Surgery and Medicine in Jerusalem in October of 1985, it was demonstrated that the light of excimer lasers can produce thermal effects in living tissue if the radiant dosage is high enough. This is not surprising in view of the energy balance depicted in Fig. 27. There it can be seen that the thermodynamic equilibrium of a population of molecules is upset in the first instance by an increase in the sum of the excited levels of this population upon receiving an inflow of radiant energy from the environment. The sum of excited levels may then be increased by photochemolysis of certain molecular species or decreased by photochemosynthesis, for endothermic reactions and vice versa for exothermic ones. If the sum of excited levels is initially increased, it will return to equilibrium in part by nonradiative transitions of energy to lower levels, which means that the energy thus given up will *increase the thermal energy of the population.* In particular, if the inflow of radiant energy from outside the population is such that the initial increase in the sum of all excited levels cannot be balanced by an increase in the sum of all chemical energy of the population, or is actually increased by a release of chemical energy into more excited levels, then the total thermal energy of the population must increase. Because photochemical effects in biological systems usually occur at low power densities of incident radiation (e.g., the photosynthesis of carbohydrates in green plants requires an irradiance of about 0.1-0.5 W/cm^2), power densities above 10 W/cm^2 are almost certain to result in some increase of thermal energy of the irradiated population.

If each photon (wavelet) of incoming radiation has sufficient energy to break an interatomic bond within a molecule, then a relatively low irradiance can cause photochemolysis. At a wavelength of 193 nm, the photon energy is $(e_p)_{193}$ = 1.029×10^{-18} J. At this wavelength, the value of z_e for fat is about 3.5×10^{-5} cm. Therefore an argon-fluoride laser beam having an average power density of only 10 W/cm^2 can deliver 2.86×10^5 J to each cubic centimeter of the absorption volume in 1 sec, or 2.78×10^{23} photons/cm^3/sec. If we assume an average density of 5×10^{22} atoms/cm^3 (probably too high for fat), then 10 W/cm^2 corresponds to 5.56 photons/sec for each atom, more than enough to cause chemical breakdown.

3. Long-Pulsed and Continuous Wave Nd:YAG Lasers

If a stationary beam from an Nd:YAG laser is directed normally at a tissue surface, it will instantly *illuminate* a volume of tissue much greater than that which is included within the geometric projection of the beam cylinder into the tissue to the extinction depth. This is so because of copious scattering of Nd:YAG rays in all soft tissue, which is due to both Mie and Rayleigh effects at λ = 1064 nm. The illuminated volume is shaped roughly like half an oblate melon, having its long axis collinear with that of the laser beam, and its small equator lying in the plane of the exposed tissue surface. The axial depth of this watermelon-shaped volume is about 1 cm (approximately z_e), and its diameter at the first surface is about 0.8 cm. Because Mie scattering has a strong forward pattern, the power density in the tissue on the axis of the beam will not be a maximum at the first surface (as with the CO_2 laser), rather at a point *below* (distal to) this surface by 2 or 3 mm. This point will rise faster in temperature than any other point within the illuminated volume. As irradiation continues, the volume of tissue bounded by an isothermal surface at 60°C will grow in size, from a small coaxial bell based in the first surface, through a series of elongated shapes resembling cucumbers, to a final watermelon shape approximating the illuminated volume.

If the total beam power is 40 W or more, this cucumber of volume will grow to about z_e in depth within 1-2 sec, and expand in its first-surface diameter to about $0.8z_e$ within 15 sec or so after the onset of irradiation. (The reader should bear in mind that as much as 40-50% of the incident power will be lost by reflection from the first surface.) If the radiant power is increased to approximately 100 W, the minor diameter of the 60°C cucumber will approximate z_e at 15 sec. At all instants of time after the start of irradiation, temperatures inside this isotherm will be higher than 60°C. When the total of incoming radiant energy (of which about half is lost by reflection) has reached a value between 700 and 1000 J, if delivered continuously or higher if delivered intermittently, the temperature of the distally displaced hot spot will reach 100°C. The histological water at this hidden point of maximum power density will then vaporize, forming a pocket of steam whose pressure will rise until it is high enough to rupture the overlying tissue and escape into the atmosphere. These phenomena have been well documented by Halldorsson and his associates (35,36), who have named the submerged explosion the *popcorn effect.*

Because a popcorn explosion can perforate the wall of a hollow organ, it should be carefully avoided by the laser endoscopist, who is well advised to apply the laser beam in short bursts not exceeding 2 sec each if the total power of the beam is 40 W or more (37). Once coagulation, whose onset is marked in vivo by a *blanching* of the tissue, has occurred, the scattering coefficient increases substantially, thus expanding the illuminated volume radially but not greatly increasing its depth in the z direction. The absorption coefficient, whose value is more dependent upon the visual color of the tissue, is greatest for dark or black substances. Hence it increases dramatically if thermal breakdown of soft tissue leads to formation of free carbon, which strongly absorbs the laser rays and then transfers this heat by *thermal conduction and radiation at longer wavelengths* to the nearby water. This set of events can cause rapid, but very imprecise, vaporization. The foregoing facts are schematically illustrated in Fig. 32.

The values of α and σ at $\lambda = 1064$ nm are not well known for many biologic materials. The values in Table 8 are derived from the studies of Halldorsson and Langerholc (36), and from those of Hillenkamp (38).

The usual means of delivering the beam of a Nd:YAG laser to a surgical target is an optical fiber made of quartz, about 3 m in length, and having a core diameter of 0.5-1.0 mm. The fiber is encased within a concentric plastic sheath of larger

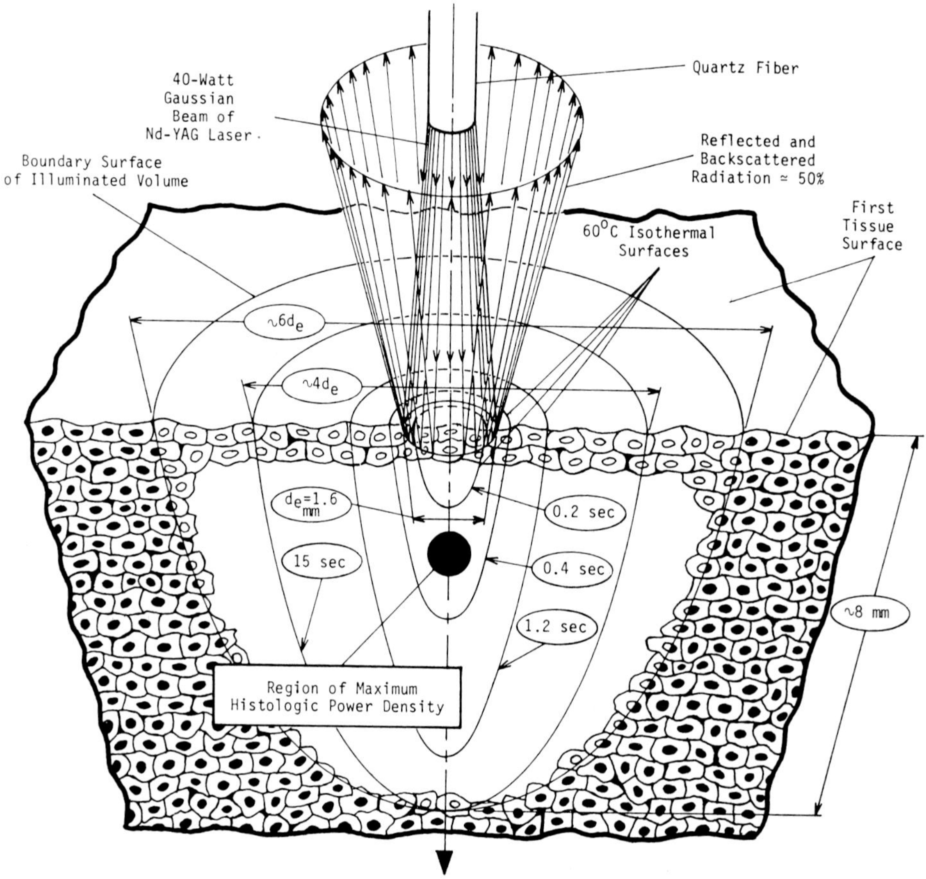

Figure 32 Irradiation of soft tissue by the rays of a Nd:YAG laser from a quartz optical fiber that does not touch the tissue. Scattering in the tissue causes the illuminated volume to be much larger in diameter than the effective diameter of the beam, and about 8 mm deep. The isothermal boundary at 60°C grows larger with time after the onset of irradiation. The point of maximum histological power density is on the beam axis below the first surface, and can be the nucleus of a popcorn explosion with prolonged irradiation. Contours shown are idealized, and the numbers are based upon the data of Halldorsson and Langerholc (36).

diameter, so that a small annular clearance exists around the surface of the fiber. Nitrogen or carbon dioxide gas is passed through this space at low speeds to cool the distal tip of the fiber and keep it free of organic debris, which can become carbonized by the laser's rays and rise to temperatures well above the melting

Table 8 Bouguer's Coefficients At 1064 nm

Substance	Melanin	HbO_2[a]	Hb[a]	Water	Albumin	Stomach or bladder
α in cm^{-1}	3.5	3.3	0.8	0.18	0.15	0.11
σ in cm^{-1}						5-10

[a]Concentration of 150 g/L.

temperature of quartz (1610-1728°C, depending upon crystalline form of the SiO_2). For protection against melting of the tip by the hot carbon, which sublimates at temperatures up to 3597°C, the tip of the quartz fiber is not brought into contact with the tissue. The rays emerging from the distal tip diverge with an included angle between 5 and 15°, thus reducing the power density as the distance between tip and tissue increases. Furthermore, the first-surface reflection at 1064 nm is about 50% for most soft tissue. These facts make vaporization of tissue by a surgical Nd:YAG laser very difficult and imprecise.

In 1984, Surgical Laser Technologies, of Malvern, Pennsylvania, introduced a family of sapphire tips with which the quartz fibers of surgical Nd:YAG lasers can be terminated. These tips are available in a wide variety of distal shapes, ranging from straight circular cylinders to long, slender, tapering needles. Because the index of refraction of sapphire (Al_2O_3) is 1.76 in the near infrared, it acts as a waveguide if surrounded by air or water. By suitably tapering a long sapphire needle, it can be made to act like a converging waveguide for light, whose rays progressively approach the critical angle of incidence on the sapphire-air interface and finally emerge into the air outside the tip with a geometric distribution of power density that depends strongly upon the geometric contour of the extreme distal end of the tip. A slender tapering needle with a tip radius of 0.1 mm concentrates the rays at the tip with very high power density at the spherical surface, from which they diverge like the radii of a sphere, quickly reducing the power density to low values within a radial distance of a few millimeters. Such a tip can be used as a photonic scalpel to cut soft tissue with a precision approaching that of the beam of a CO_2 laser, while still providing thermal coagulation of transected blood vessels up to about 2 mm in diameter. A cylindrical tip with a plane distal face perpendicular to its axis can provide a conically diverging beam for voluminous coagulation. A chisel-shaped tip can shave off thin layers of tissue. Tips for use through endoscopes in general have short lengths, more rapid tapers, and a variety of distal geometries.

Because these sapphire tips are designed to be used *in contact with the tissue,* they can perform tasks of cutting, vaporizing, and coagulating soft tissue with *less than half the total beam power required for noncontacting quartz fibers.* This saving is largely a result of the elimination of the reflection and backscatter from the

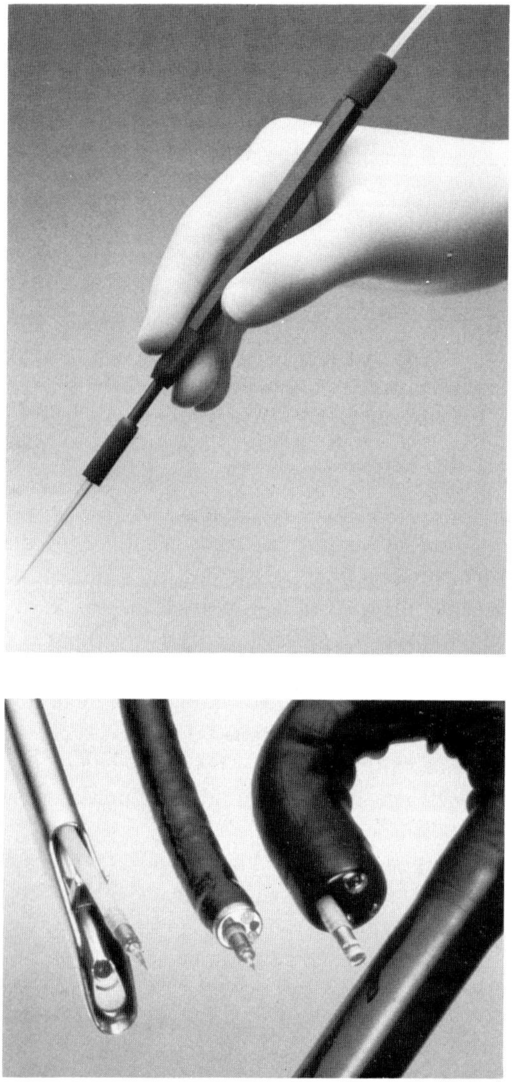

Figure 33 Sapphire tips for contact surgery with Nd:YAG lasers. The tip in the upper part of the figure is mounted in a hand-held probe holder that attaches to the distal end of the quartz fiber for use in freehand surgery. The tips in the lower photograph, from left to right, are for cutting, vaporizing, and coagulating, respectively, through either rigid or flexible endoscopes. The endoscopic tips are screwed onto a threaded ferrule which attaches permanently to the distal end of the quartz fiber (*Source*: photographs furnished courtesy of Surgical Laser Technologies of Malvern, Pennsylvania).

first tissue surface, resulting from the contact of the tip with the tissue. These unique tips, recently described in clinical detail by Daikuzono et al. and Joffe et al. (39,40), are synthetic crystalline aluminum oxide, with a melting temperature in the range of 2030-2050°C, which enables them to withstand the heat generated in tissue debris agglomerated on the tip at power densities up to 70,000 W/cm^2. If more than 20 W or so is delivered via such a tip, the distal end can be melted, but not so readily as that of a quartz fiber. Figure 33 shows a few typical shapes for both freehand surgery and endoscopic procedures.

In the opinion of the author, who has used these contacting surgical probes in animal surgery, they add an important new dimension to surgery performed with the Nd:YAG laser, enabling it to perform many of the procedures formerly accomplished only with the CO_2 laser, offering somewhat less precise removal of tissue but with considerably better first-pass hemostasis than with the carbon dioxide laser. It seems quite probable that the day of the high-powered Nd:YAG surgical laser delivering 100 W or more of output power may be coming to a close, because the sapphire tips allow the performance of all the procedures possible with noncontacting quartz fibers, but at a maximum power of 20 W or less.

4. Visible Lasers

Because of the extreme variability of α and σ in biological materials at wavelengths between 400 and 700 nm, few valid generalities can be stated for visible lasers, except that the wavelength must be chosen to suit the intended use. The values of these coefficients in water at 514 nm ($\alpha = 0.02$ cm^{-1} and $\sigma = 0.08$ cm^{-1}) are totally unreliable indicators of their values in other biomaterials. The most important single medical application of the argon ion laser is selective coagulation of ectatic vasculature, primarily at the ocular fundus and secondarily in the skin. The dye laser, bulky and expensive in output ratings of a few watts because it must be pumped by a laser of shorter wavelength (notably the argon ion), has found its most important medical use for the excitation of hematoporphyrin derivative in photodynamic therapy. In this use it represents technical overkill, because its broad tunability of wavelength is wasted in merely adjusting to the last red absorption peak of HPD. It is the author's opinion that the frequency-doubled Nd:YAG will ultimately replace the argon ion laser, and that an intense broad-banded, noncoherent source (e.g., a xenon lamp) with a bandpass optical filter to tailor its output to approximate the visible red absorption peak of HPD at 630 nm, will finally dominate this application.

V. APPENDIX

A. Maxwell's Equations

When written in modern vector notation and in the mks system of measurement, the equations of James Clerk Maxwell are as follows:

$$\nabla \times \mathbf{E} = \mu\,\mu_0 \frac{\partial \mathbf{H}}{\partial t} \qquad [63]$$

$$\nabla \times \mathbf{H} = i + K\,k_0 \frac{\partial \mathbf{E}}{\partial t} \qquad [64]$$

$$\nabla \cdot \mathbf{D} = q \qquad [65]$$

and

$$\nabla \cdot \mathbf{B} = 0 \qquad [66]$$

where the quantities in bold face type are vectors, i is the electric current density (including displacement currents) in units of (*CURRENT*) divided by (*AREA*), q is the spatial density of electric charge in units of (*CHARGE*)/(*VOLUME*), and the symbol ∇ denotes $\boldsymbol{u}_x\,(\partial/\partial x) + \boldsymbol{u}_y\,(\partial/\partial y) + \boldsymbol{u}_z\,(\partial/\partial z)$. The quantities $\boldsymbol{u}_x$, $\boldsymbol{u}_y$, $\boldsymbol{u}_z$ denote vectors of unit magnitude directed along the positive x, y, and z axes, respectively, of a cartesian coordinate system. The symbol $\times$ following the symbol ∇ denotes the *cross product* of two vectors, which is the vector whose magnitude is the product of the magnitudes of the two vectors multiplied by the sine of the angle between them, and whose direction is that of the tip of a right-handed screw that is rotated in the angular sense from the direction of the first vector to the direction of the second. The symbol $\cdot$ following the symbol ∇ denotes the *scalar product* of the two vectors, which is the product of their magnitudes multiplied by the cosine of the angle between them and has no direction (i.e., it is not a vector).

B. Values of Important Physical Constants

Symbol	Constant	Numerical value and units
c	Speed of light in free space	2.998×10^{8} m/sec
h	Planck's constant	6.626×10^{-34} J sec
κ	Boltzmann's constant	1.3807×10^{-23} J/°K
Q_e	Charge of an electron	-1.6022×10^{-19} C
m_e	Mass of an electron	9.110×10^{-31} kg
m_p	Mass of a proton	1.6726×10^{-27} kg
m_n	Mass of a neutron	1.6749×10^{-27} kg
k_0	Absolute permittivity of free space	1.11265×10^{-10} (see Table 1)

Symbol	Constant	Numerical value and units
μ_0	Absolute permeability of free space	1.00000×10^{-7} (see Table 1)
ϵ	Base of natural logarithms	2.718281828 . . .
π	Circumference/diameter ratio of a circle	3.141592654 . . .
$(e_p)_{400}$	Photon energy, λ = 400 nm	3.100 eV
$(e_p)_{700}$	Photon energy, λ = 700 nm	1.7712 eV
$(e_p)_{1064}$	Photon energy, λ = 1064 nm	1.1653 eV
$(e_p)_{10,600}$	Photon energy, λ = 10,600 nm	0.11697 eV

BIBLIOGRAPHY

1. Fisher JC: The power density of a surgical laser beam: Its meaning and measurement. *Lasers Surg. Med. 2*:301-315, 1983.
2. F. E. Jenkins and H. E. White. *Fundamentals of Optics.* McGraw-Hill, New York, 1976.
3. B. A. Lengyel. *Lasers.* Wiley-Interscience, New York, 1971.
4. F. S. Barnes (ed.). *Laser Theory.* IEEE Press, New York, 1972.
5. M. J. Weber (ed.). *Handbook of Laser Science and Technology,* Vols. 1 and 2. CRC Press, Boca Raton, Florida, 1982.
6. D. Sliney and M. Wolbarsht (eds.). *Safety with Lasers and Other Optical Sources: A Comprehensive Handbook.* Plenum Press, New York, 1980.
7. S. L. Trokel (ed.). *YAG Laser Ophthalmic Microsurgery.* Appleton-Century-Crofts, Norwalk, Connecticut, 1983.
8. E. G. Cullwick. *The Fundamentals of Electromagnetism.* The University Press, Cambridge, England, 1939; Macmillan, New York, 1949.
9. R. C. Weast (ed.). *Handbook of Chemistry and Physics.* CRC Press, Boca Raton, Florida, 1984-1985.
10. M. J. Beesley. *Lasers and Their Applications.* Taylor & Francis, London, 1978.

REFERENCES

1. Maxwell JC: A dynamical theory of the electromagnetic field. *R. Soc. Engl.* 1864.
2. Einstein A: On the quantum theory of radiation. *Physikal. Z. 18*:121, 1917.
3. Bohr N: On the constitution of atoms and molecules. *Philosoph. Mag. 26*:1, 1913.
4. Sommerfeld A: *Ann. Physik 51*:1, 1916.
5. de Broglie L: *Philosoph. Mag. 47*:446, 1924.

6. Schrödinger E: *Ann. Physik 79*:361, 489, and 734, 1926; *Physical Rev. 28*: 1047, 1926.
7. Dirac PAM: *Proc. R. Soc. A117*:610, 1928; *A118*:351, 1928.
8. Maiman TH: *Physical Rev. Lett. 4*:564, 1960.
9. S. S. Charschan (ed.). *Lasers in Industry*. Van Nostrand, New York, 1972, p. 20.
10. Ketcham AS, Hoye RC, and Riggle GC: A surgeon's appraisal of the laser. *Surg. Clin. N. Am. 47*:1249, 1967.
11. Aron-Rosa D, Aron J, and Griesman JC: Use of the neodymium-YAG laser to open the posterior capsule after lens-implant surgery: A preliminary report. *J. Am. Intraocular Implant Soc.* 352-354, 1980.
12. Fankhauser F, Roussel P, Steffen J, et al.: Clinical studies on the efficacy of high power laser radiation upon some structures of the anterior segment of the eye. *Int. Ophthalmol. 3*:129-139, 1981.
13. McKibbin LS and Paraschak DM: A study of the effects of lasering on chronic bowed tendons at Wheatley Hall Farm Limited, Canada, January, 1983. *Lasers Surg. Med. 3*:55-59, 1983.
14. Linsker R, Srinivasan R, Wynne JJ, and Alonso DR: Far-ultraviolet laser ablation of atherosclerotic lesions. *Lasers Surg. Med. 4*:201-206, 1984.
15. Choy DSJ, Stertzer SH, Myler RK, Marco J, and Fournial G: Human coronary laser recanalization. *Clin. Cardiol. 7*:377, 1984.
16. Geschwind H, Boussignac G, Tessiere B, Vielledent C et al.: Percutaneous transluminal laser angioplasty in man. *Lancet 1*:844, 1984.
17. Jacquez JA and Kuppenheim HF: Spectral reflectance of human skin in the region 235-1000 nm. *J. Appl. Physiol. 8*:212, 1956.
18. Hardy JD, Hammell HT, and Murgatroyd D: Spectral transmittance and reflectance of excised human skin. *J. Appl. Physiol. 9*:257-264, 1957.
19. T. Ohshiro. *Laser Treatment for Nevi.* Medical Laser Research Co. and Fukuin Printing Co., Ltd., Tokyo, 1980.
20. Mihashi S, Jako GJ, Incze J, Strong MS, and Vaughan CW: Laser surgery in otolaryngology: Interaction of CO_2 laser and soft tissue. *Ann. N. Y. Acad. Sci. 267*:263-294, 1976.
21. Henriques FC Jr: Studies of thermal injuries, V. The predictability and the significance of thermally induced rate processes leading to irreversible epidermal injury. *Am. J. Pathol. 23*:489-502, 1947.
21a. Henriques FC Jr and Moritz AR: Studies of thermal injury, I. The conduction of heat to and through skin and the temperatures attained therein. A theoretical and an experimental investigation. *Am. J. Pathol. 23*:531-549, 1947.
22. Castro DJ, Abergel RP, Johnston KJ, Adomian GE, Dwyer RM, Uitto J, and Lesavoy MA: Wound healing: Biological effects of Nd:YAG laser on collagen metabolism in pig skin in comparison to thermal burn. *Ann. Plast. Surg. 11*:131-141, 1983.
23. Castro DJ, Abergel RP, Meeker CA, Dwyer RM, Lesavoy MA, and Uitto J: Effects of the Nd:YAG laser on DNA synthesis and collagen production in human skin fibroblast cultures. *Ann. Plast. Surg. 11*:214-222, 1983.

24. Abergel RP, Meeker CA, Dwyer RM, Lesavoy MA, and Uitto J: Non-thermal effects of Nd:YAG laser on biological functions of human skin fibroblasts in culture. *Lasers Surg. Med. 3*:279-284, 1984.
25. S. Mihashi, S. Ueda, M. Hirano et al. Some problems about condensates induced by CO_2 laser irradiation, in *Transactions, 4th Congress of the International Society for Laser Surgery* (K. Atsumi, ed.), Intergroup Corp., Tokyo, 1981, pp. 2:25-2:27.
26. J. W. Osterhuis. *Tumor Surgery with the CO_2 Laser: Studies with the Cloudman S-91 Mouse Melanoma.* Veenstra-Visser, Gronigen, Netherlands, 1977.
27. Bellina JH, Stjernholm RL, and Kurpel JE: Analysis of plume emissions after papovavirus irradiation with the carbon-dioxide laser. *J. Reprod. Med. 27*: 268-270, 1982.
28. Grundfest WS, Litvak F, Forrester JS, Goldenberg T et al.: Laser ablation of human atherosclerotic plaque without adjacent tissue injury. *J. Am. Coll. Cardiol. 5*:929-933, 1985.
29. Isner JM, Michlewitz H, Donaldson RF, Konstam MA, and Salem DN: Laser-associated debridement of aortic valve calcium. *Am. Heart J. 109*:448-452, 1985.
30. Isner JM et al.: Optimal deployment of carbon dioxide laser irradiation in the treatment of cardiovascular disease: Importance of peak power and power density. *Abstr. 6th Cong. Int. Soc. Laser Surg. Med.*, Jerusalem, Israel, 1985, p. 37.
31. Isner JM et al.: Excimer laser irradiation of cardiovascular tissues: Standardized comparison of 193, 248, 308, and 351 nm wavelengths. *Abstr. 6th Cong. Int. Soc. Laser Surg. Med.* Jerusalem, Israel, 1985, p. 38.
32. Isner JM et al.: The use of optical fibers to deliver excimer laser energy to cardiovascular tissue sites. *Abstr. 6th Cong. Int. Soc. Laser Surg Med.* Jerusalem, Israel, 1985, p. 39.
33. Isner JM et al.: Laser assisted endocardiectomy for refractory ventricular tachyarrhythmias: In vitro findings and initial intraoperative application. *Abstr. 6th Cong. Int. Soc. Laser Surg. Med.* Jerusalem, Israel, 1985, p. 41.
34. Isner JM et al: Excimer laser irradiation of cardiovascular tissue in blood and fluosol. *Abstr. 6th Cong. Int. Soc. Laser Surg. Med.* Jerusalem, Israel, 1985, p. 42.
35. Halldorsson T, Langerholc J, Senatori L, and Funk H: Thermal action of laser irradiation in biological material monitored by egg-white coagulation. *Appl. Optics 20*:822-825, 1981.
36. Halldorsson T and Langerholc J: Thermodynamic analysis of laser irradiation of biological tissue. *Appl. Optics 17*:3948-3958, 1978.
37. J. C. Fisher. Principles of safety in laser surgery and therapy, in *Basic and Advanced Laser Surgery in Gynecology* (M. S. Baggish, ed.), Appleton-Century-Crofts, Norwalk, Connecticut, 1985, p. 97.
38. F. Hillenkamp. Interaction between laser radiation and biological systems, in *Lasers in Biology and Medicine* (F. Hillenkamp, R. Pratesi, and C. A. Sacchi, eds.), Plenum Press, New York, 1980, pp. 37-68.

39. Daikuzono N, Joffe SN et al.: Artificial sapphire probe for contact photocoagulation and tissue vaporization with the Nd:YAG laser. *Med. Inst. 19*: 173-178, 1985.
40. Joffe SN, Daikuzono N et al.: Contact probes for the Nd:YAG laser. *Proc. S.P.I.E.* (in press) 1985.

2

Operating Characteristics of Surgical Lasers and Delivery Systems

TERRY A. FULLER
Fuller Research Corporation, Vernon Hills, and Northwestern University, Evanston, Illinois

I. INTRODUCTION

Therapeutic lasers were originally developed in the early 1960s as an effective alternative to a bright (*xenon*) light source used in ophthalmology for photocoagulation of the retina. The early ophthalmic *ruby* laser was little more than a prototype commercial or scientific laser connected inflexibly to a microscope to "cross hair site" the target. Even as the earliest devices provided a magnified field, the only mechanism to deliver the laser beam was by moving the patient under the cross hair.

The second generation ophthalmic laser was the *argon* laser. Not only had a more effective wavelength been selected for photocoagulation of the retina, but a mechanism had been developed to aim a prefocused laser beam at the patient's retina. Elimination of the requirement to move the patient for focusing by introducing the ability to control the laser beam directly has had a significant impact on the human engineering advantages present in today's lasers.

In addition to ease-of-use and accuracy developments in the laser, recognition of how wavelength changes alter the therapeutic result stimulated major research that continues today.

As a result of biological and clinical research on laser-tissue interaction and as a result of improving generations of lasers, beam delivery systems, and focusing schemes, surgical lasers available to the surgeon today represent vast improvements over the early systems. The surgical laser today is a relatively sophisticated device with many wavelength choices, acceptable and improving delivery systems, and an ever-increasing list of defined clinical applications.

To have an understanding of the variety of lasers available to the surgeon today, not only is an understanding of tissue biophysics necessary (discussed in Chap. 1), but a fundamental understanding of the mode of operation of a laser and the norms and limitations on energy-delivery devices is required. Availability and flexibility of end-delivery systems are generally dependent upon the wavelength of a laser.

In this chapter, we will summarize the various lasers used in medicine and medical research and characterize them by active medium (substance or substances responsible for the radiant energy wavelength).

Next, we will define the temporal or time/function relationships of lasers. As it is not the intent of this brief chapter to explore the clinical effect created by temporal changes in laser energy, only synoptic remarks describing the principal effects caused by such changes will be made.

We will conclude with a discussion of how the laser energy, developed by the laser itself, is transmitted or delivered to the surgical site. Clearly, it is a laser's delivery system that most affects the clinician. The laser's delivery system determines the ease, control, and even the ability to perform surgery.

II. MAJOR CLINICAL LASERS

Thousands of wavelengths are emitted by a long list of solid, liquid, and gas laser mediums (Fig. 1). In the surgical use of lasers, however, only a handful of the wavelengths are of value (Table 1). The following discussion lists the major clinical lasers as well as several new devices that appear to have particular promise as a medical laser or are being actively investigated today for surgical application.

A. Carbon Dioxide Laser

The *carbon dioxide* (CO_2) laser, developed in 1964 at Bell Laboratories by C.K. N. Patel (1), initially generated approximately 1 mW of 10.6 μm (10,600 nm) energy. Today's industrial lasers are capable of delivering in excess of 5000 W of continuous wave power and millions of watts in pulsed energy.

The laser cavity of a CO_2 laser contains *nitrogen* and *helium* in addition to carbon dioxide. This gas mixture improves the efficiency of the laser by increasing the vibrational energy of the CO_2 molecule.

1. Flowing-Gas versus Sealed Systems

Conventional flowing-gas carbon dioxide lasers are designed with a constant stream of *carbon dioxide, nitrogen,* and *helium* flowing longitudinally through a hollow laser cavity (Fig. 2a). In surgical laser systems, where size is a major consideration, slow flow rates are needed to conserve gas contained in gas bottles.

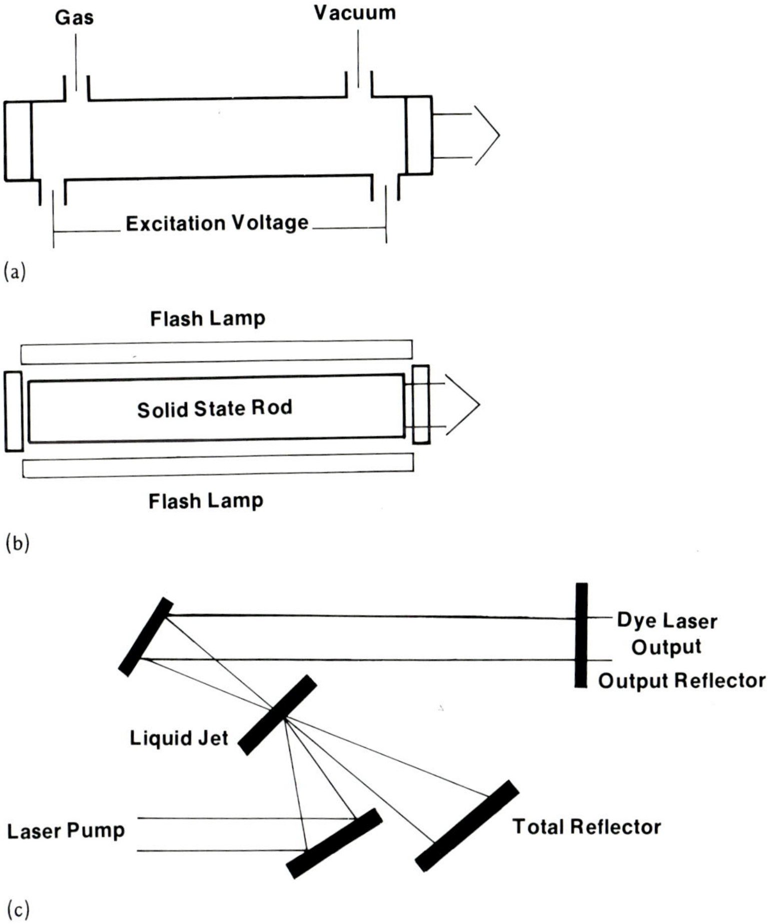

Figure 1 Three different laser-active media: (a) a gas laser showing longitudinal discharge in a flowing-gas system; (b) a solid-state media, excited by two linear flash lamps; (c) a liquid dye (CW) system.

Increasing the flow of gas through the cavity helps dissipate heat. This results in more power from the laser. The CO_2 lasers with slow axial flow rates are power limited to 75-100 W of power per meter of tube length (W/m). Industrial lasers requiring higher power achieve 200-500 W/m by rapidly passing through the cavity and dumping or recirculating the gas mixture. Even higher continuous wave (CW) power can be obtained by a process known as transverse flow (Fig. 2b). This process increases gas flow considerably, and CW power greater than 1000 W/m is possible.

Table 1 Operating Characteristics of Surgical Lasers

Phase	Laser type	Wavelength (nm)	Excitation method
Gas	Argon	488, 514	AC-discharge
	Carbon dioxide (CW)	10,600	DC-discharge
			RF-discharge
	Carbon dioxide (pulsed)	10,600	Pulsed-discharge
	Excimer	193, 248, 308	Pulsed-discharge
		351	Pulsed-discharge
	Helium:neon (CW)	632.8	AC-discharge
	Krypton (CW)	647, 676	AC-discharge
		407, 413, 515	AC-discharge
	Metal vapor		
	HeCd	442, 325	AC-discharge
	Copper	510, 578	AC-discharge
	Gold	628	AC-discharge
Liquid	Dye (CW)	400-800	Synchronously pumped argon or krypton ion laser
	Dye (pulsed)	300-900	Optically pumped by N_2, excimer, and Nd:YAG
Solid	Nd:glass	1060	Optically pumped
	Nd:YAG (CW)	1060	Optically pumped
	Nd:YAG (pulsed)	1060	Optically pumped
	Alexandrite	700-815	Optically pumped
	Ruby	694	Optically pumped

AA = articulated arm
Direct = direct mechanical connection from laser head to delivery unit
Exp. fiber = experimental fiber optic
Fiber = fiber optics
MM = micromanipulator

Typical performance	Delivery systems
0.05-10 W (blue-green)	Fiber, AA, MM
10 MW-100 W 15% eff. typ. (mid-IR)	
10 MW-100 W 15% eff. typ. (mid-IR)	AA, MM, direct, exp. fiber
<1000 J/pulse repetition rate to 2.5 kHz (mid-IR)	
0.5-20 W	AA, MM
1-1500 pps (UV)	
1-20 W	MM, fiber
1-100 pps (UV)	
0.5-1.5 MW (red)	Fiber
1-5 W (red)	Fiber, MM
1-3 W (deep blue, UV)	Fiber, MM
0.1-7 mW CW	Fiber
1-10 W (avg) rep. rate 8 kHz	AA, fiber
.01-5 W (avg) rep. rate 5-15 Hz	Fiber
Continuously tunable 0.1-4 W (visible)	Fiber
Tunable 0.01-0.1 MJ/pulse at rep. rate of pump (visible)	AA, some fiber
11-100 J (near-IR)	Direct
0.01-100 W (near-IR)	Fiber, MM
0.01-400 J/pulse, harmonics avail. (near-IR)	AA, MM
Tunable 0.01-1 J rep. rates <1 to 10 (near-IR)	Fiber, AA
>200 J long pulse (300-3000 μsec) (red)	Direct
>15 J short pulse (15-25 nsec) (red)	Direct

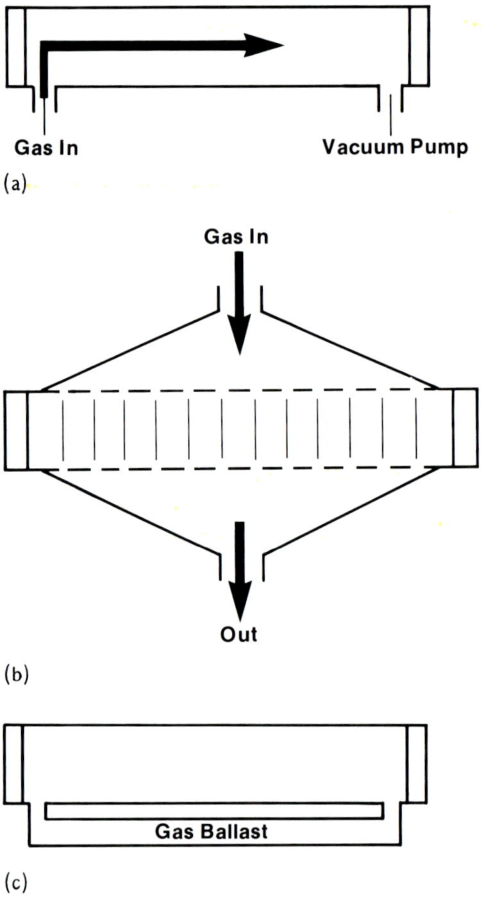

Figure 2 Three common gas laser systems: In (a) there is gas flowing from the "gas-in" port to the vacuum pump (typical of a flowing-gas CO_2 laser); (b) a transversely excited laser; (c) a sealed laser. The gas reservoir (ballast) supplies gas to the head. This is common in ion lasers and sealed CO_2 lasers.

In the low-power range, CO_2 lasers have been designed with the gas sealed in the laser tube (Fig. 2c). Sealed CO_2 lasers typically deliver less than 50 W/m. Sealed lasers differ from conventional flowing-gas systems in several respects. The most obvious is the sealing of the gas in the cavity, thus eliminating the need to replace empty gas bottles. However, there are difficulties in providing an effective seal. Containing all the elements of the gas mixture in the cavity over long periods has proved to be a challenge. The most difficult element to contain for any substantial period is *helium*; many types of glass and epoxy seals are permeable to helium. In a sealed laser there is degradation of the gas mixture with continuous use or long periods of nonuse. Thus there are both helium-containment and gas-degradation limitations in a sealed system. One approach to solving these limitations is *hard seals*. This involves an actual weld of all gas-containing elements of the cavity including the optics. Although an expensive process, hard seals do yield longer-lived lasers.

Partly because of the static nature of gas sealed in a cavity and to local heating of the gas along its small volume, the power per unit length of a sealed laser is less than in a flowing-gas system. Typically, 25-40 W/m of tube length can be achieved by sealed CO_2 lasers in contrast to the 75-100 W/m generated by conventional flowing-gas CO_2 lasers.

2. Excitation

The type of excitation that is used to stimulate the gases is another variable affecting the size, power, and mode of CO_2 lasers. The two common types of excitation employed are *direct current* (DC) and *radiofrequency* (RF). Most conventional flowing-gas CO_2 lasers are DC excited. This means a constant high voltage is applied to the ends (longitudinally) of the laser cavity (Fig. 3a). Voltages ranging from 10,000 to 30,000 V/m are typical. The current (1-30 mA) through the gas is controlled, thus regulating the resultant laser power. Although transversely

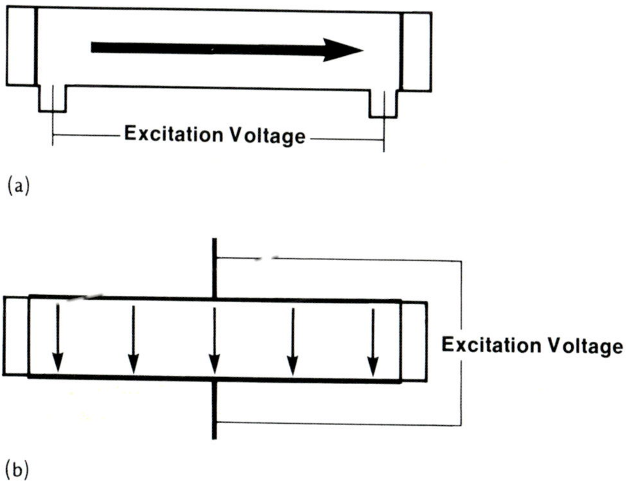

Figure 3 Two mechanisms of exciting a gas: (a) longitudinal excitation, contrasted to the transverse excitation in (b).

excited (TE) DC lasers are used in industry, where high peak powers are required, this mode of excitation has received little attention in surgery (Fig. 3b). The TE lasers develop high-power pulses with a low repetition rate.

The attractiveness of transverse excitation is principally one of reducing the required voltage of the power supply. Direct current sources cannot be effectively used on a TE laser because of problems with nonuniform arcing across the laser cavity, thus the pulsed nature of such devices. To eliminate this problem and preserve the advantages of transverse excitation, RF excitation is used. Voltages less than 300 V at frequencies ranging from 20-1000 MHz are employed. However, RF lasers currently are limited in power. Typically, 35 W is the maximum reliable power. Developing techniques of folding a laser cavity will permit more compact devices for medicine.

Most RF lasers are sealed and are optically characterized in one of two manners: unguided and waveguided. The waveguide lasers are capable of delivering higher power per meter than unguided sealed systems. Their beam diameters are very small (typically 1-3 mm) and thus have a high power density. Both RF- and DC-excited lasers can be part of an optical system of specific geometry to *guide* the wave down the cavity with minimal diffraction. This type of system is referred to as a *waveguide laser.* Because of several compelling manufacturing benefits, most RF lasers are also waveguide devices. The principal advantage of waveguiding is one of stability and mode retention. These considerations are crucial in military and certain industrial applications. Such characteristics, while not essential in a surgical laser, coupled with rugged and reliable construction, make the waveguide laser attractive as a surgical device.

B. Ion Lasers

The first observed laser emission by a gaseous ion laser was by Bell (2) in *ionic mercury* (Hg^{2+}) in 1963. The strongest emissions from the mercury laser are at 567.7 and 615.0 μm in the visible portion of the electromagnetic spectrum. Although of historic and scientific significance, perhaps the greatest contribution of the mercury laser was that its development led to several other important discoveries. Within a year, over 150 ionic emission lines were discovered. Most of the emissions came from the noble gases: *neon* (Ne), *argon* (Ar), *krypton* (Kr), and *xenon* (Xe). See the tabulation for wavelength. In that same year, Bridges (3) reported on a serendipitous discovery. While adding *argon* as a buffer gas in place of *helium,* he observed several new laser wavelengths from argon.

The blue-green emission from the argon laser was initially used for spectroscopy (4). Both Bell's and Bridges' discoveries were made through an intermittent DC-excited device. It gained wide use in *ophthalmology* for photocoagulation of the retina only after the development of a practical CW argon laser in 1966. Hundreds, perhaps thousands of ion transitions have been discovered,

Ion	Wavelength (nm)
Argon	0.488, 0.514
Helium:cadmium	0.442
Helium:neon	0.633
Krypton	0.35, 0.68
Mercury	0.568, 0.615
Xenon	0.488-0.54

but very few have been developed as practical, let alone surgical laser devices. The most common ion laser is the blue-green transition from argon, followed by the red and yellow lines of the krypton, and the ultraviolet (UV) lines from cadmium lasers.

To obtain an emission from an *ion* laser, the active medium (most frequently a gas) must be ionized by the transfer of energy. The transfer of energy can be obtained by high-voltage, pulsed discharges producing pulsed ionic laser lines. However, for applications requiring long pulses or continuous operation, high current densities (in excess of 100 A/cm^2) at relatively low voltages are required. The high densities create tremendous stress on the discharge structure, and early quartz cavities were rapidly destroyed. The progress and differentiation present in today's lasers represent approaches in handling the high current densities as well as techniques for confining the discharge to lengthen the cavity lifetime and to improve stability and efficiency of the overall device. Ion lasers require considerable dissipation of heat. Because of inherent inefficient excitation and subsequent transition of the gaseous ion, the ion lasers are faced with major engineering problems not encountered in molecular or atomic laser structures. The continuous wave ion laser has had to overcome major technical difficulties in dissipating heat while maintaining control over the output power.

In 1968, an *argon ion laser* (5) was introduced using graphite disks to contain the high current densities. The result was a laser delivering 2 W of CW power, a longer lifetime, a reduced cost, and an improved performance. In another approach to handling the required current densities and to aid in the dissipation of heat, beryllium oxide was used as a laser tube. To date, these two approaches have survived commercial use. More recent technology that promises higher reliability and a longer tube lifetime is the introduction of cooled tungsten disks to make up the laser cavity. These developments have led to greater and greater powers available from the ion lasers. Typically, the argon ion laser delivers 15-25 W.

Axial magnetic fields are commonly used to aid in the discharge of the tube, confining the fields to a narrow region along the tube axis. Inherent thermal losses and required large magnetic fields result in an overall device inefficiency and add to the requirement for cooling to prevent overheating and device destruction.

Recent innovations stimulated by the need for low-power argon lasers for industrial applications have found a home in the medical environment as well. These smaller lasers are air cooled in contrast to the traditional water-cooled devices. Although their power is typically limited to less than 5-8 W, their size and relative portability make them attractive for many applications.

C. Neodymium Lasers

Discovered in 1961 by Snitzer (6) the *neodymium:glass* (Nd:glass) laser demonstrated stimulated emission in the near-infrared portion of the spectrum using 2-6% Nd^{3+} incorporated in a host material made of barium crown glass. These early lasers were composed of thin fibers of Nd:glass, 1-7.5 cm in length. As an understanding of the process of exciting and amplifying Nd:glass took hold, larger glass rods were used. This resulted in developing powerful laser pulses. It is possible to obtain about 100 J output from a single pulse (7). Repetition rates of up to 3-5 pulses/sec (pps) have been obtained. Although great power can be obtained from the neodymium:glass laser, continuous operation is difficult because of the low thermal conductivity of glass. The main advantage of the glass structure is the size, shape, and ease at which high optical quality material can be fabricated.

Continuous operation of the neodymium laser was initiated in 1963. Neodymium-doped yttrium-aluminum-garnet (YAG) was first operated at Bell Laboratories (8). The average power of the early Nd:YAG lasers was less than 10 W. Today, average powers in excess of 100 W are common. While YAG laser technology is relatively mature, areas of improvement include reduction in size and weight.

Both the Nd:glass and Nd:YAG lasers are in production and used in a variety of fields. In medicine, the Nd:YAG laser is most popular because of the continuous wave operation and the ability to Q-switch (see Sec. III) the device.

The Nd:YAG laser cavity is made up of an optically reflective structure (typically a gold-plated cavity) that couples optical energy from a krypton arc lamp to the Nd:YAG. The krypton arc lamp creates broad-spectrum energy that excites the neodymium atoms. The concentration of neodymium in a YAG laser is typically 0.5-2%. The emission of energy is most common at 1.06 μm; however, spectral lines centered at 0.914 and 1.35 μm can also be found. In medical devices, 1.06 μm predominates, and other wavelengths are suppressed. The typical medical CW Nd:YAG laser generates 80-100 W. This device must be water cooled, most commonly from a wall source and drain. It is powered by a 230V, 30A, 3Φ electrical power source found commonly in hospital environments.

In addition to operating the neodymium lasers in a long-pulsed and continuous wave manner, the Nd:YAG laser is commonly Q-switched. Q-switching results in short (10-12 nsec) pulses with intensities of millions of watts. These pulses are used in medicine by locally ionizing tissue thereby disrupting membranes and other structures. In ophthalmology, without the requirement for opening the cornea, this device is used to create an incision in a posterior capsule that has opacified.

D. Excimer Laser

The excimer or rare-gas halide lasers were developed by AVCO Incorporated in 1975. In 1976, the first commercial device was offered, but it remained for several more years for the device to gain any substantial popularity. Several years later excimer lasers were introduced, capable of delivering up to 25 W of average power at repetition rates as high as 100 Hz (9).

Today there are four popular active gases (and thus four wavelengths) used in eximer lasers—argon fluoride (ArF), krypton fluoride (KrF), xenon chloride (XeCl), and xenon fluoride (XeF). Each gas can be stimulated to emit a different UV laser line at 193, 248, 308, and 351 nm, respectively.

Excimer lasers represented a challenge to the engineering design community. The need to ensure encapsulation of toxic gases and to provide protection from high-voltage discharges were prime concerns. Certainly in a medical environment, a gas leak would be truly unacceptable. Recently, manufacturers have reduced these concerns by appropriate housing and engineering design.

Another important area of investigation with excimer lasers is the potential of unwanted side effects from UV radiation. Concern has been expressed regarding potential mutagenic effects. The absorption peak of the DNA molecule is 248 nm. Within the medical community, excimer lasers are still in research stages. Applications are being investigated that include radial keratoplasty, to correct the shape of the cornea, and laser angioplasty, to provide debridement of occluded coronary arteries.

The exciting and novel aspect of the surgical applications of excimer laser energy is its high absorption coefficient by tissue, resulting in a clean cut with minimal thermal necrosis surrounding the lesion. Thus for applications where hemostasis is not required, the excimer laser holds substantial promise as an adjunctive surgical tool.

E. Dye Laser

The dye laser was invented at IBM Laboratories by Sorokin in 1966 (10,11). In this laser, the active dye was optically pumped by energy emitted from a ruby laser. Resultant emission of the chloro-aluminum phthalocyanine (CAP) dye was a pulsed coherent beam at 0.756 μm. Three years later, AVCO Incorporated introduced the first commercial dye laser. The device was optically pumped from a hydrogen laser. In 1970, Kodak Research Laboratories invented the first continuous wave dye laser, which was commercialized in 1971.

The active medium in a dye laser is an organic fluorescent material dissolved in a common solvent. Materials, such as rhodamines in alcohol and fluorescein in water, are typical. The color of the dyes are a result of the strong absorption bands in the visible region of the spectrum by the dyes. The excitation of the dye is by optical pumping. The optical source is typically a flash lamp (an emission arc tube or most commonly, a host laser).

There are basically two types of dye lasers—pulsed and continuous operation. Pulsed lasers are most commonly used in research involving chemical reactions and fundamental physics. These devices have also found applications in biomedical research and isotope separation programs. The most common types of pulsed dye lasers fall into four categories.

1. Flash lamp-pumped dye: A relatively inefficient laser that generates the high pulse energy with long pulse lengths but has a low repetition rate and relatively low cost.
2. Nitrogen-pumped dye: This device has low pulse energy, low repetition rates, and is relatively low in cost.
3. Neodymium-pumped dye. This laser generates high pulse energy, low repetition rates, but has a large wavelength-tuning range. This device is the most expensive and difficult to operate.
4. Excimer-pumped dye: This device is similar to the neodymium pumped dye but with lower pulse energy and higher repetition rates.

The continuous wave dye lasers have found greater acceptance in the medical community than the pulsed dye lasers. Although generally of low power (less than 4 W), they are of interest for applications requiring specific wavelengths, such as photodynamic therapy, in the treatment of malignant disease.

F. Solid State Lasers

1. Ruby

Developed in 1960 by Maiman (12), the *ruby* laser was the first device to demonstrate stimulated emission of radiation. The active medium is a pink ruby crystal made up of aluminum oxide doped with 0.05% chromium oxide (the crystal without the chromium is known as sapphire). The ruby rod is optically pumped from a flash lamp, and can be operated in either a long pulse (300-3000 μsec) or Q-switched (15-25 nsec). When long pulsed, the ruby laser is capable of delivering up to 200 J of energy. When operating in the Q-switched mode, pulsed energies of 15 J yielding peak powers of 100 MW are typical. Such high peak powers make ruby lasers advantageous for a variety of industrial applications. However, it is the long-pulsed mode that was of interest to medicine for early photocoagulation of the retina and the novel application of this pulsed energy for iritoectomies. The 694 nm red emission from the ruby laser is only able to transmit a few pulses per minute to the patient. Although this rate is low, it is suitable for creating *iridectomies* in patients with pigmented irides.

2. Alexandrite Laser

The *alexandrite* laser can be considered as a cousin to the ruby laser. Its active medium contains chromium but in a significantly different host lattice. The material is a proprietary substance of the Allied Corporation first developed in 1974. Three years later, it was discovered that the alexandrite laser had tunability characteristics. Today, this device is capable of being tuned from 700 to 815 nm in the red region of the spectrum. Continuous wave powers of up to 100 W are projected for the near future. This device is of interest to the medical community because of its tunability, but has yet not received any significant attention.

III. TEMPORAL MODES OF THE LASER

Most lasers can be operated in a variety of temporal modes. There are four such modes of operation that have interest to the medical community. They are: *continuous wave, pulsed, Q-switched,* and *mode-locked*. Each mode, independent of wavelength, will cause significantly different effects on tissue. Thus in addition to the choice of wavelength, which is an important independent variable on tissue response, controlling the rate, number, and duration of photons (energy) being absorbed by tissue will result in an accurate prediction of tissue response. Inherent in the type of laser chosen for a particular application are certain variables including power, duration, beam diameter, and wavelength. Frequently, a fixed relationship exists between several of these variates. It is common to find continuous wave lasers operating at low power and a pulsed laser operating with high peak power.

In the *continuous wave* (CW) mode, the laser is oscillating constantly, delivering power to the tissue as long as the operator activates the device. This activation is typically performed by depressing a footpedal or finger switch. Although many surgical lasers are continuous wave devices, they are rarely targeting the tissue continuously. More typically, a controlled *dose* of energy is delivered by either electronically preselecting the duration of exposure or by controlling the duration by the operator activating a switch. In either case, the control over the duration of exposure is referred to as *gating* the laser. Depending upon the type of laser being used, gating is accomplished by opening a mechanical shutter or simply turning the laser power supply on and off. Typical gated continuous wave lasers operate from a minimum duration of tens of milliseconds (10 ms = 0.01 sec) to true CW or infinite exposure. Many lasers have preselected durations. For example: a 25-W carbon dioxide laser might provide electronic gating of 0.1, 0.2, 0.5, 1 sec and CW footpedal operation. Preselected durations will vary by surgical application of the laser, as well as by the manufacturer's particular features.

In contrast with CW and gated CW lasers are a family of devices referred to as *pulsed lasers*. A pulsed laser generally has a pulsed duration that is much shorter than a continuous wave gated device. In addition, the power levels obtainable by a pulsed laser are substantially higher than can be achieved in a continuous wave laser (of the same active medium). Optical, mechanical, and electrical design considerations dictate the mode of operation of the device. Considering the continuous wave 25-W carbon dioxide laser referenced previously, a similarly sized pulsed CO_2 laser might exhibit peak powers of five to 20 times its continuous wave power or 125-500 W. The duration of such a device might typically be in the 100 msec (0.1 sec) range.

There are various types of pulsed lasers used in surgery today. Pulsed lasers of duration measured in 10s to 100s of microseconds are most frequently controlled electronically. Changing the method of pumping the active medium will result in change in the power and temporal effects described earlier. Many pulsed lasers, however, permit further compression of the pulsed width (duration) with a corresponding increase in power. The result is that the energy (power X time) is kept constant.

A clinical comparison of a lesion created by a continuous wave and pulsed laser of similar energy would reveal important differences. In CW laser lesions where the duration of exposure is long, there is sufficient time for the heat to be generated into deeper structures, thus permitting a large zone of coagulation necrosis. The zone of coagulation necrosis determines the degree of hemostasis. In contrast, the pulse laser with its short pulse duration and higher peak powers creates a lesion with less thermal damage and consequently less hemostasis.

The process referred to as *Q-switched* carries the concept of pulse compression a step farther. In this process, first proposed by Hellwarth (13), extremely short pulses can be created with intensities several orders of magnitude greater than the non-Q-switched laser. Because of the extremely high power, this porcess is commonly called *giant pulsing*. The process involves inserting a fast shutter or other optical element in the laser cavity. With the shutter closed, the excitation of the active medium can build up far above the threshold level achieved with the shutter open. After a high excitation is reached, the shutter is opened and all the excess radiation is discharged in an extremely short period. The introduction of the shutter spoils the *quality* of the laser, and thus is referred to as the *Q* of the optical cavity. Q-switching may be accomplished by changing the reflectivity of one of the mirrors, inserting a shutter that changes the transparency of a medium within the cavity, or any other process that optically interrupts the propagation in the cavity.

The process of pulse width compression can also be accomplished by changing the energy content of the modes in the laser cavity. This process, known as *mode-locking* (14) gives rise to energy bursts of very high power and short duration. The output pattern from a mode-locked laser is a series of pulses, spaced at intervals equal to the travel time across the laser cavity. Thus these devices are typically long or have a folded cavity design.

IV. DELIVERY SYSTEMS

There are many devices used to interconnect the laser with tissue. Depending upon wavelength, power, required spot size, and ability to access tissue, one or several devices may suit the purpose. The energy delivery device along with hand probes, micromanipulators, endoscopes, etc., make up a total laser delivery system. Before practical delivery systems were developed, physicians and scientists moved patients under the laser to direct the energy beam to the proper site.

Today, there are three primary schemes to bring laser energy to tissue: The first developed, and still in wide use, is an articulated arm; the second is direct optomechanical coupling of the laser and end-delivery device, and the third is via fiber optics.

A. Articulated Arms

An *articulated arm* is a series of hollow tubes and mirrors interconnected in such a way as to maintain alignment of the laser beam along the path of the arm (Fig. 4). This scheme has many features for use as a surgical device. Ideally, the energy traversing the articulated arm is simply reflected down the conduit without modifying the optical character of the beam. The spacial and temporal coherence of the beam is preserved. This means the energy exiting the arm is optically unaltered. The result is preservation of the ability to focus the beam in exactly the same manner as the incoming beam can be focused. This permits the beam to be focused to its minimal spot diameter.

The articulated arm development significantly expanded surgical usage of lasers. In the real world of surgical lasers with articulating arms, some difficulties arise out of their implementation. The most obvious is that to achieve a degree of flexibility, the rigid tubes must have several mirrors. Bumping or jarring the arm

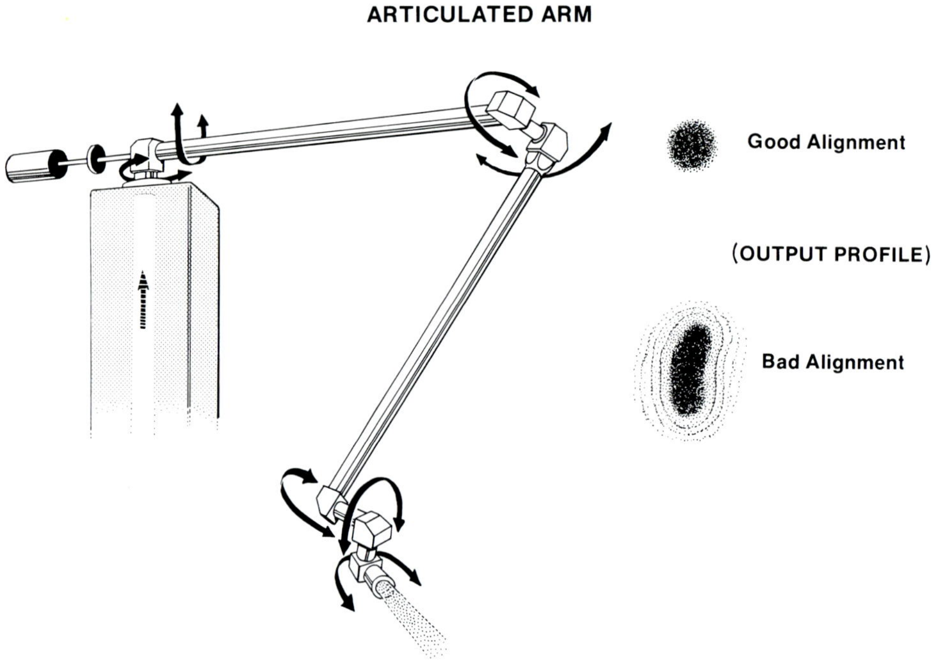

Figure 4 Configuration of an articulated arm. Note beam output profile. A good arm alignment results in a beam profile that is equal to the incoming beam profile (typically gaussian). In contrast, a poor alignment distorts and widens the beam or, in worst case, blocks the beam completely.

can cause misalignment leading to two common problems. The most common is diffraction or a widening of the beam diameter as it traverses the arm. With severe bumps, the beam can be reflected in the walls of the articulated tubes, substantially reducing or eliminating transmission of power down its path. A second problem presents itself in lasers on which two laser beams are used concurrently. Such a case is present in carbon dioxide laser systems. A harmless red helium-neon laser beam is used to *image* the invisible CO_2 laser beam (Fig. 5). When the laser is jarred, it is possible for the two lasers to lose their coaxial orientation. Thus one beam may pass properly down the articulated arms and the second may be askew. When this occurs, the surgical laser beam and the aiming beam are not together, leading to improper aiming of the beam. Although this is not a problem created by the articulated arm, it is a problem that is propagated by it. Both difficulties are managed by proper care and preventive maintenance of the arm. An articulated arm can cost as much as 6000 dollars.

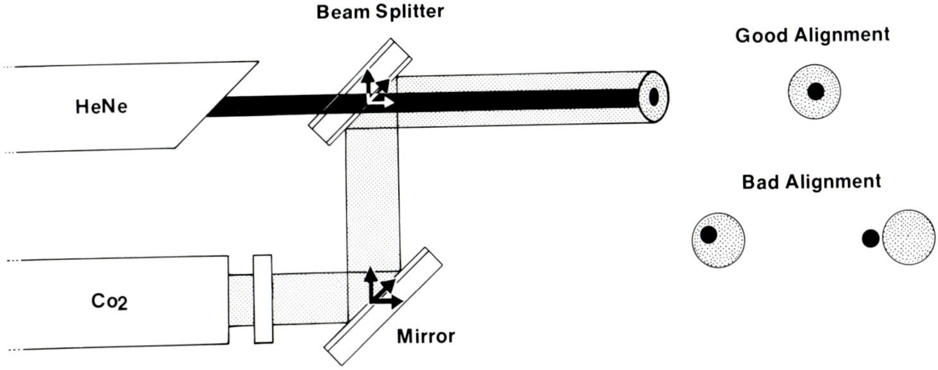

Figure 5 Alignment of an invisible (UV or IR) laser beam with a red (HeNe laser) beam. Note alignment conditions.

Perhaps the greatest limitation of an articulated arm is its flexibility. Although the typical mirror devices are well suited for coupling to other accessories and for work outside the body, work that requires flexible endoscopic delivery or sensitive tactile control are severely compromised. Tactile control is somewhat compromised by the mass or weight of the arm. The weight of the arm is usually counterbalanced upon installation. The total mass of the arm, however, commonly leads to difficulties in maneuvering. Once the arm begins to move, it continues under its own weight. Great strides have been made in improving articulating arms by developing smaller and lighter arms, but this general limitation remains.

B. Optomechanical Coupling

Several laser schemes directly couple the laser to the tissue by way of a focusing lens and beam manipulating mirror. In one sense such a coupling can be consider ed as a very short articulated arm. Direct-coupled surgical lasers are most frequently used in conjunction with operating microscopes. The instrument is designed so that the laser beam is coaxial with the optical axis of the microscope, thereby permitting visualization, illumination, and laser beam alignment. Frequently, there are two optical elements in the beam path from the laser head to the tissue. One is a reflective optic that is manipulated by a Gimball-type joystick mount. This permits the surgeon to easily move the beam around the tissue while under direct vision. Micromanipulators can be mechanically or electromechanically driven. Mechanical micromanipulators are most common for several reasons. They effectively perform the task required of the device, i.e., movement of the laser beam. In addition, they are easily manufactured and thus less expensive than the electromechanical alternative; mechanical micromanipulators physically take up less room both laterally and distally than do electromechanical devices. Depending upon the surgical application, the micromanipulator may be so arranged that when the physician removes his hand from the joystick the laser beam "jumps" to the center of the microscope's field. In other instances, the adjusting scheme permits the laser to be focused to one area of the field and remain in that area after the hand is removed from the joystick. The choice of which scheme (return-to-center or manual) is used is strictly governed by user preference.

In an electromechanical unit, as the user manipulates a micromanipulator (a device similar to the joystick of a home computer) small motors mounted orthogonally to the reflective optic surface are moved to a position proportional to the motion of the manipulator. The advantage of such a scheme is that the electronics present between the joystick and the mirrors can permit rapid and smooth changes in aspect ratio. In other words, for applications requiring high microscopic magnification, a small movement of the micromanipulator can result in a movement of the laser beam that would rapidly be out of the field of vision. In an electromechanically coupled device this ratio of movement can be easily and continuously changed. As the magnification of the microscope is changed, the appropriate adjustment for the field size and movement rate of the beam can be made. Another feature of the electromechanically coupled device is the ability to change the response time of the joystick to the mirror. In mechanical systems the relationship is 1:1, i.e., the motion of the joystick directly results in the motion of the laser beam. The response time of the electromechanically driven device is a variable. Such variance, when not used to change the aspect ratio, is a hindrance to its use but one that can be overcome with experience.

Perhaps the most valuable feature of electromechanical micromanipulators is the ability to have the micromanipulator direct the laser beam in a predefined geometry. For example, a dermatologist may wish to outline a lesion with the micromanipulator without the beam hitting the tissue. After outlining, the micromanipulator can be programmed to fill in the area with laser energy, thus providing an accurate, uniform, and predefined treatment area (providing the patient does not move).

C. Fiber Optics

Fiber optics are thin flexible devices that carry optical energy along their path (Fig. 6). In many surgical lasers, fiber optics have replaced articulated arms. This is due to their relative inexpensiveness and ease of use. Capable of carrying energy in very small diameters (typically less than 1 mm), fiber optics carry laser energy to virtually any portion of the body either by endoscopic channels or hand-held manipulation. Unlike articulated arms whose ability to carry any laser wavelength is limited only by selecting appropriate reflecting optics, fiber optics are more limited in wavelength transmissivity. The most common fibers

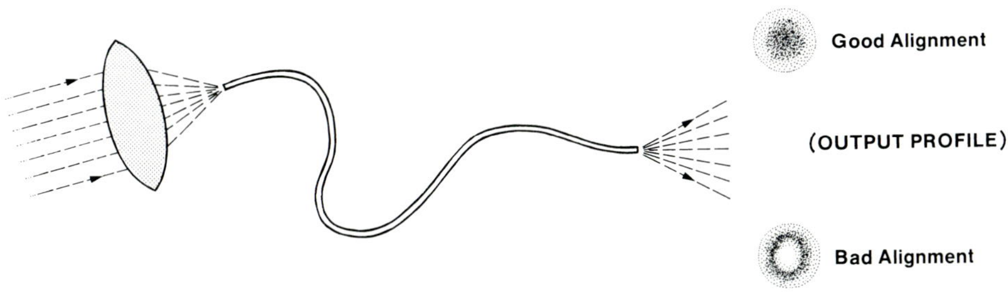

Figure 6 The geometry of a fiber optic permits flexible delivery of laser energy to tissue. Note the difference of the laser energy upon exit from the fiber.

can carry modest power from any CW laser emitting energy in the visible and near-infrared portions of the spectrum. Thus the argon-krypton, helium-neon, and Nd:YAG lasers, as well as the emission from most dye lasers, can be transmitted by inexpensive, readily available fiber optics. The wavelengths of some eximer lasers (351 nm, XeF) can also be carried by fiber optics. However shorter UV wavelengths and the emission from the carbon dioxide laser are not transmittable via traditional glass fibers. New experimental devices are emerging for the midinfrared (CO_2 laser) wavelengths. Some clinical work has already been performed using such fibers (15).

In addition to the increased flexibility and small diameter of fiber optics compared with articulated arms, once energy is *captured* in the fiber, it is transmitted down the device. Thus when multiple wavelengths are used, e.g., red HeNe and invisible Nd:YAG, the concentricity of both beams is assured.

Perhaps the greatest shortcoming of the current fiber optics used in medicine is the loss of coherence of the beam as it traverses the fiber. This results in a divergent beam of light that can, practically, only be focused to the diameter of the fiber itself. Refocusing the output of the fiber is common on hand-held fiber applications but rare on endoscopic fiber optics.

REFERENCES

1. Patel CKN, McFarlane RA, and Faust WL: Selective excitation through vibrational energy transfer and optical maser action in N_2-CO_2. *Physical Rev.* *13*:617-619, 1964.
2. Bell WE: Visible laser transitions in Hg^+. *Appl. Physics Lett.* *4*:34-35, 1964.
3. Bridges WB: Laser oscillation in singly ionized argon in the visible spectrum. *Appl. Physics Lett.* *4*:128-130, 1974. Err. *5*:39, 1964.

4. Bridges WB and Chester AM: Visible and UV laser oscillation at 118 wavelength in ionized Ne, Ar, Kr, Xe, O and other gases. *Appl. Optics 4*:573-580, 1965.
5. Coherent Radiation, Incorporated.
6. Snitzer E: Optical maser action of Nd^{3+} in a barium crown glass. *Physics Rev. Lett. 7*:444-446, 1961.
7. E. Snitzer and C. G. Young. Glass lasers, in *Lasers* (A. K. Levine, ed.), Vol. 2, Marcel Dekker, New York, 1968.
8. Johnson LF: Optical maser characteristics of rare-earth ions in crystals. *J. Appl. Physics 34*:897-909, 1963.
9. Lambda Physik, Incorporated; Lumonics, Incorporated.
10. Sorokin PP, Lankard JR, Hammond EC, and Moruzzi VL: Laser-pumped stimulated emission from organic dyes. Experimental studies and analytical comparisons. *IBM J. Res. Dev. 11*:130-148, 1967.
11. Sorokin PP and Lankard JR: Stimulated emission from an organic dye, chloro-aluminum phthalocyanine. *IBM J. Res. Dev. 10*:162-163, 1966.
12. Maiman TH: Stimulated optical radiation in ruby. *Nature 187*:493-494, 1960.
13. R. W. Hellwarth. Control of fluorescent pulsations, in *Advances in Avartun Electronics* (J. R. Singer, ed.), Columbia University Press, New York, 1961, pp. 334-341.
14. Hargrowe LE, Furk RL, and Pollack MA: Locking of HeNe laser modes induced by synchronous intra-cavity modulation. *Appl. Physics Lett. 5*:4-5, 1964.

3

Laser Surgery of the Oral Cavity and Oropharynx

R. KIM DAVIS
University of Utah School of Medicine, Salt Lake City, Utah

I. INTRODUCTION

This chapter describes recent advances in the treatment of both benign and malignant lesions of the oral cavity and oropharynx. The treatment of benign lesions will be separated from the treatment of malignant lesions because the basic principles involved are different. In addition, attention will be given to the potentially best use in this area of the CO_2 laser, the low-power argon laser, and the neodymium:YAG laser. The greatest emphasis will be placed on use of the CO_2 laser in malignant lesions because the greatest amount of research and current publication deals with this particular laser. Because the treatment of benign vascular lesions of this area is now becoming better understood, attention also will be focused on this particular aspect.

II. TREATMENT OF MALIGNANT LESIONS

Carcinoma of the oral cavity and oropharynx represents a broad spectrum of disease extending from solitary carcinoma in situ, usually associated with areas of leukoplakia or erythroplakia, to large tumors infiltrating the deep musculature of the tongue, the palatine arch, the mandible, or the neck skin. The most commonly accepted approach to oropharyngeal squamous cell carcinoma in the last several decades has been that early cancer (stages I and II) is curable in 70-90% of the patients by either surgery or irradiation therapy. The choice of which form of therapy to use is determined by the age of the patient, the location of the tumor, the functional outcome anticipated, the complications from either procedure, the cost-effectiveness, and the clinicians' biases.

Another factor strongly influencing the choice of treatment has been the concept of field cancerization as originally proposed by Slaughter (1). Tobacco smoking and alcohol consumption are considered important causative factors inducing widespread dysplasia and carcinogenesis. The incidence of second primary cancers arising in this field of dysplasia typically is reported to range from 14 to 20%.

Recommended therapy for patients with clinically dysplastic epithelium who have small intraoral cavity lesions has typically been surgery. One of the best current uses of the CO_2 laser has been in the excisional biopsy of early carcinoma in this area (2,3). The laser has allowed a precise excision of the lesion and the involved mucosa that provides an excellent specimen for histological verification of surgical margins. The complications from laser excision have been minimal, so that tracheostomy is not needed. The area seldom needs to be reconstructed, and patients can almost always be discharged the day after surgery.

Stage III and stage IV cancers have usually been treated by a combination of surgery and either preoperative or postoperative radiation therapy. The rationale for combining radiation with surgery in patients with larger primary cancers or metastatic disease of the neck is based on the fact that disease-free margins are often difficult to obtain. Even radical surgical procedures can fail at the margins, leaving nonresectable diffuse microscopic disease. Conversely, the failure to eradicate all malignant cells despite high dosages of radiation therapy is thought to be due to the large number of hypoxic cells in a fungating tumor. The concept that surgery can remove all but microscopic disease has been coupled with the concept that irradiation will successfully eradicate diffuse microscopic disease.

A critical analysis of the therapy results in oral cavity disease is very distressing. Table 1, which represents a large review of European patients, shows a 3-year no evidence of disease (NED) rate of only 68% in stage I patients and an NED rate of 17% in stage IV patients (4). Table 2 shows the 3-year NED status of a large group of treated American patients as recently reported (5-8). These studies suggest that when surgery or irradiation is used as a single entity, unsatisfactory cure rates result. In general, irradiation therapy has failed at the primary site, while surgery has failed both at the primary site secondary to inadequate margins and in the neck where in early disease usually no treatment has been given.

An interesting finding, reported by Platz et al. (4), was that patients with $T_1N_0M_0$ oral cavity cancer who showed any evidence of infiltration of adjacent structures had a 3-year NED rate of 48% compared with a 3-year NED rate of 66% in patients without infiltration (4). Yamamoto et al. (9) used Jacobsson's classification of the mode of invasion, as outlined in Table 3, to evaluate the success of therapy in their patients with oral cavity cancer. In

Table 1 Treated Oral Cavity Cancer: 3-Year NED Status, European

Stage	No. patients	%
I	161	68
II	241	52
III	231	39
IV	169	17

Table 2 Treated Oral Cavity Cancer: 3-Year NED Status, American

Stage	Surgery (%)	XRT[a] (%)	Both (%)
I	48	52	82
II	33	52	74
III and IV	31	15	46

[a]XRT = radiation therapy.

Table 3 Mode of Invasion by Jacobsson's Criteria

Grade	Definition
1	Well-defined borderline
2	Cords, less-marked borderline
3	Groups of cells, no distinct borderline
4	Diffuse invasion
	cordlike invasion
	diffuse-type invasion

this report the mode of invasion of oral cavity cancer was investigated to determine the clinical course and sensitivity to bleomycin chemotherapy. Bleomycin sensitivity correlated with the mode of invasion, with the best sensitivity found in the less invasive forms of cancer. Even more important, however, primary and secondary metastases were found in nine of 12 $T_1 N_0 M_0$ cancers that had grade 4 invasion (diffuse invasion) versus only 4 of 28 (14%) in all

Laser excisional biopsy (clear margins)

- No further therapy (grades 1, 2, 3)
- Radiation therapy (grade 4—No neck involvement)
- Neck dissection and irradiation (any grade with $N_{1\text{-}3}$ neck involvement)

Figure 1

other grades. The overall survival at 3 years follow-up was 93% NED in patients with grade 1-3 infiltration versus only 50% disease-free survival in patients with grade 4 infiltration.

The problem encountered with traditional surgery for $T_1 N_0 M_0$ oral cavity cancer is well emphasized in the study by Marks et al. (8). Initial surgery controlled floor-of-mouth lesions in only three of seven patients. With secondary surgical procedures all seven were controlled. Three of the four initial failures, however, needed ultimate treatment by segmental mandibulectomy, a procedure associated with substantially higher morbidity than is a simple transoral excision. One reason cited in this report for the failure to gain local control was that margins were difficult to visualize secondary to excess blood in the field. Use of an electrocautery will decrease this bleeding problem at the local site but it has led to substantial patient pain, edema, and to potential marginal artifacts on review of the histological specimen.

Use of the CO_2 laser coupled with careful review of the mode of invasion as described by Jacobsson has circumvented many of the problems previously mentioned. Figure 1 shows an approach to stage 1 oral cavity cancer utilizing a precise laser procedure. After initial excellent exposure of the lesion, the CO_2 laser is used to circumscribe the lesion with wide margins. This initial circumferential incision is carried down to the level of the mylohyoid muscle. This surgery is performed with the CO_2 laser coupled to the Zeiss operating microscope to provide microscopic precision during tumor removal. Sutures are then placed on the edges of the lesion, and the lesion is removed using the CO_2 laser with countertraction applied by the sutures. The specimen is then analyzed pathologically. If the mode of invasion is either Jacobsson's grades 1, 2, or 3, no further therapy is given. If the mode of invasion is Jacobsson's grade 4, then full-course radiation therapy is suggested to the patient. If the patient initially presents with any nodal disease, the initial excisional surgery is coupled with neck dissection and full-course postoperative radiation. Using this approach, our NED rate at a median follow-up period of 22.8 months is 94% (16 of 17 patients).

Table 4 shows the complications that developed in patients treated by laser in

Table 4 Laser Related Complications, 17 Patients

Complication	No. patients	%
Sialadenitis	2/17	12
Submandibular gland excision	1/17	6
Need for oral cavity reconstruction	3/17	18

this manner. During the initial excision, no attempt is made to preserve Wharton's duct (the submandibular duct) on either side. The duct is simply transected if it is within the field of cancer excision. Only two of 17 patients (12%) have developed sialadenitis requiring therapy. In this same group, only one patient has required a subsequent excision of the submandibular gland. In this case, the patient presented 6 months after his initial excision with induration and firmness in the left submandibular gland. This gland was removed to rule out the presence of metastatic disease, and, in fact, the biopsy showed only sialadenitis. Fourteen of seventeen patients required no reconstruction of the surgical defect and were discharged from the hospital no later than the third postoperative day. The three patients who had reconstruction had undergone en bloc excisions of the floor of mouth with inner table mandibulectomies secondary to proximity of the cancer to the mandibular inner table periosteum.

Excisional biopsy of T_1 oral cavity lesions with the CO_2 laser has several definite advantages: the specimen obtained can be readily analyzed pathologically, as stated earlier, and no difficulty with marginal artifacts has been seen during pathological review. Use of the operating microscope allows precise intraoperative examination of surgical margins. When gross tumor approaches the tumor margin, the margins can be enlarged to more safely encompass the tumor. Because of the hemostatic effect of the laser, bleeding is no longer the problem it was with other methods. Morbidity for patients with use of this approach compared with use of the hot knife is decreased because the frequency of sialadenitis and subsequent excision of submandibular glands is minimal. Postoperative pain is definitely less than in patients treated with electrocautery. Reconstruction of the surgical defect after laser excision is seldom necessary, and indeed, no patient showed clinical evidence that would suggest any lingual disability secondary to subsequent scar tissue formation in our series. The CO_2 laser is believed to offer a definite advantage to patients with T_1 lesions of the anterior floor of mouth or tongue.

The basic principles stated for oral cavity cancer also apply to the palatine arch and retromolar trigone. Lesions that are T_1 in size and are readily visualized can be safely excised for biopsy with the CO_2 laser. This applies to primary tumors as well as very limited postirradiation recurrences. Laser excisional biopsy of these lesions is an important step in definitive staging of the tumors. When any palatine arch or retromolar trigone lesion extends to involve the base of tongue, excision with the laser is inappropriate. Safe visualization in the base of tongue area cannot be obtained to allow laser excision.

The studies earlier quoted definitely show that patients who present with stage II and stage III disease have better survival and local regional control when treated by surgery and radiation therapy. Patients with T_2 oral cavity disease can benefit by surgical excisions using the laser initially to outline the lines of excision about the lesions. These patients generally have generally been treated by pull-through operations with an in continuity neck dissection. The initial outlining of the lesion with adequate margins under microscopic control has been helpful in defining exactly where the lesion is "pulled through" the oral cavity. In other words, careful in continuity margins are better maintained when the lesion is initially outlined in the oral cavity and then dropped out of the mouth in continuity with the neck dissection specimen along the lines of the earlier laser circumferential outline.

Stage IV oral cavity cancer has been managed poorly by all current therapeutic approaches. No survival advantage has been documented when surgery has been coupled with irradiation therapy in this advanced disease. Several regimens of chemotherapy have been used in treatment of stage IV disease, and these regimens have shown no current survival advantage. Adjuvant chemotherapy has been of definite benefit in other tumor systems and has been tried in advanced head and neck cancer. The Head and Neck Contracts Program sponsored by the National Cancer Institute represents the only large prospective randomized study that has been evaluated (10). One branch of this study consisted of induction chemotherapy with bleomycin and cisplatin followed by surgery and postoperative irradiation. Following this, maintenance chemotherapy was offered as an adjuvant toward improvement of survival. Notably, only 13 of 156 patients in this branch of the study were able to complete maintenance chemotherapy. The major problem preventing completion of therapy was the lack of patient compliance with the regimen. Figure 2 shows a potential use of laser cytoreduction in stage IV inoperable patients. It has been noted anecdotally that when the CO_2 laser has been used to debulk cancer, patients have been able to proceed immediately to radiation therapy or chemotherapy. The complications of CO_2 laser cytoreduction for the patient with advanced disease has not prevented the immediate application of either chemotherapy or radiation. This is in definite contrast to the conventional approach in which patients undergo extensive nonlaser operations.

The concept of treating patients with advanced inoperable disease with cytoreduction coupled with other modalities definitely merits further consideration.

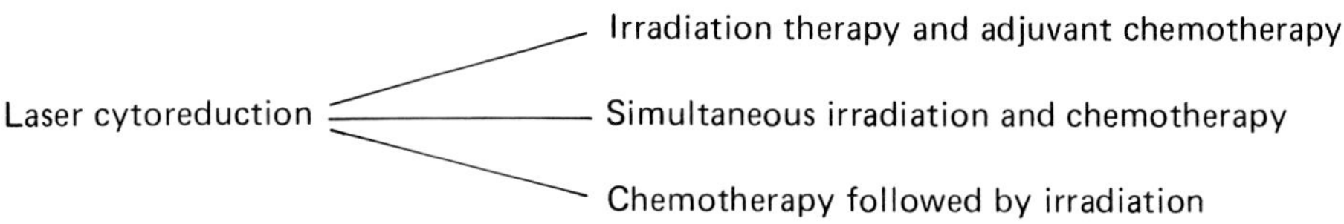

Figure 2 A potential use of laser cytoreduction in stage IV inoperable patients.

The prognosis of these patients is already dismal and the application of laser therapy will not interfere with treatment by current standard therapy. Laser surgery in this group offers definite palliation.

One difficulty with attempting early cytoreduction of large tumors by using the CO_2 laser alone is the frequent lack of significant hemostasis in treating these lesions. Coupling of the neodymium:YAG laser to coagulate the tumor initially would allow subsequent cytoreduction with the CO_2 laser to proceed more expeditiously. This concept needs thorough evaluation in a large prospective institutional study.

III. TREATMENT OF BENIGN LESIONS

The first application of laser surgery to the oral cavity was done with premalignant lesions. Excisional biopsies of areas of dysplasia and erythroplasia were accomplished using the laser to the great advantage of patients. Such treatment was basically free of pain, caused almost no bleeding, and did not require reconstruction. The biopsy specimens obtained could be readily analyzed. Application of the CO_2 laser was also extended to other benign oral cavity soft-tissue lesions. It was discovered early that the CO_2 laser conferred no advantage in treating benign bony lesions of the oral cavity. In fact, attempted laser excision of bony lesions led to unnecessary thermal damage of adjacent tissue, and this practice, fortunately, never gained favor. Some attempts were made to extend laser excision to common benign oropharyngeal problems such as tonsillectomy or adenoidectomy. Here again, laser excision was more tedious and certainly less cost-effective than traditional methods. Although there may be a decreased loss of blood at the time of palatine tonsillectomy, there is no change in the postoperative potential for bleeding and clearly no advantage to the patient.

One area in which the CO_2 laser offers a great advantage in the oropharynx is its application to lingual tonsillectomy. When lingual tonsillectomy is performed by either the hot or cold knife technique, patients have a notable morbidity and somewhat prolonged recuperative period. It was found that the CO_2

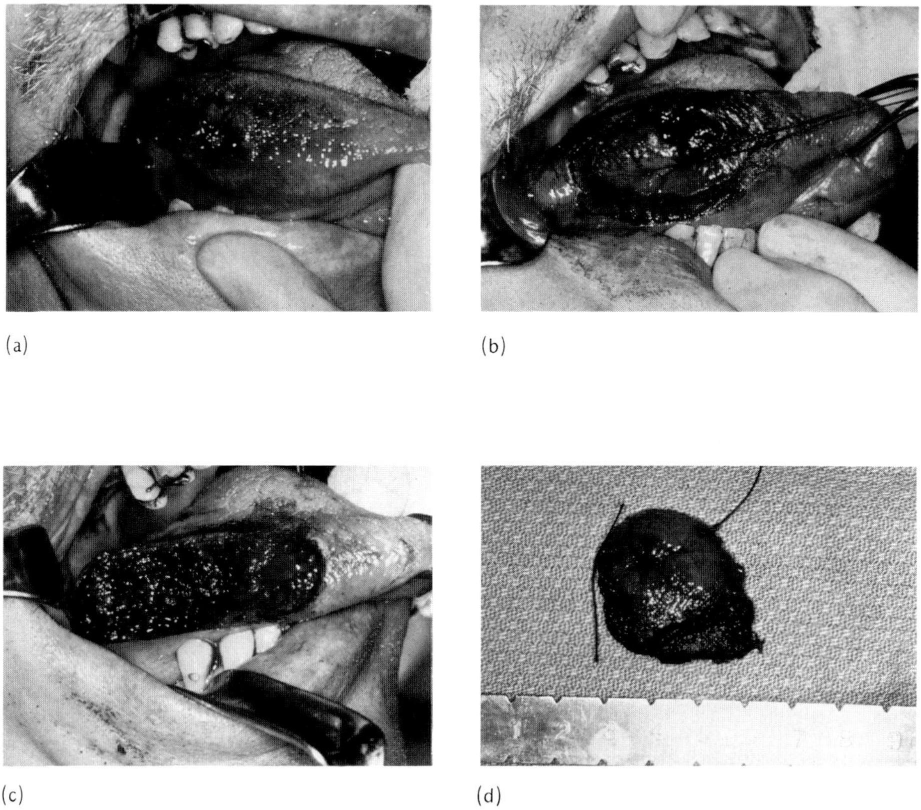

(a) (b) (c) (d)

Figure 3 (a) A 2-cm carcinoma of the lateral tongue. (b) With adequate margins the tongue cancer has been outlined with the laser and tension applied with silk sutures. (c) Operative site after excisional biopsy of tumor. This is allowed to granulate. Closure is not necessary. (d) Excisional biopsy specimen removed from tongue. It must be stressed that use of the laser in oral cavity cancer must be excisional biopsy.

laser could be used to perform a lingual tonsillectomy with a great decrease in adverse effects. This procedure can be done by either directly visualizing the lingual tonsil using the large-bore laryngoscope and excising the tonsil in this manner or by using a steel or nickel mirror to reflect the laser beam, thereby ablating the lingual tonsil. In performing any of these procedures a portion of the tonsil must be saved and sent as an incisional biopsy specimen. The rest of the lesion can safely be vaporized. When surgery is performed in this manner, the patient must be prepared for the use of electrocautery if any vessel larger than that easily controlled by the laser is encountered. These vessels can then be gently grasped with a forceps and electrocoagulated.

Another area of effective use of a CO_2 laser is in the treatment of superficial capillary hemangiomas or lymphangiomas of the tongue that cannot reasonably be completely removed. In both of these examples the CO_2 laser provides adequate hemostasis and allows excellent debulking of these lesions without inducing complications for the patient.

Hemostasis cannot be adequately assured using the CO_2 laser for other vascular neoplasms.

An additional area in which the CO_2 laser has recently been used is in performing a uveopalatopharyngoplasty. This operation can certainly be accomplished using conventional techniques, but it often leaves the patient with notable soreness and several days of poor swallowing related to pain. When the CO_2 laser is used to do the palatal portion of this procedure, the degree of pain is definitely decreased. In patients with residual tonsils, most are best treated by performing a conventional tonsillectomy. If the tonsils are small and the point of dissection is very clear, the CO_2 laser can also be used to excise the tonsil or redundant soft tissue, i.e., mucosa and submucosa of the lateral pharynx. This indication for the laser currently is being further evaluated.

With the advent of the argon and Nd:YAG lasers, the ability to treat vascular neoplasms of the oral cavity and oropharynx has been greatly enhanced. The argon laser is particularly helpful in the photocoagulation of superficial hemangiomas. When these lesions are symptomatic because of recurrent bleeding or difficulty with feeding or speech related to the mass effect, the argon laser can be used to selectively decrease the tumor size and prevent bleeding. This application is somewhat limited in the massive hemangiomas, which tend to have a larger cavernous vascular component. More recently the Nd:YAG laser has been used to treat these lesions. Notable palliation of bleeding and the mass effect from these lesions has been seen by using the Nd:YAG laser at approximately 30-40 W power for a duration of 0.3-0.5 sec. Areas of laser impact are carefully spaced 3-4 mm apart. Closer placement of laser impacts will lead to substantial tissue necrosis because the depth and spread of energy are much higher than with the argon laser. If areas are treated in this manner, notable palliation of major cavernous hemangiomas and even limited AV malformations has been seen. Currently, work in this area has been somewhat limited, but the potential benefit of this application remains very high.

IV. SUMMARY

Careful application of the CO_2, low-power argon, and Nd:YAG lasers has allowed major advances in the treatment of benign and malignant lesions of the oral cavity and oropharynx. It is important to stress that the basic principles of treatment are notably different between benign and malignant lesions. When incisional biopsy confirms the benign nature of a lesion, it can then be either vaporized or photocoagulated. The choice of CO_2, argon, or Nd:YAG laser will depend upon the degree of vascularity. In general, the CO_2 laser treats nonvascular lesions very effectively. The argon laser has its best use in superficial small vascular lesions. The Nd:YAG laser is currently being applied to vascular lesions with cavernous components, with initial good success.

When malignant lesions are treated, there is almost no role for a simple vaporization or photocoagulation. These lesions are best treated by excisional biopsy with adequate margins using the CO_2 laser as a staging procedure. In very rare instances, when no other treatment is possible, the combined application of lasers may allow some palliation.

REFERENCES

1. Slaughter DP: Multicentric origin of intraoral carcinoma. *Surgery 20*:133-146, 1946.
2. Strong MS, Vaughan CW, Healy GB, Shapshay SM, and Jako GJ: Transoral management of localized carcinoma of the oral cavity using the CO_2 laser. *Laryngoscope 89*:897-905, 1979.
3. Strong MS, Vaughan CW, Jako GJ, and Polanyi T: Transoral resection of cancer of the oral cavity: The role of the CO_2 laser. *Otolaryngol. Clin. N. Am. 12*:207-218, 1979.
4. Platz H, Fries F, Hudec M, Tjoa AM, and Wagner RR: Carcinoma of the oral cavity: Analysis of various pretherapeutic classifications. *Head Neck Surg. 5*:93-107, 1982.
5. Newman AN, Rice OH, Ossoff RH, and Sisson GA: Carcinoma of the tongue in persons younger than 30 years of age. *Arch. Otolaryngol. 109*:302-304, 1983.
6. Leipzig B, Cummings CW, Chung CT, Johnson JR, and Sagerman RH: Carcinomas of the anterior tongue. *Ann. Otol. Rhinol. Laryngol. 91*:94-97, 1982.

7. Holm LE, Lundquist PG, Rudén BI, Silfverswärd C, Sobin A, and Wersäll J: Combined preoperative radiotherapy and surgery in the treatment of carcinoma of the anterior two-thirds of the tongue. *Laryngoscope 93*:792-796, 1983.
8. Marks JE, Lee F, Smith PG, and Ogura JH: Floor of mouth cancer: Patient selection and treatment results. *Laryngoscope 93*:475-480, 1983.
9. Yamamoto E, Kohama G, Sunakawa H, Iwai M, and Hiratsuka H: Mode of invasion, bleomycin sensitivity, and clinical course in squamous cell carcinoma of the oral cavity. *Cancer 51*:2175-2180, 1983.
10. Baker SR, Makuch RW, and Wolf GT: Preoperative cisplatin and bleomycin therapy in head and neck squamous carcinoma: The prognostic factors for tumor response. *Arch. Otolaryngol. 107*:683-689, 1981.

4

Laser Treatment of the Nasal Cavities and Nasopharynx

JAMES L. PARKIN
University of Utah School of Medicine, Salt Lake City, Utah

I. INTRODUCTION

With the development of new therapeutic instruments and modalities, significant temptation exists to utilize expensive instruments where conventional methods perform well. For example, common nasal polyps can be removed with lasers or nasal polyp snare wires. Unless medical advantages exist, the snare wire is a more economical instrument. In the early experiences with the laser or any new instrument or drug, experimentation is necessary to establish usefulness and cost-effectiveness.

The applications of the laser to diseases of the nose and nasopharynx have led to the identification of disease states that are beneficially managed with this modality. The variety of available lasers enlarges the scope and results of laser treatment. The discussion in this chapter will be limited to the carbon dioxide (CO_2), argon, and neodymium:yttrium-aluminum-garnet (Nd:YAG) laser systems.

II. EXTERNAL NOSE

The CO_2 laser has proved to be very useful in the treatment of rhinophyma (Figs. 1-3) (1). Excellent results have been obtained using the CO_2 laser in the treatment of this problem. It is my preference to use the CO_2 laser coupled to the microscope. The benefit of microscopic control is the assessment of depth of tissue removal. The hypertrophied cutaneous epithelium is carefully planed with the CO_2 laser with contouring of the nasal surface. It is important not to excise the dermal appendages to their full depth. These dermal appendage

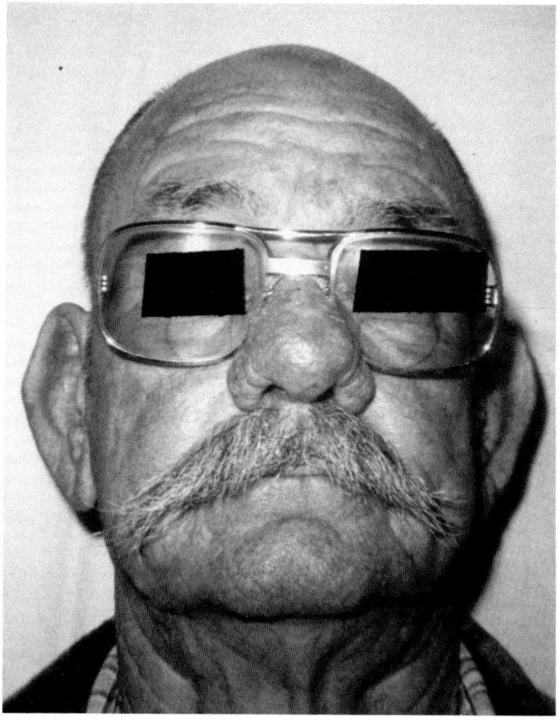

Figure 1 Pretreatment rhinophyma.

remnants serve as sources for new epithelium to resurface the external nose.

The procedure is performed under general anesthesia with dual suction set up to keep up with the removal of the smoke and vapors. The laser is used at a 8-10 W power setting on the continuous mode. The skin is prepared with iodine-based solution and then cleansed with alcohol. Systemic antibiotics are not routinely used.

A light nonadherent dressing can be applied for the first several postoperative days. Following dressing removal, the patient is instructed in gentle cleansing and application of topical antibiotic preparations to prevent secondary infection.

Vascular lesions of the external nose have been successfully treated with the argon laser and with the Nd:YAG laser (2). Small or solitary lesions are treated under local anesthesia. Large or multiple lesions are treated under general anesthesia. Examples of such lesions are telangiectasias, hemangiomas, and small arteriovenous malformations (Fig. 4). I prefer to utilize the argon laser in the

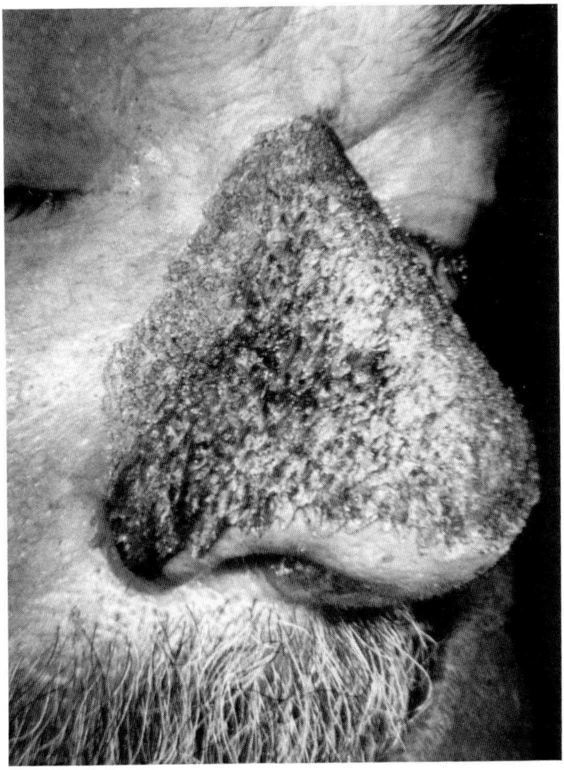

Figure 2 Immediate posttreatment appearance of rhinophyma.

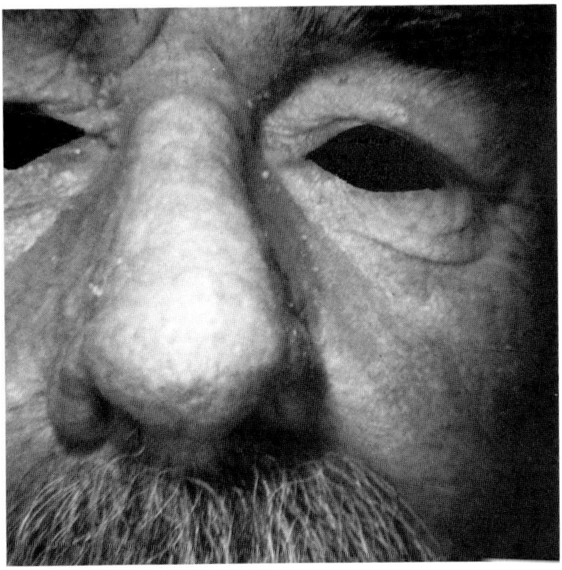

Figure 3 One-year posttreatment appearance of rhinophyma treated with CO_2 laser.

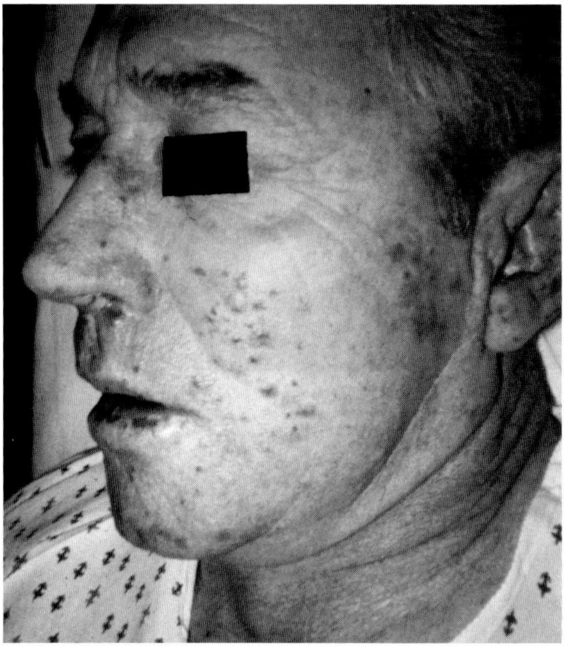

Figure 4 Facial telangiectasia in patient with Osler-Weber-Rendu syndrome.

treatment of these lesions. In my experience, the long-term results are better with the argon laser. When the argon laser is used for the treatment of cutaneous telangiectasias, lesion recurrence in the same location is rarely experienced. However when the Nd:YAG laser has been used, I have seen recurrent telangiectasias develop in the center of the lightened areas from previous treatments.

In the treatment of facial port-wine stains of the external nose and face, the area of the nasolabial fold and upper lip is particularly vulnerable to posttreatment scarring (3). Treatment of hemangiomas in these areas should not be aggressive. In the treatment of telangiectasias and hemangiomas, the argon laser energy is delivered through a fiberoptic bundle utilizing power settings in the range of 2 W. It is important to treat the areas only once at one sitting to avoid deep thermal injury and resultant untoward scarring.

III. INTERNAL NOSE

Epistaxis is a distressing symptom that brings many patients to visit their physician. Fortunately, epistaxis usually originates from minor anterior septal problems easily controlled with cautery, or packing, or both. Hereditary hemorrhagic telangiectasia (Osler-Weber-Rendu syndrome) is a cause of frequent severe epistaxis. The treatment modalities utilized in this disorder are legion. The mucosal telangiectasias can be successfully treated with either the argon or the Nd:YAG laser or with both (4,5). The CO_2 laser has also reportedly been utilized in the treatment of this disorder (6). However, the CO_2 laser is a less-effective photocoagulator of vascular lesions than the argon or Nd:YAG laser. Experience in the treatment of many patients with hereditary hemorrhagic telangiectasia and secondary epistaxis has indicated good control of nasal bleeding from the mucosal telangiectasias associated with this disease. There is cartilage sparing by the argon and the Nd:YAG lasers allowing treatment of both sides of the nasal septum simultaneously. Very large telangiectasias may not be well controlled with laser therapy and may still require a procedure such as a septodermoplasty for control.

Argon laser treatment of Osler-Weber-Rendu syndrome is accomplished under general anesthesia to allow more complete treatment of the multiple lesions and also to handle the bleeding encountered with less distress to the patient.

Self-retaining nasal specula and trimming of the vibrissae aid in the visualization of intranasal lesions. Small fiberoptic scopes also facilitate lesion identification. Two suctions are set up. One suction is connected to a red rubber catheter that is placed into the nasopharynx via the nares not being treated. The second suction is used with a nasal tip by the surgeon or assistant to remove smoke and/or blood from the side being treated.

The laser is used in the continuous mode with a power setting of 2-4 W. The fiber tip is held 3-5 mm from the lesion. If the center of the telangiectasia is treated first, bleeding often ensues. I treat each lesion by a circular motion beginning around the periphery and working toward the center. If a vigorous bleeder develops, I control it with electrocautery and then photocoagulate the area with the laser. If an electrocautery must be used, care is taken not to treat the same area on the opposite side of the septum.

Posttreatment nasal packing is rarely required. Treatment can be accomplished as an outpatient unless blood replacement is necessary or other associated medical conditions mandate hospital admission. Treatment results are indicated in Table 1.

Congenital choanal atresia has, in the past, required early oral airway maintenance while allowing the patient to grow sufficiently to enable operative correction of the problem. The CO_2 laser allows the early successful treatment of patients with congenital choanal atresia with excellent visualization and low morbidity (7). Even premature infants in the first several days of life have undergone successful correction of the problem with the CO_2 laser.

When the diagnosis has been confirmed, the procedure is performed under general anesthesia. Topical decongestants are placed in the nasal cavity. The operating microscope is utilized. An aural speculum is then placed in the external cho-

Table 1 Laser Treatment Experience in Hereditary Hemorrhagic Telangiectasia–25 Consecutive Patients

Age: 16-79 yr (mean: 54.5 yr)	
Sex: 9 males, 16 females	
Lasers used	
argon	21/25
Nd:YAG	2/25
none	2/25
Prelaser treatments	
cautery	25/25
septodermoplasty	9/25
arterial ligation	2/25
arterial embolization	2/25
Results: laser better than earlier treatment	22/23

ana. The area of atresia is identified. A saline-soaked cottonoid is placed in the nasopharynx to allow identification of that space and prevent damage to the posterior nasopharyngeal wall. Beginning inferomedially, a small opening is created into the nasopharynx. A CO_2 power setting of 6-8 W is sufficient for membranous atresia, but over 15 W is necessary to vaporize bony plates. Once the nasopharynx is identified, the atresial plate is completely removed. The posterior end of the nasal septum is also removed. The necessity for stenting such patients is controversial. If a stent is to be used, it is recommended that a soft, individually tailored Silastic stent be used (Fig. 5). The parents are instructed in postsurgical nasal hygiene. Atresia recurrence may require subsequent retreatment.

Neoplasms of the nasal cavity can be removed with the laser. A small angiosarcoma of the nasal septum was expeditiously and safely removed using the CO_2 laser, with few aftereffects. Theoretically, nasopharyngeal angiofibromas could lend themselves to removal utilizing the various laser forms. Nasal polyps could be excised with lasers; however, the polyp snare is an effective way to manage these lesions.

Turbinate hypertrophy has been treated on an outpatient basis with the laser. Decreased blood loss and precision of removal are advantages of lasers in the management of turbinate hypertrophy.

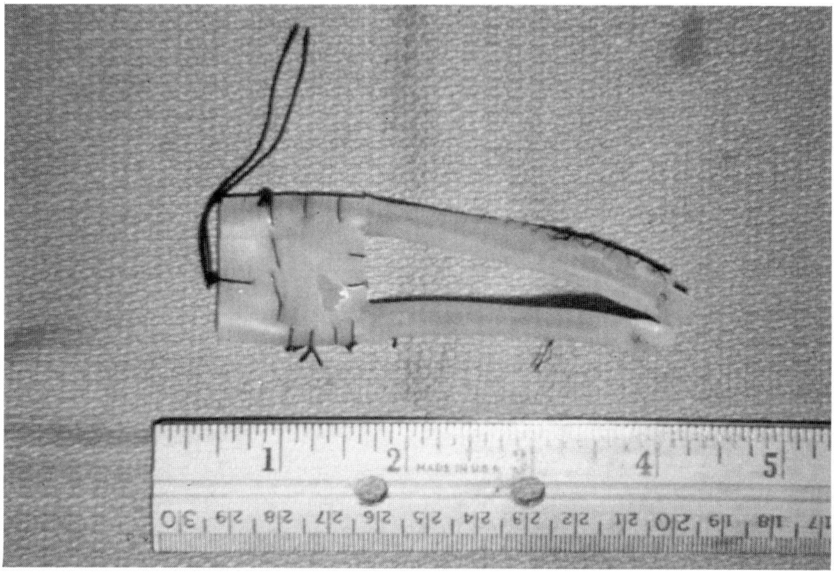

Figure 5 Tailored nasopharyngeal stent of reinforced silicone rubber sheeting for choanal atresia.

IV. NASOPHARYNX

The most common surgical procedure of the nasopharynx is adenoidectomy. Laser removal of the adenoidal tissue is technically possible, but offers no apparent advantage over the classic adenoidectomy techniques.

Nasopharyngeal stenosis is a complication rarely encountered following adenotonsillectomy (T&A) and other pharyngeal procedures. I have utilized the CO_2 laser for the correction of a post-T&A nasopharyngeal stenosis. The early result was very satisfactory; however, restenosis developed within 6 months following laser surgery.

Nasopharyngeal carcinoma is usually considered an inoperable lesion by classic cancer removal principles. Because of its location, the surgeon is not able to remove the tumor plus adequate surgical margins. The CO_2 laser is effective in debulking nasopharyngeal carcinomas before institution of irradiation therapy. The CO_2 laser has also been used to remove nasopharyngeal carcinoma. The underlying base of skull bone can be melted and sealed with laser energy.

V. SUMMARY

The introduction of laser therapy in the treatment of rhinological disease has been a major advance in certain disease entities. Utilization of the laser has been demonstrated to be cost-effective with improved results and few complications in the treatment of rhinophyma, vascular abnormalities of the external and internal nose, and congenital choanal atresia. The addition of laser technology has not resulted in abandonment of classic treatment techniques for adenoidectomy, nasal polypectomy, nasal septal reconstruction, etc. The judicious use of laser technology in the management of rhinological disease results in patient and surgeon satisfaction.

REFERENCES

1. Shapshay SM, Strong MS, Anastasi GW et al.: Removal of rhinophyma with the carbon dioxide laser. *Arch. Otolaryngol. 106*:257, 1980.
2. Parkin JL and Dixon JA: Argon laser treatment of head and neck vascular lesions. *Otolaryngol. Head Neck Surg. 93*:211, 1985.
3. Dixon JA, Huether S, and Rotering R: Hypertrophic scarring in argon laser treatment of port-wine stains. *Plast. Reconst. Surg. 73*:771, 1984.
4. Parkin JL and Dixon JA: Laser photocoagulation in hereditary hemorrhagic telangiectasia. *Otolaryngol. Head Neck Surg. 89*:204, 1981.
5. Shapshay SM and Oliver P: Treatment of hereditary hemorrhagic telangiectasia by Nd:YAG laser photocoagulation. *Laryngoscope 94*:1554, 1984.
6. Ben-Bassatt M, Kaplan I, and Levy R: Treatment of hereditary hemorrhagic telangectasia of the nasal mucosa with the carbon dioxide laser. *Br. J. Plast. Surg. 31*:157, 1978.
7. Healy GB, McGill T, and Jako GJ: Management of choanal atresia with the carbon dioxide laser. *Ann. Otol. Rhinol. Laryngol. 87*:658, 1978.

5

Laser Surgery in the Larynx

GEZA J. JAKO
Boston University School of Medicine, Boston, and Melrose Hospital, Melrose, Massachusetts

I. HISTORICAL BACKGROUND

In 1962 I developed the concept and instrumentation for soft-tissue microsurgery (1). This work was directed toward endoscopic surgery because of my clinical involvement in this area. Microsurgical techniques had been used in ear surgery since the early 1950s, but most of this effort involved bone microsurgery. In the early 1960s soft-tissue microsurgical techniques were also being developed for anastomosing small blood vessels and for procedures in the eye. Magnification to improve detection of pathological changes in vivo were first used in otology and then in gynecology. This latter application led to the development of clinical colposcopy. In 1960, Kleinsasser (2) used a monocular magnifying microscope with direct laryngoscopy to improve detection of laryngeal pathology. Following my work in soft-tissue microsurgery of the larynx, Kleinsasser became a major proponent of this technique in Europe—contributing new techniques and instrumentation.

In 1961, at the first International Congress of Biophysics in Stockholm, I witnessed Townes' first presentation on the application of "optical masers" (lasers) in biology. Since then, I have been interested in how the interaction of an electromagnetic beam with biomatter, such as the laser, could be used to improve precision in surgical removal, cutting, and coagulation of tissues. In 1965, I began my experiments with the neodymium:glass laser in the cadaver larynx, and by 1967 this work led to my application of the CO_2 laser to surgery and microsurgery (3). The subsequent clinical application of the CO_2 laser by myself and my colleagues at the Boston University School of Medicine, including Drs.

Strong (4,5), Vaughan (6), and later Healy and Shapshay (7), and Simpson, laid the groundwork for laser surgery in otolaryngology.

Currently, many basic aspects and clinical applications of CO_2 and neodymium: yttrium-aluminum-garnet (Nd:YAG) lasers for the detection and removal of benign pathological lesions have been established.

II. OVERVIEW OF LASER APPLICATIONS

In experienced hands CO_2 laser microsurgery provides precision that can be easily applied to endoscopic procedures. This capability made the development of the microsurgical treatment of cancer possible. The ability to view a dry surgical field under binocular magnification opened a new frontier in intraoperative visualization of fine details, such as in the differentiation between normal and cancerous tissues. This improved by order of magnitude the detection and removal of cancer cells. In localized laryngeal cancer, endoscopic procedures can make the detection and removal of most or all cancer cells possible. In the future, the effectiveness of endoscopic procedures can improve when combined with external microsurgery. Major portions of pathological laryngeal tissue can be removed as long as the valve action of the larynx is maintained. Laser surgery, initially used in the larynx, has led to the expansion of this modality into many surgical subspecialties. Now macroscopic and microscopic laser surgery is utilized for the treatment of both benign and malignant disease. With the aid of a new generation of laser surgical systems, major improvement will be made in the treatment of cancer. This will be the result of improved intraoperative detection and of precision in the cytoreduction of cancers.

III. ENDOSCOPIC AND LASER INSTRUMENTATION FOR LARYNGEAL MICROSURGERY

Indirect laryngoscopy, as it has been practiced since 1858, provides for practical observation of laryngeal pathology. It may also be used for some crude surgical procedures, such as biopsies, polyp removals, etc. In addition, it is useful for still and video photography under local anesthesia. Direct laryngoscopy, initiated in 1894 by Kirstein, provides a satisfactory approach to diagnosis and treatment. However, along with indirect laryngoscopy, it has been largely outmoded because of lack of precision (8). The rare exception is a situation in which the larynx is anatomically inaccessible because of the increased size of microlaryngeal endoscopes.

Fiberoptic laryngoscopy is helpful for diagnosing pathological and functional disorders of the larynx. It is mainly useful when the patient has a vigorous gag reflex or when the larynx cannot be viewed with a mirror for anatomical reasons.

None of these techniques provide the necessary access and resolution required in the endoscopic diagnosis, treatment, and postoperative follow-up of laryngeal cancer.

IV. MICROLARYNGOSCOPY

A. Endoscopes

The two main types of microlaryngoscopes are the Jako type, which provides superior distal illumination using fiberoptics, and the Kleinsasser type, in which the illumination is provided proximally by the light source of the operating microscope. There are different-sized microlaryngoscopes available, such as adult, adolescent, child, and infant sizes. Other modifications are useful for exposing the anterior commissure when certain anatomical limitations are present. An endoscope to view the subglottic area (subglottiscope) was developed by Healy. An endoscope with a built-in suction tube was developed for laser surgery by the author. I also designed a longer scope for laser surgery of the hypopharynx and esophagus. In connection with these endoscopes, various holders and suspension systems are used.

B. Microlaryngeal Instrumentation

Forceps, scissors, probes, knives, hooks, suction tubes, and suction coagulators are available in various sizes and shapes. Most of them were originally developed by the author.

C. Accessory Instrumentation for Laser Surgery

Various metal protectors and vocal cord retractors connected with a suction-tube handle were developed by Andrews (9) and modifications have been made by others.

V. INSTRUMENTATION FOR LARYNGEAL LASER SURGERY

After the first surgical laser was built by Polanyi and colleagues, a laser micromanipulator system was developed at my suggestion and with my cooperation. This was later published in 1972. As a result of these developments, clinical laser microsurgery became practical (10,11). Because of its precise cutting and hemostatic effects (for capillary-sized blood vessels), the CO_2 laser is the instrument of choice for almost all endoscopic treatment of laryngeal cancers. The CO_2 lasers come in different power ranges. A lower-power laser tube (up to 25 W) can be attached directly to the operating room microscope. The other parts of the equipment are connected by an umbilical cord to the laser tube (12-14).

A one-unit system that includes the microscope and utilizes a flowing-gas-type laser tube has recently been developed by the author.

Higher-power CO_2 laser surgical systems with maximum powers from 30 to 45 W and 60 to 80 W use an articulated delivery arm and can be connected to a micromanipulator on the microscope. For most endoscopic laryngeal surgery, 5-20 W is satisfactory. Up-to-date laser surgical equipment provides a single-mode laser beam with a variable spot size (0.8-2 mm).

VI. SAFETY AND PRECAUTIONS

The action of infrared lasers, such as the CO_2 laser, is purely thermal. Any combustible material, such as surgical drapes or flammable anesthetic gases, can be ignited. If a surgeon accidentally places his hand in the path of the laser beam, skin or fingernail will be burned. Because the hand will be pulled away immediately, no major damage will result. If the surgeon is wearing rubber gloves, the rubber will contain the heat. Therefore, immediate removal of the rubber glove will shorten the burning action. It is important that areas around the inserted laryngoscope, such as lips and face, be protected by soaked gauze or towel, which should be rewet periodically. The eyes should be covered separately by wet gauze. Operating room personnel should wear protective eyeglasses to avoid accidental laser injury to the eye from laser beams reflected off metal instruments. The surgeon's eyes do not need to be protected while he is using the microscope (glass optics will not transmit the CO_2 laser). While operating, the laser beam can be reflected from the inner surface of the laryngoscope or instruments within it, causing unnecessary burns on the patient. The beam, while evaporating tissues, creates steam and smoke inside the larynx, which is removed by suction. The removal of the steam and smoke is necessary to maintain visualization and to minimize unnecessary first- and second-degree burns of adjacent tissues. These burns do not become visible until several hours after the operation. Most patients' larynxes will not be seen under indirect laryngoscopy by the surgeon for about a week. Therefore, this additional damage, such as swelling or fibrin formation, is not always detected. These injuries can be avoided by using metal protectors with suction or by water-soaked microsurgical pads (surgical pattis; 0.5, 0.75, and 1-in. squares). The string that is attached to these pads should also be soaked. Otherwise, the string can easily be burned. Intact strings facilitate removal of the pads.

The most important and imperative consideration is the protection of the endotracheal tube. An endotracheal tube through which oxygen is flowing can be ignited if perforated by a laser beam. This can cause a blowtorchlike fire leading to permanent disabling injury or fatality. Plastic tubes must not be used because they ignite more easily and burn more briskly. Red rubber tubes wrapped with self-adhesive aluminum tape are used. It is wrapped starting from the cuff of the endotracheal tube until a length of about 12-15 cm of tube is covered. Each spiral of tape should overlap about one-fourth the width of the previous spiral, (this 0.25 in.-wide aluminum tape is available in the United States from 3M Company as an electrical tape). It is important that the tape is an aluminum tape and not an aluminized self-adhesive plastic tape such as Mylar. The inflated cuff of the endotracheal tube is protected by 1-in.^{2} (25 mm^2) water-soaked microsurgical pads. These pads should be periodically resoaked.

During the time that the laser is in use, some anesthesiologists prefer to use 25% oxygen in their air mixture. Details for anesthesia used in laryngeal laser microsurgery have been described in greater detail (15-19). During laryngeal microsurgery the anesthesiologists and surgeons share the same airway necessitating close cooperation.

VII. INTERACTION OF SOFT TISSUE AND LASER BEAMS

The 10.6 μm infrared radiation of the CO_2 laser beam is strongly absorbed and vaporizes the intracellular and extracellular water of tissues. Combustion of the organic elements is incomplete, yielding CO_2, water, and some carbonized particles. Higher-power lasers seem to create less carbonization than lower-power laser settings. Higher-power densities (using higher power and smaller spot sizes) are less effective for coagulating and providing hemostasis of larger blood vessels (0.2-0.5 mm diameter). However, lower power densities (lower power and larger spot size) provide relatively more hemostasis, especially when the blood vessels are compressed by an instrument or suction tip.

The CO_2 laser beam, in contrast with the Nd:YAG beam, is so well absorbed by water-containing soft tissues that the heat damage to the surrounding area is very minimal; thus, postoperative edema is greatly diminished. This is one reason why there is considerably less postoperative pain.

Pulsed or the so-called superpulsed CO_2 laser beams are a feature on most modern equipment. According to some authors, cutting with the pulsed laser beam causes less damage to the surrounding tissues. However, cutting with the pulsed or superpulsed beam creates acoustic shock effects in tissues and increases the likelihood of dislodging cells. For this reason, I prefer the use of continuous wave (CW mode of the laser) in cancer surgery.

VIII. TECHNIQUE

The operation of the microsurgical laser system should be tested by the surgeon or a technician before placing the patient under general anesthesia. This should include testing the delivery of the beam on a piece of wood, such as a wooden tongue blade. The patient is then put under general anesthesia, using the protected endotracheal tube. The largest-bore endoscope that can be inserted into the patient should be used. All parts of the larynx and surrounding areas should be checked with the laryngoscope, and the pathological area should be identified. Then, the laryngoscope is placed in the appropriate position and is held in place with the holder or suspension apparatus. The larynx is systematically inspected under the microscope, and color photographs are taken for documentation.

When an excisional biopsy is required, the tissue is grasped by an upcurved cup forceps and excised with the laser beam or with upcurved scissors, according to the surgeon's preference. When the laser beam is used, there is usually no bleeding. If scissors are used, the bleeding is first controlled with a small 1:1000 epinephrine (Adrenalin)-soaked pad. The bleeding is then controlled with the laser beam. Further tissue removal is achieved by laser evaporation; the smoke and vapor are removed by aspiration. Adjacent tissues are shielded by metal protectors or microsurgical pads. Beginners are advised to use the pulsed modes with repeated application of 0.2- to 0.1-sec pulses. More experienced surgeons use the CW mode. On occasions when less heat damage is preferred, e.g., cutting through scarred and stenotic areas, the pulsed mode is preferred.

In laryngeal surgery, 5-20 W of power is generally satisfactory. For small lesions or polyps, 5-10 W and a smaller spot size are preferred. For debulking of larger cancers and papillomas, 10-20 W of power with a larger spot size provides more effective hemostasis. In the event of brisk bleeding from a small laryngeal artery, the bleeding is stopped more easily by using epinephrine packing and then a suction coagulator.

If the lesion to be removed is in the posterior commissure of the larynx, the endotracheal tube is temporarily deflated. Then the laryngoscope is repositioned, using the lip of the laryngoscope to push the endotracheal tube into the anterior commissure, after which the cuff is reinflated.

In most cases, a biopsy specimen should be obtained. In cancer patients, the orientation of the specimen is important. When removing small vocal cord nodules, it is sometimes impractical to remove a biopsy specimen because the wound created by the removal could be considerably larger than the specimen itself. Under these circumstances, photographs should be taken for documentation and legal purposes.

Satisfactory exposure of most anatomical areas can be obtained by positioning the laryngoscope and externally moving the larynx. For areas with an obstructed view, it is possible to redirect the laser with a steel mirror. Such mirrors are available as part of the accessorial instrumentation for laryngeal microsurgery.

During laser surgery, various degrees of tissue carbonization are generated. Most of the carbon particles should be cleansed from the tissues with a soaking-wet pad. These particles could be embedded during healing. After healing, any remaining carbon particles can be visible under the epithelial surface. Histological sections have shown mild foreign-body-type reactions to these particles. However, these reactions have not caused clinical problems.

Precise surgery with the CO_2 laser requires a magnified view of the operative area, which also should be dry. If there is excess mucus or saliva on the tissues, it should be removed by suction before cutting begins. In the laryngeal tissues and vocal cords, the blood vessels are relatively small (capillary size). Hemostasis can be increased if the tissues are held under tension while cutting with the laser.

In a small larynx, such as in a child or infant, an endotracheal tube can obstruct the view and access to the operative side. Under these conditions, a Venturi type of injection anesthesia can be used by an experienced anesthesiologist (20,21).

In patients with laryngeal papilloma, the pathological lesion can extend into the subglottic area or trachea. In this situation, a subglottiscope or a laser bronchoscope can provide access (22).

IX. LARYNGEAL PHOTOGRAPHY AND VIDEO RECORDING

The photographic recording of the laryngeal disease before surgery is useful for clinical and research documentation and for legal purposes (23). Recently when biopsy specimens were not available, as in the removal of small laryngeal nodules, some insurance companies have begun to require photographs for clinical documentation. Pictures are also taken postoperatively. Video recording is useful for documentation and educational purposes.

Using a modern automatic reflex camera, 35-mm still photography is practical with the available photo attachment of a Zeiss operating microscope. When choosing the camera, it is important that the photosensing element be located in the center of the view. This would correspond to the field actually seen through the microscope. Taking pictures and getting good exposure with modern automatic single lens reflex (SLR) cameras is not difficult. On the Zeiss operating microscope, when using a 400-mm lens and the photo attachment, the 25X magnification setting will fill the frame. Larger fiberoptic bundles will provide more illumination. If a high-intensity light source, such as color-corrected halogen or xenon, is utilized, higher-speed daylight films such as Kodak Ektachrome 200, 400, or 1000 ASA can be used with shorter exposure times. This minimizes distortion resulting from movement of patients, instruments, or microscope. With incandescent or quartz iodine light sources, an indoor-type film such as Kodak Ektachrome B ASA is used. An information camera back simplifies picture identification. Cine or video attachments available with the Zeiss microscope allow simultaneous color video recording.

X. LASER SURGERY OF BENIGN LESIONS

Small vocal cord nodules, polyps, and cysts can be removed by grasping them with small cup forceps and excising them with the laser. After photodocumentation, a small vocal cord nodule is evaporated by "grazing" it with the laser beam. With this technique, the excess laser beam is absorbed by a soaked pad placed under the vocal cord to protect the distal trachea and the inflated cuff of the endotracheal tube. Taking a biopsy specimen in such a situation would unnecessarily remove healthy tissue.

In localized hyperkeratosis (leukoplakia), as well as in its more diffuse manifestation, the superficial pathological tissue can be shaved off with the laser (24). A number of clinical cases suggest that after removal of the pathological tissue, a more normal epithelium will regenerate if the previous irritation (e.g., smoking) is avoided.

In an edematous vocal cord, a wedge can be resected. While the tissue heals, some of the fluid in the tissue can drain out. There is minimal scarring, and the healed vocal cord has normal elasticity. The voice greatly improves.

In patients with bilateral vocal cord polyps, especially when there is involvement of the anterior commissure, only one vocal cord can be operated on at a time. The opposing epithelium should not be denuded because of the danger of web formation. Thus, the opposite polypoid cord must be protected by a soaked, wet pad.

A. Laryngeal Papilloma

In the past 10 years, laser surgical management of laryngeal papilloma has become the treatment of choice. The removal of papillomatous lesions is greatly simplified by the CO_2 laser, which provides hemostasis. This makes microscopic surgery more precise, thus decreasing the likelihood of damaging healthy tissue such as the vocal cords. Also, postoperative edema is minimal. This is expecially important in small children who have a much smaller laryngeal lumen. Larger areas of papillomatous tissues are held with a forceps and removed by laser excision. In the initial operation, this larger specimen is submitted for histological examination. Smaller papilloma pieces are removed by evaporation. The heat of the laser beam also helps destroy the papilloma virus. The hemostatic effect prevents virus-contaminated blood from flowing into the trachea and bronchi, thus decreasing the likelihood of spread. During 15 years of experience with laser surgical management of laryngeal papilloma in adults, considerably longer remissions have been seen than with previous techniques. In addition, cures were more common. Experiences with children at Boston University Medical Center and Boston Children's Hospital demonstrated that noncontact removal minimizes scarring and laryngeal damage, thus preserving better function. The recurrence in children is still frequent, indicating that laser surgery provides for mainly airway management, rather than cure. During laser papilloma surgery, as in conventional surgery, care must be taken not to denude opposing surfaces of vocal cords in the anterior commissure to avoid web formation.

B. Wedge Resection of the False Vocal Cords

Resecting a wedge in the mid-third of the false vocal cords results in the improvement of dysphonia plicae ventricularis. This relatively bloodless procedure results in a smaller false vocal cord after healing. Because the false vocal cords cannot approximate effectively, the true cords once again are forced into use, thus greatly improving the voice.

C. Laser Surgical Treatment of Bilateral Abductor Paralysis

Endoscopic arytenoidectomy with the CO_2 laser has been described by Ossoff with favorable results. It has also been claimed that partial lengthwise removal of the vocal cords on both sides improves the airway and maintains satisfactory vocal cord function.

XI. SURGICAL LASER TREATMENT OF LARYNGEAL CANCERS

Twelve years of experience using endoscopic laser surgery resulted in considerable improvement in the cure of early vocal cord cancers. The treatment became simplified and cost-effective. In less-malignant cancers of the larynx, such as verrucous cancer and carcinosarcoma, endoscopic laser excision became the treatment of choice. In the treatment of intraepithelial carcinoma or carcinoma in situ, simple or multiple endoscopic laser surgeries provide the most effective and simple treatment (25). In addition, it can be done repeatedly.

In cases of T_1 cancers of the vocal cord in which radiation treatment failed, endoscopic laser cordectomy can result in long-term cure and avoidance of hemi- or total laryngectomy. Multiple small cancers, including recurring and new cancers that involve both vocal cords, can now be treated effectively with repeated laser surgical excisions while avoiding laryngectomy.

In more-advanced vocal cord cancers, such as T_2 or T_3, the likelihood of cervical lymph node metastases is greater (26-28). Nevertheless, in a number of these patients endoscopic laser excision of localized cancers can still be accomplished. Experience also shows that when the primary tumor is controlled by removal, cervical metastases that manifest themselves at a later date can be handled separately. Thus, en bloc removal of the larynx and cervical lymph nodes is not necessary. In more advanced cancers, when the surgeon feels that he did not remove all cancerous tissues from the larynx, as proved by multiple biopsies, postoperative chemotherapy and radiation therapy or radiation therapy alone may effect a cure.

In early cancers of the false vocal cord, excision biopsy can also be curative. In more advanced cases, precision cytoreduction of cancerous tissues can provide local control. If cervical lymph node involvement subsequently manifests itself, it can be controlled by excision of the lymph node.

Endoscopic laser surgery controlled the primary lesion in one case of supraglottic cancer that involved the aryepiglottic folds on both sides, the epiglottis, and a small area in the midportion of the base of the tongue (T_4 supraglottic cancer). Subsequent cervical metastases, which developed 6 months later on the left side and 3 years later on the right side, were controlled by surgical removal of the lymph nodes. Three years after the original operation, the patient developed a small lung lesion that had grown slowly over a 6-month period. The patient denied a biopsy procedure for this lung lesion. Three and a half years after the initial operation another physician recommended chemotherapy, which resulted in major immune depression and multiple liver metastases. At the time of his death, 4 years and 2 months after the initial operation, his larynx and neck were free of cancer.

A. Clinical Observations Concerning Laryngeal Cancer Biology and Laser Surgical Treatment

Cancers of the vocal cords can vary from each other biologically. Exophytic cancers tend to grow into the lumen with minimal infiltration into deeper tissues such as the vocal cord muscle. The removal of this type of cancer is easier, and the results are better. Other types of squamous cell cancers tend to infiltrate deeper into the vocal cord causing diminished vocal cord movement and eventual fixation. There are also increased chances for regional lymph node involvement. It appears that anterior commissure lesions tend to be infiltrative. They can involve both sides and can extend sub- or supraglottically. Occasionally these cancers can perforate the thyroid cartilage, transforming a T_{1b} cancer into a T_4. The extension of the cancer through the thyroid cartilage cannot be detected reliably on laminagrams or computerized tomographic scanning of the larynx. The extent of the tumor can only be followed by layer-by-layer laser excision of the cancer.

Vocal cord cancers are frequently diagnosed early because voice changes can be caused by as few as 10 million cancer cells (10^7 cells). In these early cases, the local host tissue reaction against the tumor diminishes the migration of cancer cells into the regional lymph nodes. As the number of cancer cells increases, the local host resistance is overwhelmed, thus allowing greater infiltration and spread of cancer cells to regional lymph nodes.

In heavy smokers, the entire respiratory mucosal lining is diseased owing to longstanding irritation. However, there is variation in the degree of change of the mucosa in different areas. Clinical observation supports the hypothesis that once a squamous cell cancer of the vocal cord becomes invasive (infiltrating through the basal membrane) it can induce transformation of the dysplastic cells around it. In this manner, a small infiltrating cancer can become more extensive by both multiplication of existing cancer cells and by inducing transformation of surrounding dysplastic cells. This phenomenon may also be responsible for inducing malignant change in the apposing vocal cord epithelium with which it sometimes makes contact. Clinical observation also seems to show that recurrence of cancer in a previously laser-excised area is more common in patients who continue to smoke after surgery.

The most significant advantage of laser surgery, which is especially evident in the larynx, is a dry surgical field. This enables viewing the area under binocular magnification. With experience, normal and cancerous tissues in laryngeal structures can be differentiated to a reasonable degree. A magnification of 16-25X occasionally allows the detection of a few hundred to a few thousand cancer cells. The clinical observation of tissues under a microscope, combined with biopsies, fosters the development of this skill of differentiation. Microsurgical laser techniques improve by two to three orders of magnitude the minimum number of cancer cells that can be detected intraoperatively. Further improvement can be anticipated if fluorescent intraoperative tumor cell markers are developed (29). The advantages of laser surgery in cancer treatment are as follows:

1. Dry surgical field allows precision detection under magnification.
2. Precision removal of cancer cells; maximum normal tissue preservation.
3. Total kill in comparison with chemotherapy, biological therapy, and radiotherapy.
4. Repeatable (in contrast with radiotherapy).
5. Minimal trauma surgery, causing minimal immunodeficiency and minimal postoperative edema (minimal postoperative pain and rapid healing).
6. No mechanical effect to dislodge cancer cells.
7. Sealing of lymph vessels; diminished lymphatic metastasis.
8. Ability to sterilize the surgical field of tumor cells, bacteria, and viruses.
9. Ease of delivery through endoscopes, fiberoptics, and sapphire tips.
10. Hemostatic effect; diminished need for transfusion.

B. Excisional Biopsy by Layers

With localized laryngeal cancers, precision removal in the form of an excision biopsy can effect a cure. In the case of vocal cord cancer, an initial specimen is removed by grasping it with cup forceps and excising it with the laser beam or scissors. With the latter technique, the bleeding is controlled by epinephrine packing. The specimen obtained is then carefully marked for orientation and placed on a fixed cucumber slice which is then sent to the pathologist. Another layer of tissue is then removed with the laser and the dry, cut surface is carefully observed under 16X and 25X magnification. Nine biopsies are then taken with cup forceps. These biopsies are taken in the following order: 1, anterior; 2, mid; and 3, posterior part of the superior edge of the vocal cord's excisional bed. Similarly, three more biopsies are taken from both the midportion (specimens 4,5,6) and inferior edge (specimens 7,8,9) of the vocal cord. This is followed by epinephrine packing before another layer is excised with the laser. All biopsy specimens are then carefully marked in separate containers. If there is further suggestion of cancerous tissues, biopsies are taken followed by further layer-by-layer removal. If the cancer is extensive and deeper, the excision is carried to the thyroid cartilage, and the cartilage is examined for tumor penetration.

Depending upon the surgeon's experience, the following endoscopic partial laryngectomy procedures can be performed as was previously described by Jako in 1979.

1. Partial cordectomy
2. Subtotal cordectomy
3. Total cordectomy
4. Epiglottidectomy
5. Anterior vertical partial laryngectomy
6. Lateral vertical partial laryngectomy
7. Extended frontolateral vertical hemilaryngectomy
8. Supraglottic hemilaryngectomy
9. Supraglottic laryngectomy

When the cancer is more extensive, the chances for lymph node metastases are greater. In T_1 glottic tumors, the incidence of lymph node metastases is less than 1%. In T_2 glottic tumors this incidence is 6.6-16.7%. With T_3 tumors the incidence of metastases is slightly higher. Thus, in most cases of T_1, T_2, and T_3 glottic tumors (which because of early detection are those usually found in the United States) the patient has no palpable cervical lymph nodes. Patients without palpable nodes are primarily treated as having localized cancers. These are removed by laser excision. If the patient later develops evidence of lymph node metastases, they are dealt with separately using surgery or radiotherapy. If local control of the tumor could not be achieved with the initial excision as proven by the results of multiple biopsies, further options are available.

1. Endoscopic reexcision of the tumor (This can be done repeatedly.)
2. Standard external partial or total laryngectomy
3. Radiotherapy or adjuvant chemotherapy followed by radiotherapy

The third option can be a viable alternative in a T_3 cancer when an experienced surgeon is confident that 95% or more of the tumor cells have been removed endoscopically.

C. General Considerations

Before operating on patients with laryngeal cancer, the surgeon must have considerable experience with microlaryngeal surgery and the use of the microsurgical laser. Experience with laser dissection of human cadaver larynges and learning surgical anatomy via layer-by-layer removal of laryngeal tissues are important. Methodology for performing cordectomies in human cadaver larynges has been described by Davis et al. (30). Adequate endoscopic exposure of the larynx is essential. When performing microsurgery for cancer, the surgeon should make every attempt to remove all pathological and suspicious tissues. For cancer surgery, the CW mode of the laser should be used. Electrocoagulation equipment with a suction tip should always be available to control bleeding from larger blood vessels. There are larger blood vessels entering the vocal cords in the anterior commissure and posteriorly at the muscular process of the arytenoid.

Frozen-section biopsies can be used, but permanent histological sections are preferred.

D. Endoscopic Partial Laryngectomies

The following classification was described by myself in 1979, in conjunction with case histories (31). Since then, there has been more time to follow-up patients who have had these treatments. In addition, new patients have received these treatments. Partial, subtotal, and total cordectomies provide long-term cure. Ninety-six percent of T_1 cancers result in long-term cures, and the patient is left with a reasonable voice because a secondary or pseudo vocal cord develops. Epiglottidectomy is a practical procedure and provides cure for small cancers involving the free edge. The other procedures are used for extensive cancer. They can provide local control alone, in combination with radiotherapy, or in combination with chemotherapy followed by radiotherapy.

1. Partial Cordectomy

One to two-thirds of the vocal cord is removed; this could include part of the vocal cord muscle (thyroarytenoid).

2. Subtotal Cordectomy

This procedure includes removal of more than two-thirds of the vocal cord, including a large part of the vocal cord muscle, removal of the vocal process, and partial removal of the ventricles. Part of the false vocal cord could be included for better visualization.

3. Total Cordectomy

The entire vocal cord on one side is removed, including the vocal process of the arytenoid. The anterior commissure on one side, including the anterior commissure tendon, is removed. The ventricle and the vocal cord muscle are removed. The subglottic soft tissue, conus elasticus laryngis, is removed 5-10 mm below the free edge of the cord. Part of the false cord can be removed for better visuaization.

4. Epiglottidectomy

The epiglottis is removed in one or two pieces with the laser.

5. Anterior Vertical Partial Laryngectomy

The anterior one-third of both vocal cords is removed, including the anterior commissure tendon (32). Part of the thyroid cartilage can be removed from the laryngeal side with the laser. If the thyroid cartilage is calcified, both laser and scraping should be used. Part of the anterior false vocal cord could be removed for better visualization. The removal could include all or part of the epiglottis.

6. Lateral Vertical Partial Laryngectomy

This procedure includes subtotal or total cordectomy with removal of the false vocal cord and the ventricle on one side.

7. Extended Frontolateral Vertical Hemilaryngectomy

The procedure is similar to lateral vertical partial laryngectomy, except that it includes removal of the anterior third of the opposite vocal cord.

8. Supraglottic Hemilaryngectomy

This procedure includes removal of the false vocal cord, the aryepiglottic fold on one or both sides, the epiglottis, and exposure of the hyoid bone.

E. Postoperative Care and Follow-Up

Patients generally tolerate endoscopic partial laryngectomies well and have relatively minimal postoperative pain. No tracheotomies were needed with any of these procedures. There was also no problem with aspiration in the less extensive procedures. In more extensive procedures, aspiration can occur. This requires intravenous fluid feeding and nasal tubes. Periodic cold vapor inhalation for 12-24 hr after the operation is recommended. Breathing exercises, performed for 2-3 weeks are helpful. In the case of aspiration, antibiotics should be given to prevent pneumonia. In addition, careful reintroduction of oral feeding should be initiated. One patient experienced hemorrhage 3 weeks postoperatively. This was controlled by intubation.

All patients should undergo follow-up microlaryngoscopy 2-3 months after the initial operation. At this time, if there is no cancer, the vocal cords and other excised areas should be covered with normal epithelium. In patients in whom large segments of the vocal cord had been removed, a smaller and bowing vocal cord is present. It generally takes 6-12 months for a secondary vocal cord to develop. During this time, the voice continually improves. Microlaryngoscopy follow-up with photographic documentation is recommended at 6, 12, and 24 months, and at 5 years after the original operation. All laryngeal cancer patients should have follow-up chest x-rays at the time of each microlaryngoscopy. Development of secondary cancers in the lung or esophagus has been reported in up to 25% of patients. Careful checking of the neck for lymph node enlargement should be performed weekly by the patient. In the event of an unhealed or suspicious area in the patient's larynx, biopsies and further laser excisions are mandatory. If this follow-up biopsy is positive for the first specimen, but deeper layers are negative, the patient undergoes microlaryngoscopy according to the previously mentioned schedule. If there is any question, the patient should have a second follow-up microlaryngoscopy in 2 months. If cancer is still present at the time of this second follow-up, the patient should be then offered radiation therapy or external surgery. If the patient refuses these treatments, repeated endoscopic laser excisions could be an alternative. One should especially consider this option when the patient is over 80 years old, has severe cardiac or other disabling diseases, or has a second inoperable cancer, such as lung, esophagus, or colon cancer with liver metastases (see Fig. 1 for a summary of this follow-up procedure).

XII. FUTURE CONSIDERATIONS

As the next generation of laser surgical instruments become available, combined endoscopic and external laser microsurgical techniques will be developed (Fig. 2). They will provide local control for more extensive laryngeal cancer, such as those that involve extension into or through the thyroid cartilage. This surgery will be performed by a team of two surgeons; one working endoscopically and the other externally, both using microscopes.

This new instrumentation will provide more accurate and precise removal of cervical lymph nodes to which metastases have spread. This will prevent leakage of cancer cells from lymph nodes and vessels, and will clear the wound area of contaminating cancer cells. The surgeon will also be able to differentiate the malignant tissue from normal tissue, such as nerves and blood vessels, with much greater precision.

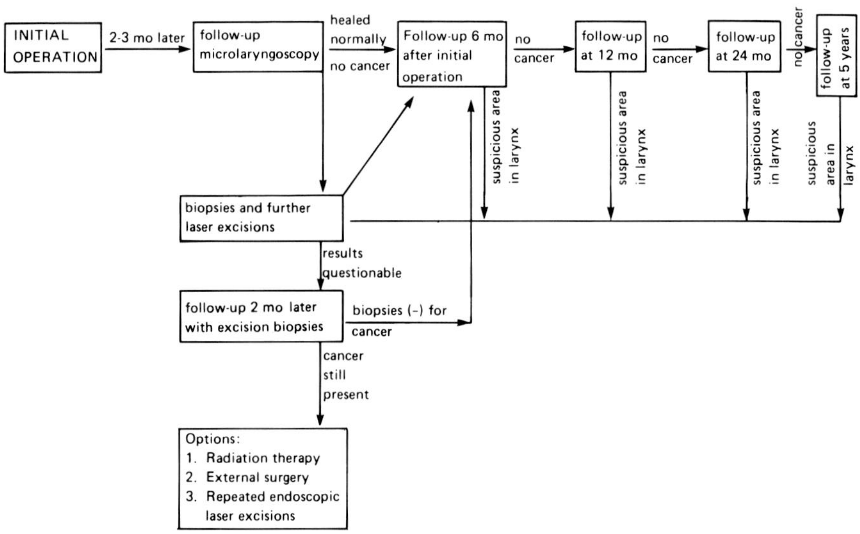

Figure 1 Excision biopsy of laryngeal cancer: follow-up flow chart.

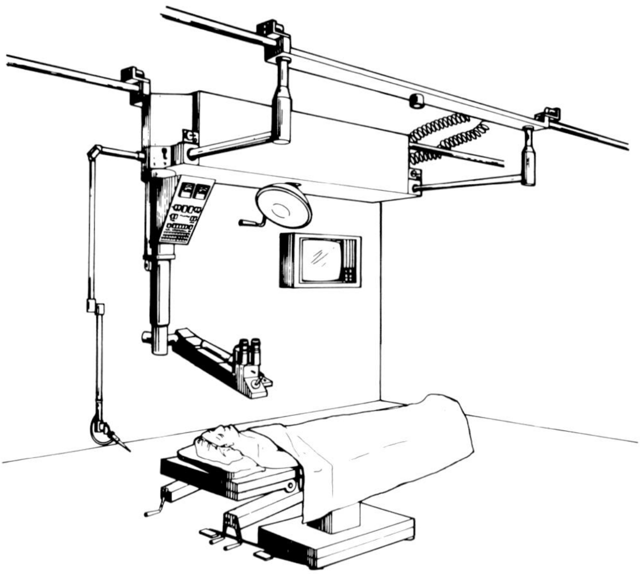

Figure 2 Next generation of multipurpose and multiwave length operating room laser. To be used for endoscopic, external head and neck, and general surgery (Jako, U.S. Patent #4,503,854).

The indications for partial and total laryngectomies vary considerably in different parts of the world. At the present, I estimate that endoscopic laser surgery offers a temporary or permanent alternative to partial or total laryngectomies in 20% of patients with laryngeal cancer. Current state-of-the-art laser surgical instrumentation and techniques could save thousands of larynges each year. In turn, this could provide major reductions in pain and suffering as well as in disability payments. The new generation of laser surgical equipment will contribute to the development of even more conservative surgical procedures. In combination with adjuvant cancer therapies, one could expect a 50% reduction in the number of laryngectomies by the end of this century.

ACKNOWLEDGMENT

I would like to express my appreciation to Howard Herman, a medical student at Boston University, for his help in compiling this chapter.

REFERENCES

1. Jako J: Laryngoscope for microscopic observation, surgery and photography. *Arch. Otolaryngol. 91*:196, 1970.
2. Kleinsasser O: Mikrochirugie im Kehikopf. *Arch. Ohrenheilk. 183*:428, 1964.
3. Ketham AS, Hoye RC, and Riggle CC: A surgeon's appraisal of the laser. *Surg. Clin. N. Am. 47*:1249, 1967.
4. Strong MS and Jako GJ: Laser surgery in the larynx. *Ann. Otol. Rhinol. Laryngol. 81*:791, 1972.
5. Strong MS, Jako GJ, Polanyi TG, and Wallace RA: Laser surgery in the aerodigestive tract. *Am. J. Surg. 126*:529, 1973.
6. Vaughan CW: Use of the carbon dioxide laser in the endoscopic management of organic laryngeal disease. *Otolaryngol. Clin. N. Am. 16*:849-864, 1983.
7. Shapshay SM and Simpson GT: Lasers in bronchology. *Otolaryngol. Clin. N. Am. 16*:879-886, 1983.
8. G. J. Jako. Laryngeal endoscopy and microlaryngoscopy. In *Otolaryngology* (M. M. Paparella and D. A. Shumrick, eds.), Vol. 3, 2nd Ed., W. B. Saunders Co., Philadelphia, 1980, pp. 2410-2430.
9. Andrews AH Jr. and Moss HW: Experience with the carbon dioxide laser in the larynx. *Ann. Otol. Rhinol. Laryngol. 83*:462, 1974.
10. G. J. Jako and T. G. Polanyi. Carbon dioxide laser surgery in otolaryngology. In *Laser Surgery* (I. Kaplan, ed.), Israel Academic Press, 1976.
11. G. J. Jako and R. A. Wallace. Carbon dioxide laser microsurgery and its applications in laryngology. In *Laser 77: Opto-Electronics: Conference Proceedings* (W. Waidelich, ed.), Science and Technology Press, Schenectady, NY, 1977

12. G. J. Jako. Laser biomedical engineering: Clinical applications in otolaryngology. In *The Biomedical Laser* (L. Goldman, ed.), Springer-Verlag, Berlin, 1981.
13. Jako GJ: Future directions and developments in laser surgery. *Otolaryngol. Clin. N. Am. 16*:821-828, 1983.
14. Jako GJ: Microscopic and endoscopic laser surgery. *Md. Instr. 17*:407-408, 1983.
15. Snow JC, et al.: Anesthesia for carbon dioxide laser microsurgery on the larynx and trachea. *Anesth. Analg. 53*:507, 1974.
16. Konchiegei HH and Shaker MH: Anesthesia for intra-laryngeal surgery. *Can. Anaesth. Soc. J. 21*:343, 1974.
17. Snow JC and Norton ML: Lasers and anesthesia. *J. Am. Assoc. Nurse Anesth. 43*:464, 1975.
18. Norton ML: Anesthesia for laser surgery in laryngobronchoesophagology. *Otolaryngol. Clin. N. Am. 16*:785-792, 1983.
19. Hermens JM, Bennet MJ, and Hirshman CA: Anesthesia for laser surgery. *Anesth. Analg. 62*:218-229.
20. Davis RK and Simpson GT: Safety with the carbon dioxide laser. *Otolaryngol. Clin. N. Am. 16*:801-814, 1983.
21. Healy GB: Complications of laser surgery. *Otolaryngol. Clin. N. Am. 16*: 815-820, 1983.
22. Simpson GT and Strong MS: Recurrent respiratory papillomatosis: The role of the carbon dioxide laser. *Otolaryngol. Clin. N. Am. 16*:887-894, 1983.
23. Jako GJ and Strong MS: Laryngeal photography. *Arch. Otolaryngol. 96*: 268, 1972.
24. Strong MS: Laser management of pre-malignant lesions of the larynx. *Can. J. Otolaryngol. 3*:560, 1975.
25. Strong MS: Laser excision of carcinoma of the larynx. *Laryngoscope 85*: 1286, 1975.
26. Cocke EW: Cancer of the larynx: Surgery. *CA 26*:201-211, 1976.
27. Kirchner JA: One hundred laryngeal cancers studied by serial section. *Ann. Otol. Rhinol. Laryngol. 78*:689, 1969.
28. Kirchner JA: Staging as seen in serial sections. *Laryngoscope 85*:1816, 1975.
29. Jako GJ, Strong MS, and Vaughan CW: Identifying and excising tumor. *Arch. Otolaryngol. 110*:824, 1984.
30. Davis RK, Jako GJ, Hyams VJ, and Shapshay SM: Anatomic limitations of CO_2 laser cordectomy. *Laryngoscope 92*:980-984, 1982.
31. G. J. Jako. Microsurgery of the larynx with the CO_2 laser. In *Otolaryngology,* Vol. 3. Harper and Row, Hagerstown, MD, 1979, pp. 1-20.
32. Broyles EN: The anterior commissure tendon. *Ann. Otol. Rhinol. Laryngol. 52*:342, 1943.

6

Applications of the Neodymium:Yttrium-Aluminum-Garnet Laser in Bronchology

JEAN-FRANCOIS DUMON
Salvator Hospital, Marseilles, France

I. INTRODUCTION

Like so many discoveries and inventions, the laser originated from research and development aimed primarily at military ends. This technology continues to be the focus of considerable effort by the military for arms systems. The word *laser* often conjures up images of awesome, frightening, destructive power. This idea is reinforced by science fiction literature and especially motion pictures. Recently in the film *Star Wars,* for instance, the laser beam was once again presented as a death ray. And what fan of James Bond films can forget the laser with which the fiendish Goldfinger tried to cut Agent 007 in half! This negative perception nonetheless is not wholly deserved, because the laser has been used peacefully for industrial purposes for a number of years. Unfortunately, these latter applications were not newsworthy, i.e., commercially exploitable for the mass media. At present, however, the laser is finding its way into everyday life. Discotheque and concertgoers enjoy laser light shows, and record collectors now listen to their favorite music on laser disks. As a result, the laser is starting to shed its sinister reputation for a more positive one.

One domain in which lasers have found many constructive uses is medicine, where they have been put to use by specialists in a wide range of fields. In endoscopy, the cumbersome mirror arm delivery system of the carbon dioxide (CO_2) laser was a major obstacle to extensive utilization. With the arrival of the neodymium:yttrium-aluminum-garnet (Nd:YAG) laser, however, it became possible to deliver laser energy through an optical fiber at levels high enough for resection or hemostasis or both. This development opened the way to a host of endoscopic applications, including endobronchial Nd:YAG laser surgery, a technique that I pioneered more than 5 years ago at Salvator Hospital in Marseilles, France. Presently, other endoscopists in Europe and the United States are also practicing the Salvator method, and many others are receiving training at regularly scheduled courses in France, Italy, and the United States. This chapter describes the main features of that method whose reliability now appears unequivocally established.

II. SPECIAL MATERIALS REQUIREMENTS

Laser therapy is not a diagnostic procedure and therefore must not be attempted in a noninstitutional environment. Most hospital endoscopy units are well equipped, and no major transformations are required for the practice of laser surgery. Two notable additional materials requirements are the laser and the rigid bronchoscope. A recovery room is also indispensable.

A. The Nd:YAG Laser

Unlike the CO_2 laser, the Nd:YAG laser can be delivered through an optical fiber, making it more suitable for endoscopic applications. For other reasons, the Nd:YAG laser is better for this type of therapy, but it has a few disadvantages.

The interaction between a laser and living tissue depends upon several factors, especially the wavelength of the laser's beam. A CO_2 laser is ideal for cutting because its energy is almost fully absorbed at the point of impact, with little scattering to adjacent tissue. Consequently, the CO_2 laser is unequaled as an optical scalpel. As will be explained more completely later, removal of the tissue causing tracheobronchial obstruction through an endoscope relies more on controlled coagulation than on incision and ablation, and the CO_2 laser is not well suited to that purpose (1). The Nd:YAG laser, on the other hand, is superior when it comes to heating large volumes of tissue to below 100°C to induce hemostasis and necrosis. Unlike the CO_2 laser, its energy scatters out around the point of impact. In fact, in heavily pigmented soft tissue, such as that of skin, liver, and spleen, scattering is about twice that of absorption. Because of this property, the Nd:YAG laser can be used to "cook" malignant or benign neoplasms into a friable state to enable removal by sloughing or suction. Similarly, the Nd:YAG laser is efficient for control of hemorrhage.

In connection with this scattering effect, a word of caution is needed. It is extremely difficult to judge the depth of penetration, especially where pale tissue is concerned. The surface appearance of the irradiated zone does not necessarily indicate the degree of underlying involvement. Under these conditions, the inexperienced, uninformed operator can cause serious damage by continuing to irradiate the area. The worst hazard is the dreaded *popcorn* effect in which a pocket of steam forms below the surface and eventually explodes leaving a large cavity. To avoid this type of injury, irradiation should be carried out at low power settings (40-45 W)—in short, repeated bursts (1 sec or less)—while paying close attention to surface changes. The exact technique will be described, but continuous exposure at high power outputs is both senseless and dangerous.

Another consideration in the comparison between the CO_2 and the Nd:YAG lasers is that the former demands total immobility of the operating field. When that field is in the airways, this is difficult to accomplish because it entails a heavy general anesthesia and usually jet ventilation. Both these techniques are incompatible with our technique and increase the inherent risks of tracheobronchial resection first by reducing visibility and second by making suction impossible.

The Nd:YAG laser also compares favorably with the argon laser for endoscopic work in tracheobronchial regions. Although it, too, can be delivered through fiber, the argon laser can only be used to control bleeding from small vessels because it lacks the power to stop massive bleeding or penetrate large tissue masses. On the other hand, the argon dye laser system has been used indirectly to achieve photoactivation of hematoporphyrin dye injected into neoplasms. This technique results in chemical reactions that effectively kill the malignant cells. Unfortunately, given the risk of perforation and hemorrhage with extensive invasion of the cartilage or mediastinum, the success of this therapy hinges on early detection of the growth. The Nd:YAG laser is thus the system of choice for endoscopic tracheobronchial therapy.

A word is needed about the laser fiber. Fibers are composed of a quartz core surrounded by silicone rubber cladding and a Teflon cover. On standard models these three components are enclosed in a polyethylene catheter, which guards the fiber against fouling and allows the passage of gas for cooling or clearing the treatment site. Fibers are the most expendable part of a laser system; they can and do wear out. The mean time between replacement differs from one fiber to another and is dependent upon the conditions and frequency of use and on maintenance.

B. Rigid Bronchoscope System

Rigid open-tube bronchoscopy is not widely used for diagnostic purposes. Most endoscopy suites are equipped with flexible fiberscopes, such as the Machida 6TL II, Olympus BFI, and Pentax 17C. As they all have large working channels, any one of these instruments can, in theory, be used to perform laser surgery. For high-risk procedures, however, fiberscopes do not permit rapid and effective response to major complications, and thus advantages in terms of patient comfort, endoscopist convenience, and extended visibility are largely offset. A mandatory requirement for any department planning to perform laser resection

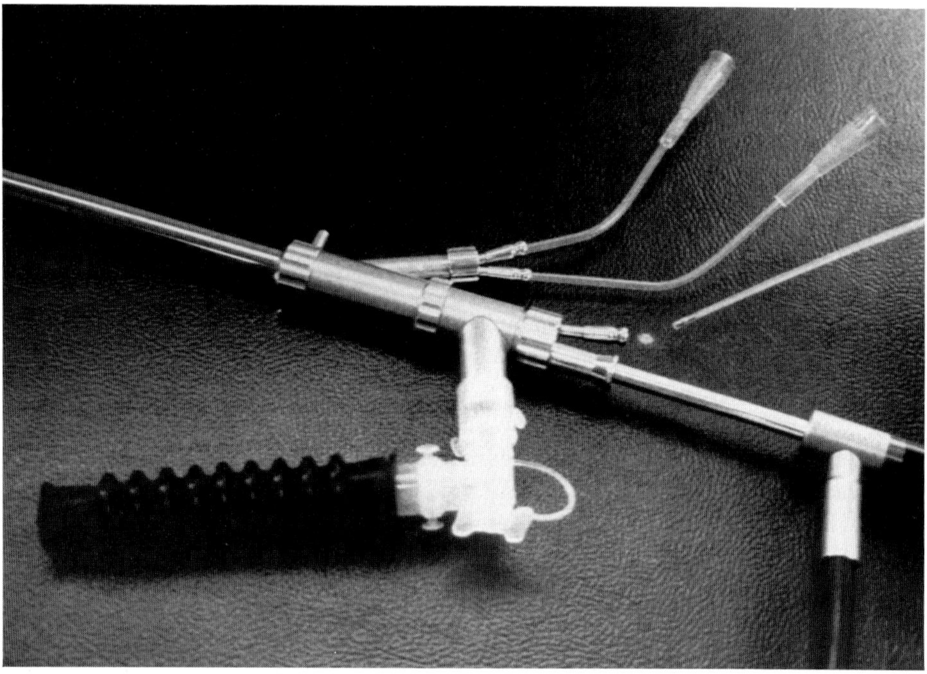

Figure 1 Specially designed rigid laser bronchoscope (Dumon-Harrell) with side-port ventilation on a swivel adapter. Note separate ports for two semirigid plastic suction catheters and the Nd:YAG laser fiber. The light source is attached to a 0° telescope. A set of bronchoscope barrels (different lumen calibers) is available to attach to the proximal laser ventilation fitting.

on a regular basis is a rigid bronchoscope with special features for laser resection.

Working in collaboration with Dr. James Harrell (San Diego, California) and the Efer Company (distributors are the Bryan Corporation, Woburn, Massachusetts), we have developed a third-generation universal bronchoscope system (Fig. 1). It is suitable for all purposes, including laser resection. It has a headpiece and comes with a set of barrels of various sizes. The barrels can be clipped on and off the headpiece quickly and easily. The headpiece has all the design features of the previous model, i.e., two side-entry ports, rubber cuffs on all channel openings, and a rotating ventilation connector for simultaneous lasing and suctioning under closed-circuit conditions. An innovation of this new system is that the side-entry ports are mounted on a cap that, when removed, leaves an opening large enough for passage of large tumor fragments. The bronchoscope is designed for use with a straight viewing lens.

Suction catheters play a critical role in laser resection. Their purpose is threefold: palpation of the target area, removal of debris, and cleansing. A second catheter is often inserted alongside the telescope directly down the working channel. Given the heat factor in proximity to the laser fiber, our catheters are made of heat-resistant plastic, and we have never experienced a problem with ignition or melting.

C. Recovery Room

An endoscopy unit offering laser surgery must have an especially equipped recovery room at its disposal. This facility should contain an adjustable hospital bed, an Ambu bag system as well as a compact ventilator, oxygen supply and suctioning capability, an intubation setup with a laryngoscope and full set of blades, equipment to monitor vital signs, a cardioscope, and a defibrillator. The drugs on hand should include corticosteroid agents, epinephrine, bronchodilators, buffers, respiratory stimulants, and antiarrhythmic agents. Two specially trained staff members should be assigned to duty in the recovery room.

III. METHODOLOGY

Any surgical procedure requires thorough preparation, skilled execution, and conscientious follow-up. Each phase is essential to a successful lasting outcome. Laser resection unfolds along similar lines. The preparatory phase includes screening, admission, and workup. The operative phase consists of anesthesia, intubation, endoscopic inspection, resection, coagulation, and extubation. The postoperative phase covers recovery, discharge, and periodic inspection (follow-up).

A. Preoperative Workup

Patients are often seen in extremis so that almost no time is available for extensive examination. However, because they are usually inoperable, are surgical failures, or have recurrent tumor, their disease is already well documented. The endoscopist can thus obtain most of the data that are needed from the referring physician, such as information about complicating illnesses (essentially cardiovascular and respiratory deficiencies), previous therapy (including current medications), results of recent medical testing (cardiovascular, pulmonary, neurological, blood, and urine), and standard radiography (chest and localized views).

For patients not already hospitalized, admission should be demanded no less than 8 hr before treatment. For a morning procedure, it is best to have the patient admitted the evening before. The main purpose of early admission is to ensure that the patient does not eat or drink, but it also gives the anesthesiologist or the endoscopist or both an opportunity to conduct last-minute testing.

An interview with the patient is also helpful for patient and physician. If informed consent is necessary, this too can be obtained.

Preoperative workup for laser treatment is essentially aimed at assessing the dangers that may arise intraoperatively and not at screening. The absolute contraindication is extrinsic compression. Most patients should be brought into the endoscopy room for direct visual examination. If purely extrinsic airway compression is present, the laser is useless, and treatment is aborted. If the narrowing is partially or completely intraluminal, laser treatment is initiated and continued to the safe limits of the patient's condition. The endoscopist should keep in mind that the effects of laser treatment are not cumulative, and therefore repetitive procedures are possible. In an emergency, we do not hesitate to deal first with the immediate life or death situation in the initial session and after the patient's condition has improved to reschedule new sessions to achieve more complete removal.

Another purpose of the preoperative workup is selecting the appropriate anesthesia. This decision can be rapidly made based on the size and location of the lesion. Any sizable lesion in a major airway must be treated with the open, rigid bronchoscope and thus under general anesthesia. Local anesthesia as well as the fiberscope must be used only for small or peripheral lesions. Cardiovascular and respiratory deficiencies, although they may complicate the procedure (some patients need continuous monitoring of blood gases), are not absolute contraindications because general anesthesia for Nd:YAG laser surgery can be light. In this regard, and unlike the CO_2 laser, the Nd:YAG laser does not require complete immobility, and therefore jet ventilation and curarization are not necessary. Light general anesthesia can be achieved with many drugs. As for the cough reflex, fentanyl is a good inhibitor. For many years we obtained satisfactory anesthesia using a combination of diazepam (Valium) and alfadione (alfaxalone/alfadolone, 3:1 mixture) for induction with reinjection of alfadione and fentanyl during the procedure as needed. Because of the high frequency of anaphylactic reaction, alfadione is currently unavailable, and we have replaced it with sodium pentothal. This drug works well but necessitates higher doses of fentanyl to suppress the cough reflex and leads to a longer recovery period. At any rate, a variety of anesthetic protocols are undoubtedly possible. We advise against gases because of leakage, and nitrous oxide, which is flammable, must never be used. In connection with anesthesia, it is important to stress the key role played by the anesthesiologist, not only in designing an anesthetic protocol for each patient but also in carefully monitoring the patient's vital signs throughout the procedure.

B. Operative Procedures

The operative period unfolds in a logically ordered manner as follows (2):

The patient is placed on the table in the supine position for an open-tube procedure with the head in hyperextension. For a fiberscopic procedure on one lung, the patient is turned so that the side containing the tumor is down on the table. This position prevents flooding of the healthy lung in the event of massive hemorrhage.

The electrodes for electrocardiography and, if necessary, a blood gas probe are applied, and transnasal oxygen, 4 L/min, is administered. Diazepam is given first followed by the anesthetic agent, presently sodium pentothal. When induction has been achieved, a local anesthesia of the glottis using lidocaine is obtained, and the patient is manually ventilated by means of an Ambu bag for a few moments. Induction obviously is only included in a general anesthetic procedure; for a local anesthetic procedure, this step consists of application of the appropriate topical agent.

The patient is intubated through the oral route. Intubation with the rigid, open tube is easily mastered with a little practice. At this point, endoscopic inspection of the whole tracheobronchial tree, not just the pathological area, should be performed. Special attention must be given to determining the limits of intraluminal growth and the route of the airway. For this purpose, the suction catheter serves well to palpate the target area. When it is used, the open tube serves conveniently as a conduit for the fiberscope during the preresection inspection. It is also wise to take a biopsy specimen at this time.

With the cooling system (nitrogen) on high flow, the laser fiber is inserted down the working channel until it can be seen about 1 cm outside the distal end of the bronchoscope. Lasing is initiated by aiming at close range (but not too close to avoid soiling the fiber, which leads to premature deterioration). If the tissue is very pale, the laser beam may not be well absorbed, and it may be necessary to scratch the surface lightly to produce some bleeding. The endoscopist should start in the center of the lesion working gradually in larger circles out to the edge. Under no circumstances should the laser be fired perpendicularly into the wall. Resection of tumor must obligatorily end with a thorough laser coagulation of the treated area. This is the best insurance against hemorrhage in the postoperative period. Also, after coagulation, a careful cleansing of the whole tracheobronchial tree must be carried out to remove secretions, blood, and any remaining tissue fragments. In this way, the risk of infection is greatly lowered, and corticosteroid agents are not needed.

The major concern of the endoscopist during laser application is to limit laser exposure while achieving removal of the tumor completely and quickly. These goals may seem contradictory, but with the proper approach they are easily attained. A basic safeguard is to keep the power down; more than 40-50 W is seldom needed. As for time exposure, 0.5-1.0 sec is the average range; the continuous mode is never indicated. At settings within these limits, coagulation

—the Nd:YAG laser's forte—is efficiently achieved. Excision of the tumor is accomplished by slicing off fragments of coagulated tissue layer by layer with the beveled end of the open tube. As they are detached, these tissue fragments are removed through the open tube using the suction catheters. When the base of the tumor is reached, purely "laser" resection may be performed to eliminate any remnants of endoluminal growth.

The patient is extubated on the operating table. Depending upon the respiratory status, mechanical ventilation with an Ambu bag or oxygen mask may be required. Reintubation is avoided if possible because this may cause hemorrhage from the tumor bed.

C. Postoperative Care

The postoperative period is biphasic in that it consists of immediate follow-up and long-term surveillance. Immediate follow-up includes the time spent by the patient in the recovery room and the rest of the hospitalization. This phase of treatment must not be underestimated; most of our complications occurred after resection. In view of this danger, patients should be sent to the recovery facility which, as previously stated, is an integral part of the laser-capable endoscopy unit. The length of transit in recovery depends on several factors, including the type of anesthesia used, the endoscopist's skill and meticulousness, and the patient's overall condition. The decision to release the patient from the recovery room is made on the basis of clinical observation. We rely on the *Aldrette scale,* which rates the patient's limb motor activity, respiration, circulation, consciousness, and skin color on a scale from 0 to 2 for each category. When a score of 10 is attained, transfer to a hospital room is authorized. Given the lingering effects of anesthesia and the, albeit, remote risks of latent complications, at least one night in the hospital is recommended. After treatment and release from the hospital, periodic checkup visits should be made.

Follow-up study depends largely upon the nature of the disease (extent of involvement, degree of malignancy, growth rate), adjuvant treatment undertaken, and the success of initial resection. Further laser resection may be performed whenever needed; in some patients, several procedures may be required to achieve complete resection. In practice, long-term surveillance is often problematic because it presupposes the cooperation of both the patient and the referring physician(s) as well as geographic proximity.

IV. POTENTIAL DANGERS AND RESPONSES

The principal hazards of laser resection are hypoxemia, hemorrhage, perforation, and fire (3). A few other dangers, such as tension pneumothorax, pneumopericardial tamponade, and infection, loom as remote possibilities if proper precautions are not taken. Most of these problems can arise postoperatively as well as intraoperatively and must be carefully watched for by the recovery room staff.

A. Hypoxemia

Hypoxemia can be brought on by anesthetics and, as stated previously, the main concern of the anesthesiologist is to anesthetize the patient such that spontaneous respiration is preserved. For a skilled anesthesiologist, this is not usually a problem, and only occasional brief respiratory assistance is necessary. In practice, most hypoxemia is procedure related. Many patients are treated as an emergency, and they are barely compensating ventilationwise. It follows that introduction of any instrument into the airway can lead to decompensation, not only by cutting down on airflow but also by increasing and accumulating secretions. The first act after intubation often consists of oxygenating the patient. Immediate oxygenation is, of course, the only way of halting hypoxemia before the appearance of cardiovascular complications. Consequently, at the slightest sign of hypoxemia, laser treatment should be interrupted to lessen this danger. Immediate counteraction against hypoxemia greatly depends upon cooperation between the anesthesiologist and endoscopist. The former must closely monitor the patient's vital signs using blood gas analysis when required. The latter must be ready and willing to interrupt treatment without hesitation to ventilate the patient under closed-circuit conditions, if necessary.

To achieve effective oxygenation, the endoscopist must take into account the location of the obstructing lesion. When the stenosis involves the trachea, the first objective is to establish an oxygenation route. With the open tube, this can be done quickly by pushing the beveled tip through the narrowing before beginning oxygenation (Fig. 2). In bilateral obstruction involving the entrance to both mainstem bronchi, the procedure is the same except that the endoscopist must choose the least-obstructed bronchus before proceeding distally (Fig. 3). When only one bronchus is diseased, the endoscopist should withdraw the tip of the bronchoscope into the trachea so that the air supply can reach the healthy lung (Fig. 4). In all patients, tracheobronchial cleansing should be performed before treatment is resumed.

B. Hemorrhage

Death by exsanguination is not a major risk; in fact, it is extremely rare. Actually, hemorrhage is dangerous because by filling the airways it can result in hypoxemia and its attendant cardiovascular sequelae. Bleeding is, by far, the number 1 cause of hypoxemia during laser resection. A massive hemorrhage obscuring the operative field obviously imposes instant attention, but a slow hemorrhage can become life-threatening if it is ignored too long. Therefore, it is good policy to take any hemorrhage seriously, regardless of its rate.

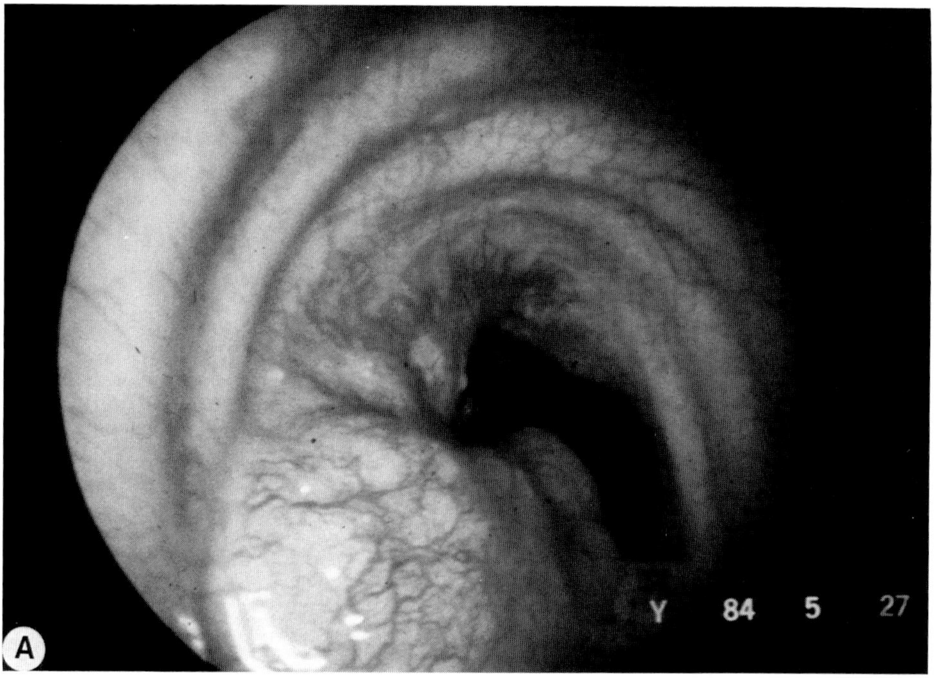

Figure 2A Endoscopic photograph showing an adenoid cystic carcinoma of the distal trachea extending into the left main bronchus with 90% obstruction. This patient had already received radiotherapy and was not considered a surgical candidate.

Given the difficulty in controlling the depth of penetration, it behooves any endoscopist practicing Nd:YAG laser resection to bear in mind the anatomical surroundings of the tracheobronchial tree. A convenient wall chart should be designed for display in the endoscopy room. In the trachea, the most notable close vascular relations are the innominate artery, which runs diagonally up the front, and the aorta, which crosses on the left-hand side at the base. The aorta is also in close contact with the left mainstem bronchus in the front and across the top. The bifurcation of the left and right pulmonary arteries presses hard against the back of the left mainstem bronchus at its commencement. As for the right mainstem bronchus, the right pulmonary artery transverses the top at the point where the upper-lobe bronchus arises.

The chance of these dangers is usually small because in most patients the route of the airway is readily recognizable. In the pulmonary lobes, however, disease often results in great distortion. The pulmonary arteries and veins are in close

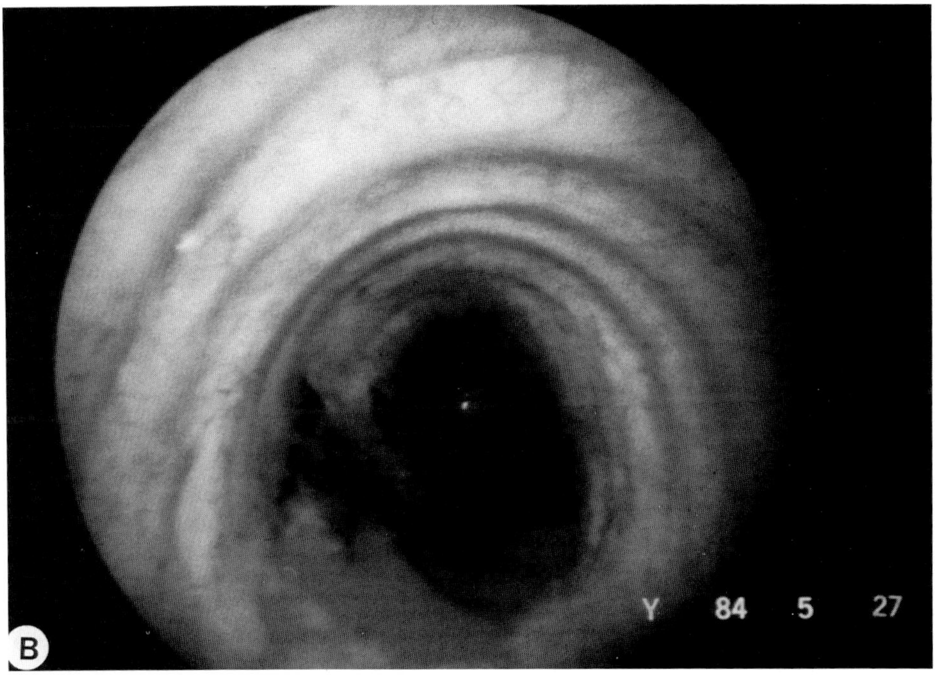

Figure 2B The same patient immediately after Nd:YAG laser photocoagulation and removal of tumor. The carina is distorted, but the opening into the left main bronchus is adequate. Tumor has been left on the posterior wall as this is a danger zone because of its proximity to the esophagus.

contact with the lobar bronchi in many places, and the endoscopist may unwittingly fire directly into the wall of the lumen and accidentally rupture a major vessel. Hence, particular caution is demanded in the upper lobe, which is almost beyond the reach of laser therapy.

Through the preoperative workup, the endoscopist should know the degree of vascularity present in the pathological tissue. Tumors have neovasculature, which sometimes branches from major arteries. Such is true for many carcinoid tumors, and the risk of hemorrhage is high.

Unfortunately, prevention cannot guarantee that hemorrhage will not occur, and the endoscopist must be ready to cope with this complication. As I have said, controlling hemorrhage hinges on the open-tube system, which permits simultaneous use of two suction tubes and the laser fiber. Because the main danger with hemorrhage is hypoxemia, priority should be given to securing ventilation. Therefore, the first measure to take in response to bleeding is often oxy-

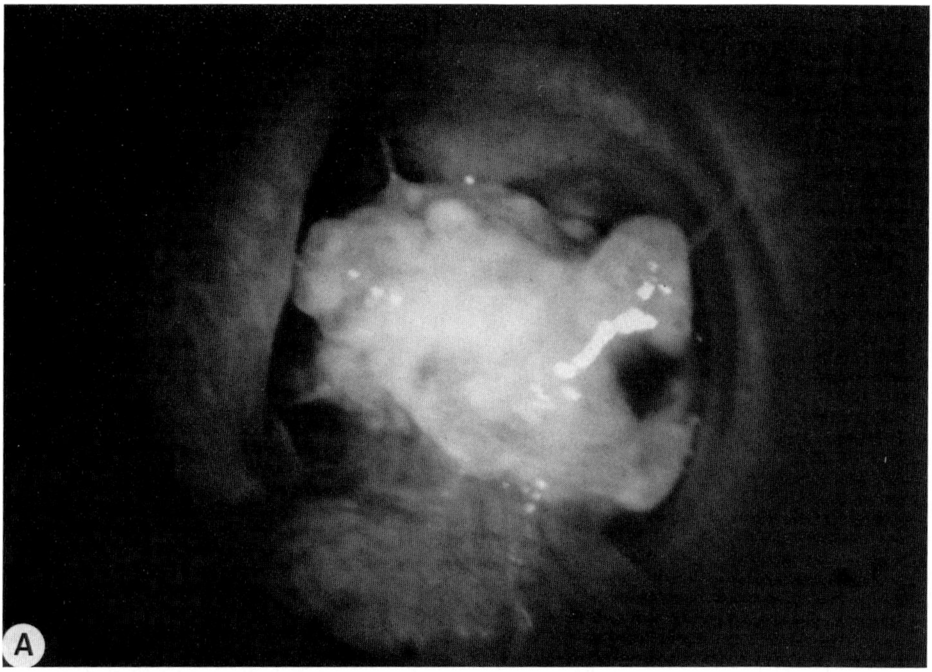

Figure 3 A Endoscopic photograph showing a large exophytic squamous cell carcinoma based on the carina partially obstructing both main bronchi. This patient presented with severe dyspnea and a history of hemoptysis.

genation and tracheobronchial cleansing, which should be carried out as described earlier, depending upon the location of the lesion. If massive bleeding occurs, the patient should be turned on the side containing the tumor, thus preventing blood from entering the healthy bronchus. After oxygenation, the endoscopist can secure hemostasis. For tracheal or bilateral lesions, when the tip of the bronchoscope has been passed distal to the stenosis, hemostasis should be carried out from distal to proximal airway as the open tube is withdrawn. In this way, bleeding is stopped by tamponade, and the operative field remains clear. Hemostasis should be performed circumferentially around the edges of the bleeding area and then directly into the source. One suction tube should be kept in contact with the operative field to suction blood out of the field.

C. Perforation

Because perforation is irreversible, every possible effort must be made to avoid its occurrence. As with large vessels, it is well to know the main anatomical dan-

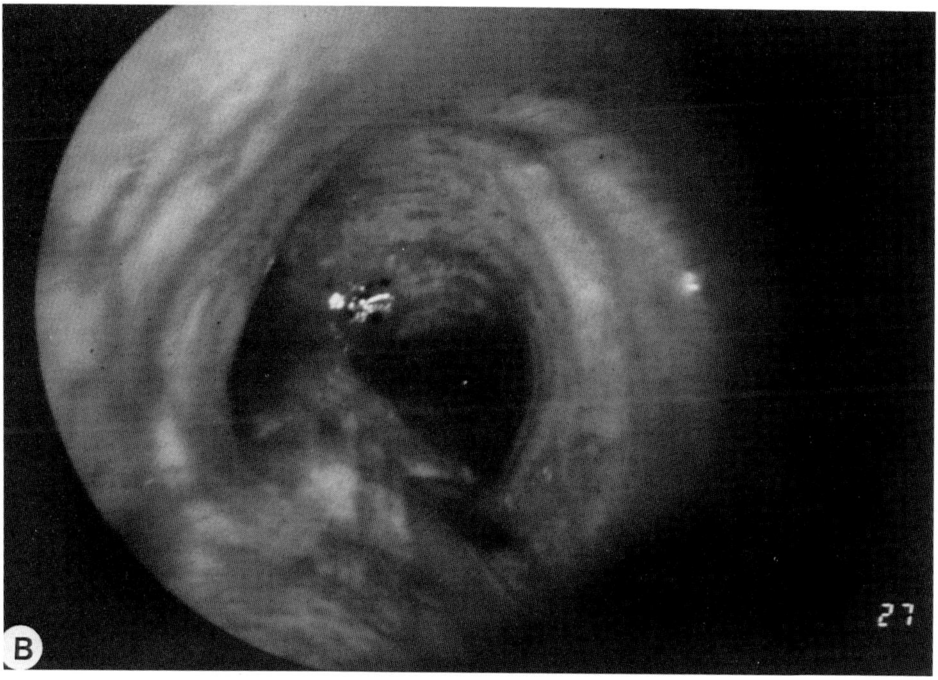

Figure 3B Photograph taken immediately after Nd:YAG laser photocoagulation and mechanical removal. Note the presence of some charring on the carina and widely patent main bronchi. The patient received radiotherapy after laser treatment.

gers. The back of the trachea and the first inch of the left mainstem bronchus are intimately bound to the front of the esophagus, and no cartilage is present in this wall to act as a buffer. Before lasing a tumor on the esophagus partition, a wise precautionary measure is to perform esophagoscopy to be sure that the tumor does not communicate with, or originate from, the esophagus. Work in lobar bronchi can also be hazardous if, as often happens, the lumen is not clearly distinguishable. In all of these areas it is especially important to minimize exposure by using short, low-powered bursts and by being more conservative in the amount of tissue resected.

D. Fire

One incident, which has been reported (4) during laser treatment, is fire. We did not experience this problem and believe that this hazard has been overstated. Almost invariably, ignition took place when the laser was being used through the

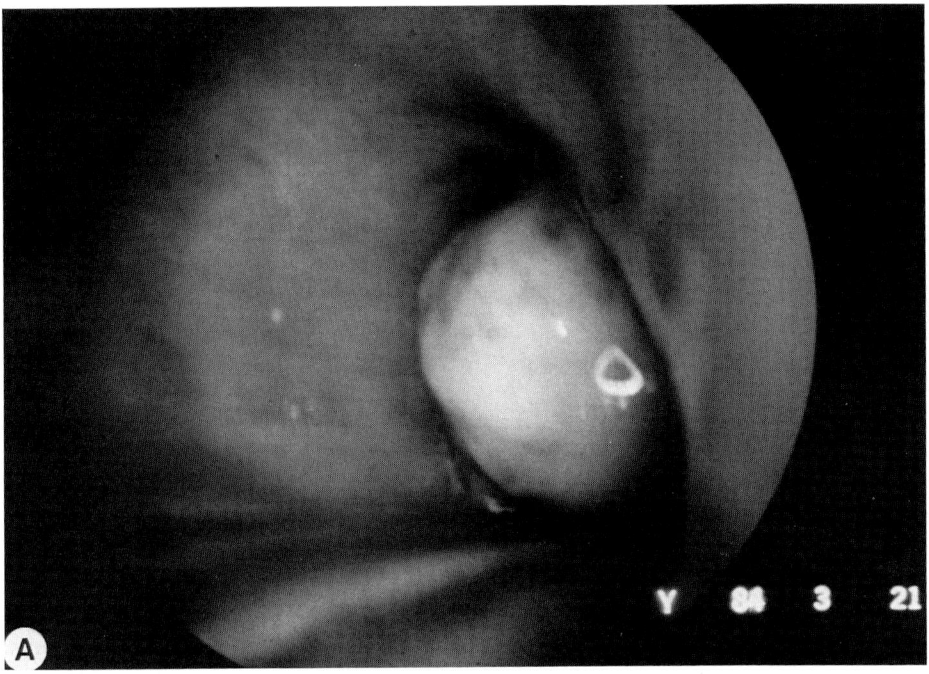

Figure 4A Endoscopic photograph showing complete obstruction of the right main bronchus by an exophytic adenocarcinoma. The patient's symptoms were dyspnea and partial lung collapse (postobstructive pneumonitis).

fiberscope at high power settings. This argues strongly against the fiberscope or any other unnecessary flammable materials, such as endotracheal tubes. Our suction tubes are made of hard, heat-resistant plastic, and ignition would require continuous high power exposure. Also, it is not advisable to use the laser in an oxygen-rich atmosphere. We recommend that power be kept below 45 W and that during firing the oxygen content be no higher than 50%.

E. Miscellaneous Dangers

Other complications can also happen during or after a laser procedure. Tension pneumothorax has occurred when jet ventilation has been used, but when the patient is breathing spontaneously, this fatal complication has not been encountered in our unit. The same can be said for pneumopericardial tamponade. Another possible complication is postoperative infection, which is best avoided by removing any remaining debris or secretions, or both, from the tracheobronchial tree before extubation.

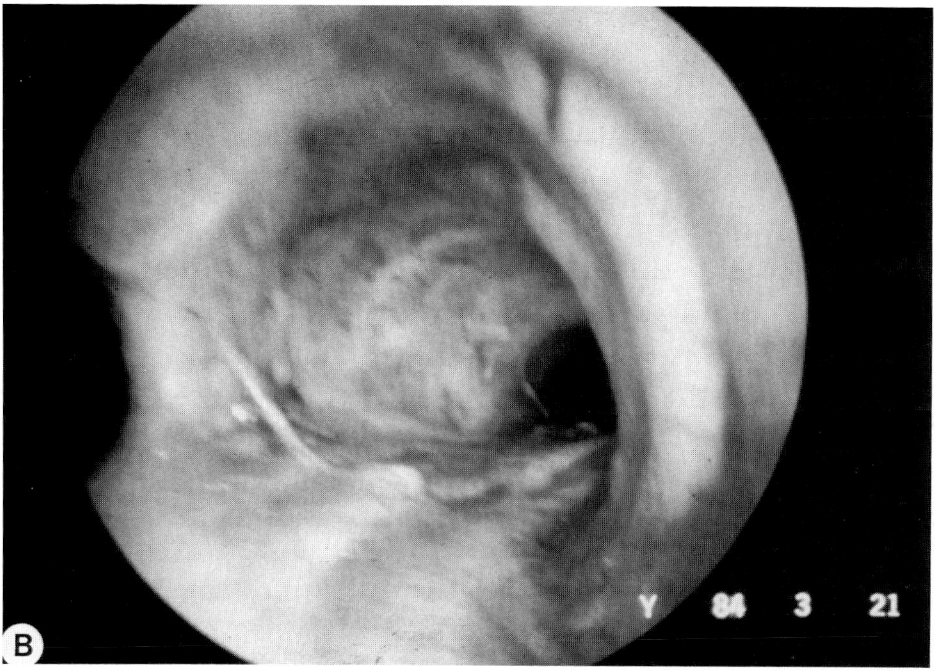

Figure 4B Endoscopic photograph taken immediately after Nd:YAG laser application. Note patency of the right main bronchus and some extrinsic narrowing of the lumen resulting from metastatic tumor in the mediastinum.

V. CLINICAL ASPECTS OF Nd:YAG LASER SURGERY

The exact clinical situation--tumor, stenosis, or miscellaneous conditions (i.e., foreign-body removal or hemostasis)—calls for some further commentary (2,5).

A. Tumor

The two factors that greatly determine the success of laser treatment for tumors are location and histology. Location is certainly the most important for immediate results because laser resection of endoluminal growth is always possible provided the neoplasm is accessible. Access is excellent in the trachea and right mainstem bronchus whose deviation from the trachea is slight. It follows that excision of lesions in these regions is almost fail proof. The left mainstem bronchus, which because its centerline deviates more radically from that of the trachea, is more difficult to enter, and the chances of success decrease to 75%. As for the lobar bronchi, the influence of accessibility is equally important. The bronchus intermedius on the right permits easy entry, and the success rate is high, but in other lobar areas the chances of success drop drastically. A factor that further complicates treatment in the lobes is that the degree of distortion is great with sizable tumors, and anatomical landmarks are obscured. The left-upper lobe bronchus is by far the most hazardous region for laser treatment. It is pertinent to repeat that, although the fiberscope can reach the upper lobes, it also reduces the response potential in the event of complications—most importantly, hemorrhage.

The second factor influencing success in laser surgery for tumoral obstruction is histology. As mentioned, my experience with benign tumors has been good enough to recommend the laser as the modality for initial treatment in many patients. Even papillomas, which have a marked tendency to recurrence, have responded well. Other benign entities occurring in the tracheobronchial tree include fibromas, leiomyofibromas, lipomas, angiomas, and chondromas, all of which are usually found in the trachea or mainstem bronchi. Fiberscopic resection is frequently possible in benign tumors, which are usually small, but a large lesion in a major airway, even if it is benign, requires the use of the open-tube system.

Most tumors encountered in the tracheobronchial tree are malignant to some degree. Carcinoid, adenoid cystic, and spindle cell tumors, all of which can occur in the tracheobronchial tree, are classified as mildly malignant lesions. All three have a high hemorrhagic potential and thus demand the twin suction tube technique. In inoperable carcinoid lung cancer, the Nd:YAG laser can be a second-choice modality used for removing airway obstruction. Intramural carcinoid growth frequently constitutes only the visible tip of an extensive parenchymatous mass. The endoscopist must not violate the airway wall. Checkup visits are necessary every 4-6 months, depending upon growth rate. Only localized adenoid cystic tumors are resectable; laser surgery does not manage the diffuse variety any better than conventional surgery. Spindle cell carcinoma is only slightly sensitive to radiotherapy so that when surgery is unavailable, laser use is a logical alternative therapy. Initially, multiple sessions are needed to establish patency of the airway, and given the tumor's rapid growth rate, checkup visits with further laser treatment must be scheduled at close intervals.

Squamous cell carcinoma, adenocarcinoma, and oat cell carcinoma are the most common highly malignant entities encountered in pulmonary medicine. Of these three cancers, squamous cell carcinoma is by far the most frequent. In stage I of squamous cell carcinoma, treatment with radiotherapy may be curative, but in later phases, the Nd:YAG laser can be used only for palliation. Recurrence is often rapid. Localized adenocarcinoma is a slower-growing cancer, and laser resection brings more lasting results. Oat cell carcinoma is usually accompanied by marked extrinsic compression, and laser treatment is seldom effective. Highly malignant tumors usually occur in high-risk patients (those with cardiac and chronic lung disease), cause a great amount of airway distortion, and involve some degree of extraluminal growth. The open-tube system is thus mandatory for control of the airway and hemorrhage.

B. Stenosis

The most frequent cause of tracheal stenosis is hospital-related injury. Prolonged intubation invariably results in tracheal injury. Extended tracheostomy, especially with a balloon cuff tracheal tube, produces mechanical irritation. Healing after any one of these procedures can lead to narrowing of the lumen. With these lesions the degree of involvement of the tracheal wall is variable in terms of circumference, depth, and length. The simplest form would be a solitary granuloma on a suture thread or in the vicinity of the tracheostomy tube. Solitary granulomas can usually be destroyed quickly and easily under local anesthesia through a fiberscope. If a suture thread is at the base of the granuloma, the laser can be used to cut and remove the suture provided at least 14 days has elapsed since the original procedure. Before granulomas are resected, it is prudent to obtain a biopsy specimen to be sure the lesion is benign. Inflammatory stenosis is an ongoing process implying some degree of involvement of the tracheal wall. At the onset, this involvement is only mucosal or submucosal, and laser therapy is potentially an effective therapeutic method, provided the pathological section extends circumferentially rather than longitudinally. In some cases of concentric stenosis, a membranous outgrowth gradually closes off the airflow. The endoscopist should begin resection on the inside edge of the lesion and work circularly out to the wall.

In the later stages of inflammatory stenosis, the tracheal wall becomes ulcerated, exposing cartilage, which ultimately softens. This leads to sagging and collapse of the airway. With time, fibrosis sets in, and the lesion is irreversible. The length of the section involved further complicates such a pathological process. Sometimes several foci of inflammation are seen at different levels. For this type of lesion, called bottleneck stenosis, laser treatment alone is not usually effective. My first attempts to apply laser resection to this condition were largely unsuccessful, and I now take a somewhat more subtle approach, deciding on which therapeutic strategy to follow after the first laser session.

The four alternatives include no further treatment, one or two additional laser treatments, sleeve resection, or insertion of a T-tube stent for 6 months at the end of which time a choice is made among the other three alternatives. We believe that this strategy has been effective. The laser has been successful for the simple membranous type of stenosis.

C. Miscellaneous Uses

I have used the Nd:YAG laser successfully for other purposes—mainly for removal of foreign bodies and control of hemorrhage. Occasionally, a foreign body may be so impacted that it is impossible to dislodge it with conventional tools, such as biopsy forceps. The Nd:YAG laser can be used in such circumstances either to free the object by removing granulation tissue or to carve a handle on the object for stronger attachment of the forceps used to remove a foreign body.

Although other modalities are certainly to be tried first, the laser can be called on for rapid management of tracheobronchial bleeding caused by traumatic aspiration or by a biopsy. Obviously, it is not effective in massive bleeding subsequent to tuberculosis or radiotherapy.

VI. CONCLUSION

The goal of Nd:YAG laser therapy is removal of tracheobronchial obstruction. For some applications, it is possible to speak of curative treatment, as in certain cases of tracheal stenosis, benign tumors, and foreign-body removal. With high-grade malignancy, laser treatment alone is only palliative, but then no currently available treatment of end-stage cancer, including surgery, can guarantee long-term stability. Thus, it would not be fair to use long-term results as a criterion for evaluating the usefulness of this modality.

Actually, laser resection, not having lingering or cumulative aftereffects, can be performed in conjunction with a wide variety of treatments. For example, for some forms of cancer a cure may be possible by removing the bulk of the tumor by means of the laser and then using localized radiation therapy (brachytherapy) to irradiate only the pathological zone. This strategy would achieve the elimination of malignant tissue without damaging surrounding structures. Another good illustration of this point is tracheal stenosis, which in many patients can be cured. Laser treatment leaves all therapeutic options open to the patient, and it is available to them after other modalities have failed. No other treatment can make the same claim.

This is not to say that we recommend laser resection as a primary treatment for malignant lung cancer. We do not argue that surgical excision can, under certain circumstances, achieve definitive resolution. It would be irresponsible to refuse surgery to a patient whose chances of long-term survival are high, but it would also be irresponsible to inflict the ordeal of surgery on one whose chances are slim. Quality survival is better than needless suffering.

In this respect, the laser is being called on belatedly in many patients. In short, although a number of factors must be taken into account in deciding on an appropriate therapeutic approach to each patient (histology, stage of disease, and the patient's physical condition and previous treatment), the laser deserves to be given serious consideration in the management of patients with lung disease. We do not believe that this is presently the case.

A further advantage of laser treatment is that with proper precaution it is safe in almost all patients. In our studies, we performed 1500 treatments on 817 patients with only two deaths. Overall mortality in most of the high-risk patients was only 0.36%.

To reiterate, the main principles of our technique, which we believe are directly responsible for our excellent safety record, include the following:

Light general anesthesia, i.e., the patient breathes spontaneously with only occasional assistance, is crucial because by eliminating the need for curarization and jet ventilation it reduces the dangers associated with anesthetizing high-risk patients. At the least sign of decompensation, treatment should be interrupted at least long enough to reestablish favorable respiratory status. In certain patients, blood oxygenation should be monitored. Teamwork between the endoscopist and the anesthesiologist is essential.

Rigid bronchoscopy using a specially designed rigid bronchoscope for laser resection is a major asset for three reasons. First, the beveled end of the open tube can be used to shave off coagulated tissue fragments, thus permitting a great economy in laser exposure both in terms of time and power. Second, the tactile feedback provided by palpation of the target area with a suction tube is highly valuable in gauging the safe limits of treatment. Third, this instrumentation is the endoscopist's best insurance in case of hypoxemia or hemorrhage because it enables simultaneous use of the laser and suction tubes.

An analogy, which well illustrates our feelings about the rigid bronchoscope system, can be made with automobile seat belts. You may not have an accident every time you get into your car, but the possibility is there, and it is therefore safer to have your seat belt fastened. A flexible fiberscope may be used safely in certain situations, but accidents can occur during short or minor procedures. If general anesthesia is given, why not go ahead and use the rigid bronchoscope system?

Thorough laser coagulation and complete tracheobronchial cleansing at the end of resection are mandatory to prevent postoperative complications. Our two fatal complications at Salvator Hospital occurred after the procedure. Once again we emphasize that a recovery room is a requirement for an endoscopy facility offering laser therapy.

ACKNOWLEDGMENT

I would like to express my gratitude to A. Corsini, who put my ideas into English words, and to N. Macciocu for secretarial assistance.

REFERENCES

1. Shapshay SM, Dumon J-F, and Beamis JF Jr: Endoscopic treatment of tracheobronchial malignancy: Experience with Nd-YAG and CO_2 lasers in 506 operations. *Otolaryngol. Head Neck Surg. 93*:205-210, 1985.

2. J-F. Dumon. *YAG Laser Bronchoscopy,* Surgical Science Series. Vol. 5. Praeger Publishers, New York, 1985.
3. Dumon J-F, Shapshay S, Bourcereau J, Cavaliere S, Meric B, Garbi N, and Beamis J: Principles for safety in application of neodymium-YAG laser in bronchology. *Chest 86*:163-168, 1984.
4. Casey KR, Fairfax WR, Smith SJ, and Dixon JA: Intratracheal fire ignited by the Nd-YAG laser during treatment of tracheal stenosis. *Chest 84*:295-296, 1983.
5. Dumon J-F, Reboud E, Garbe L, Aucomte F, and Meric B: Treatment of tracheobronchial lesions by laser photoresection. *Chest 81*:278-284, 1982.

7

Tracheobronchial Tree: Carbon Dioxide Laser

ROBERT H. OSSOFF
Vanderbilt University Medical School, Nashville, Tennessee

Laser technology has provided us with new options for the endoscopic management of patients with selected diseases of the tracheobronchial tree. Polanyi et al. (1) reported the first endoscopic delivery system for carbon dioxide (CO_2) laser surgery of the tracheobronchial tree in 1970. Three years later, Strong and his colleagues (2) successfully coupled this prototype endoscopic adapter to a standard, rigid, ventilating bronchoscope. Problems with this first adapter included its size and rather cumbersome arrangement for coupling to the bronchoscope; therefore, a smaller and lighter endoscopic adapter that coupled easily to the bronchoscope was developed in 1974. Strong et al. (3) presented their initial experiences with these two endoscopic adapters for bronchoscopic CO_2 laser surgery at the 1974 Annual Meeting of the American Broncho-Esophagological Association. This report represented a logical extension of the clinical application of lasers in medicine and surgery. Specifically, use of the CO_2 laser coupled to a rigid, ventilating bronchoscope allowed the surgeon to perform "hands-off" endoscopic surgery through the relatively long and narrow operative field of a ventilating bronchoscope. The advantages of precision, hemostasis, and decreased perioperative edema helped to facilitate visualization and control over the final result of the operation. However, lack of an aiming beam that would have allowed the surgeon to visualize the precise area of tissue to be vaporized by the invisible CO_2 laser limited the growth of bronchoscopic CO_2 laser surgery with this device. Laforet et al. (4) in 1976, Healy et al. (5) in 1979, Kullman (6) in 1980, Andrews and Horowitz (7) in 1980, McElvein and Zorn (8,9) in 1981 and 1983, Simpson et al. (10) in 1982, and Shapshay et al (11,12) in 1983 reviewed their experience with this bronchoscopic laser system in the management of patients with benign and malignant diseases of

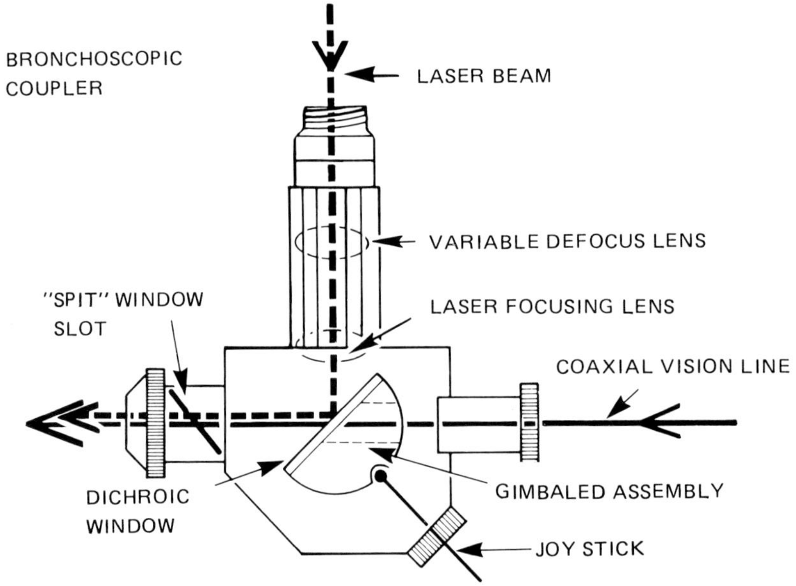

Figure 1 Diagram of the bronchoscopic coupler (*Source*: reprinted with permission of *Otolaryngology and Head and Neck Surgery,* C. V. Mosby, Publisher).

the tracheobronchial tree. Limitations with this laser bronchoscope system prompted research and development of a new endoscopic coupler for bronchoscopic CO_2 laser surgery as well as investigation of a different wavelength for bronchoscopic laser surgery, the neodymium:yttrium-aluminum-garnet (Nd:YAG) laser.

The universal endoscopic coupler for bronchoscopic CO_2 laser surgery was presented as a new instrument at the 1982 Annual Meeting of the American Broncho-Esophagological Association (13). Certain safety features were incorporated into this endoscopic coupler that were previously unavailable. At the time of presentation, this coupler had been used extensively in the animal laboratory but had not been tried clinically. However, a later report in 1985 by Ossoff and his colleagues (14) described their experience with this instrument.

The universal endoscopic coupler was designed with a gimbaled, partially reflective dichroic beam window to facilitate direct, coaxial vision with the invisible CO_2 laser beam (Fig. 1). The dichroic window was treated with special optical coatings that allowed the helium-neon aiming laser beam of the CO_2 surgical laser to be transmitted down the lumen of the bronchoscope with minimal distortion. This feature permitted the surgeon to visualize the precise area of tissue to be vaporized by the invisible CO_2 laser beam. A continuous variable defocus lens system was included to facilitate focus/defocus settings for incision, vapori-

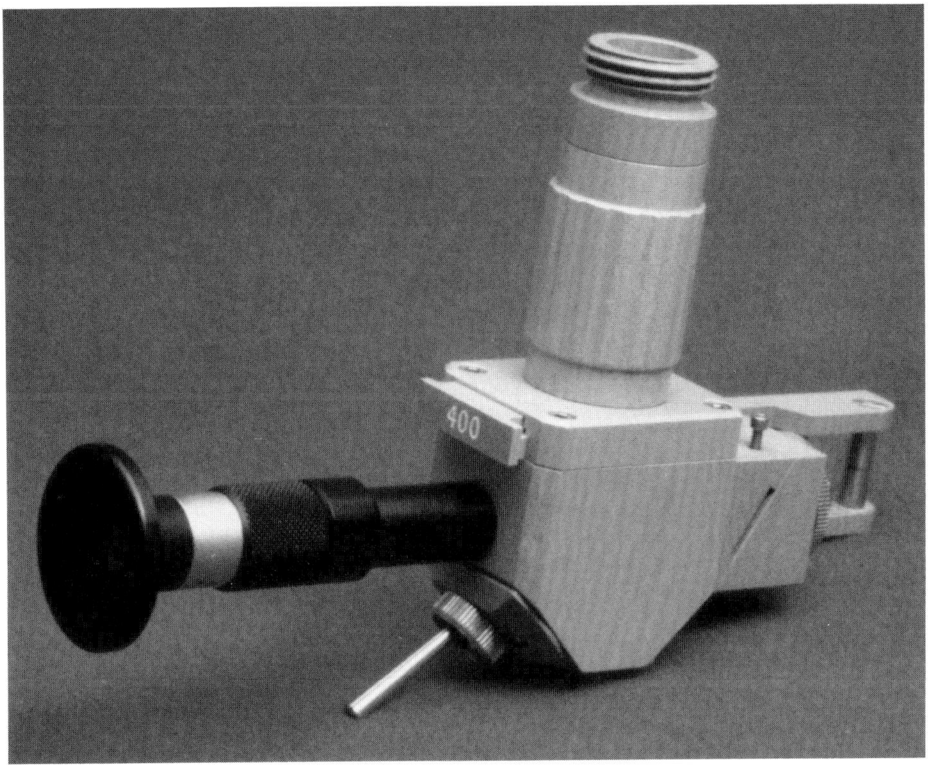

Figure 2 Bronchoscopic coupler showing joystick (*Source*: reprinted with permission of *Otolaryngology and Head and Neck Surgery,* C. V. Mosby, Publisher).

zation, and coagulation. By using this lens system, the surgeon could vary the spot size of the laser beam according to his surgical needs.

When using the previous generation of endoscopic adapters, the laser beam had to be centered within the lumen of the bronchoscope before the bronchoscope was introduced into the patient's tracheobronchial tree. Because standard, rigid, ventilating bronchoscopes are slightly flexible, slight bending of the bronchoscope after insertion into the patient could cause the laser beam to reflect off its inside wall, causing subsequent loss of power and possible heating of the bronchoscope itself. The trachea, larynx, pharynx, and oral cavity could be burned as a result of such an event. Inclusion of the gimballed mirror and "joystick" into the design of the universal endoscopic coupler allowed the surgeon to manipulate the helium-neon aiming laser within the lumen of the bronchoscope (Fig. 2). This feature allowed the aiming beam to be centered visually while the bronchoscope was in the

Figure 3 Bronchoscopic coupler with attached bronchoscope showing spit window (black arrow) and thumb screw attachment (white arrow) (*Source*: reprinted with permission of *Otolaryngology and Head and Neck Surgery*, C. V. Mosby, Publisher).

patient's tracheobronchial tree. In addition, the laser could be aimed anywhere within the optical field of the bronchoscope.

The first generation universal endoscopic coupler required a disposable "spit" window to prevent moisture present in the airway from fogging the dichroic window. These disposable windows, however, were too delicate, and the newer couplers have been shipped with a glass window that serves the same function as was originally planned for the disposable one. These newer windows require gas autoclaving between uses.

A quick on/off centering screw was designed to facilitate coaxial alignment and dislodgment of the bronchoscope from the coupler. The knurled thumb screw on the coupler had to be turned only 57° to accomplish this (Fig. 3). The endoscopic coupler, itself, weighed approximately 1 lb, and this feature enabled the surgeon to grasp the coupler easily with the attached bronchoscope in one hand and perform bronchoscopic laser surgery in the same fashion as conventional bronchoscopy (Fig. 4).

A set of custom-designed laser bronchoscopes was used with the universal endoscopic coupler (Fig. 5). The proximal end of these bronchoscopes was

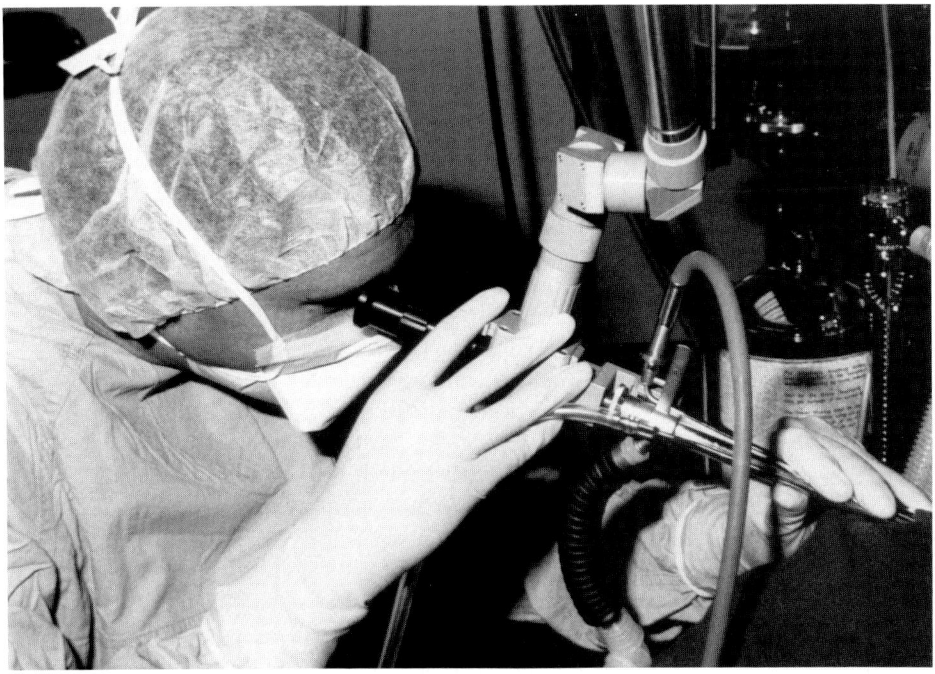

Figure 4 Bronchoscopic coupler with attached bronchoscope being used in lower trachea (*Source*: reprinted with permission of *Otolaryngology and Head and Neck Surgery,* C. V. Mosby, Publisher).

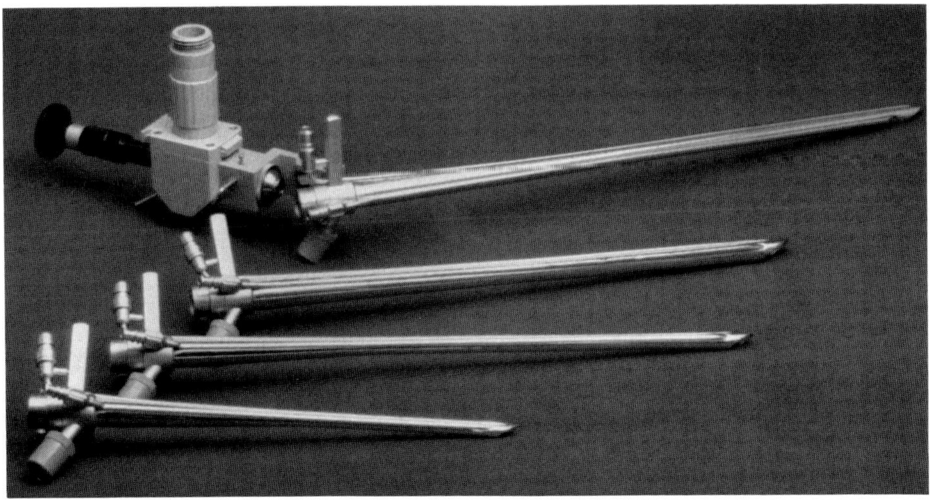

Figure 5 Bronchoscopic coupler with set of custom bronchoscopes (4 X 250, 6 X 350, 8 X 450, and 9 X 350 mm) (*Source*: reprinted with permission of *Otolaryngology and Head and Neck Surgery,* C. V. Mosby, Publisher).

machined optically true to facilitate coaxial alignment with the quick on/off centering screw of the endoscopic coupler. A smoke-evacuating channel was built into the wall of the bronchoscope to provide for the evacuation of laser-induced steam and smoke. An on/off bracket attached to the side of the proximal end of the bronchoscope facilitated quick attachment and dislodgment from the bronchoscopic coupler. Two measures were taken to reduce the potential for direct or reflected laser beam irradiation damage to the surrounding tissues. First, the inner lumen of the bronchoscopes was sandblasted to disperse any reflected laser beam energy, and second, the side ventilation ports were eliminated (15). In later generations of the bronchoscopes, the side ventilation ports were again added. Here, however, they were extremely small and angled to prevent any direct or indirect laser irradiation from escaping through them.

Because the rigid bronchoscope is utilized with the universal endoscopic coupler, general anesthesia is required. Several options exist including the choice of spontaneous versus controlled ventilation, squeezing the reservoir bag of a standard anesthesia circuit, or using jet ventilation at conventional frequencies with a Sanders injector or at high frequencies with appropriate equipment and whether anesthesia will be provided totally by the intravenous route or whether inhaled drugs will be used. If jet ventilation is chosen, there is no satisfactory way to scavenge the system, and total intravenous anesthesia must be provided (16).

Carbon dioxide lasers produce light with a wavelength of 10.6 μm which is in the invisible range of the spectrum. The CO_2 laser has a very high water absorption coefficient and is therefore readily absorbed by all soft tissues. Furthermore, its absorption is independent of tissue color with virtually no scattering. This high degree of absorption in soft tissue with limited lateral damage is the property that makes the CO_2 laser a precise surgical instrument for use in vaporizing tissue with a hands-off technique. The hallmark of the CO_2 laser is, then, its surgical precision. With current technology, the lateral zone of tissue damage extends less than 0.5 mm from the impact point. Because of the precise localization of the CO_2 laser beam effect, as well as the sealing effects of the beam, surrounding tissue exhibits minimal edema and scarring.

Current indications for bronchoscopic CO_2 laser surgery include the following diseases of the trachea and bronchi: (1) recurrent respiratory papillomatosis, (2) stenosis, (3) granulation tissue, (4) adenoma, (5) web, and (6) palliation of obstructing carcinoma. Contraindications to bronchoscopic CO_2 laser surgery are few. If patients suffer from tracheal stenosis with tracheomalacia or loss of cartilage, use of the laser bronchoscope must be discouraged. Here the danger of tracheal perforation is too great. Laser bronchoscopy must not be performed in patients with airway obstruction caused by extrinsic compression of the tracheobronchial tree.

Bronchoscopic CO_2 laser excision of recurrent respiratory papillomatosis in patients with tracheobronchial involvement is the best method of treatment at present. It must be emphasized that use of the CO_2 laser does not in any way reduce the frequency of recurrence. However, larger excisions can be performed with the aid of the laser bronchoscope than with conventional techniques.

Bronchoscopic laser surgery should be limited to the initial management of those patients with tracheal stenosis when one or more of the following factors associated with a poor result are found: (1) loss of cartilaginous support; (2) stenosis length of more than 1 cm; (3) circumferential scarring; and (4) carinal involvement. When possible, open surgical procedures should be performed in this group of patients. All patients with strictures of the trachea who failed endoscopic laser surgery, in my experience, had at least one of these predisposing factors. The inability to cure only about 50% of patients with tracheal stenosis with bronchoscopic CO_2 laser surgery, then, is not an indictment of this technology but rather a significant limitation of the disease process itself (17).

Patients with tracheal granulation tissue can be treated successfully with the CO_2 laser. Hemostasis was not a problem in any of these cases.

The principal advantages of bronchoscopic CO_2 laser surgery include precision, excellent vaporization, hemostasis of the microcirculation, decreased perioperative edema, and excellent healing.

Certain disadvantages are associated with using the CO_2 laser bronchoscope system for the management of patients with selected tracheobronchial diseases. First, hemostasis is limited to the microcirculation, vessels 0.5 mm and smaller. This raises difficulty with some of the moderately vascular tracheobronchial neoplasms often encountered when performing bronchoscopic laser surgery; however, when treating patients with other indications for bronchoscopic laser surgery, hemostasis was not a problem. In those patients with obstructing carcinomas, and in one patient with a bronchial carcinoid (Fig. 6), hemostasis would frequently have to be obtained by placing epinephrine-saturated pledgets on the tumor or by utilizing a bipolar bronchoscopic suction cautery. In either case, extra time and equipment were required. Second, lack of a fiber delivery system for bronchoscopic CO_2 laser surgery precludes the utilization of the new rod lens telescopes and associated photodocumentation equipment currently used with the Nd:YAG laser and limits bronchoscopic CO_2 laser surgery to a rigid delivery system. It is felt that this limitation is only a relative drawback, because the rigid ventilating bronchoscope is the delivery system of choice for airway control even when utilizing the Nd:YAG laser, which can be passed through a flexible fiberoptic bronchoscope.

The advantages of a rigid delivery system (rigid, ventilating bronchoscope) include ventilatory control of the compromised airway, multiple distal suction capabilities simultaneous with laser application, palpation of the tumor/cartilage interface, rapid evacuation of tumor fragments and debris to prevent hypoxemia, use of the bronchoscope tip as a "cookie cutter," and use of the bronchoscope tip to compress a bleeding tumor bed for hemostasis. Disadvantages of the rigid

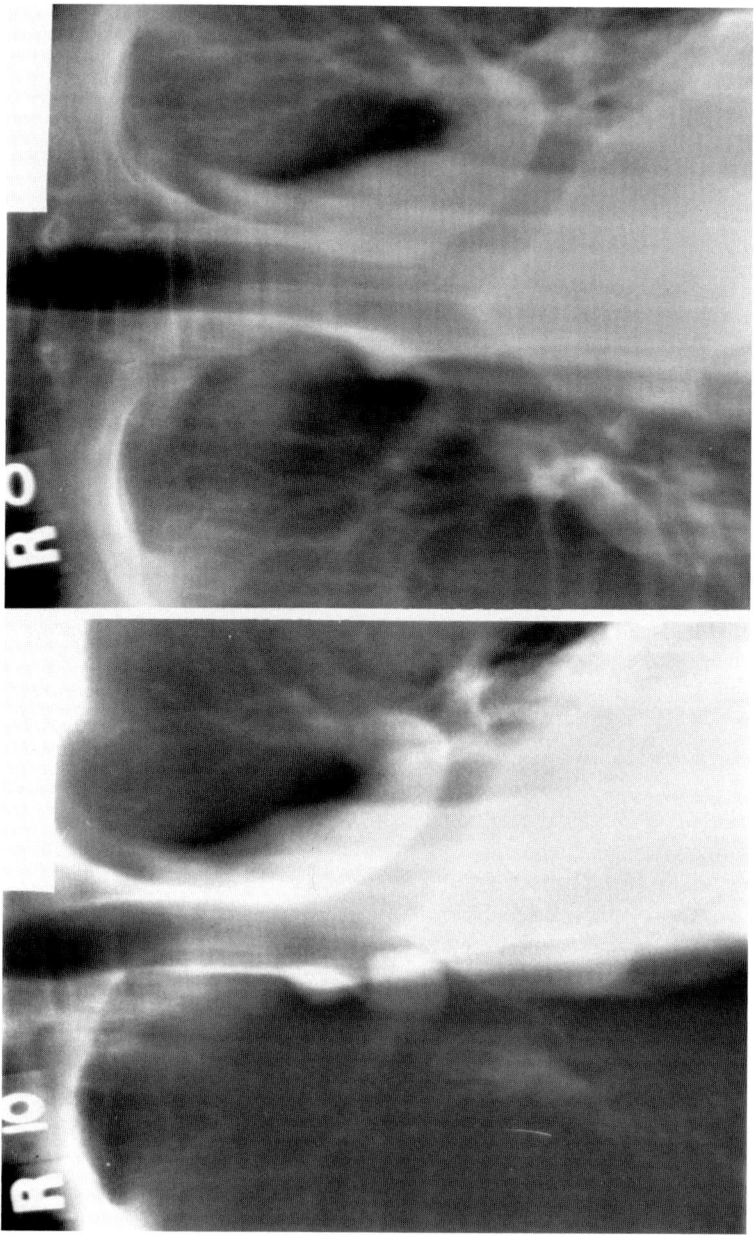

Figure 6 Preoperative (left) and postoperative (right) polytomogram of patient with bronchial adenoma showing obstructive lesion in the preoperative film at the right mainstem bronchus (*Source*: reprinted with permission of *Otolaryngology and Head and Neck Surgery*, C. V. Mosby, Publisher).

delivery system include the need for general anesthesia for airway control and patient comfort, ability to reach and treat only tracheal and proximal endobronchial lesions, and a learning curve with the rigid bronchoscopic instrumentation for those not already trained in its use. Despite these disadvantages, major obstructive disease of the tracheobronchial tree is best approached with the rigid delivery system and telescopic, visual control.

Advantages of a flexible fiberoptic bronchoscope system include the ability to perform the operations easily with local (topical) anesthesia and the ability to reach upper lobe and distal endobronchial lesions. Disadvantages with the flexible delivery system include lack of ventilatory control of the compromised airway, lack of multiple distal suction capabilities simultaneous with laser application, lack of ability to palpate the tumor/cartilage interface, slow evacuation of tumor fragments and debris when compared with the open, rigid system, the inability to compress a bleeding tumor bed with the tip of the bronchoscope, and the risk of a fire (sheath). Minor conditions such as granulation tissue or nonobstructive distal tumor can be approached with safety using the flexible fiberoptic bronchoscope.

Certain precautions must be taken when performing CO_2 laser surgery in the tracheobronchial tree through a rigid, ventilating bronchoscope. First, to reduce the risk of ocular damage, the patient's eyes should be protected by a double layer of moistened eye pads, and all operating room personnel should wear protective glasses. Second, all of the patient's exposed skin in the immediate area of the surgical field should be protected by a double layer of moistened surgical towels. Knowledge of the anatomy beyond the lesion is necessary to avoid inadvertent tracheal perforation. The long axis of the bronchoscope should parallel the long axis of the airway. Finally, the cartilaginous wall of the trachea or bronchus should not be exposed.

Potential complications associated with this procedure are few but impressive. They include tracheal perforation, tracheal burns, and postoperative airway obstruction. All three can be minimized by judicious use of the instrument and by atraumatic technique.

REFERENCES

1. Polanyi TG, Bredemeir HC, and Davis TW Jr: A CO_2 laser for surgical research. *Med. Biol. Eng. 8*:541-548, 1970.
2. Strong MS, Jako GJ, Polanyi TG, and Wallace RA: Laser surgery in the aerodigestive tract. *Am. J. Surg. 126*:529-533, 1973.
3. Strong MS, Vaughan CW, Polanyi TG, and Wallace RA: Bronchoscopic carbon dioxide laser surgery. *Ann. Otol. Rhinol. Laryngol. 83*:769-776, 1974.
4. Laforet EG, Berger RL, and Vaughan CW: Carcinoma obstructing the trachea: Treatment by laser excision. *N. Engl. J. Med. 291*:941, 1976.

5. Healy GB, McGill T, Simpson GT, and Strong MS: The use of the carbon dioxide laser in the pediatric airway. *J. Pediatr. Surg.* *14*:735-740, 1979.
6. Kullman GL: Laser palliation of bronchogenic carcinoma. *J. Fla. Med. Assoc.* *67*:566, 1980.
7. Andrews AH and Horowitz SL: Bronchoscopic CO_2 laser surgery. *Lasers Surg. Med.* *1*:35-45, 1980.
8. McElvein RB: Laser endoscopy. *Ann. Thorac. Surg.* *32*:463-467, 1981.
9. McElvein RB and Zorn GL: Indications, results, and complications of bronchoscopic carbon dioxide laser therapy. *Ann. Surg.* *199*:522-525, 1984.
10. Simpson GT, Strong MS, Healy GB, Shapshay SM, and Vaughan CW: Predictive factors of success or failure in the endoscopic management of laryngeal and tracheal stenosis. *Ann. Otol. Rhinol. Laryngol.* *91*:384-388, 1982.
11. Shapshay SM, Davis RK, Vaughan CW, Norton M, Strong MS, and Simpson GT: Palliation of airway obstruction from tracheobronchial malignancy: Use of the CO_2 laser bronchoscope. *Otolaryngol. Head Neck Surg.* *91*:615-619, 1983.
12. Shapshay SM, Healy GB, Davis RK, and Strong MS: Endoscopic management of airway obstruction from tracheobronchial neoplasia: Use of the carbon dioxide laser. *Lahey Clin. Found. Bull.* *32*:29-38, 1983.
13. Ossoff RH and Karlan MS: Universal endoscopic coupler for carbon dioxide laser surgery. *Ann. Otol. Rhinol. Laryngol.* *91*:608-609, 1982.
14. Ossoff RH, Karlan MS, Duncavage JA, Toohill RJ, Gluckman JL, Keane WM, Adkins JP, Norris CW, and Tucker JA: The universal endoscopic coupler for bronchoscopic carbon dioxide laser surgery: A multi-institutional clinical trial. *Otolaryngol. Head Neck Surg.* *93*:824-830, 1985.
15. Ossoff RH and Karlan MS: A set of bronchoscopes for carbon dioxide laser surgery. *Otolaryngol. Head Neck Surg.* *91*:336-337, 1983.
16. Eisenman TS and Ossoff RH: Anesthesia for bronchoscopic laser surgery. *Otolaryngol. Head Neck Surg.* (in press).
17. Ossoff RH, Duncavage JA, Toohill RJ, and Tucker GF Jr: Limitations of bronchoscopic carbon dioxide laser surgery. *Ann. Otol. Rhinol. Laryngol.* *94*:498-501, 1985

8

Brochoscopic Phototherapy with Hematoporphyrin Derivative and Argon Dye Laser

DENIS A. CORTESE
Mayo Clinic and Mayo Foundation and the Mayo Medical School, Rochester, Minnesota

I. INTRODUCTION

Lipson and Baldes reported in 1960 (1) that a derivative of hematoporphyrin (HpD) was preferentially concentrated by malignant tissue to a greater degree than pure hematoporphyrin (2). This derivative serves as a cancer marker by producing a characteristic red fluorescence when excited by light of appropriate wavelength (3-10). Furthermore, it absorbs light and uses the photon energy in a series of chemical reactions that eventuate in the death of the cell. Cytotoxic agents are generated by the transfer of energy from the excited porphyrin (sensitizer) to oxygen. The major cytotoxic agent is believed to be singlet oxygen, an excited oxygen species (11,12). Superoxide and the hydroxyl radical are also produced but are believed to play only minor roles (13,14).

The therapeutic effectiveness of HpD phototherapy (HpD-PT) was first reported in a patient with cancer of the bladder (15). Subsequent pioneering work by Dougherty and associates demonstrated the potential for this technique in the treatment of various solid, malignant tumors of the skin and subcutaneous tissues (16-19). Recent reports have indicated the possible application of this technique in the treatment of carcinoma of the tracheobronchial tree.

In principle, any source of light of appropriate spectrum and power output may be used for HpD-PT. The spectral output of the light must closely match the absorption maxima of HpD to produce the desired photochemical cytotoxic effect. The total energy density irradiating the cancer cells containing HpD must be sufficient to activate the chemical reactions. (Energy density, in joules per square centimeter, is the product of power density, in watts per square centimeter, times irradiation time, in seconds.) The main problem is focusing light of sufficient power density into the proximal end of a suitable optical fiber. It is not technically possible to focus enough power from conventional extended light sources, such as incandescent lamps and mercury arcs, into this fiber. However, the high power density and small beam diameter of lasers make them ideally suited for this task. The optical fiber can be inserted into a working channel of an endoscope to deliver the activating light to the tracheobronchial tree.

Hematoporphyrin derivative phototherapy produces cytotoxic effects through two possible mechanisms: one is energy-dependent, and the other is power-dependent. The energy-dependent mechanism is a result of photon absorption and energy transfer by HpD. The ensuing photodynamic chemical reactions cause cellular death. The power-dependent mechanism is the direct thermal effect of absorption of the light by the tissue. The cellular death in this instance is related to the power of the light and is independent of the photochemical reaction mediated by HpD.

Experimental results show that the pure photodynamic effect of HpD-PT may have a role in the treatment of small cancers (20,21); clinical experience indicates that phototherapy without the introduction of significant hyperthermia of tissue is ineffective as primary treatment of obstructing lesions (22-26). The therapeutic success with this type of therapy is limited to lesions that are within the reach of the flexible fiberbronchoscope and invade the bronchial wall to a depth of no more than 5-10 mm. This means that most bronchial carcinomas, perhaps more than 95%, are not amenable to this type of therapy because of extrabronchial extension of tumor or metastatic spread.

Particularly if accompanied by hyperthermia, HpD-PT has potential for palliative therapy in patients with larger, obstructing carcinomas of the tracheobronchial tree who have failed to respond to other standard therapy. The recent introduction of the neodymium:yttrium-aluminum-garnet (Nd:YAG) laser, which can provide dramatic and rapid palliation by debulking large endobronchial tumors, has limited the role of HpD-PT in palliation of tracheobronchial malignancy (27-34). However, HpD-PT does offer a unique alternative to surgical therapy in carefully selected patients with in situ or early invasive squamous cell carcinoma of the bronchus. These patients are not candidates for high-power laser resection with the Nd:YAG laser but would be suitable candidates for superficial phototherapy mediated by HpD. Being able to offer HpD-PT would be extremely helpful, particularly because patients with early squamous cell carcinoma are frequently at high surgical risk because of coexistent chronic obstructive lung disease or ischemic cardiovascular disease. Most important, they are at significant risk for the development of subsequent primary lung cancers at a rate of approximately 5%/yr (35-37).

This report summarizes the clinical therapeutic experience in 29 patients who received HpD-PT for cancer of the tracheobronchial tree from December 1980 to October 1984 at the Mayo Clinic.

II. PROCEDURES

A. Selection of Patients

All patients had biopsy-proved malignant tumors involving the tracheobronchial tree. Each patient's condition was evaluated by a bronchoscopist, a thoracic surgeon, and an oncologist. All patients studied either had undergone a pulmonary resection for another lung cancer or were considered to be inoperable for medical or technical reasons. Furthermore, all patients either had been rejected for, or had failed to respond to, conventional therapy with irradiation and chemotherapy. Once patients met these criteria, they were considered candidates for this new experimental treatment. Informed consent was obtained from each patient.

B. Hematoporphyrin Derivative

Hematoporphyrin derivative was prepared by a published (23) modification of the technique of Lipson and associates. All batches were tested in a murine tumor model system for effectiveness. The HpD was prepared for human use by passage through a Millipore filter and then sterilization and packaging by the hospital's pharmacy. A sample from each bottle was cultured to detect contamination before intravenous administration.

C. Source of Light for Irradiation

A medical-grade quartz optical fiber, 5 m long and having a core diameter of 400 μm (Quartz et Silice, Paris), was used to carry the laser beam through a bronchoscope to the endobronchial cancer. The exit beam diverged approximately 20°. The fiber was passed through the open channel of a flexible fiberoptic bronchoscope so that the distal end protruded from the tip of the bronchoscope.

During treatment, the distal tip of the fiber was maintained between 1.0 and 1.5 cm from the surface of the tumor. Light was directed to the surface of the tumor in all cases; in addition, some of the larger cancers received irradiation by intralesional implantation of the fiber probe. The beam had a diameter of approximately 10 mm and a centrally located intense spot of 4-5 mm. In some cases in which the tumor was 2-3 cm^2 in area, a small microlens was provided at the distal tip of the fiber, and in this way the intense spot could be spread over a surface diameter of 15-20 mm. The size of the light spot was considered in calculating the power density and the energy density (see earlier discussion).

The proximal end of this fiber was attached to a positioning device that was coupled to the output of a continuous-wave dye laser. The wavelength produced with the dye laser was 630 nm for all patients. The light beam was focused directly into the optical fiber. Rhodamine B dye was circulated in the

Figure 1 Argon pump-dye laser system used in bronchoscopic phototherapy.

dye laser. A continuous-wave argon ion laser was used to pump the dye laser (Figure 1). The power output from the optical fiber was measured with a calibrated, flat-response, continuous-wave power meter.

D. Treatment

All patients received HpD intravenously (2.0-5.0 mg/kg of body weight) 3-6 days before irradiation with light. General anesthesia was used for most of the procedures. Occasionally, topical tracheal anesthesia was sufficient to prevent coughing and to allow adequate treatment. Photographs, required for adequate documentation during follow-up, were taken before treatment and at each subsequent bronchoscopic procedure. Biopsies and brushings were performed either several days before the administration of HpD or immediately after the laser irradiation, to eliminate the possibility that traumatic bleeding would interfere with the absorption of the laser light. Power densities ranged from 170 to 500 mW/cm^2 and energy densities ranged from 90 to 675 J/cm^2 for tumors treated with surface irradiation. Power ranged from 200 to 500 mW and energy from 54 to 600 J for tumors treated with intralesional insertion of the fiber probe.

E. Follow-up and Interpretation of Results

All patients understood the need for close follow-up, including multiple bronchoscopic procedures, to document the therapeutic response. Results are reported in terms of the following definitions: complete response, no demonstrable tumor by biopsy *and* brushing; less than complete response, cancer still present by biopsy *or* brushing at the time of bronchoscopy.

III. RESULTS

There were 30 cancers in the 29 patients. In all, 48 sessions of HpD-PT were completed. Follow-up data range from a minimum of 3 months to a maximum of 46 months. Table 1 presents information on those patients in whom a complete response was obtained. Table 2 presents similar information on those patients who had less than a complete response. The data on all of these patients are complete as of October 1984.

A. Complete Response

In the nine cases in which response to HpD-PT was complete, one cancer was small cell carcinoma and nine were squamous cell carcinoma. The lesions were distributed throughout the tracheobronchial tree but all were within easy reach of the flexible fiberoptic bronchoscope (see Table 1). Seven of the lesions responded completely after a single session of HpD-PT; the remaining three responded after a second session (Figure 2). (In no patient was a complete response obtained if three or more sessions of HpD-PT were required.) The follow-up ranged from 3 to 38 months (mean, 25.9 months; median, 29 months).

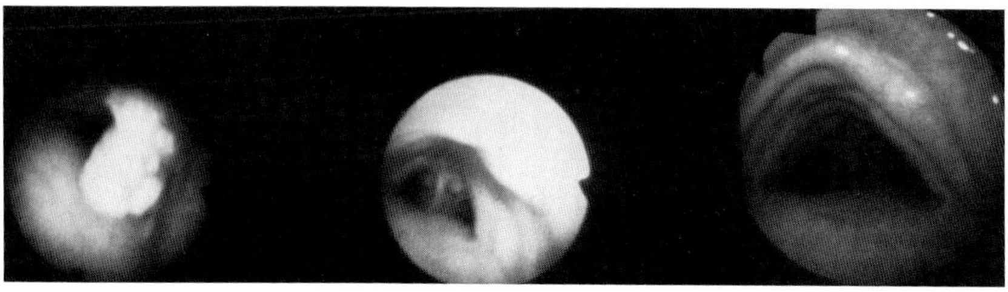

Figure 2 Squamous cell carcinoma of the anterior tracheal wall before treatment in case 9. Left, prior to first HpD-PT session. Middle, three months later cancer is still present by biopsy. Right, after second HpD-PT session, no cancer cells present on biopsy, brushing, and washing

Table 1 Summary of Cases With Complete Response[a] to Hematoporphyrin Derivative Phototherapy

Case	Site[b]	HPD-PT treatments (no.)	Tumor area (cm^2)	Follow-up	
				Months	Status[c]
1	RML	1	<1	38	A,NLR
1	LLL	2	2	35	A,LR
2	RUL	1	<1	29	D,2nd ca,NLR
3	Trachea	1	3	29	D,2nd ca,NLR
4	Carina	1	1	24	D,LR at 12 mo
5	BI	1	1-2	34	A,NLR
6	Carina	1	1-2	20	A,NLR
9	Trachea	2	2	35	A,NLR
22	Carina	2	3	12	A,2nd ca,NLR
27	RB6	1	1-1.5	3	A,2nd ca,NLR

[a]All cases were squamous cell carcinoma except for case 5 which was small cell carcinoma. In all patients, chest radiographs were negative.
[b]RML, right-middle lobe; LLL, left-lower lobe; RUL, right-upper lobe; BI, bronchus intermedius.
[c]A, alive; ca, carcinoma; D, dead; LR, local recurrence; NLR, no local recurrence.

In case 4 the patient died of a cancer that initially had shown a dramatic, complete response. It recurred locally 12 months after phototherapy, and the patient died 12 months after that. In cases 2 and 3, the patients died of a second primary lung cancer but had no local recurrence at the treatment site of their first cancers.

Second primary lung cancers developed in cases 22 and 27 but with no local recurrence of the original tumors treated with HpD-PT. In case 1, the patient had simultaneous cancers; one tumor completely disappeared and has not recurred, but the second cancer recurred locally at 35 months after an initial complete response.

The patients are alive with no known cancer in cases 5, 6, and 9.

All patients who eventually had a complete response to HpD-PT had had pretreatment chest roentgenograms that were negative for that cancer. Furthermore, the cancers were quite small on endoscopic examination before treatment: surface areas were approximately 3 cm^2 in two of the lesions and smaller in eight.

B. Less Than Complete Response

Eighteen of these 20 cancers were squamous cell carcinoma, one was a fibrosarcoma, and one was a tracheal cylindroma. The latter two cell types showed no response, even after several sessions of HpD-PT.

Three patients (cases 16, 17, and 25) died of hemoptysis which occurred between 1 and 4 weeks after HpD-PT (see Table 2). In cases 16 and 17, the hemoptysis most likely was due to tumor necrosis and bleeding related to HpD-PT; in each case the tumors were large, obstructing, and visible on roentgenograms. In case 25, the patient died 31 days after treatment because of hemoptysis, likely related to the cancer, but whether or not it was precipitated by the HpD-PT is unclear.

The other deaths were as follows: case 13, extensive fibrosarcoma that produced bilateral airway obstruction and involved the pericardium; cases 18, 19, and 21, metastatic disease; case 14, metastasis and malignant airway obstruction; case 15, local disease and obstruction.

In case 10, there was incomplete response to two sessions of HpD-PT and mild obstruction of the distal trachea and right mainstem bronchus was developing. This area was then treated with three separate sessions of Nd:YAG laser phototherapy. The most recent examinations show no evidence of residual cancer on visual examination or in biopsy tissue, bronchial brushings, and bronchial washings.

The remaining 11 patients are alive with residual local cancer and are receiving additional available experimental therapy, such as endobronchial brachytherapy and Nd:YAG laser phototherapy.

It should be noted that the surface area of these tumors was estimated to

Table 2 Summary of Cases With Less Than Complete Response[a] to Hematoporphyrin Derivative Phototherapy

Case	Site[b]	X-ray[c]	Tumor area (cm^2)	Follow-up	
				Months	Status[d]
7	Trachea	–	3	46	A,HPD-PTx4,YAG,Brac,RC
8	Trachea	+	4	46	A,HPD-PTx3,YAG,RC
10	Trachea	–	2	27	A,HPD-PTx2,YAGx3 to CR,NLR
11	RLL	–	2	26	A,Brac,RC
12	Trachea	+	3	19	A,HPD-PTx3,YAGx2,Brac,RC
13	L main	+	5	9	D,HPD-PTx2,RC
14	BI	+	5	9	D,HPD-PTx2,Ob&Mets,RC
15	R main	+	4	16	D,RC
16	LLL	+	10	6 days	D,Hem
17	RUL	+	12	11 days	D,Hem

18	LUL	+	3	8	D,Mets
19	Trachea	–	4	2	D,Mets
20	R main	+	2	16	A,HPD-PTx2,YAGx2,RC
21	LB6	–	<1	9	D,HPD-PTx2,Mets,RC
23	LLL	–	3	16	A,RC
24	RB3	–	<1	12	A,HPD-PTx2,YAG,Brac,RC
25	RUL	+	2	31 days	D,Hem
26	RUL	–	2	8	A,HPD-PTx2,RC
28	Carina	+	4	3	A,YAG,RC
29	Trachea	+	3	3	A,RöRx,RC

[a]All cases were squamous cell carcinoma except: case 8, cylindroma; case 13, fibrosarcoma.
[b]RLL, right-lower lobe; LUL, left-upper lobe; also see footnote in Table 1.
[c]Finding on chest radiograph: –, no evidence of cancer; +, cancer detectable.
[d]A, alive; HPD-PT, hemoporphyrin derivative phototherapy; Brac, endobronchial brachytherapy; CR, Complete response; D, dead; Hem, hemoptysis; Mets, metastases; NLR, no local recurrence; Ob, airway obstruction; RC, residual cancer; RöRx, radiation therapy; YAG, Nd:YAG laser.

range from less than 1 cm^2 to 12 cm^2. Overall, these lesions were larger in estimated surface area than those in patients who had a complete response. Furthermore, the cancer was visible on the chest roentgenogram in 12 of these 20 patients. Clearly, these lesions were more advanced than the lesions of patients who had a complete response.

C. Complications

Three patients experienced significant sunburn reaction. In two the reaction involved the face, and in one it involved the hands. The degree of reaction ranged from first- to second-degree burn and responded promptly to symptomatic treatment with antihistamines and steroids. In four patients, cough that was productive of blood-tinged sputum developed. Gray, necrotic material was frequently expectorated with the cough; usually, the symptoms resolved within 2 weeks after treatment. Four patients had temporary airway obstruction that was thought to be related to posttreatment bronchial edema. In each case, this responded to intravenous bronchodilator therapy and corticosteroids.

IV. DISCUSSION

Hematoporphyrin derivative phototherapy of bronchogenic carcinoma now appears to be a new addition to the standard modalities of surgical resection, radiation therapy, and chemotherapy. The role of HpD-PT is limited because the maximal therapeutic depth of penetration of the activating light is no more than 2 cm, and probably much less than 1 cm, when the fiberoptic bronchoscope is used. Therefore the major therapeutic use is in treatment of bronchogenic carcinoma at the in situ or intramural invasive stage. The likelihood of nodal metastasis from this type of tumor is negligible, and the risk of perforation through the bronchial wall is minimal (37). The results described in this summary support this conclusion as do the data of others (22,24,26). Small tumors, particularly if they have not eroded through the bronchial wall, seem to respond best. Major complications after treatment with HpD-PT, such as hemoptysis, have occurred only in patients with very large obstructing tumors. Other complications—such as retained secretions, airway obstruction due to accumulation of exudate, fever, pneumonia, and abscess formation—have been reported after HpD-PT for advanced-stage tumors.

Large obstructing tumors of the tracheobronchial tree will respond partially to HpD-PT. Obstructed airways can be opened with this technique. However, the degree of response is unpredictable, particularly because the maximum effect of therapy occurs several days after treatment. Edema and an exudative response frequently develop and can result in temporary complete obstruction of a previously partially obstructed airway. These larger obstructing cancers are better treated by phototherapy with a Nd:YAG laser, a powerful laser that exerts its therapeutic effect directly through thermal coagulation/necrosis of the tumor. This results in vessel coagulation and tumor shrinkage, permitting relatively safe and bloodless removal of an obstructing tumor. The operator can debulk the tumor directly and improve the airway at the time of bronchoscopy. The success of treatment is obvious immediately. However, Nd:YAG laser therapy is considered to be palliative and cannot be used for aggressive treatment of lesions that are eroding into and through the bronchial wall.

The current state of knowledge about HpD-PT indicates that it does have the potential to become an alternative to surgical resection as the primary treatment in patients who have bronchogenic cancers that are smaller than 3 cm^2 in surface area and that appear to be superficial and nonobstructing at the time of bronchoscopy. Whether the long-term, curative potential of this treatment will be equivalent to that of surgical resection must await further studies and longer follow-up.

ACKNOWLEDGMENTS

This investigation was supported in part by Research Grant CA-28923 from the National Institutes of Health, Public Health Service.

REFERENCES

1. Lipson RL and Baldes EJ: The photodynamic properties of a particular hematoporphyrin derivative. *Arch. Dermatol. 82*:508-516, 1960.
2. Figge FHJ, Weiland GS, and Manganiello LOJ: Cancer detection and therapy: Affinity of neoplastic, embryonic, and traumatized tissues for porphyrins and metalloporphyrins. *Proc. Soc. Exp. Biol. Med. 68*:640-641, 1948.
3. Lipson RL, Baldes EJ, and Olsen AM: The use of a derivative of hematoporphyrin in tumor detection. *J. Natl. Cancer Inst. 26*:1-10, 1961.
4. Gregorie HB Jr, Horger EO, Ward JL, Green JF, Richards T, Robertson HC Jr, and Stevenson TB: Hematoporphyrin-derivative fluorescence in malignant neoplasms. *Ann. Surg. 167*:820-828, 1968.
5. Kinsey JH, Cortese DA, and Sanderson DR: Detection of hematoporphyrin fluorescence during fiberoptic bronchoscopy to localize early bronchogenic carcinoma. *Mayo Clin. Proc. 53*:594-600, 1978.
6. Cortese DA, Kinsey JH, Woolner LB, Payne WS, Sanderson DR, and Fontana RS: Clinical application of a new endoscopic technique for detection of in situ bronchial carcinoma. *Mayo Clin. Proc. 54*:635-641, 1979.
7. Doiron DR, Profio E, Vincent RG, and Dougherty TJ: Fluorescence bronchoscopy for detection of lung cancer. *Chest 76*:27-32, 1979.
8. Hayata Y, Kato H, Konaka C, Ono J, Matsushima Y, Yoneyama K, and Nishimiya K: Fiberoptic bronchoscopic laser photoradiation for tumor localization in lung cancer. *Chest 82*:10-14, 1982.

9. King EG, Man G, LeRiche J, Amy R, Profio AE, and Doiron DR: Fluorescence bronchoscopy in the localization of bronchogenic carcinoma. *Cancer 49*:777-782, 1982.
10. Kato H, Aizawa K, Ono J, Konaka C, Kawate N, Yoneyama K, Kinoshita K, Nishimiya K, Sakai H, Noguchi M, Tomono T, Kawasaki S, Tokuda Y, and Hayata Y: Clinical measurement of tumor fluorescence using a new diagnostic system with hematoporphyrin derivative, laser photoradiation, and a spectroscope. *Lasers Surg. Med. 4*:49-58, 1984.
11. Weishaupt KR, Gomer CJ, and Dougherty TJ: Identification of singlet oxygen as the cytotoxic agent in photo-inactivation of a murine tumor. *Cancer Res. 36*:2326-2329, 1976.
12. Dougherty TJ, Gomer CJ, and Weishaupt KR: Energetics and efficiency of photoinactivation of murine tumor cells containing hematoporphyrin derivative. *Cancer Res. 36*:2330-2333, 1976.
13. Buettner GR and Oberley LW: The apparent production of superoxide and hydroxyl radicals by hematoporphyrin and light as seen by spin-trapping. *FEBS Lett. 121*:161-164, 1980.
14. Hariharan PV, Courtney J, and Eleczko S: Production of hydroxyl radicals in cell systems exposed to hematoporphyrin and red light. *Int. J. Radiat. Biol. 37*:691-694, 1980.
15. Kelly JF and Snell ME: Hematoporphyrin derivative: A possible aid in the diagnosis and therapy of carcinoma of the bladder. *J. Urol. 115*:150-151, 1976.
16. Dougherty TJ, Kaufman JE, Goldfarb A, Weishaupt KR, Boyle D, and Mittleman A: Photoradiation therapy for the treatment of malignant tumors. *Cancer Res. 38*:2628-2635, 1978.
17. Dougherty TJ, Lawrence G, Kaufman JH, Boyle D, Weishaupt KR, and Goldfarb A: Photoradiation in the treatment of recurrent breast carcinoma. *J. Natl. Cancer Inst. 62*:231-237, 1979.
18. Dougherty TJ: Photoradiation therapy for cutaneous and subcutaneous malignancies. *J. Invest. Dermatol. 77*:122-124, 1981.
19. Dougherty TJ, Thoma RE, Boyle DG, and Weishaupt KR: Interstitial photoradiation therapy for primary solid tumors in pet cats and dogs. *Cancer Res. 41*:401-404, 1981.
20. Kinsey JH, Cortese DA, and Neel HB: Thermal considerations in murine tumor killing using hematoporphyrin derivative phototherapy. *Cancer Res. 43*:1562-1567, 1983.
21. Svaasand LO, Doiron DR, and Dougherty TJ: Temperature rise during photoradiation therapy of malignant tumors. *Med. Phys. 10*:10-17, 1983.
22. Hayata Y, Kato H, Konaka C, Ono J, and Takizawa N: Hematoporphyrin derivative and laser photoradiation in the treatment of lung cancer. *Chest 81*:269-277, 1982.
23. Cortese DA and Kinsey JH: Endoscopic management of lung cancer with hematoporphyrin derivative phototherapy. *Mayo Clin. Proc. 57*:543-547, 1982.

24. Vincent RG, Dougherty TJ, Rao U, Boyle DG, and Potter WR: Photoradiation therapy in advanced carcinoma of the trachea and bronchus. *Chest 85*: 29-33, 1984.
25. Cortese DA and Kinsey JH: Hematoporphyrin derivative phototherapy in the treatment of bronchogenic carcinoma. *Chest 86*:8-13, 1984.
26. Hayata Y, Kato H, Konaka C, Amemiya R, Ono J, Ogawa I, Kinoshita K, Sakai H, and Takahashi H: Photoradiation therapy with hematoporphyrin derivative in early and stage I lung cancer. *Chest 86*:169-177, 1984.
27. Toty L, Personne C, Colchen A, and Vourc'h G: Bronchoscopic management of tracheal lesions using the neodynium yttrium aluminium garnet laser. *Thorax 36*:175-178, 1981.
28. Dumon J-F, Reboud E, Garbe L, Aucomte F, and Meric B: Treatment of tracheobronchial lesions by laser photoresection. *Chest 81*:278-284, 1982.
29. Hetzel MR, Millard FJC, Ayesh R, Bridges CE, Nanson EM, Swain CP, and Williams IP: Laser treatment of carcinoma of the bronchus. *Br. Med. J. 286*:12-16, 1983.
30. McDougall JC and Cortese DA: Neodymium-YAG laser therapy of malignant airway obstruction: A preliminary report. *Mayo Clin. Proc. 58*:35-39, 1983.
31. Dumon JF, Shapshay S, Bourcereau J, Cavaliere S, Meric B, Garbi N, and Beamis J: Principles for safety and application of neodymium-YAG laser in bronchology. *Chest 86*:163-168, 1984.
32. Arabian A and Spangnolo SV: Laser therapy in patients with primary lung cancer. *Chest 86*:519-523, 1984.
33. Wolfe WG, Cole PH, and Saviston DC Jr: Experimental and clinical use of the YAG laser in the management of pulmonary neoplasms. *Ann. Surg. 199*: 526-531, 1984.
34. Gelb AF and Epstein JD: Laser in treatment of lung cancer. *Chest 86*:662-666, 1984.
35. Martini N and Melamed MR: Occult carcinomas of the lung. *Ann. Thorac. Surg. 30*:215-221, 1980.
36. Cortese DA, Pairolero PC, Bergstralh EJ, Woolner LB, Uhlenhopp MA, Piehler JM, Sanderson DR, Bernatz PE, Williams DE, Taylor WF, Payne WS, and Fontana RS: Roentgenographically occult lung cancer: A ten-year experience. *J. Thorac. Cardiovasc. Surg. 86*:373-379, 1983.
37. Woolner LB, Fontana RS, Cortese DA, Sanderson DR, Bernatz PE, Payne WS, Pairolero PC, Piehler JM, and Taylor WF: Roentgenographically occult lung cancer: Pathologic findings and frequency of multicentricity during a 10-year period. *Mayo Clin. Proc. 59*:453-466, 1984.

9

Laser Therapy for the Upper Gastrointestinal Tract

DAVID FLEISCHER
Georgetown University Hospital, Washington, D.C.

I. INTRODUCTION

The two major applications for endoscopic laser therapy (ELT) in the upper gastrointestinal (UGI) tract are for the treatment of bleeding and for neoplastic disease. Acute hemorrhage and lesions that have bled but are not actively bleeding have been treated. Benign and malignant tumors of the pharynx, esophagus, stomach, duodenum, and ampulla of Vater have been lased. For neoplasms, the treatment is generally palliative to ameliorate symptoms and to improve the quality of life. Curative therapy has been accomplished on rare occasions in the United States, but it is uncommon. In Japan, where the diagnosis of early gastric cancer is more commonly made, there is a growing experience with curative treatment by laser. In China, where esophageal cancers may be detected earlier, there has been some application of the laser for curative therapy. In addition to the laser's utility for bleeding and neoplastic diseases of the UGI tract, it has also been used endoscopically to treat some benign strictures of the esophagus secondary to reflux, postanastomotic benign strictures, esophageal and duodenal webs and rings, and to drain pseudocysts of the pancreas (Table 1).

The first laser used for gastrointestinal purposes in man was the argon laser. When the neodymium:yttrium-aluminum-garnet (Nd:YAG) laser was first applied, there was a great deal of controversy over which laser was "best." As is often the case, one may work better in one situation than another and vice versa. However, in both the United States and other countries, the Nd:YAG is used far more commonly. In this country it is the laser most frequently applied in more than 90% of gastroenterology units. It is preferred because of its broader versatility. Because of its wavelength and power output, it is more effective in destroying tissue

Table 1 Gastrointestinal Applications of Lasers

Coagulation
acute hemorrhage
active bleeding
recent bleeding (e.g., visible vessel)
recent hemorrhage—not currently bleeding
angiodysplasia
varices
hemorrhoids
Vaporization
neoplastic disease
palliation
curative treatment
ancillary (e.g., placement of esophageal prosthesis)
benign stricture or web
biliary disease
infundibulotomy
strictures
fracture "gallstones"
cyst drainage

than the argon laser. Additionally, it has proved to be very effective for treating GI bleeding. Initial fears that its greater depth of penetration would lead to an unacceptable level of perforation have not been borne out. There has been a great interest in applying the carbon dioxide (CO_2) laser endoscopically in the GI tract. Delivery systems that permit its satisfactory and routine application with flexible endoscopes have not yet been developed.

II. LASER THERAPY OF UPPER GASTROINTESTINAL BLEEDING

With the possible exception of an aortoenteric fistula, literally every bleeding site in the UGI tract has been approached by laser therapy to control blood loss. There are several points of clarification that should be addressed. First, laser therapy is only one of numerous modalities that can be applied via the endoscope to arrest bleeding (Table 2). Second, there is relatively little comparative data available to help one decide which mode of therapy is best. Although much more data have

Table 2 Endoscopic Therapy for Gastrointestinal Bleeding

Topical	
	tissue glue
	clotting factors
	collagen hemostat
	ferromagnetic tamponade
Injection	
	variceal sclerosis
	nonvariceal injection
Mechanical	
	balloon tamponade
	clips
	sutures
Thermal	
	electrocoagulation
	monopolar
	electrohydrothermal
	bipolar (multipolar)
	laser photocoagulation
	argon
	Nd:YAG
	heater probe
	cryotherapy

been published about the laser than about other modalities, the question is far from decided. Again, it is likely that no one modality will be best for all situations. The laser is currently more expensive and less portable than some alternative methods. Third, a distinction must be made between treatment of an actively bleeding lesion (e.g., spurting duodenal ulcer) and a lesion thought to be the site of recent bleeding that is not currently bleeding (e.g., an ulcer with stigmata of recent hemorrhage). There may be differences in therapeutic approaches as well as technique. Fourth, discrete lesions (e.g., a single ulcer) are generally regarded as ideal lesions for ELT. However, there is controversy over whether or not a diffuse lesion, such as hemorrhagic gastritis, is an appropriate condition for ELT. Fifth, most endoscopists agree that nonvariceal lesions, especially those that are discrete (ulcer, erosion, Mallory-Weiss tear, bleeding tumor, angiodysplasia, etc.), should be treated by laser coagulation. Although there is a moderately large body of data suggesting that lasers can effectively stop acute variceal bleeding, endoscopic injection sclerotherapy is by far the most commonly employed type of treatment, and even the advocates of laser therapy for acute variceal bleeding generally feel that if the laser is used to arrest the initial bleeding, some more definitive follow-up therapy is required.

A. Acute Hemorrhage

The first reports of endoscopic laser therapy in humans emerged in the mid-1970s. At about the same time, Dwyer et al. (1), Nath et al. (2), and Frühmorgen et al. (3) performed successful hemostasis in humans. The initial reports were from uncontrolled studies but are worthy of notice. In 1979, Kiefhaber (4) reported the entire international experience regarding endoscopic therapy of UGI bleeding. At that time 37 centers (three in the United States) had used lasers for this purpose. A total of 1729 patients had been treated. Successful initial hemostasis was achieved in 84% (range 70-100%) of 196 patients treated with argon laser in seven centers. In 90% (range 74-100%) of 1533 patients treated with the Nd:YAG laser in 31 centers, initial hemostasis was accomplished. This survey does not address the prevalence of rebleeding and was not intended to address follow-up. By 1985, more than 300 centers were using lasers for gastrointestinal bleeding. The most experience is that of Kiefhaber (5). He has reported the treatment of 994 acute bleeding incidents in 625 patients. Lesions included varices, Mallory-Weiss tears, ulcers, erosions, vascular anomalies, and tumors. His overall success rate for achieving initial hemostasis is 94%. He reports a reduced mortality for acute and chronic ulcer bleeding when laser results are compared with those before laser therapy was available in his hospitals. Other European investigators such as Schonekas and Sander report initial hemostasis success rates of greater than 90% with uncontrolled patient series of greater than 300. Dwyer in the United States reports initial control of hemorrhage in about 90% of more than 300 bleeding patients. In sum, there is a large amount of experience from several investigators that suggests that endoscopic laser therapy effectively produces initial hemostasis for acutely bleeding lesions.

As impressive as the above-mentioned information is, it must be stressed that little data are available from controlled studies. To that end a sense of critical caution must be maintained. The reports of high success rates for laser therapy of UGI bleeding must be interpreted by considering that approximately 70-80% of all UGI bleeding episodes are self-limited and abate without any specific therapy. Although laser therapy effectively halts bleeding, the frequency of rebleeding, morbidity, transfusion requirements, and complications, and the number of hospitalization days, and ultimate patient outcome must be considered. To this end, controlled, randomized studies are essential to establish the overall value of laser therapy for treatment of UGI bleeding. The exciting and encouraging uncontrolled reports must be followed by controlled evaluation. To date, there is information available from 10 controlled studies assessing the efficacy of lasers in the setting of acute UGI bleeding. Seven of these studies used the Nd:YAG laser, and three used the argon laser. There are some differences in design and considerable differences concerning the interpretation. The seven Nd:YAG laser studies will be discussed first.

Rutgeerts et al. (6) divided 152 patients with UGI hemorrhage into three groups: the 23 patients in group 1 had acute spurting arterial bleeding; the 86 patients in group 2 had actively, bleeding, oozing, but not spurting lesions; and those in group 3 had recent bleeding as determined by stigmata of recent hemorrhage (SRH) but no active bleeding. The authors did not feel it was ethically justifiable to use a control population in the group 1 patients with spurting arterial bleeding. Initial hemostasis was achieved in 20 of the 23 patients. In nine of the 20, the hemostasis was permanent. In 11 patients rebleeding occurred. Fourteen of the 23 patients (61%) required surgery. The overall mortality was 30%, with three of the seven deaths (two postoperatively) the result of uncontrolled bleeding or shock. In the 86 patients with group 2 lesions with active oozing, 46 were treated by laser and 40 were in the control group. Significantly more patients continued to bleed in the control group than in the laser group, and there was a numerical reduction of rebleeding after initial hemostasis as well as a reduction of the surgical requirement in the patients receiving laser therapy. The overall mortality was similar in both groups. In the 43 patients with inactive bleeding but with SRH, there was a reduction in rebleeding, the requirement for an operation was not significant, but there was no reduction in mortality.

Ihre et al. (7) evaluated 135 patients with UGI bleeding. His study considered bleeding lesions of all types (including varices) and both active and inactive bleeding. The laser group included 23 acute bleeders and the control group 19. Eight of the 23 chosen randomly for laser therapy could not be treated for technical reasons. Fourteen of 15 patients who were treated with the laser stopped bleeding, but seven rebled. Overall, those patients with acute bleeding treated by laser, fared no better than the control group with regard to continued or recurrent bleeding, need for emergency surgery, or mortality. Additionally there was no difference in these variables in the 43 patients in the laser group and the 50 patients in the control group who were not bleeding at endoscopy.

A study by Escourrou (8) enlisted 83 patients with active acute UGI bleeding. Forty-one patients were placed in the control group. Forty-two were randomly assigned to laser therapy, but eight of these were excluded because endoscopic therapy was not possible. There was no difference between the two groups with regard to permanent hemostasis, need for surgery, or mortality.

Rohde et al. (9) randomized 105 patients with hematemesis and/or melena to receive either ELT or no endoscopic therapy. The need for urgent surgery and the mortality were similar in the two groups.

Macleod et al. (10) randomized 45 patients in their study. Those that had spurting arteries or visible vessels in the ulcer fared considerably better than the patients who did not receive laser therapy.

The study by Swain et al. (11) deserves special emphasis because it is so well done. Of 465 patients presenting to two London hospitals with UGI bleeding, 232 had peptic ulcers, and 147 of the ulcers were either bleeding or showed stigmata of recent hemorrhage. Twenty-four patients were excluded, and of the 123 who were included in the study, 62 were treated with the Nd:YAG laser and 61 served as controls. Bleeding in the laser-treated group was more effectively controlled ($P < .002$) and the mortality in the laser-treated group was less ($P < .05$).

Fleischer's study (12) differs from the other six in that it focuses exclusively on patients with active bleeding from esophageal varices. Endoscopic laser therapy controlled acute variceal hemorrhage in 70% of the patients, but none in the sham group stopped spontaneously. However, rebleeding was a problem so that statistical benefit could not be demonstrated in the overall course for those who received laser therapy. He concluded that it may be considered an interim solution but does not provide definitive therapeutic benefit.

The first argon laser trial was reported by Vallon et al. (13) in 1981. One hundred thirty-six patients with bleeding peptic ulcers were divided into three groups and randomized to laser therapy or control. Group 1 included 28 patients with actively bleeding ulcers; group 2 included 35 patients with nonbleeding visible vessels, and group 3 included 73 patients with nonbleeding ulcers without visible vessels but with a "central spot." Initial hemostasis was obtained in 10 to the 15 laser-treated patients, and four of the 13 controls stopped bleeding. Two of the laser-treated patients rebled bringing to seven the number in the laser group who required surgery. Although fewer patients in the laser group required emergency surgery, the difference was not significant. In both groups of patients with SRH but no active bleeding, laser therapy did not appear to confer benefit with regard to prevention of rebleeding or need for surgery.

The study by Swain et al. (14) culled 76 patients with an actively bleeding peptic ulcer, or SRH accessible to laser therapy, from a group of 330 patients admitted with acute UGI hemorrhage. They divided these 76 patients into three groups. Group 1 comprised patients with visible vessels; patients in group 2 had other SRH, and in group 3 had overlying clots. All three groups contained patients with both active and inactive bleeding. Laser therapy in the group of patients with visible vessels reduced rebleeding or continued bleeding. Additionally, mortality was less in the laser-treated group 1 patients. There was no difference between group 2 and group 3 patients.

The last of the three argon-controlled trials is a small series reported by Jensen et al. (15). Permanent hemostasis was greater and the need for emergency surgery diminished in the group of ulcer patients treated with the laser.

Information about complications is available in seven of the nine studies. No perforations are reported with either the Nd:YAG or argon lasers. (Reports from uncontrolled series generally give an incidence of 1-2%.) There is general agreement that laser therapy paradoxically can increase the severity of bleeding. If the laser beam causes vaporization of the vessel, increased hemorrhage is possible. In one patient with variceal bleeding, aspiration and death occurred. This may have related to a prolongation of the endoscopic procedure so that therapy could be delivered. The patient was a Child's classification C cirrhotic for whom the prognosis is generally dismal. There was one case of severe gaseous overdistention leading to respiratory compromise. The coaxial gas that is used with most laser therapy is still problematic for many laser investigators. The risk of complications from endoscopic laser therapy increases with the amount of energy delivered and the length of the procedure. As with any therapeutic modality, the benefit/risk ratio must be assessed for that patient, and availability of alternative methods of therapy must be reviewed. If this is done, avoidable complications will be minimized.

Controlled trials can best establish the efficacy and safety of specific treatment modalities. What can we conclude from the controlled laser studies performed in the setting of acute UGI bleeding? It is apparent that only a very limited amount of data exist. To date only about 800 patients have been included in these 10 controlled studies and the population is heterogeneous. Of this number, some patients were actively bleeding; some were not. The firmest data relate to bleeding ulcers. Little information is available for nonulcer bleeding. There is a trend, in those studies that focus upon specific lesions with either active bleeding or visible vessels, suggesting that laser therapy may confer some clinical benefit. In these studies complications of endoscopic therapy were few with both Nd:YAG and argon lasers. This is particularly noteworthy because the patient population is generally at high risk for complications.

B. Angiodysplasia

When angiodysplastic lesions of the upper GI tract are seen endoscopically, they are small (1-10 mm), discrete, red, and either flat or minimally elevated (Fig. 1). They may be single or multiple and the borders may be smooth, spidery, or fernlike. In a large series enumerating the causes of UGI bleeding found at endoscopy, they were identified in 0.5% of the patients (16). It is likely that this number underestimates the true incidence. If the endoscopy is performed while there is blood in the stomach, the lesions may be covered. If the patient is vasoconstricted or volume depleted, they may not be "red enough" to be seen. The

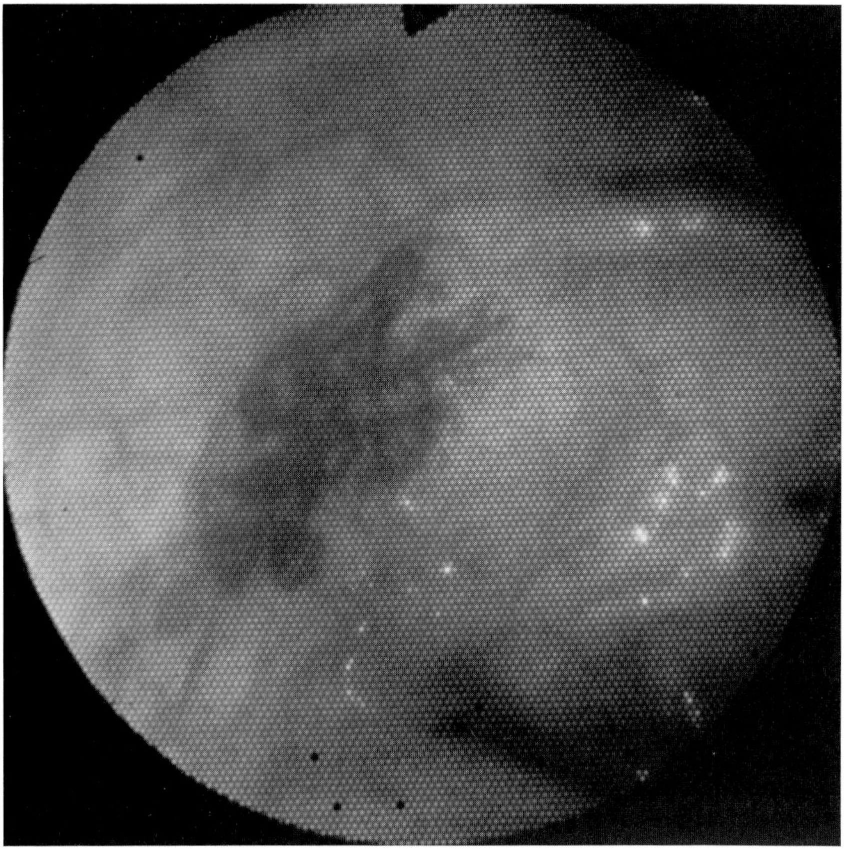

Figure 1 Angiodysplastic lesion in stomach. It has the appearance of a cherry-red birthmark (actual size 7 mm).

lesions are small and may be hidden between gastric folds. Their appearance is similar enough to nasogastric tube erosions or endoscopic trauma that they may be mistaken for these entities. Angiography, as an alternative diagnostic approach to endoscopy, may fail to demonstrate angiodysplastic lesions. Additionally they may not be apparent during surgical exploration. A high index of suspicion is necessary to make the diagnosis.

The etiology and pathogenesis of angiodysplastic lesions of the UGI tract are not understood. A recent review discusses the possibilities (17). Histologically they comprise numerous compact small dilated blood vessels within the superficial lamina propria. The vessels are of capillary and small venular size lined by a thin endothelium (17). Although some authors attempt to segregate them from the gastrointestinal lesions of hereditary hemorrhagic telangiectasia (HHT), they are indistinguishable in endoscopic and histological appearance and may have a similar clinical course. Patients differ from those diagnosed as having HHT in that there is no familial association and no evidence of extragastrointestinal lesions.

Therapy for bleeding upper gastrointestinal angiodysplasia is usually supportive. No pharmacological treatment is of proved benefit. The initial excitement following reports of decreased bleeding in patients with HHT treated with estrogen has not been sustained (18,19). There is no evidence that histamine-2 antagonists or other measures that reduce gastric acidity are of value. The use of prostaglandins, based on the logic that their cytoprotective effect may be beneficial, is an avenue worthy of pursuit but is untested. A report describing the temporary benefit of gastric radiotherapy for a patient with HHT and von Willebrand's disease who had gastrointestinal bleeding has recently been published (20). Surgical therapy has been undertaken as a last resort.

All of the endoscopic thermal modalities (laser, monopolar and bipolar electrocoagulation, heater probe) and injection techniques have been used to obliterate angiodysplasia. In most instances, endoscopic treatment was effective in destroying the lesion. There is no clear evidence that one method is superior to the others. Success seems to be more dependent upon establishing as clearly as possible that the lesion(s) treated is(are) the source of the blood loss, rather than just an incidental finding, and additionally, upon carrying out a compulsive surveillance program, because these lesions may tend to recur.

Several studies using laser therapy to treat angiodysplasia have been published. Both the argon and Nd:YAG lasers have been used. The most detailed report is from Jensen et al. (21) who use the argon laser. Thirty patients had 830 lesions of the esophagus, stomach, duodenum, and colon treated endoscopically with this laser. Ten patients with Osler-Weber-Rendu (OWR) syndrome had transfusions with more than 460 units of blood before treatment for 659 angiodysplastic lesions. The mean hematocrit level rose from 23 to 37 in 2 years of follow-up. Transfusion requirements and hospitalizations were dramatically reduced. Twenty patients with non-OWR angiodysplasia had 171 lesions treated. In 2 years of follow-up, mean hematocrit level rose from 24 to 35, and there was a decrease in transfusion requirements and hospitalizations. Other investigators, such as Waitman et al. (22), Bowers and Dixon (23), Johnston (24), and Fleischer (25), report similar results.

III. LASER THERAPY OF GASTROINTESTINAL NEOPLASMS

Because the fiberoptic endoscope can place the physician in close proximity to neoplasms of the gastrointestinal tract, an opportunity exists for local oncolytic therapy. Neoplasms of the oral cavity, esophagus, stomach, duodenum, ampulla, colon, and rectum have been treated with endoscopic laser therapy. For the most part, the therapy has been palliative, but on some occasions curative treatment has been intended and achieved.

Endoscopic management of gastrointestinal cancer has several appealing aspects: (1) it averts the need for surgery and general anesthesia with their attendant complications; (2) it considerably diminishes the likelihood of systemic side effects; (3) it can be performed under direct vision; (4) unlike radiotherapy, there is no maximum dose, so that if the tumor recurs in the same area, retreatment can be performed. It is limited in that it does not affect pathological tissue outside the gastrointestinal lumen, and in that regard it is generally palliative.

When palliative treatment is applied, the goal is usually to relieve obstruction or to reduce blood loss. Attempts to eliminate pain have generally proved unfruitful, presumably because pain is related to neoplastic spread outside the gastrointestinal lumen. Therapy of bleeding may either be achieved by coagulating the bleeding sites on the tumor or by destroying the tumor and thereby eliminating the vasculature that is bleeding. Treatment for bleeding is generally not effective in patients whose tumors are excessively large (e.g., greater than 8 cm in any dimension). The author has successfully reduced or eliminated bleeding from primary esophageal, gastric, and colorectal neoplasms. In addition, bleeding metastatic lesions to the GI tract (e.g., renal cell carcinoma to the stomach) have also been treated successfully.

Most of the experience with upper gastrointestinal tract tumors has been with neoplasms of the esophagus and esophagogastric junction. To date, there have been publications from 11 different centers describing the efficacy of this procedure (26-36). Although there are some minor variations in technique, they are for the most part similar to the original method described by Fleischer et al. (26), in which the Nd:YAG laser is used to vaporize tumor tissue.

Endoscopic laser therapy is performed by the following technique (Fig. 2): The endoscope is advanced to either the proximal or distal margin of the tumor. In some instances there is no option because luminal narrowing is so marked that even the smallest endoscopes cannot traverse the tumor. The quartz waveguide that carries the laser beam is passed out of the biopsy channel of a conventional endoscope (Fig. 3). Often a two-channel endoscope is used. The laser energy is preselected such that when the beam hits the tissue, vaporization will occur. With the technique used by Fleischer et al. (26), the Nd:YAG laser is generally set at 90-100 W for 2 sec or more at a 1.0 cm distance from the tissue. The beam is aimed at the neoplastic tumor closest to the lumen, and treatment progresses in increasingly larger concentric circles toward, but not to, the wall of the GI organ in which the treatment is being delivered. If the lumen is wide enough to enable passage of the instrument, treatment may be carried out at various levels. If it is

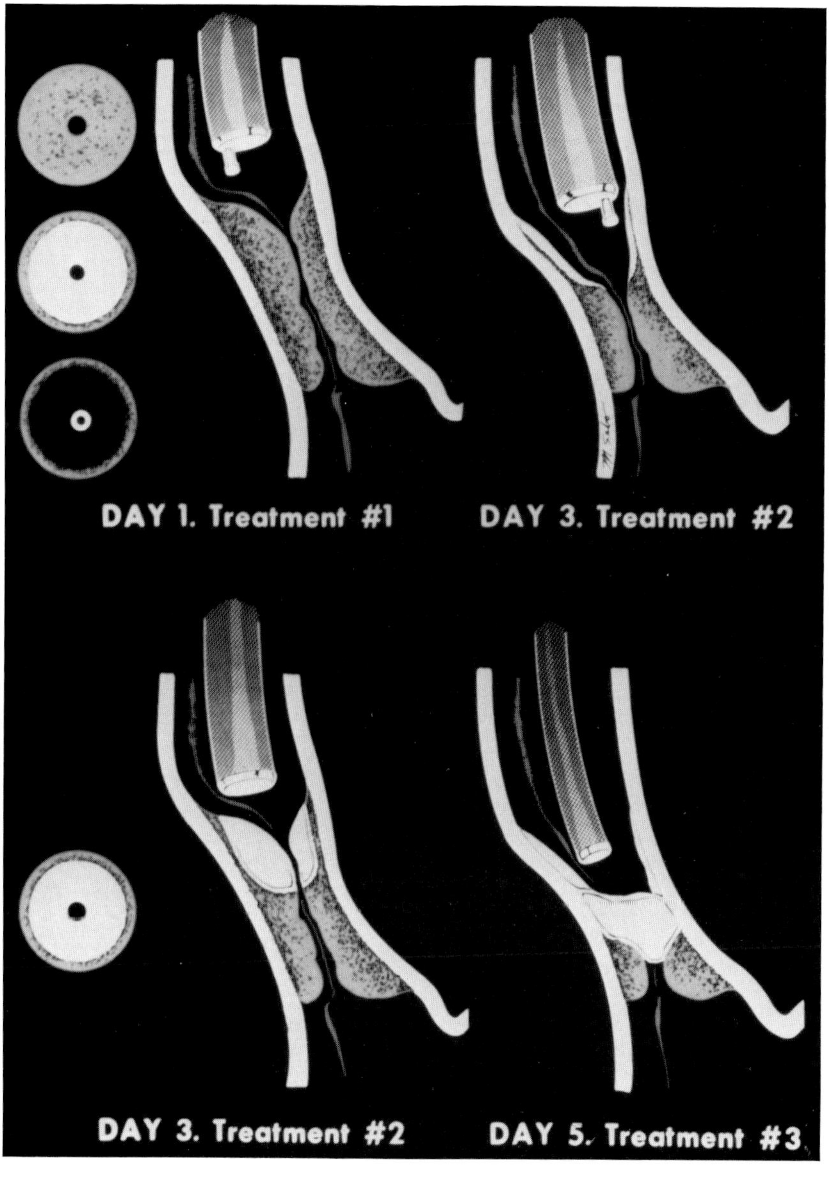

Figure 2 Treatment technique for endoscopic laser therapy of esophageal carcinoma: Day 1, endoscope advanced to superior margin of esophageal cancer. Laser waveguide passed through biopsy channel aiming at tumor. Treatment 2 shows schematic appearance 48 hr later; laser-treated area is puddinglike and will need to be removed; day 3, appearance after necrotic tumor tissue removed; day 5, before third treatment.

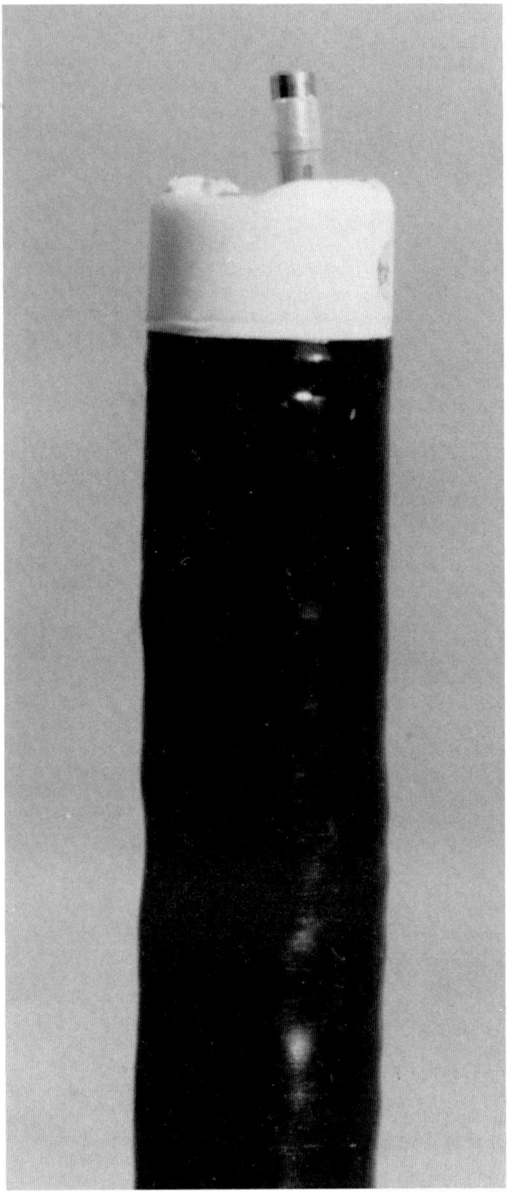

Figure 3 Quartz waveguide projecting through distal tip of endoscope.

not, then treatment is completed for that day when the cross-section of tumor has been vaporized. Subsequent laser treatments are then carried out approximately every other day until the goals of laser therapy are achieved. The period of 48 hr between treatments generally permits maximal tissue necrosis and is well tolerated by patients. At the beginning of each laser session after the first, the previously treated neoplastic tissue that plugs the lumen must be removed so that progressive "coring-out" of the new alimentary canal can be done.

Using this technique, the following results were obtained. In the first report, Fleischer's group treated five patients with squamous cell carcinoma who had marked dysphagia (26). In all patients luminal opening could be achieved, clinical improvement was seen, and no complications ensued (Fig. 4). That experience was expanded and, in addition to treating patients with squamous cell carcinoma of the esophagus, patients with adenocarcinoma of the esophagogastric junction were treated (32). As my overall experience has grown to approximately 100 patients, a retrospective review has been undertaken to see if pretreatment factors could be defined that might determine how well or poorly a patient might respond to endoscopic laser therapy and to determine what is the likelihood that complications will develop. Four factors were assessed: (1) histological type, (2) endoscopic appearance, (3) location, and (4) length of tumor.

There was no difference in tissue response between squamous cell carcinoma and adenocarcinoma. This is not surprising because tissue effects are achieved by gross thermal destruction. Although I have not personally treated patients with primary melanoma of the esophagus, it is conceivable that the darkly pigmented lesion may absorb the 1.06-μm beam of the Nd:YAG laser differently than "pinkish" carcinoma tissue.

The endoscopic appearance is of vital importance. Better results are obtained when the tumor is exophytic and polypoid rather than submucosal or extrinsic. Risk of perforation is less, and it is technically easier to decide what tissue should be vaporized and what tissue should be left. It is unusual for the patient to experience pain with laser treatment if the tumor is exophytic and polypoid but common when the tumor is submucosal.

Location of the tumor is also of critical importance. If the tumor is located in a relatively straight segment of esophagus, it is technically easier to treat and the risk of perforation is diminished, because the beam is directed along the axis of the esophagus. After the obstruction is relieved, gravity will help carry the food through the rigid, aperistaltic tumorous segment (particularly if the patient is upright), and food will traverse the esophagus more readily. Conversely, if the narrowed segment lies horizontally (e.g., at the esophagogastric junction or in a transverse anastomosis in the chest), it will be technically more difficult to treat and the food will pass less easily after luminal patency is achieved. The least

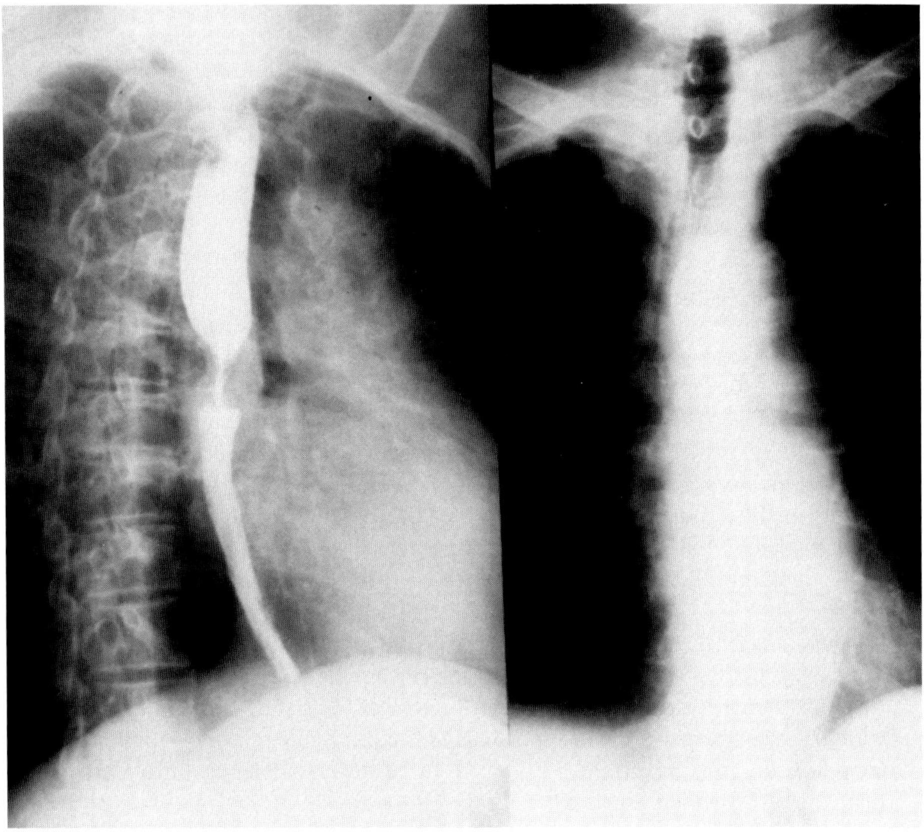

Figure 4 Radiographic appearance shown on barium swallow of patient with esophageal cancer before and after laser therapy: left = before; right = after.

favorable results of ELT for esophageal malignancy have been found when the obstruction is high in the cervical esophagus, just below the cricopharyngeus. These are technically difficult or impossible, and the procedure can only be done safely if the superior margin of the tumor is greater than 3-4 cm below the cricopharyngeus. The risk of aspiration during treatment is great. Perhaps most disconcerting is that even when the lumen can be opened by ELT to admit an 11-mm or 13-mm endoscope, dysphagia often persists and the patient is no better off than before the treatment. Presumably an underlying neuromuscular disorder explains the dysphagia, because the obstruction has been removed.

Not surprisingly, shorter tumors (less than 6 cm) can be destroyed in fewer sessions than longer tumors (more than 10 cm) and they tend to recur less quickly.

Similar observations were made by Lambert et al. (29), who reported treating 76 nonoperative patients with squamous cell carcinoma of the esophagus. Best results were obtained with intraluminally obstructing tumors. The results were less satisfactory with submucosally infiltrating tumors. In six patients early perforation occurred. In studies involving fewer patients, Buset et al. (30), Krasner and Beard (33), Dwyer (34), and Kiefhaber et al. (35) have testified to technical feasibility of this method.

In addition to reporting the technical efficacy of Nd:YAG laser therapy for esophageal and esophagogastric cancers, Mellow and Pinkas (27) addressed two other important aspects. First, they pointed out that technical success (achieving luminal patency) did not correlate with functional success (ability to leave hospital for home and ingest all necessary calories). In a series of 30 consecutive patients, technical success was achieved in 29, but functional success in only 21. The discrepancy was caused by severe anorexia or poor performance status that limited intake and activity, even if the lumen was patent; radiation-associated pharyngeal dysphagia was present; or there were treatment complications. They felt that poor performance status at accession correlated closely with poor functional outcome irrespective of technical outcome.

In a second paper, Mellow and Pinkas (28) look at survival when laser-treated patients with esophageal cancer were compared with a similar group of patients who did not receive laser therapy, but who received radiation therapy. The laser-treated patients' survival was significantly increased, whether measured from time of onset of radiotherapy (36 vs. 17 weeks, $P = .02$) or from time of recurrent symptoms after radiotherapy (25 vs. 8 weeks, $P < .05$).

The flip side of benefit is risk. What can be said about the complications? Perforation is the most worrisome concern. In most series the incidence has been 5-10%. Risk of perforation is increased if the tumor is high in the cervical esophagus; if the tumor is preponderantly submucosal; if the lumen is so narrow that an endoscope cannot traverse it; or if the obstructing tumor occurs in a horizontal segment at the gastroesophageal junction. As the experience of the laser endoscopist increases, technical and judgmental expertise grows, reducing the likelihood of perforation. Bleeding is only rarely seen. Clinical sepsis is also unusual. Bacteremia occurs in about 30% of patients, but it is associated more with passage of the endoscope through the necrotic tumor than with the treatment per se (37). Pain with treatment is unusual with exophytic tumors but common when submucosal cancer is treated.

There are several reports of laser therapy for gastric cancers. Fleischer and

Sivak (32) described 15 patients with recurrent adenocarcinoma at the cardia treated by Nd:YAG laser therapy. Relief of symptoms was obtained in all patients in one to three sessions (mean 2.2). No complications were encountered. If the patient had had a previous esophagogastrectomy, laser therapy was technically very easy because the area of tumor recurrence was so short. Swain et al. (38) described the palliative relief of gastric outflow obstruction obtained with argon laser treatment. The patient was able to return home and could swallow minced solids. Several Japanese investigators have described endoscopic laser therapy for gastric cancers using Nd:YAG lasers. Imaoka et al. (39) treated 15 patients with gastric neoplasms. In some patients with early gastric cancer, curative therapy was described. In some patients with stenosing gastric adenocarcinoma, treatment was effective in about half the patients. In addition to adenocarcinoma they describe effective treatment of gastric adenomas and leiomyomas. Iwasaki et al. (40) claimed clinical benefit in 80% of patients treated for stenosis at the gastric cardia. Ichikawa et al. (41) reported variable effects when gastric cancers were treated by Nd:YAG laser. They suggested that the histological tumor type affected the results and that undifferentiated adenocarcinoma, signet ring cell carcinoma, and mucinous adenocarcinoma responded less well. Mizushima et al. (42) reported beneficial effect of Nd:YAG laser therapy for gastric polyps. Ito et al. (43) reported on the treatment of 18 "borderline" neoplasms and 16 early gastric cancers and concluded that the laser effect may be beneficial. Brunetaud et al. (44), in France, have used both argon and Nd:YAG laser for palliative treatment of gastric carcinomas. Richey and Dixon (45) used both argon and Nd:YAG laser treatment to ablate premalignant gastric polyps in one patient. In none of these studies are complications of laser therapy reported.

Isolated reports describe the use of laser therapy for neoplasms in other parts of the UGI tract. Hutton described a case of a tumor of the soft palate that recurred after surgical excision and was causing difficulties with respiration. Laser ablation provided distinct relief (personal communication). Tumors of the duodenum and ampulla of Vater are rare, but there are reports of laser treatment for such tumors that were causing symptoms of obstruction or blood loss. Both Bowers and Sivak (personal communications) have treated such patients. In the case treated by Sivak, the patient had had recurrent bouts of pancreatitis secondary to a tubulovillous tumor of the ampulla of Vater. Surgical treatment was not curative and duodenoscopic laser therapy was undertaken with hopes of decreasing bouts of pancreatitis and destroying potentially premalignant tissue. At short-term follow-up (6 months) both goals had been realized.

The method of therapy used in the cases just described depended upon gross thermal destruction. Quite simply, high intensities of energy are focused on the neoplastic tissue, and it is vaporized. In the future it is possible that the laser might be utilized in other ways. Some substances such as hematoporphyrins have a high affinity for neoplastic tissue. When given intravenously they preferentially tend to localize there. Afterward if light of an appropriate wavelength (e.g., argon laser, 630 nm) is directed to this tissue it can be selectively destroyed. Photodynamic therapy of lasers is discussed in Chap. 17. Alternatively, there is evidence that extremely low laser energy can disrupt neoplastic cells by destroying cellular organelles. This entire area is just beginning to be explored.

IV. FUTURE CONSIDERATIONS

The future of endoscopic laser therapy for the upper GI tract is dependent upon several factors.

The following can be said about therapy for gastrointestinal bleeding. It is effective treatment for many patients with severe, persistent hemorrhage. Like other endoscopic therapies it has the practical appeal that therapy can be piggy-backed to diagnostic endoscopy, which has emerged as the best test to determine the cause of hemorrhage. At this juncture, laser therapy has the most established track record for efficacy and safety. However, the equipment costs more, and it is less portable (it cannot be taken to the intensive care unit) than other modalities. The lack of comparative data simply leaves the issue of "which is best" unresolved at this time. From my personal experience with all commercially available modalities, I prefer the Nd:YAG laser (if I could just have one device). With all modalities, including the laser, initial hemostasis has been high, but rebleeding has occurred in 15-20% of the cases. We need to better understand why rebleeding occurs in some patients and alter our approaches to try to decrease this occurrence.

When considering the therapy for UGI tract neoplasms, several issues are unresolved. It is uncertain which subgroups of upper gastrointestinal cancers are best treated with lasers. Dosage of laser energy is imprecisely measured. Operator judgment plays a major role in selecting treatment variables and endpoints. There is little information that enables the physician to decide what sequence of combined therapy is best for a given tumor—or for that matter, that combined therapy is better and/or safer. Should an endoscopic prosthesis be placed after an esophageal lumen has been widened with the laser? Should ancillary chemotherapy or radiotherapy or both be used? What timing is appropriate? Little data exists about prognostic indicators, and there is only minimal information about the effect on survival. These concerns are not so much problems as questions.

Clearly, the future is exciting to ponder. The technology is growing rapidly and the entire field is in its infancy. Numerous developments now in their conceptual stages might greatly influence the future use of lasers for treatment of upper gastrointestinal diseases. Only a few wavelengths are being used. It is conceivable that many more wavelengths can be utilized, and these other wavelengths may have specificity, which may allow them to be used more safely and more effectively. Tunable (selectable) lasers that allow the physician to select a wavelength as one does channels on a television should be available in the future. Lasers may have a role in diagnostic evaluation, either in association with fluorescent dyes (e.g., for earlier diagnosis of premalignant conditions such as dysplasia in inflammatory bowel disease) or in totally separate methodologies. Computer application should assist with a more precise delivery of dosage. As mentioned, tissue sensitizing agents may render abnormal tissue more susceptible to laser destruction. Advances can be expected with both waveguides and endoscopic delivery systems to increase access to the pathological process. Videoendoscopes, already in use, may provide new diagnostic and therapeutic capability when this technology is wedded to laser technology (46). It will be most exciting to follow the evolution of endoscopic laser therapy for diseases of the gastrointestinal tract.

REFERENCES

1. Dwyer RM, Yellin AE, Craig J, Cherlow J, and Bass M: Gastric hemostasis by laser phototherapy in man: A preliminary report. *J. Am. Med. Assoc. 236*:1383-1384, 1976.
2. Nath G, Gorisch W, and Kiefhaber P: First laser endoscopy via fiberoptic transmission system. *Endoscopy 5*:208-213, 1973.
3. Frühmorgen P, Reidenbach H-D, Bodem F, Kaduk B, and Demling L: Experimental examinations on laser endoscopy. *Endoscopy 6*:116-122, 1974.
4. P. Kiefhaber. International experience with lasers for gastrointestinal bleeding, in *Proceedings of the International Laser Congress.* Detroit, 1979.
5. P. Kiefhaber. Laser endoscopic experience, in *Brussels Fifth International Symposium on Digestive Endoscopy.* Brussels, 1982.
6. Rutgeerts P, Vantrappen G, Broeckaert L, Janssens J, Coremans G, Geboes K, and Schurmans P: Controlled trial of YAG laser treatment of upper digestive hemorrhage. *Gastroenterology 83*:410-416, 1982.
7. Ihre T, Johansson C, Seligson U, and Törngren S: Endoscopic YAG-laser treatment in massive upper gastrointestinal bleeding: Report of a controlled randomized study. *Scand. J. Gastroenterol. 16*:633-640, 1981.
8. J. Escourrou. Nd:YAG laser therapy for acute gastrointestinal hemorrhage, in *Laser Tokyo 81* (K. Atsumi and N. Nimsakul, eds.). Intergroup Corp. Tokyo, 1981.
9. Rohde H, Thon K, Fischer M, Stoltzing H, Elsasser P, Haibach L, Ohmann C, and Lorenz W: Results of a defined therapeutic concept of endoscopic neodymium-YAG-laser therapy in patients with upper gastrointestinal bleeding. *Br. J. Surg. 67*:360-376, 1980.
10. Macleod IA, Mills PR, Mackenzie JF, Russell RI, and Carter DC: Neodymium YAG laser photocoagulation for major acute upper gastrointestinal haemorrhage. *Gut 23*:A905, 1982.

11. Swain CP, Bown SG, Salmon PR, Kirkham JS, and Northfield TC: Controlled trial of Nd:YAG laser photocoagulation in bleeding peptic ulcers (abstr.) *Lasers Surg. Med. 3*:111, 1983.
12. Fleischer DE: Endoscopic Nd:YAG laser therapy for acute esophageal variceal bleeding: A randomized controlled trial. *Gastrointest. Endosc. 31*:4-9, 1985.
13. Vallon AG, Cotton PB, Laurence BH, Armengol Miro JR, and Salord Oses JC: Randomized trial of endoscopic argon laser photocoagulation in bleeding peptic ulcers. *Gut 22*:228-233, 1981.
14. Swain CP, Storey DW, Northfield TC, Bown SG, Kirkham JS, and Salmon PR: Controlled trial of argon laser photocoagulation in bleeding peptic ulcers. *Lancet 2*:1313-1316, 1981.
15. Jensen DM, Machicado GA, Tapia JI, Beilin DB, and Mautner W: Endoscopic argon laser photocoagulation of patients with severe upper gastrointestinal bleeding. *Gastrointest. Endosc. 28*:151, 1982.
16. Silverstein FE, Gilbert DA, Tedesco FJ, Buenger NK, and Persing J: The national ASGE survey on upper gastrointestinal bleeding. I. Study design and baseline data. II. Clinical prognostic factors. *Gastrointest. Endosc. 27*:73-79, 80-93, 1981.
17. Weaver GA, Alpern HD, Davis JS, Ramsey WH, and Reichelderfer M: Gastrointestinal angiodysplasia associated with aortic valve disease: Part of a spectrum of angiodysplasia of the gut. *Gastroenterology 77*:1-11, 1979.
18. Koch HJ Jr, Escher GC, and Lewis JS: Hormonal management of hereditary hemorrhagic telangiectasia. *J. Am. Med. Assoc. 149*:1376-1380, 1952.
19. Harrison DFN: Familial haemorrhagic telangiectasia: 20 cases treated with systemic oestrogen. *Q. J. Med. 33*:25-38, 1964.
20. Schwartz JT, Patton GA, Graham DY, and Cain GD: Gastric radiotherapy as treatment for hereditary hemorrhagic telangiectasia. *Am. J. Gastroenterol. 77*:53-54, 1982.
21. Jensen DM, Machicado GA, and Silpa ML: Treatment of GI angioma with argon laser, heater probe, or bipolar electrocoagulation. Paper presented to American Gastroenterological Association Plenary Session, Digestive Disease Week, New Orleans, May 23, 1984.
22. Waitman AM, Grant DZ, and Chateau F: Argon laser photocoagulation treatment of patients with acute and chronic bleeding secondary to telangiectasia. *Gastrointest. Endosc. 28*:153, 1982.
23. Bowers JH and Dixon JA: Argon laser photocoagulation of vascular malformations in the GI tract: Short-term results. *Gastrointest. Endosc. 28*:126, 1982.
24. Johnston JH: Complications following endoscopic laser therapy. *Gastrointest. Endosc. 28*:135, 1982.
25. Fleischer DE: Nd:laser photocoagulation for upper gastrointestinal angiodysplasia (abstr.). *Gastrointest. Endosc. 27*:122, 1981.
26. Fleischer D, Kessler F, and Haye O: Endoscopic Nd:YAG laser therapy for

carcinoma of the esophagus: A new palliative approach. *Am. J. Surg. 143*: 280-283, 1982.
27. Mellow MH and Pinkas H: Endoscopic laser therapy for malignancies affecting the esophagus and gastroesophageal junction: Analysis of technical and functional efficacy. *Arch. Intern. Med. 145*:1443-1446, 1985.
28. Mellow MH and Pinkas H: Endoscopic therapy for esophageal carcinoma with Nd:YAG laser: Prospective evaluation of efficacy, complications, and survival. *Gastrointest. Endosc. 30*:334-339, 1984.
29. Lambert R, Sabben G, Chavaillon A, Guyot P, and Descos F: Results of Nd YAG laser treatment in epidermoid esophageal cancer (abstr.). *Lasers Surg. Med. 3*:340, 1983.
30. Buset M, Dunham F, Baize M, de Touef J, and Cremer M: Nd-YAG laser, a new palliative alternative in the management of esophageal cancer. *Endoscopy 15*:353-356, 1983.
31. Swain CP, Bown SG, Edwards DAW, Kirkham JS, Salmon PR, and Clark CG: Laser recanalization of obstructing foregut cancer. *Br. J. Surg. 71*:112-115, 1984.
32. Fleischer D and Sivak MV: Endoscopic Nd:YAG laser therapy as palliative treatment for advanced adenocarcinoma of the gastric cardia. *Gastroenterology 87*:815-820, 1984.
33. Krasner N and Beard J: Laser irradiation of tumours of the oesophagus and gastric cardia. *Br. Med. J. 288*:829, 1984.
34. Dwyer R: Single session palliation of obstructing esophageal cancer (abstr.). *Lasers Surg. Med. 3*:172, 1983.
35. Kiefhaber P, Kiefhaber K, and Huber F: Endoscopic Nd-YAG laser therapy of tumors in the gastrointestinal tract (abstr.). *Lasers Surg. Med. 3*:368, 1984.
36. Mosquet L, Brunetaud JM, Mudry J, Cortot A, and Paris JC: Are lasers useful in palliative treatment of digestive cancers? (abstr.). *Lasers Surg. Med. 3*:340, 1984.
37. Wolf D, Fleischer D, and Sivak MV: Incidence of bacteremia with elective upper gastrointestinal endoscopic laser therapy. *Gastrointest. Endosc. 31*: 247-250, 1985.
38. C. P. Swain, S. G. Bown, and J. Edwards. Neoplastic gastric outflow tract obstruction relieved by argon laser at endoscopy, in *Laser Tokyo 81* (K. Atsumi and N. Nimsakul, eds.). Intergroup Corp., Tokyo, 1981.
39. W. Imaoka, J. Okuda, K. Ida et al. Treatment of digestive tract tumors with laser endoscopy, in *Laser Tokyo 81* (K. Atsumi and N. Nimsakul, eds.). Intergroup Corp., Tokyo, 1981.
40. M. Iwasaki, M. Sasako, T. Konishi et al. Clinical application of Nd:YAG laser endoscopy, in *Laser Tokyo 81* (K. Atsumi and N. Nimsakul, eds.). Intergroup Corp., Tokyo, 1981.
41. T. Ichikawa, S. Nakosawa, and Y. Ema. The effects of Nd:YAG laser irradiation on gastric cancers, in *Laser Tokyo 81* (K. Atsumi and N. Nimsakul, eds.). Intergroup Corp., Tokyo, 1981.

42. K. Mizushima, R. Marada, M. Namik et al. Endoscopic therapy of the YAG laser in early gastric cancer and gastric polyp, in *Laser Tokyo 81* (K. Atsumi and N. Nimsakul, eds.). Intergroup Corp., Tokyo, 1981.
43. Y. Ito, M. Sugura, T. Kano et al. Endoscopic laser treatment of borderline lesions and early gastric cancers, in *Laser Tokyo 81* (K. Atsumi and N. Nimsakul, eds.). Intergroup Corp., Tokyo, 1981.
44. J. M. Brunetaud, P. Houche, J. S. Delmotte et al. Laser in digestive endoscopy, in *Laser Tokyo 81* (K. Atsumi and N. Nimsakul, eds.). Intergroup Corp., Tokyo, 1981.
45. Richey GD and Dixon JA: Ablation of atypical gastric mucosa and recurrent polyps of endoscopic application of laser. *Gastrointest. Endosc. 27*:224-225, 1981.
46. Sivak MV Jr and Fleischer DE: Colonsocopy with a VideoEndoscope: Preliminary experience. *Gastrointest. Colonoscopy 30*:1-5, 1984.

10

Laser Therapy for the Lower Gastrointestinal Tract

STEPHEN N. JOFFE
University of Cincinnati Medical Center, Cincinnati, Ohio

I. INTRODUCTION

Lasers have been produced employing various designs and base materials, and new technology is continuing at a rapid rate to find medical applications. The three main types of lasers used in the lower gastrointestinal (GI) tract are carbon dioxide (CO_2), argon, and neodymium:yttrium-aluminum-garnet (Nd:YAG) lasers (Table 1). The different wavelengths determine the visibility of the beam, depth of tissue penetration, and ultimately, the desired tissue effect to be obtained clinically. This varies from coagulation, vaporization, cutting, localized hyperthermia, and photodynamic therapy. Each of these biological actions has specific applications in the lower tract (Table 2), which are related to the interaction of laser light on the living tissue (Table 3).

II. APPLICATIONS

Gastrointestinal diseases were initially treated by laser endoscopy in the mid-1970s. The spectrum of diseases being treated and the number of patients has expanded exponentially since then: from 12 medical centers in the United States using lasers for gastrointestinal disease in 1981 to 200 by 1983 (1) and 1200 by 1986. Lasers provide a multidisciplinary and multispecialty modality and increased therapeutic options where such choices either did not exist or were previously limited. The relative safety, ease of use, efficiency, reduction of morbidity and mortality, and cost-effectiveness are important reasons for this growth. The following discusses the clinical applications of lasers in the lower gastrointestinal tract. The continuous wave Nd:YAG laser is the laser of choice

Table 1 Specifications of Lasers Used in the Lower Gastrointestinal Tract

Laser	CO_2	Argon	Nd:YAG
Type	Gas	Gas	Crystal
Wavelength (μm)	10.6	0.5	1.06
Absorbed by	Water	Hemoglobin and melanin	Tissue and protein
Penetration (mm)	0.5-1	2-3	5-7 (noncontact) 0.5-1 (contact)
Delivery system	Mirror	Fiberoptic	Fiberoptic
Coagulation	Low	Fair	High
Cutting	High	Low	Fair/High
Endoscopy	Rigid	Flexible	Flexible

Table 2 Application of Lasers in the Lower Gastrointestinal Tract

- Bleeding
  - angiodysplasia
  - AV malformations
  - diverticular disease
  - tumors
  - inflammatory bowel disease
- Tumors
  - benign
    - polyps
      - adenomatous
      - villous
  - malignant
    - primary (adenocarcinoma)
    - secondary
- Strictures
  - benign
  - malignant
- Perianal
  - warts, fissures, hemorrhoids

Table 3 Biological Action of Laser Energy in Tissue

Energy density (J/cm^2)	Biological effect
<4	Biostimulation
>4	Biosuppression
40	Phototherapy (nonthermal) with sensitizing agents
400	Photocoagulation (thermal)
4000	Vaporization (thermal)

Source: Adapted from Ref. 26.

in most of these applications. This laser is used to coagulate bleeding, vaporize tumors, excise polyps, and treat cancer by local hyperthermia. It is used in either the conventional noncontact mode or with the more precise and safer contact probes.

A. Bleeding

The therapy for massive lower gastrointestinal hemorrhage includes surgery, angiographic intervention with vasopressin infusion or embolization, and most recently, endoscopic laser photocoagulation. The major advantage of the laser technique is that the bleeding site can be directly visualized and photocoagulated with minimal risk to the patient. The therapeutic modality is carried out immediately and during the same diagnostic and colonoscopic procedure.

Extensive experimental studies have been carried out to assess the merits of various endoscopic probes, which include monopolar and bipolar electrocautery, liquid electrodes, and heater probe. Little information is available regarding their use in colonic bleeding, and there have been no controlled studies comparing them with lasers.

An experimental study comparing monopolar electrocautery and lasers on the normal canine colon found the maximal thermal tissue damage to be delayed for 6 days (2). Transmural necrosis occurred with the argon laser, Nd:YAG laser, and electrocautery at mucosal ablation energies of 6, 2, and 1 times, respectively. The energy density necessary to cause mucosal ablation resulted in less transmural injury with the two lasers compared with electrocautery. Furthermore, there was no statistical difference in depth of necrosis between the argon or Nd:YAG lasers at similar mucosal ablative power densities. This animal data suggested that lasers were safer and more effective than electrocautery in the treatment of mucosal lesions of the colon.

Kiefhaber and coworkers treated 22 patients with multiple angiodysplasia oc-

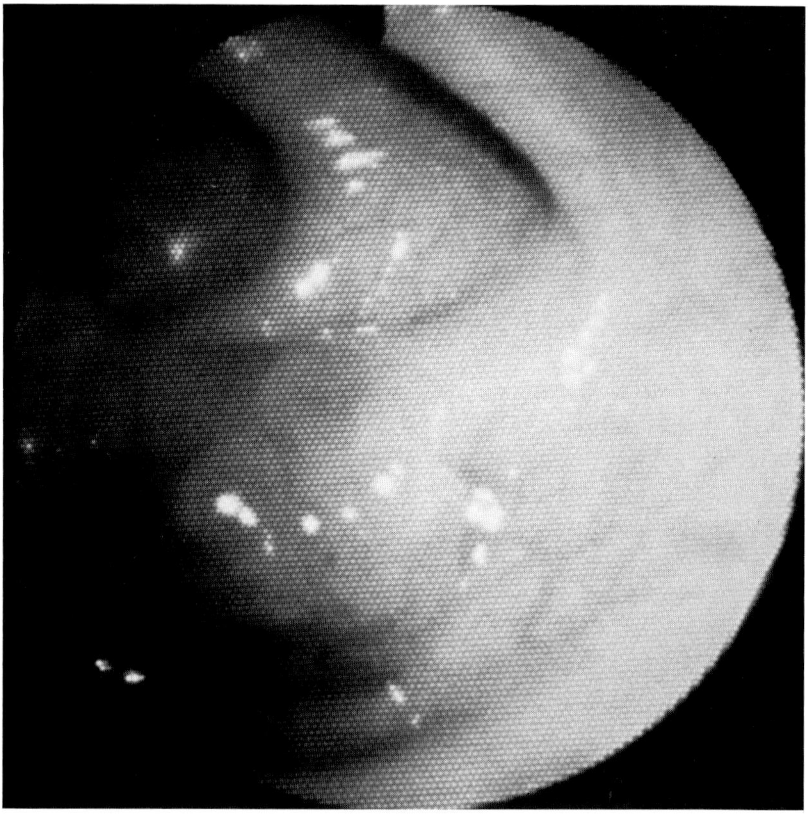

(a)

Figure 1 Protruding arteriovenous vascular malformation in colon before treatment (a), during treatment with contact endoprobe and Nd:YAG laser (b), and 2 weeks after therapy (c). Note: total disappearance of vascular malformation.

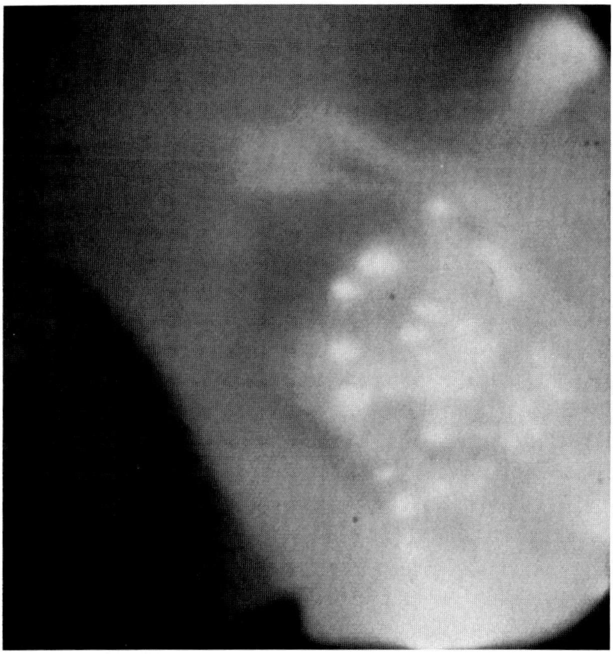

(b)

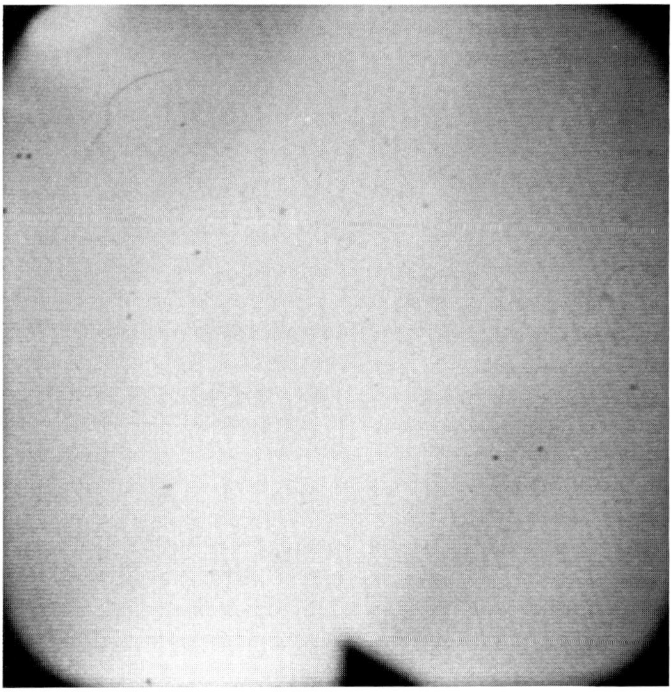

(c)

curring predominantly in the cecum (3). Except in the first patient in whom the coaxial CO_2 gas flow was too high (4.5%) no other perforations occurred, no further operations were necessary, and there was no mortality. Because of the thinness of the colonic wall, excessive insufflation must be avoided, and lower powers need to be used with conventional noncontact Nd:YAG and argon lasers when stopping colonic bleeding.

Hunter et al. endoscopically treated 15 colonic bleeds by laser photocoagulation (4). The argon laser was used primarily for the arteriovenous malformations, which tended to be small and superficial, and the Nd:YAG laser for larger tissue or those lesions more actively bleeding. In the 27 patients with GI arteriovenous malformations, transfusion requirements fell from a mean of 17 (± 5.9) units in the year before laser therapy, to 1 (± 0.8) units in the year after laser phototherapy began ($P < .01$).

We have had surprisingly good results in eradicating multiple arteriovenous vascular malformations in a young female using the contact endoprobes (Fig. 1).

Laser Doppler velocimetry is a new technique for measuring intestinal mucosal blood flow (5). It utilizes the Doppler shift principle, and measures blood flow in terms of milliliter per minute per gram of tissue. Attempts to pass the laser Doppler pole endoscopically would be an important aid in the treatment of angiodysplasia and hemangiomas affecting the colon. It would enable selective pinpointing of arterial inflow for photocoagulation of supply blood vessels.

Rectal bleeding from endometriosis coli was treated using the Nd:YAG laser in three patients (4). The laser reduced rectal bleeding, but the benefit was only partial and lasted only until further endometrial implants appeared. There were no complications or deaths. The Nd:YAG laser has also been used to treat a patient bleeding from the rectosigmoid junction as a result of radiation injury related to therapy for cervical carcinoma (6).

A disadvantage of attempts at endoscopic laser therapy during colonoscopy for emergency lower GI bleeding is the inability to visualize the bleeding point because the colon is filled with blood. Attempts to prevent this include colonic lavage either by use of purgatives, whole gut irrigation via a nasogastric tube using saline lavage, or colonic irrigation per rectum. During active bleeding this is often inadequate, and a combined approach offers some advantages. Angiography is performed to rapidly localize the bleeding site. This is followed by a vasopressin infusion into the superior mesenteric artery for right-sided colonic lesions or into the inferior mesenteric artery for left-sided lesions to lessen or stop the bleeding. Embolization is not performed because of the risk of bowel infarction and the possible increased risk of perforation after laser photocoagulation caused by decreased colonic perfusion. With the bleeding somewhat decreased, the exact bleeding site is identified. Fiberoptic colonoscopy is then more readily and effectively performed, and laser photocoagulation achieves the desired hemostasis. This combined, low-risk approach offers an alternative treatment in which neither modality alone is satisfactory in high-risk surgical patients.

B. Carcinomas

1. Advanced Colorectal Carcinoma

In the United States, carcinoma of the colon and rectum is the third most common tumor in men and the second most common in women. Despite advances in screening techniques—such as stool Hemoccult testing, air contrast barium enema, fiberoptic colonoscopy—many patients with colorectal carcinoma have a tumor that has spread beyond the colon at the time of diagnosis. Approximately 35% of all patients coming to surgery have metastases to regional lymph nodes. A sizable proportion of these patients have widespread intraabdominal, hepatic, and/or pulmonary metastases with a median survival time from laparatomy of 8 months. Local resection of the involved colon is still considered the treatment of choice in these patients.

The rationale for the surgical treatment of the primary colonic lesion in far-advanced disease includes relief or prevention, or both, of colonic obstruction; prevention or treatment, or both, of bleeding, and debulking the tumor mass. Advanced age, coexistent medical disease, and frequently, poor nutritional status contribute to a relatively high morbidity and mortality for palliative surgery in many patients with this advanced colorectal disease. In addition, a number of patients undergoing surgical exploration with far-advanced disease are unresectable. In others, if resection is technically possible, a temporary or permanent colostomy is required.

The consequences of colostomy are often devastating, both to the patient and family. Until recently, there has been no acceptable alternative to operative management in the patient with advanced colorectal cancer. Success with the laser used endoscopically has been reported in establishing an adequate lumen in previously obstructed malignant lesions involving the esophagus, gastroesophageal junction, and in a few cases, the stomach (7).

Fiberoptic colonoscopy similarly offers the ability to vaporize malignant tumors using the Nd:YAG laser and to restore patency of the lumen. This can be used palliatively in the nonoperative management of an obstructing colorectal cancer in patients with advanced metastases. A case reported by Mellow and McCoy (8) achieved the surgical goals of relief of obstruction, prevention of bleeding, and tumor debulking using an Nd:YAG laser and local anesthesia in such a clinical setting. Four treatment sessions were required over 7 days. The Nd:YAG laser was used at powers of 85-100 W. At 3-month follow-up, the lumen remained widely patent with no major increase in tumor bulk or bleeding (8).

Bowers (9) reported 14 patients treated with the Nd:YAG laser for colonic obstruction. All were free of obstruction or bleeding at the time of death or last follow-up. Treatment was given in three to six outpatient procedures 3-5 days apart. The 4% perforation rate reported by Lambert and Sabben (10) in Nd:YAG laser treatment of rectosigmoid tumors has not been confirmed and is currently at a much lower rate.

Groisser (11) has treated eight patients with colorectal cancer using the Nd:YAG laser over a 1-year period. Treatment was to reduce obstruction and/or bleeding. Six had a carcinoma in the rectosigmoid region, one had a large exophytic tumor in the rectum, and one was a sigmoid carcinoma. Five were treated for partial obstruction and three for bleeding. The reported results were excellent.

Kiefhaber and colleagues (3) have treated 13 patients with obstructing colorectal tumors. They were the first to point out that an obstructing colonic and rectal cancer can be vaporized with the Nd:YAG laser through an unmodified colonoscope. If recanalization is achieved preoperatively, then a primary resection can be performed with immediate anastomosis as opposed to the two- or three-stage operation requiring a primary colostomy.

A subsequent report from Utah stated that seven cases of colorectal cancer for obstruction or bleeding were treated using the Nd:YAG laser with successful palliation and required one to six Nd:YAG phototherapy sessions. All patients were free of obstruction or hemorrhage at the time of death. No complications, especially perforation or death resulting from the endoscopic laser procedures, were reported (4).

For large bulky tumors, protruding into the lumen of the GI tract, the Nd:YAG, because of its deeper tissue penetration, will vaporize the mass more rapidly than will the argon laser (2). The contact laser endoprobes have been used successfully, at much lower powers (10-15 W), to recanalize obstructing colorectal tumors.

2. Early Colorectal Cancer

In the United States, the treatment of cancer by laser vaporization (ablation) would be reserved for the medically unfit who could not undergo a colectomy. In Japan, however, 22 cases of early colonic cancer have been treated by radical tumor vaporization using the Nd:YAG laser and colonoscopy (Fig. 2). In 14 cases followed for more than 1 year, results of multiple biopsies have been negative with no recurrent or persistent tumors (12).

C. Benign Strictures

To date, nonneoplastic stenoses of the GI tract have been treated either conservatively with bougies, balloon dilation, electrocautery, or operatively by resection of the stenotic area. Bougienage is often unsuccessful in immediately widening

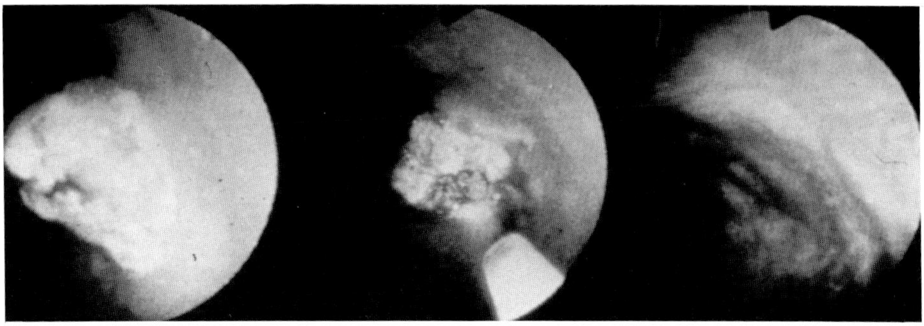

Figure 2 Early colonic cancer before treatment (left), during treatment with contact endoprobe (middle), and 9 months later (right) with normal-appearing mucosa and normal biopsy specimens.

the lumen of a narrowed organ. Furthermore, there is the risk of perforation at the site of dilation, and the effect of dilation is often temporary. Surgery is associated with its own morbidity and mortality especially in high-risk patients.

The Nd:YAG laser has been used to restore patency of a stenosed anastomosis at the rectosigmoid junction following a previous colectomy with good results in two patients (13). Follow-up showed an ulcer that had healed by 3 weeks with reepithelialization. In addition, Sasako et al. successfully treated two postoperative strictures with the laser (14). The procedure can be performed by either vaporization of the surrounding tissue with the Nd:YAG laser to sufficiently open a scarred ring or, using the contact cutting endoprobes, to incise the stenosis at two to four areas on the circumference. These stenotic areas tend to be short following surgical resections and can be cured in a single session with no further treatment necessary.

Following use of the stapling instruments for colorectal surgery in the formation of end-to-end and end-to-side anastomoses, strictures may occur at the anastomotic site. The Nd:YAG laser has been successfully used endoscopically to restore patency (15). The major complication of perforation can be avoided by careful selection of patients and avoidance of excessive treatment at the stricture site. Other strictures that could be treated in the lower GI tract include chronic inflammatory Crohn's disease, and those following radiotherapy. However, more information is required.

D. Benign Mucosal Lesions

1. Familial Polyposis Syndrome

This syndrome is usually treated by a subtotal colectomy, ileoproctostomy, and close surveillance of the remaining rectal segment. Biopsy specimens of polyps remaining in the rectal stump are taken, and the polyps usually are vaporized with monopolar electrocautery. This technique carries the risk of perforation and major bleeding. In Utah, 11 patients with a total of more than 500 polyps in the rectal stump have been treated with the argon laser without complication as an outpatient procedure (16). Despite successful polyp control in 92% of the patients, the argon laser did not cure the disease, and it is possible that some early cancers were missed by laser ablation of the polyps. However, no malignancy developed in any patient during follow-up. The Nd:YAG laser is more effective because of its depth of penetration and greater ability to destroy tissue. Proctectomy is indicated when the polyp growth is uncontrollable or when early malignant change or generalized mucosal dysplasia is found in biopsy specimens. Contact endoprobes provide greater control.

2. Adenomatous Polyps

Pedunculated polyps can usually be easily removed with the snare biopsy, electrocautery, or the contact laser endoprobe with the Nd:YAG laser. Biopsy specimens of polyps over 4 mm should be taken first to exclude malignancy. Sessile sporadic adenomatous polyps can be ablated easily and successfully without complication with Nd:YAG laser (Fig. 3).

3. Villous Adenomas

Large villous adenomas in the colon and rectum have been successfully treated with the Nd:YAG laser after partial debulking using polypectomy snares to reduce the tumor mass (3). Fifteen patients have been treated successfully by Kiefhaber with no recurrence. Of these 15 patients, three required several treatments with subsequent normal healing of the overlying mucosa (3).

4. Hemorrhoids

Hemorrhoids represent a normal anatomical state and occur in all adults. Only when they become enlarged and symptomatic is treatment indicated. Internal hemorrhoids are a plexus of superior hemorrhoidal veins above the mucocutaneous junction, which are covered by mucosa. They thus represent a vascular cushion in loose areolar tissue of the lower rectum and occupy the three primary positions of right-anterior, right-posterior and left-lateral areas. Hemorrhoids are classified as first-degree (bright, painless rectal bleeding), second-degree (protrude on straining and reduce spontaneously), and third-degree (protrude and need manual reduction). Conservative treatment requires ingestion of a high-fiber diet and regulation of bowel habits. If unsuccessful, treatment varies from injection therapy, rubber-band ligation, cryosurgery, infrared photocoagulation, anal dilation, or surgical excision. In the United States, over 150,000 formal hemorrhoidectomies are performed each year, with an average inpatient hospitalization stay

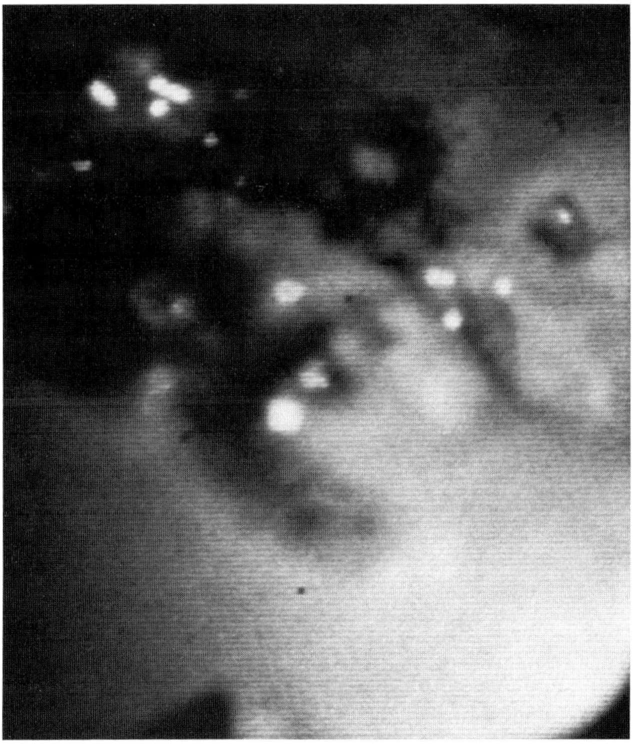

Figure 3 Adenomatous polyp following treatment with the contact endoprobe using the Nd:YAG laser. The polyp had recurred three times following electrocautery. Note mucosal defect immediately after treatment. The polyp has been totally ablated and the surrounding area was subsequently treated to prevent a recurrence.

of 5.9 days (17). The Nd:YAG laser offers an alternative method of treating hemorrhoids as an outpatient procedure, with or without local or regional anesthesia. Unless a large number of patients are being treated, it would be difficult to justify the purchase of a laser for this single indication.

5. Anal Fistula

Anal fistula, in which the tract passes through or above the anal sphincters, presents a dilemma in treatment. Adequate excision of the fistula may cause fecal incontinence. Using the CO_2 laser, the fistulous tract can be cored out. The small and accurate focus of the laser allows excision of the fistula with minimum bleeding while maintaining sterility of the wound. Studies indicate good healing without sphincter damage. The CO_2 and contact Nd:YAG laser scalpel can be used for low anal fistula and perianal fissure excision as well as for performing an anal sphincteroplasty.

6. Perianal Condyloma Acuminatum

Condyloma acuminatum is caused by a human DNA papillomavirus (HPV), usually sexually transmitted. The incidence of the disease in the Western world is increasing, and in the United States it is the fourth most frequently transmitted disease. The most common form of treatment is local application of podophyllin, which cures only 24-64% of patients (18). Cytotoxic drugs, such as 5-fluorouracil and bleomycin, are toxic drugs for a benign disease. Operative methods, such as surgical excision, electrocoagulation, and cryotreatment, have not proved to be effective and lead to bleeding and disfiguring scars.

The introduction of laser techniques appears to offer an alternative and more satisfactory form of treatment. Several reports indicate an effectiveness of the CO_2 laser in the treatment of condyloma acuminatum. Further advantages for laser treatment would be during pregnancy and in patients with extensive and multiple lesions. The latter conditions are often found in immunosuppressed and diabetic patients.

Most of these patients are seen by gynecologists and dermatologists and the perianal condyloma acuminatum is often associated with genital lesions and carcinoma in situ changes of the cervix. In a recent report by Baggish (19) of 53 patients (71% female) with condyloma acuminatum, 38% were genital, 9% perianal, and 53% had lesions in both regions. Most patients had failed treatment using conventional methods, and the disease had been present in 90% for more than 5 years. Treatment was usually carried out under general anesthesia using a CO_2 laser with powers at 15-20 W, either as a single procedure (70%) or on multiple occasions if the warts were extensive.

Complications of treatment are usually minimal and include minor bleeding and infection. However, a rectal perforation has been reported in a diabetic patient, with the development of an ischiorectal abscess. Results at follow-up indicate a high cure rate (83%) with recurrences occurring in 25%. The Danish results showed a 75% cure rate with CO_2 lasers, of which there was a 75% cure with a single treatment (20). Scott and Castro (21) had an even lower recurrence rate of 2%.

Recent work suggests a strong association between human papillomavirus (HPV) and cancer of the female genitals. This means that there is an increasing need for early vigorous and more effective treatment of these lesions with lasers. The Nd:YAG laser because of its greater depth of penetration will be associated with a higher cure, lower recurrence rate, and less tissue damage if used at lower powers.

E. Phototherapy of Colorectal Tumors

Light intensity in the visible or infrared region of the spectrum, below that producing thermal effects in living tissue, has little effect. Pretreatment with photosensitizing agents can enable light of this intensity to produce a cytotoxic effect for tumor therapy. Neoplastic tissue has an affinity for porphyrins and can be detected by fluorescence under ultraviolet light. The selectivity for malignant tumors is improved using a derivative of hematoporphyrin (HPD). The mechanism of the cytotoxic effect occurs by the activation of HPD by incident laser light that converts the HPD to triplet O_2 to excited singlet O_2 which is cytotoxic to the cell membrane. The lifetime of singlet O_2 is too short for it to move more than a few cell diameters from the site of production, and hence the effect is localized to the areas of high HPD concentration. Selective destruction of a malignant tumor is possible in areas in which the surrounding normal tissue has a low affinity for HPD. Quantitative studies on the treatment of solid tumors in animals is difficult to carry out, and few have been reported (Fig. 4).

Clinical studies from several centers report encouraging results of phototherapy of malignant tumors including bronchial cancers (22), breast cancers (23), malignant melanomas, and even gliomas (24). However, the treatment regimens used have been empirical. There is no good evidence available to determine the optimum values for the dosage of HPD, the time after administration before exposure to light, and the light intensity and exposure time, although rough guidelines can be found in the results of animal experiments. Little data are available on the extent and severity of the necrosis produced around each treatment point under given conditions in normal or neoplastic tissue (25).

In bronchial cancers treated with flexible bronchoscopes, tumors of less than 3 cm have responded well; only two tumors over 4 cm in one reported series of 10 patients bled massively at 6 and 11 days after treatment (22). It was thought that extensive tumor necrosis in an already necrotic cancer resulted in further hemorrhage and death. Similar problems could be anticipated in the treatment of colorectal tumors with bleeding and perforation with extensive necrosis.

It has been suggested that long exposure times may deposit enough energy to produce some tissue heating, thereby producing thermal biological effects, as well as those caused by singlet O_2. Nevertheless, the relatively low energies involved mean that the light can be transmitted to the fibers using the external (no contact) or contact and interstitial technique.

Selective tumor phototherapy is therefore possible in the lower GI tract by sensitization with HPD. Most current phototherapy with HPD sensitization uses red light of a wavelength between 625 and 630 nm. The HPD has several absorption peaks from the ultraviolet region of the spectrum to the visible red. The peak at 625 nm is the weakest and yet the longest wavelength at which a peak is seen. Most animal tissues absorb light in the blue and green regions,

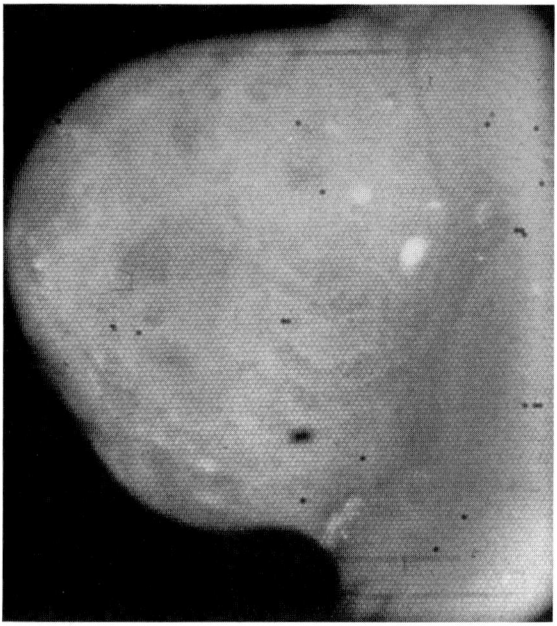

(a)

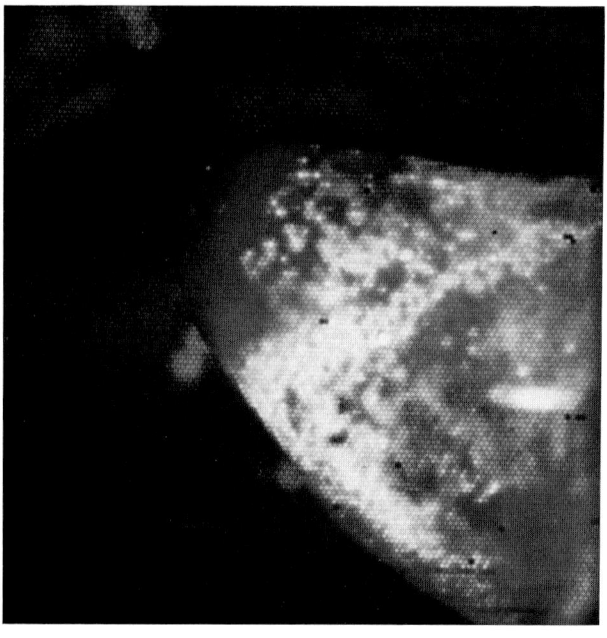

(b)

Figure 4 (a) Experimentally produced adenocarcinoma in the stomach of a dog, at 1 year; (b) during photodynamic therapy with the argon dye laser and the interstitial contact probe.

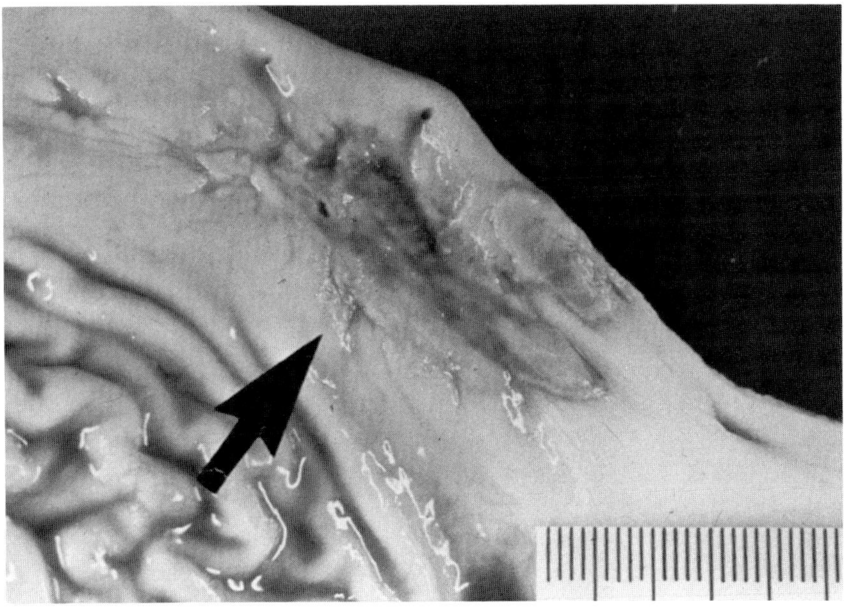

(c)

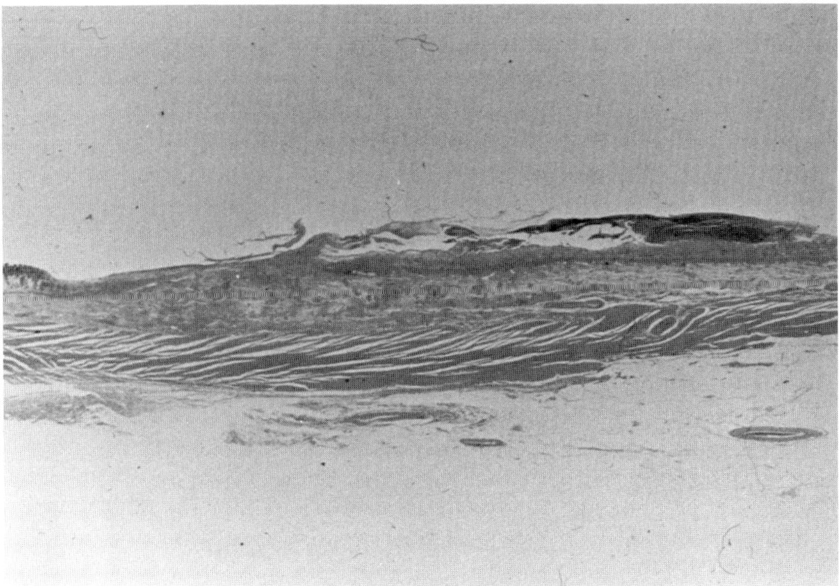

(d)

Figure 4 (continued) (c) resected stomach 10 days later, showing small necrotic ulcer; (d) light microscopy after photodynamic therapy with no evidence of malignancy, only necrosis and fibrosis.

which severely limits the penetrating power of light at these wavelengths. Absorption of red light in unsensitized tissues is less, making the peak at 625 nm the best choice. The most efficient source of light at this wavelength is a rhodamine B dye laser. The argon or copper vapor laser enables the light to be transmitted via a flexible fiber. Nonlaser light sources (e.g., xenon arc lamp or quartz halogen lamp with suitable filters) can be used if fiber transmission is not required, as with rectal lesions.

Phototherapy with HPD is extremely promising in colorectal tumors, but problems remain to be overcome. Although its chemistry is reasonably well understood and the cytotoxic effects on cell cultures are well documented, studies on whole animals and patients are largely empirical, and treatment of internal tumors may carry a risk of complications. The selectivity of HPD for malignant tumors is only relative and does not apply in all organs. The use of a laser and fiber transmission system with the colonoscope does provide a separate degree of selectivity of defining the area to which the light is applied. Hematoporphyrin derivative is the best photosensitizing agent currently available, but it is not ideal because it has a weak absorption peak in a suitable part of the spectrum (25). It is possible endoscopically to place multiple fibers with interstitial contact endoprobes to obtain a more uniform light intensity through a larger tissue volume. Tumor selectivity using thermal techniques appears to occur only if the pH level in the center of tumor is lower than that in the surrounding tissue.

F. Interstitial and Local Hyperthermia Therapy for Colorectal Tumors

In conventional use of the argon or Nd:YAG laser, the beam is transmitted via a sheath of fibers with the tip of the fiber being held at a distance of 0.5-1 cm from the target tissue (external therapy). Cellular debris and blood on the tip of the fiber absorb the energy and raise the temperature sufficiently to destroy it. By inserting the bare quartz or glass fiber tip into tissue, the direct thermal contact with a larger volume of tissue conducts the heat away (26). This interstitial therapy can heat tissue adjacent to the fiber to 100°C for a prolonged period. The extent of irreversible tissue damage depends upon the laser power and exposure time for each type of tissue. Treatment of cutaneous deposits from various malignant tumors at powers of 5-20 W for 10-15 sec (75-200 J) have been carried out by Bown (26). There is immediate vaporization followed by delayed sloughing with replacement by fibrous tissue.

To date, interstitial therapy has not been used for lower GI tumors. However, because of the accessibility of colorectal tumors to a single fiber and with the predictability of tissue damage without loss of mechanical integrity caused by scar formation, this treatment may find an increasing role in lower GI tumors both for primary and recurrent cancers and for large sessile polyps.

Local hyperthermia has been considered a useful modality for many years. In the last decade interest has increased and it can be used synergistically with chemotherapy or radiotherapy (27). The main heating techniques include microwaves, radiofrequency, ultrasound, and lasers. Each has advantages and disadvantages, with the objective to maximize damage to tumor cells and minimize effects in normal areas. Reports that malignant cells are more heat-sensitive than normal may be incorrect, and there is also no consistent difference between the heat response of hypoxic and oxic cells. A low pH level inside tumors may enhance thermal sensitivity, and hyperthermia should be limited to the tumor area.

Superficial tumors are easy to heat by all techniques, but problems arise for the lower GI tract for microwave and radiofrequency heating. Even using an array of applicators, adequate localization of the effects, other than near the skin, is extremely difficult. Using contact endoprobes placed centrally in tumors and controlling the power density, tumors can be heated to 44°C, giving a uniform heating to a radius of 1-2 cm from the treatment point (Fig. 5). These artificial sapphires can be directly attached to fiberoptic delivery systems of the Nd: YAG laser and be placed endoscopically (Fig. 6). A computer-controlled system is currently being developed to be used in conjunction with thermocouples to control the temperature for the local hyperthermia.

G. Open Surgery—Colon Resection

One-third of the mortality after colonic surgery is due to leakage at the anastomosis. The reported incidence of leakage from the suture line varies from the high rate of a radiological diagnosis of 6-35% and obvious clinical leakage of 1-18%. Factors influencing this leakage rate include operative technique, integrity of the suture line, nutritional status, blood supply, and bowel preparation. The contact Nd:YAG laser scalpel (Fig. 7) would prove useful as a cutting tool based on its accuracy, bactericidal, and hemostatic properties.

An experimental study in rats was undertaken to evaluate the CO_2 laser in colorectal surgery (28). The strength of an end-to-end anastomosis of the rat colon was assessed following resection with a CO_2 laser, conventional surgical scalpel, and diathermy. The bursting pressure was measured on the third, fifth, and seventh postoperative days. The CO_2 laser resulted in a significant increase of colonic bursting pressure at the anastomosis site. Diathermy was more harmful in its thermal damage. The use of lasers such as CO_2 and contact Nd:YAG may reduce the frequency of colonic anastomosis leakage and be useful in excision of liver metastases.

H. Intractable Rectal Pain

1. Nonmalignant

The surgical treatment of intractable pain has an important role in the neurosurgical armamentarium. Operations are usually ablative and directed at the pain-

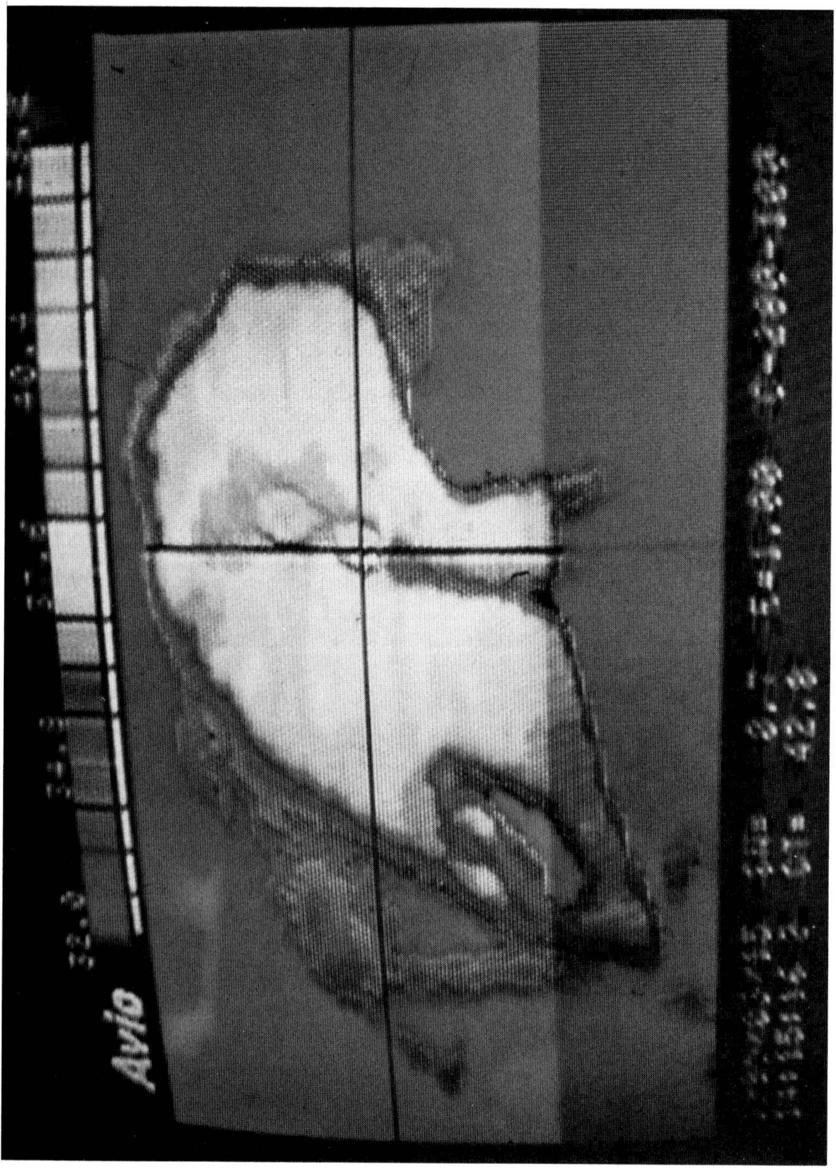

Figure 5 Local hyperthermia showing radius of heat distribution as measured by thermography in a tumor being treated.

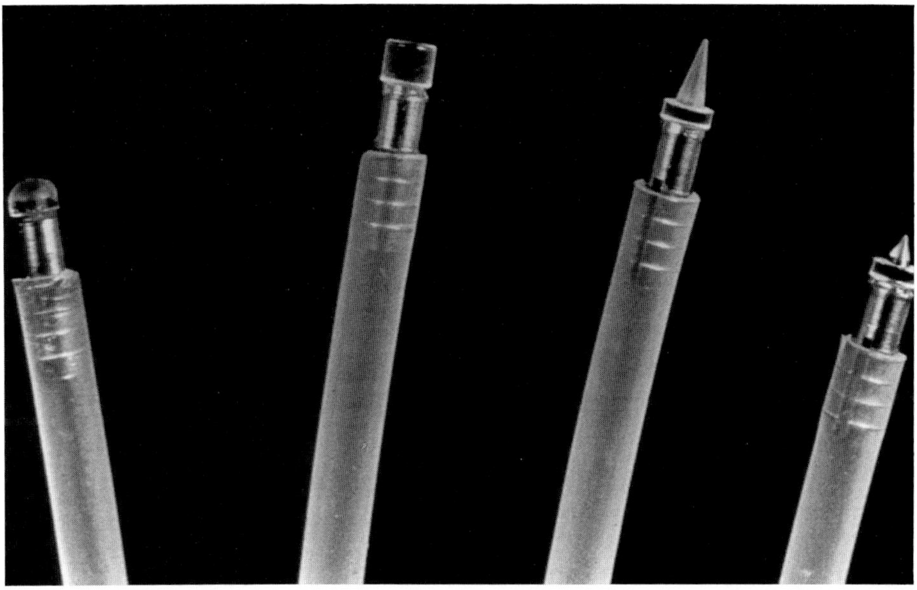

Figure 6 Contact laser endoprobes to be used with the Nd:YAG and argon lasers. These include probes for cutting, coagulating, vaporizing, local hyperthermia, and photodynamic therapy. A universal holder with a screw-on mechanism allows easy changing of each synthetic ceramic probe.

conducting pathways of the nervous system, the risk being between pain relief and loss of other functions. The CO_2 laser with the operating microscope has been used to perform a commissural myelotomy in a female patient with chronic intractable rectal pain not associated with any obvious organic pathology in the rectum (29). The procedure was successful in relieving the pain without any impairment of function of the lower extremity or sphincteric control. Surgical lasers will thus probably improve neuroablative procedures for intractable pain, but more experience is required.

2. Malignant

Laser neuroablative procedures have a role in the treatment of intractable pain associated with colorectal cancer. Microsurgical techniques with lasers can selectively denervate pelvic dermatomes and perform a myelotomy. Good results should be obtained, avoiding the complications of leg weakness, posterior column deficit, or change in bladder and bowel function.

Figure 7 Contact laser scalpel for open anal and colorectal surgery.

III. CONTACT VERSUS NONCONTACT LASER TREATMENT

The conventional method of applying either Nd:YAG or argon laser therapy for coagulation of bleeding, vaporization of tumors, or excision of polyps is using a noncontact quartz or glass fiber. Disadvantages include the tip burning, which requires recleavage, imprecision of the surgery, deep penetration with tissue necrosis, and the production of smoke. By using the newly described technique of attaching a synthetic sapphire to the end of the fiber contact laser surgery can now be performed (30). The tips do not melt, tissue damage is less, and a much lower laser energy is required (Figs. 6 and 7). The geometric shape of the endoprobes determines the light dispersion in the tissue (Fig. 8).

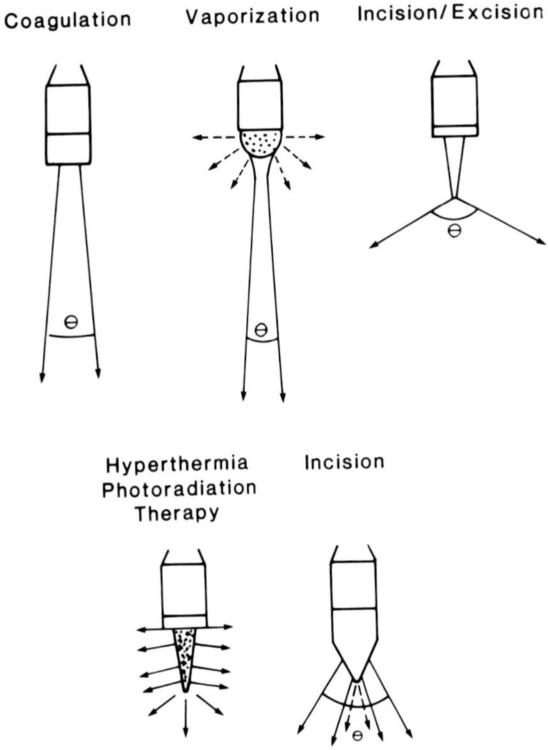

Figure 8 Geometric shape of the contact laser endoprobes, their applications, and light dispersion in tissue.

IV. LASER SAFETY

Hazards associated with the use of surgical laser systems in the lower gastrointestinal tract require appropriate safety precautions and policies (31). In addition to the electrical hazards, safety precautions center around eye safety and care of the endoscopes. The Nd:YAG laser presents the most serious visual hazards of the three types of lasers. When the laser is used in the colon or rectum in closed applications, such as through an endoscope, operating room personnel need not wear protective eyewear. For the endoscopist's protection, a near-infrared filter is placed over the eyepiece of the endoscope. If a teaching head is attached, it will also require a filter. These may be either permanently glued to the eyepiece or be removable as a cap.

When filters are being used on the colonoscope and personnel are not wearing eye protection, a hazard exists if the laser fiber or entire endoscope is removed from the patient and fired in the open room. It is imperative that the laser be disabled before and immediately after the closed procedure is completed.

It is possible to cause extensive damage to the flexible colonoscopes by firing

Table 4 Results of Questionnaire Reported to Washington Laser Symposium Regarding Laser Treatment of Lower Gastrointestinal Bleeding and Cancer

Lower GI bleeding (treated by 19 laser users)	
angiodysplasia (nonbleeding)	196
(bleeding)	88
neoplasms	112
radiation colitis	45
ulcerative colitis	8
hemorrhoids	25
other	3
Total treated	477
Lower GI cancer (treated by 21 laser users)	
rectum	759
colonic	106
Total treated	865

Source: D. Fleischer (1985), personal communication.

the laser with the fiber not fully extended from the channel. As a policy, the physician must always see the tip of the fiber in his field of view before firing the laser.

Because of the high degree of backscatter with the Nd:YAG laser and the black color of the plastic tip of most endoscopes it is possible to melt the tip of the endoscope and cause a small fire if high concentrations of flammable gas are present. Adequate bowel preparation for colonic laser surgery is important to prevent an explosion. To date this complication has not been reported.

For open colonic, rectal, and perianal laser surgical procedures, all personnel and the patient must wear appropriate safety glasses of the correct optical density for the Nd:YAG, argon, or CO_2 laser. It is also prudent to consider the installation of door interlocks that automatically stop the laser's action while the door is open. Windows need to be covered and appropriate warning signs should be placed on all entries to the surgical suite. These signs should state the type and class of laser being used and its maximum power output.

Applications of lasers, especially the Nd:YAG laser, used endoscopically are increasing at a rapid rate and are completely revolutionizing our current method of treatment (Table 4). Safety and efficacy are being proved in many centers, and lasers will be a standard tool in the endoscopist's armamentarium.

ACKNOWLEDGMENTS

Messrs. N. Daikuzono and D. Joseph of Surgical Laser Technologies, Malvern, Pennsylvania, for use of contact endoprobes and laser scalpels; Dr. H. Tajiri of the National Cancer Institute, Tokyo, Japan, for Figure 4; Ms. S. Hansen for preparation of the manuscript.

REFERENCES

1. Fleischer D: The current status of gastrointestinal laser activity in the United States. *Gastrointest. Endosc. 28*:157-161, 1982.
2. Hunter JG, Bowers JH, Burt RW, Sullivan JJ, Stevens SL, and Dixon JA: Lasers in endoscopic gastrointestinal surgery. *Am. J. Surg. 148*:736-741, 1984.
3. P. Kiefhaber, K. Kiefhaber, F. Huber, and G. Nalb. In *Neodymium-YAG Laser in Medicine and Surgery* (S. N. Joffe, ed.). Elsevier Publishing Co., New York, 1983, pp. 5-14.
4. Hunter JG, Burt RW, Becker JM, Lee RG, and Dixon JA: Colonic mucosal lesions: Evaluation of monopolar electrocautery, argon laser, and neodymium:YAG laser. *Curr. Surg. 41*:373-375, 1984.
5. Feld AD, Fondacaro JD, Holloway GA Jr, and Jacobson ED: Laser Doppler velocimetry: A new technique for the measurement of intestinal mucosal blood flow. *Gastrointest. Surg. 30*:225-230, 1984.

6. Leuchter RS, Petrilli ES, Dwyer RM, Hacker NF, Castaldo TW, and Lagasse LD: Nd:YAG laser therapy of rectosigmoid bleeding due to radiation injury. *Obstet. Gynecol. 59*:65S-67S, 1982.
7. Fleischer D, Kessler F, and Haye O: Endoscopic Nd:YAG laser therapy for carcinoma of the esophagus: A new palliative approach. *Am. J. Surg. 143*: 280-283, 1982.
8. Mellow MH and McCoy GA: Endoscopic laser therapy in the palliative treatment of colorectal carcinoma: A case report. *J. Okla. State Med. Assoc. 77*:383-386, 1984.
9. J. Bowers. In *Therapeutic Laser Endoscopy in Gastrointestinal Disease* (D. Fleischer, D. Jensen, and P. Bright-Asare, eds.). Martinus Nijhoff, Boston, 1983, p. 139.
10. Lambert R and Sabben G: Laser therapy in colorectal tumors: Early results (abstr.). *Gastroenterology 84*:1222, 1983.
11. Groisser VW: YAG laser therapy of gastrointestinal tumors. *Gastrointest. Endosc. 30*:311-312, 1984.
12. Y. Oguro and H. Tajiri. Abstract, in *2nd International Nd:YAG Laser Conference*. Munich, Germany, 1985.
13. Sander R, Posl H, and Spuhler A: Management of non-neoplastic stenoses of the GI-tract—A further indication for Nd-Yag laser application. *Endoscopy 16*:149-151, 1984.
14. Sasako M, Iwasaki M, Konishi T, Maruyama Y, and Wada T: Clinical application of the Nd:YAG laser endoscopy. *Lasers Surg. Med. 2*:137-147, 1982.
15. Chen P-C, Wu C-S, Chang-Chien C-S, and Liaw Y-F: YAG laser endoscopic treatment of an esophageal and sigmoid stricture after end-to-end anastomosis stapling. *Gastrointest. Endosc. 30*:258-260, 1984.
16. Dixon JA, Burt RW, Rotering RH, and McCloskey DW: Endoscopic argon laser photocoagulation of small sessile colonic polyps. *Gastrointest. Endosc. 28*:162-165, 1982.
17. Commission on Professional and Hospital Activities (CPHA): *Length of Stay by Operation, United States 1982.* CPHA, Ann Arbor, MI, 1983, p. 44.
18. Hahn GA: Carbon dioxide laser surgery in treatment of condyloma. *Am. J. Obstet. Gynecol. 141*:1000-1008, 1981.
19. Baggish MS: Carbon dioxide laser for condylomata acuminata venereal infections. *Obstet. Gynecol. 55*:711-715, 1980.
20. Kryger-Baggesen N, Falck Larsen J, and Hjortkjaer Pedersen P: CO_2-laser treatment of condylomata acuminata. *Acta Obstet. Gynecol. Scand. 63*: 341-343, 1984.
21. Scott RS and Castro DJ: Treatment of condyloma acuminata with carbon dioxide laser: A prospective study. *Lasers Surg. Med. 4*:157-162, 1984.
22. Cortese DA and Kinsey JH: Endoscopic management of lung cancer with hematoporphyrin derivative phototherapy. *Mayo Clin. Proc. 57*:543-547, 1982.

23. Dougherty TJ, Kaufman JE, Goldfarb A, Weishaupt KR, Boyle D, and Mittleman A: Photoradiation therapy for the treatment of malignant tumors. *Cancer Res. 38*:2628-2633, 1978.
24. Forbes IJ, Cowled PA, Leong ASY, Ward AD, Black RB, Blake AJ, and Jacka FJ: Phototherapy of human tumours using haematoporphyrin derivative. *Med. J. Aust. 2*:489-493, 1980.
25. Bown SG: Phototherapy of tumors. *World J. Surg. 7*:700-709, 1983.
26. S. G. Bown. In *Neodymium:YAG Laser in Medicine and Surgery* (S. N. Joffe, ed.). Elsevier Publishing Co., New York, 1983, pp. 51-58.
27. Ultrasound microwave and radiofrequency radiations: The basis for their potential in cancer therapy. Proceedings of the 10th L. H. Gray Conference: Oxford 13-16 July 1981. *Br. J. Cancer 45*(suppl. 5):1-257, 1982.
28. Ferulano GP, Danzi M, Abate S, Fresini A, Vanni L, and Califano G: Effects of CO_2 laser in the healing of experimental colon anastomoses. *Eur. Surg. Res. 16*:127-130, 1984.
29. Fink RA: Neurosurgical treatment of nonmalignant intractable rectal pain: Microsurgical commissural myelotomy with the carbon dioxide laser. *Neurosurgery 14*:64-65, 1984.
30. N. Daikuzono and S. N. Joffe. *Medical Instrumentation. 19*:173-178, 1985.
31. G. T. Absten and S. N. Joffe. *Lasers in Medicine: An Introductory Guide.* Chapman and Hall, London, 1985, pp. 63-69.

11

Lasers in Hepatobiliary and Pancreatic Surgery

STEPHEN N. JOFFE and M. Y. SANKAR
University of Cincinnati Medical Center, Cincinnati, Ohio

I. INTRODUCTION

Numerous types of lasers have been produced employing various designs and base materials, and development of new laser technologies is continuing at a rapid rate. The three main types of lasers that have been used in general surgery are carbon dioxide (CO_2), argon, and neodymium:yttrium-aluminum-garnet (Nd:YAG) lasers. The different wavelengths determine the visibility of the beam and the depth of tissue penetration (Table 1).

II. APPLICATIONS

Gastrointestinal diseases were initially treated endoscopically using lasers in the mid-1970s. The spectrum of diseases being treated and the number of patients have expanded exponentially since then. In 1981, less than 12 medical centers in the United States were using lasers in the treatment of gastrointestinal (GI) disease. By the end of 1983, there were more than 200, and this number is increasing (1). There are several reasons for such proliferations, including increased applications of the use of lasers in GI problems and the relative ease with which it can be used. Lasers provide a multidisciplinary and multispecialty therapeutic modality with increased therapeutic options where previously such choices either did not exist or were limited. More than anything, lasers are relatively safe to use, efficient, reduce morbidity and mortality, and are cost-effective. The following discusses experimental and clinical applications of lasers in hepatobiliary and pancreatic surgery with emphasis on the continuous wave Nd:YAG laser.

Table 1 Specifications of Lasers Used in Biliary and Pancreatic Surgery

Laser	CO_2	Argon	Nd:YAG
Type	Gas	Gas	Crystal
Wavelength (μm)	10.6	0.5	1.06
Absorbed by	Water	Hemoglobin and melanin	Tissue and protein
Penetration (mm)	0.5-1	2-3	5-7 (noncontact) 0.5-1 (contact)
Delivery system	Mirror	Fiberoptic	Fiberoptic
Coagulation	Low	Fair	High
Cutting	High	Low	Fair

III. EXPERIMENTAL STUDIES USING LASERS IN HEPATOBILIARY AND PANCREATIC TISSUES

A. Liver Tissue

Three types of laser radiation were compared relative to their capabilities for hepatic resection in New Zealand rabbits (2). Hemostasis after widely performed resection was possible with both argon and Nd:YAG laser. Bleeding from hepatic vessels up to 4-5 mm in size was controlled using only the laser without adding suture ligation. The argon laser radiation, at a wavelength of 0.5 μm, is selectively absorbed by hemoglobin, and hemostasis was obtained by a lower power output than with the Nd:YAG laser. At a wavelength of 1.06 μm, the Nd:YAG laser has the capability for deeper penetration as well as greater tissue damage and protein denaturation. Several disadvantages are seen when comparing CO_2 and Nd:YAG lasers for experimental liver resection. The CO_2 laser radiation, at a low-energy output of 15 W, is unsuitable for hemostasis even from small-diameter vessels and is a slow and tedious operation. The Nd:YAG laser radiation causes additional necrotic tissue, and therefore it theoretically increases the risk of postoperative infection. The articulating arm of the CO_2 laser is also difficult to manipulate during surgery. The conventional noncontact Nd:YAG laser is also slow. Liver resection is possible with all three types of lasers. The argon laser induced superficial lesions at all power densities. The CO_2 laser produced a cutting defect, free of excessive tissue necrosis, but vessels with an internal diameter of more than 1 mm were not sealed (3,4). The Nd:YAG laser caused a deeper penetration that was variable in depth. The thermal effects extended to vessels in the vicinity, resulting in coagulation of blood in veins and sealing off of arteries. Controlled studies of liver resection in dogs done by Nakajima have not substantiated the superiority of the CO_2 laser technique over the conventional "finger-fracture" or electrocoagulation technique (5). Tranberg et al. (6), in controlled studies, have clearly shown several points in favor of using the Nd:YAG laser in hepatic resections in dogs. The Nd:YAG laser used as a nontouch technique in liver resections was compared with the conventional "finger-fracture" procedure and a technique using ultrasonic surgical aspirator with suture or clip ligations of vessels. It was concluded that in elective liver resections, both the Nd:YAG laser and ultrasonic surgical aspirator reduced the operating time and blood loss significantly. Similarly, the likelihood of postoperative infection was consequently less when compared with the conventional procedures (6). In addition, the use of both the modalities, i.e., Nd:YAG laser and ultrasonic surgical aspirator, had no adverse effects. Durtschi et al. (7) in their study compared the Nd:YAG laser with the standard technique, e.g., suture ligation, stainless steel clips, and electrocautery, for performing partial hepatic lobectomy in dogs after a standardized injury. They feel that hemostasis remains a major technical problem in surgery of the liver and spleen. A high-powered Nd:YAG laser has been coupled with a fiberoptic delivery system and quartz blade designed to yield maximal hemostasis and minimal tissue injury. They were unable to demonstrate a significant advantage of the use of Nd:YAG laser in such experimental studies. They also noted that increasing the power density of the Nd:YAG laser consistently resulted in an increased depth of tissue damage, but they did not mention the duration of laser exposure. In our laboratory, we have found that increasing power density in liver resections produces deeper penetration without additional lateral scattering, provided the time duration was fixed. A normal, but slow, healing process follows, probably because of the presence of necrotic tissue.

A contact laser scalpel has been developed for use with the delivery system of the Nd:YAG laser (Fig. 1). A specially designed artificial sapphire rod is mounted onto the laser fiber, which gives the advantage of directly touching the liver tissue during dissection and considerably reduces the backscattering of the laser beam. This enables the surgeon to get the tactile sensation of the tissue, and a lower power density can be used to achieve cutting comparable with the CO_2 laser but with the advantage of obtaining the hemostatic effects of the high-power Nd:YAG laser. Preliminary studies in our laboratory in performing experimental liver resection on rats have shown that the touch technique using the laser scalpel at low power density is associated with less bleeding, less laser "plume," decreased tissue necrosis, and fewer postoperative adhesions. Operating time is also significantly reduced by using this touch Nd:YAG laser technique (Fig. 2).

In their studies of hepatic resection with Nd:YAG laser in pigs, Godlewski et al. concluded that the laser must be considered a complementary and safe tool in the treatment of liver trauma (8).

B. Experimental Cholangioscopic Lithotomy

A human calcium stone was placed in a beaker filled with a saline solution. The stone was irradiated by using a laser beginning at a distance of 0.5 cm from the stone. At a power density of 70 W for 0.5 sec, the stones fractured in smaller

Figure 1 Contact laser scalpel.

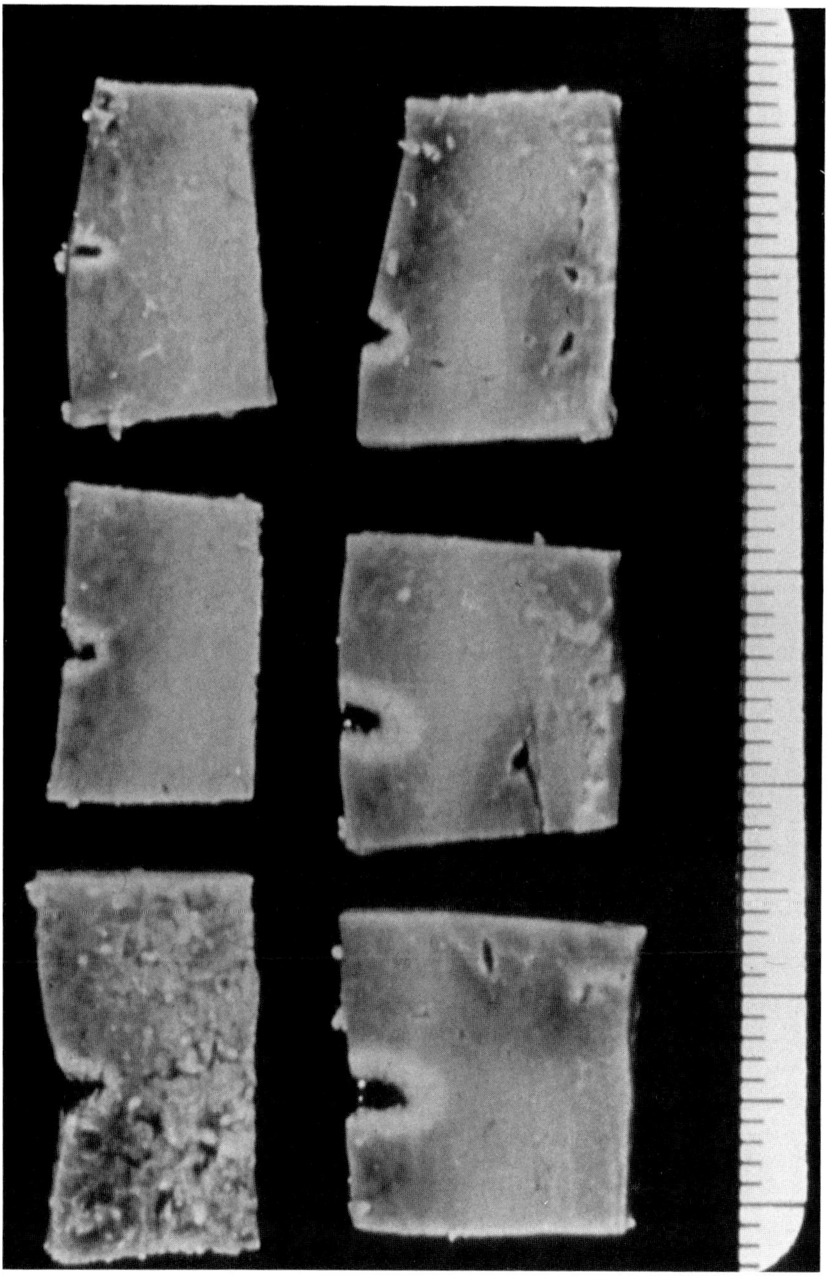

Figure 2 Liver resection: the upper row shows the effects of contact (Nd:YAG) laser scalpel on liver tissue using the touch technique (20 W/1 sec), and the lower row shows the effects of noncontact laser (80 X/1 sec).

pieces. Cholesterol stones were more resistant to disintegration than bilirubin stones, but as the irradiation time increased to 0.8 sec, these stones also disintegrated, similar to the melting of lava (9). Kouzu and Sato conducted experiments in adult mongrel dogs (9). The common bile duct was ligated to produce ductal dilatation. After a week, following relaparotomy, a human calcium bilirubinate stone was inserted into the bile duct to simulate a retained gallstone. The choledochoscope was then introduced into the dilated common bile duct and a laser fiber inserted through the choledochoscope to the surface of the stone. They concluded that the Nd:YAG laser was able to disintegrate the gallstone. Using a saline perfusion to cool the stones, the procedure could be conducted safely with no irreversible histological damage being noted in the wall of the bile duct around the stone, provided the stones were irradiated under direct vision.

C. Pancreatic Tissue

1. Dogs

Experimental studies have been carried out in our laboratory on dogs and rats with the aim of evaluating the potential use of the Nd:YAG laser in surgical procedures involving pancreatic resection. In dogs, either a partial or total pancreatic resection has been performed with duodenal preservation. The hallmark of the latter operation is a precise and meticulous dissection of the pancreas from the duodenal wall along the common blood supply of these two organs. Assessment included blood loss, operating time, graft survival, duodenal loop viability, infection, and histology. The laser surgery had many advantages over the conventional procedure of Eloy et al. (10), especially with regard to microsurgically separating the vessels to be coagulated and those to be divided and yet preserving the vascular arcade to the duodenum. The Nd:YAG laser was found to be effective and significantly reduced the operating time ($P < .01$). Insufficient amounts of energy diffusely applied to the target vessel resulted in failure of coagulation and bleeding when the vessel was divided. Delivery of an excessive amount of energy resulted in extensive damage to the surrounding tissues ranging from thrombosis to evaporation and perforation of the critical main vessel on the underlying bowel (11). The laser beam was delivered to the operative field via a flexible fiberoptic system with an inner glass core diameter of 400 μm. The beam could be aimed at the desired structure in the abdominal cavity with free movements in all directions without obstructing the vision (Fig. 3). Addition of a focusing device or contact laser surgery is expected to improve the performance and the versatility of the Nd:YAG laser. With the direct-touch technique, the amount of energy delivered to the tissue could be tailored to meet the desired effect of dissection and discriminate the coagulation, allowing minimal damage to adjacent tissues. Such work is currently in progress.

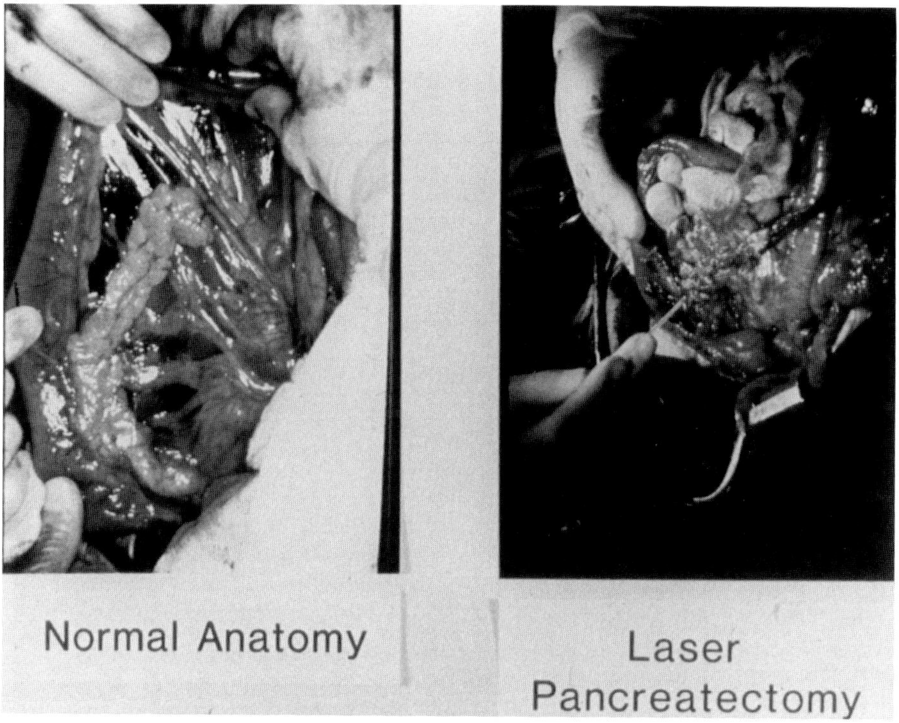

Figure 3 Experimental pancreatectomy with preservation of duodenal loop using Nd:YAG laser nontouch technique in the dog.

2. Rats

The direct effects of the Nd:YAG laser on rat pancreatic tissue after partial pancreatectomy is currently being studied. Histological observation of the rat pancreas on the day of surgery reveals the presence of four zones of tissue damage caused by the laser resection procedure (Fig. 4).

The zones are as follows: On the superficial resected surface, there was a thin layer of dense structureless protein coagulum containing carbonized tissue debris. Immediately beneath this surface was a zone of cavitation consisting of large fluid-filled cavities up to 100 μm in diameter separated by either protein coagulum nearest the surface or elongated, condensed acinar cells with pyknotic nuclei. These two layers extended for 0.1-0.15 mm. The next layer consisted of a wide band of acidophilic cells with condensed cytoplasm and nuclei varying from being pyknotic in the superficial cell layers to swollen forms in those near-

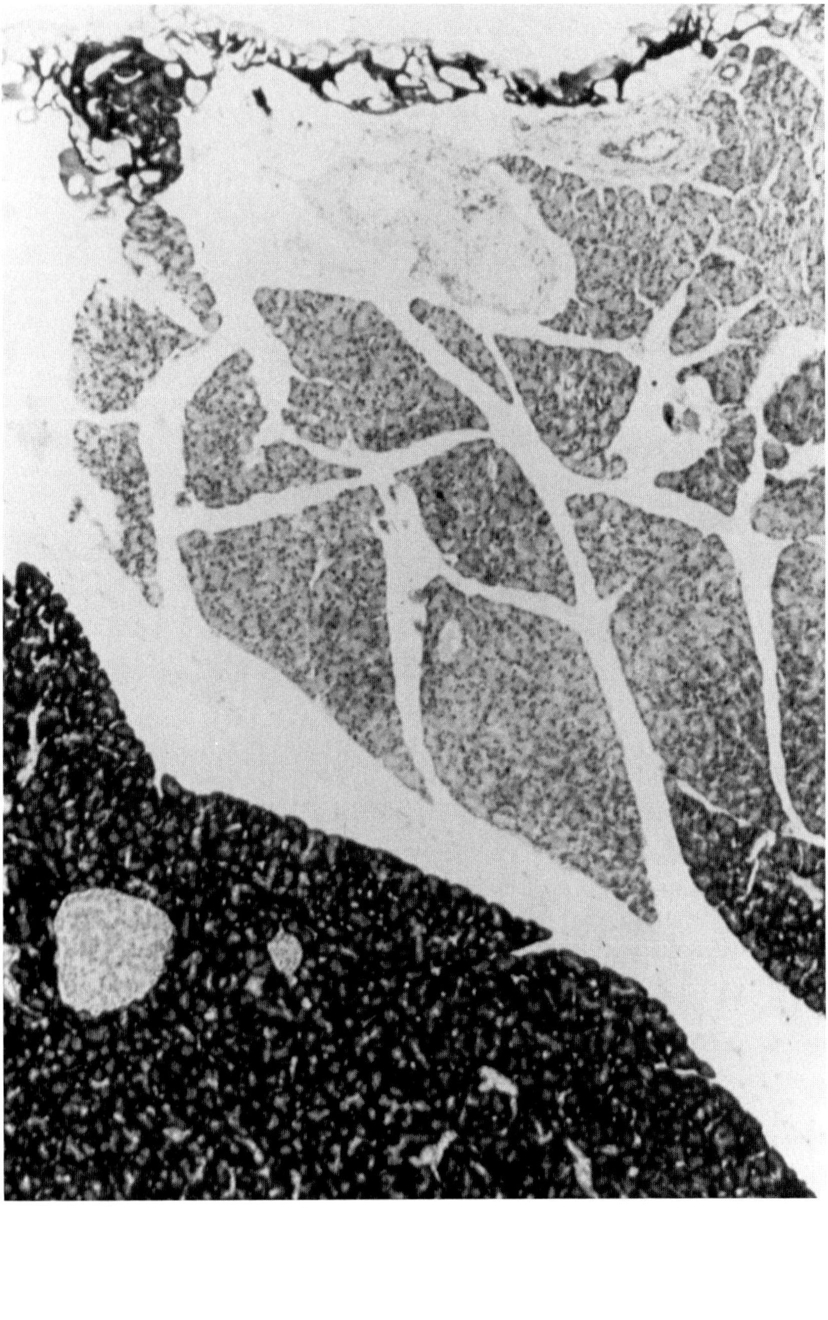

Figure 4 Histological appearance of the rat pancreas after subtotal pancreatectomy using Nd:YAG laser (70 W/1 sec) on day 0.

est the normal tissue. This extended an additional 0.3-1.3 mm into the tissue. Interlobular edema was present in this layer. The deepest layer was a narrow band (0.1-0.2 mm) of acinar cells with basophilic cytoplasm but degenerative or possibly autolytic changes including cytoplasmic vacuolation and nuclear swelling. There was no obvious boundary between normal and affected tissue. The total depth of lateral penetration was not substantially changed by increasing the power density from 50 to 100 W.

Postoperatively, the pancreatic tissue was examined at weekly intervals following laser resection. In the first and second week, all samples of pancreas showed various degrees of acute pancreatitis. At power densities of 50-80 W, edema, focal hemorrhage, vacuolation of acinar cells, and infrequently, areas of necrosis were seen (Fig. 5). At a power density of 90-100 W, much larger areas of necrosis and severe pancreatitis were found. At 14 days at these higher power levels, only a small amount of normal pancreas remained. A large area of advanced necrosis, massive cellular infiltrate, and encystment of the organ by a dense fibrocellular capsule were present. At 3 weeks, all samples from tissues resected at a power density of 50-80 W were either normal or recovering from the induced pancreatitis (Fig. 6). From these experimental studies, lower power density appears to be advantageous when performing a pancreatectomy using the Nd:YAG laser. Perhaps contact laser surgery has more to offer in this field and is currently being evaluated.

IV. CLINICAL APPLICATIONS OF ND:YAG LASER IN HEPATOBILIARY AND PANCREATIC SURGERY

A. Hepatobiliary Surgery

Hepatic resection, regardless of whether it is performed for trauma or for cancer, is a difficult procedure. The intraoperative and postoperative mortality remains at 10%. The main reason for this is the complex liver anatomy making it difficult to avoid bleeding or necrosis, or both, as well as biliary leakage. All these factors increase the risk of infection. In current hepatic surgery, resection with the laser should be considered as an adjuvant safety factor for adequate hemostasis (7). We have used the Nd:YAG in the surgical approach to a high intrahepatic bile duct stricture that was associated with cholangitis and intrahepatic choledocholithiasis. To approach the strictured portion of the hepatic duct, the Nd:YAG laser beam was attached to the laser scalpel contact probe, and the liver was split into its right and left lobes with minimum blood loss and tissue damage. The strictured portion of the hepatic duct was excised and a hepaticojejunostomy constructed. The advantages of using the laser in such situations were to reduce blood loss and operating time and to prevent biliary

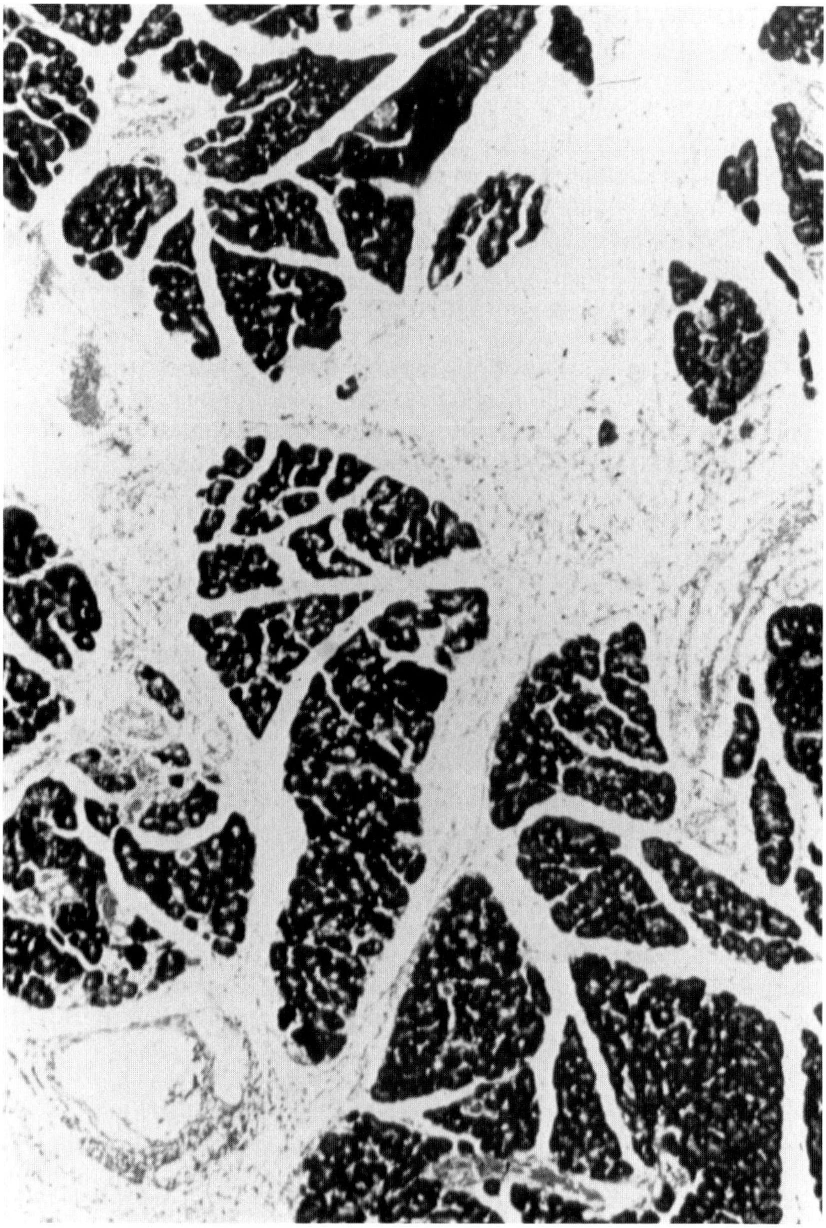

Figure 5 Histological appearance of the rat pancreas after subtotal pancreatectomy using Nd:YAG laser (70 W/1 sec) on day 7.

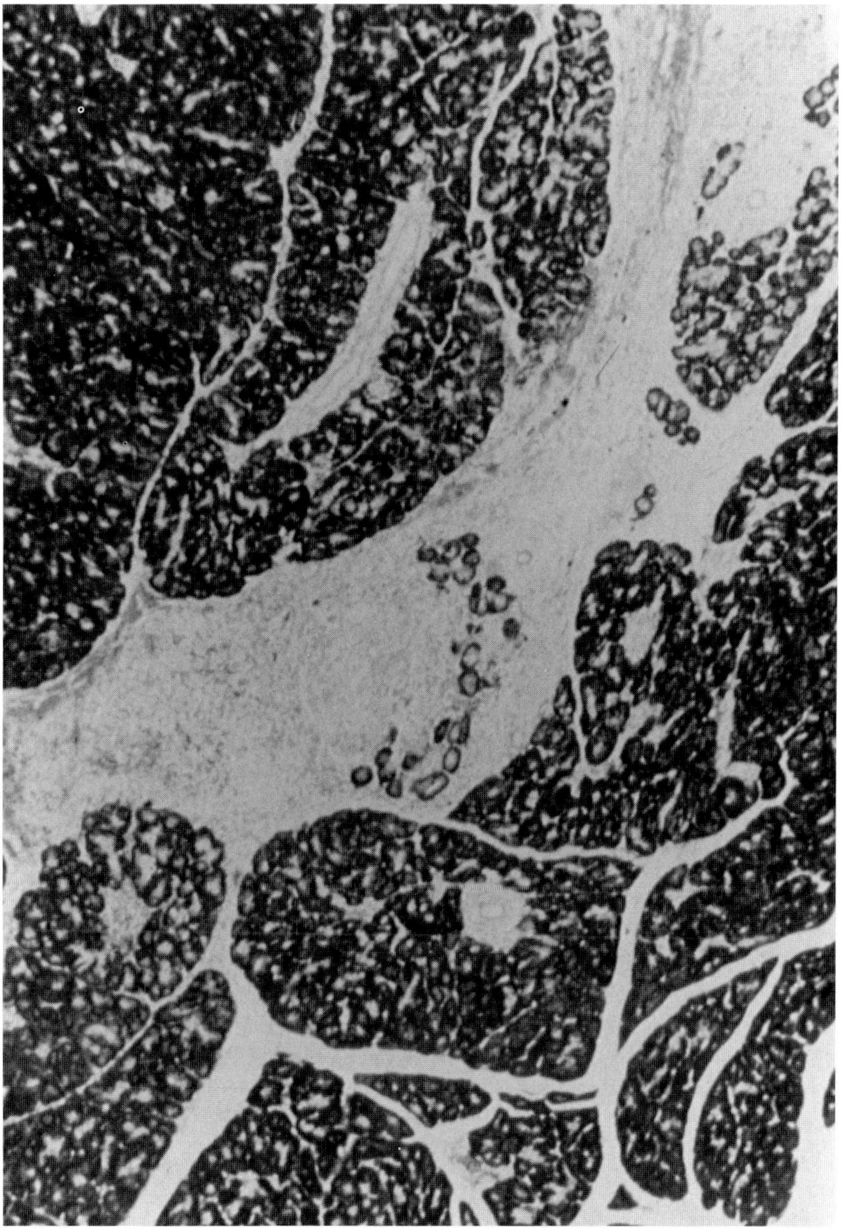

Figure 6 Histological appearance of the rat pancreas after subtotal pancreatectomy using Nd:YAG laser (70 W/1 sec) on day 21.

leakage. All of these factors are contributory, especially in the prevention of postoperative infection.

We have used the Nd:YAG laser in vaporizing and excising metastatic secondary liver tumors and for the palliative treatment of pain and associated bleeding. Godlewski et al. (2) express interest in metastatic liver tumor destruction, and having first used the CO_2 laser, they feel that the Nd:YAG laser radiation would be of better use in these situations because of its depth of penetration, hyperthermic effects, and the blocking by thrombosis of the peripheral venous system. Nims and McCaughan (13) have concluded in their study that the CO_2 laser could be used effectively as a vaporizing instrument in the management of primary and metastatic liver tumors, and the laser was used as an adjunct to other surgical procedures. The Nd:YAG laser can also be used to obtain tissue for histological diagnosis of primary and secondary liver tumors either at open surgery or via a laparoscope.

Percutaneous transhepatic biliary decompression has been used for the palliation of malignant biliary obstruction with less morbidity and mortality than with surgery. Complications of catheter placement include hemorrhage, sepsis, and pancreatitis. Treatment of arterial bleeding in this situation is difficult. The management of this complication has been limited to intraarterial infusion with vasoconstrictive agents and selective hepatic artery embolization, which may cause massive liver necrosis resulting in death. As an alternative method to embolization, Carpenter et al. (14) have successfully used Nd:YAG laser photocoagulation for control of arterial bleeding from a malignant intrahepatic bile duct tumor. They introduced a modified flexible fiberoptic bronchoscope along the catheter tract and photocoagulated the bleeding tumor using a fiberoptic waveguide under direct vision. This is another example that demonstrates the potential application of laser photocoagulation to achieve hemostasis and improve biliary drainage.

Laser photoradiation may be helpful in the treatment of obstructing carcinomas of the papilla of Vater that could not be adequately treated by other methods (15). Classen (16) feels that this mode of therapy is not the treatment of choice in all cases, although it may be indicated in patients who are inoperable or refuse surgery. This may change and become the treatment of choice as experience accumulates.

As the conventional 600 μm quartz fiber with its outer sheath cannot be introduced through the biopsy channel of a choledochofiberscope, Orii et al., in the late 1970s, tried the much smaller-diameter quartz fiber, used in the communication industry, to disintegrate stones in the bile duct (17). They found that they were able to break up bilirubin stones bathed in a solution of sodium chloride using the Nd:YAG laser, but disintegration of cholesterol stones was more difficult. The treatment of intrahepatic stones is a different and complicated situation when compared with extrahepatic calculi. It requires difficult and protracted surgery that often includes a partial liver resection when the stones are impacted and are situated distal to the common hepatic ducts. Orii et al. (18) have been successful in removing these impacted intrahepatic bilirubin stones by using the Nd:YAG laser with a choledochofiberscope (Fig. 7). This is performed using the percutaneous transhepatic route under local anesthesia. Kouzu and Sato (9) suggest delaying the lithotomy of intrahepatic stones when found during surgery because it is a time-consuming procedure with complications. They emphasize the ease of removal of these stones by postoperative percutaneous transhepatic cholangioscopic lithotomy. When the intrahepatic stones are removed in the conventional way at open surgery, the success rate is a low 64.1%, compared with 91% for choledocholithotomy. The technique of ultrasound and computer tomography allows stones in the intrahepatic ducts to be located without difficulty. The percutaneous transhepatic removal of the stones can then be performed using ultrasound guidance. Stones 5-6 mm or more in size cannot be extracted through the dilated biliary drainage tracts unless they

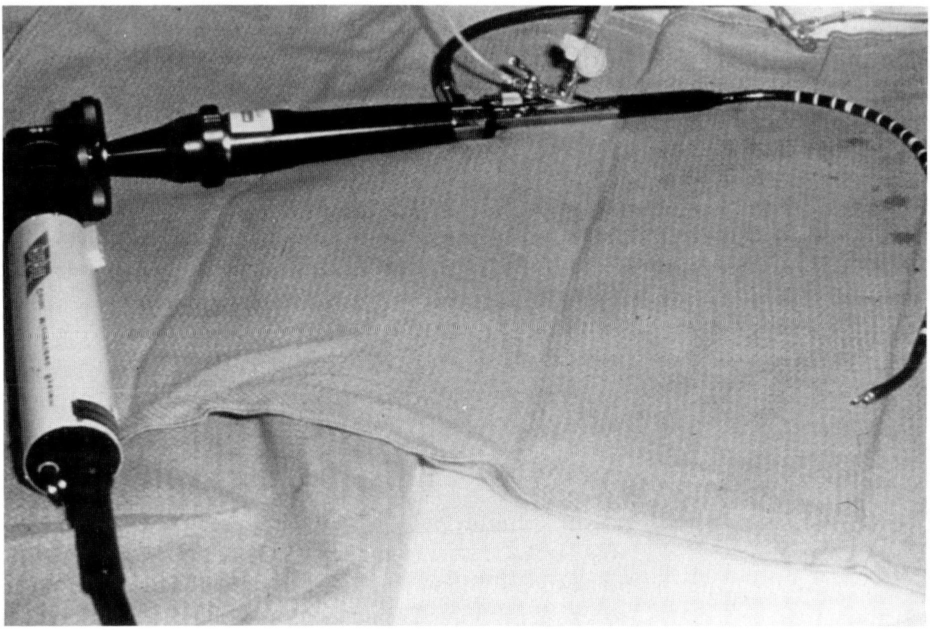

Figure 7 Nd:YAG laser fiber introduced through the channel of the choledochofiberscope.

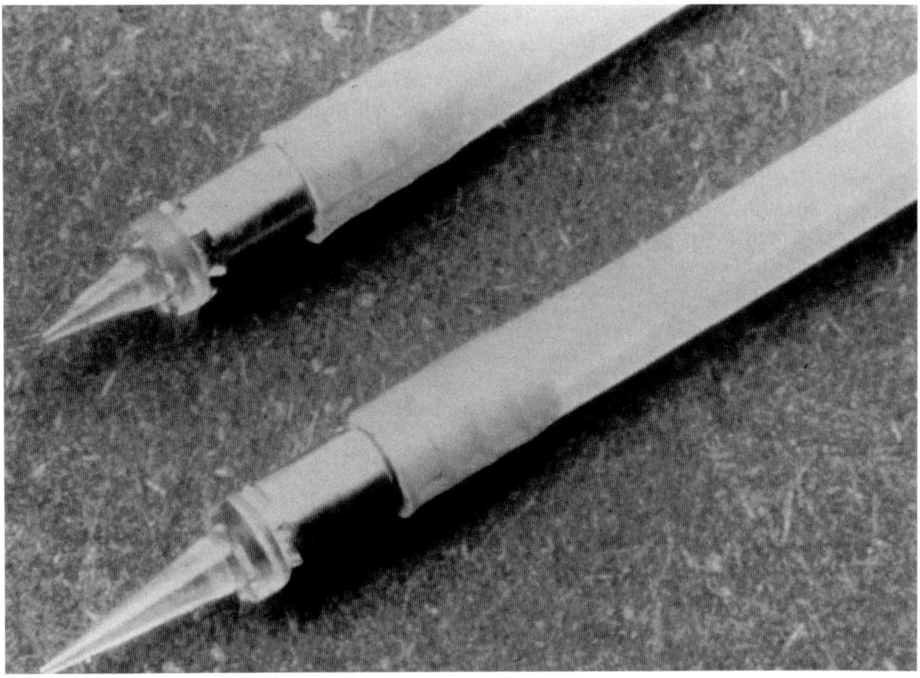

Figure 8 Synthetic ceramic contact laser endoprobes.

are crushed. Application of Nd:YAG laser to cholangioscopic lithotomy causes the disintegration and fracturing of bilirubin stones. The lithotomy then becomes an easy, safe, and effective procedure compared with removal by conventional methods. Laser irradiation under cholangioscopic visualization is also now being used for dilatation and recanalization of a strictured anastomosis following hepaticojejunostomy. This prevents major reoperations and can be done as an outpatient procedure.

Recently Kouzu (19) has tried irradiating cholesterol stones in the biliary tract by using the newly developed artificial ceramic contact laser endoprobe (Surgical Laser Technologies, Inc., Philadelphia, Pennsylvania) (Fig. 8). By changing the technique of using the Nd:YAG laser from the nontouch to the contact technique, he improved the results of lithotomy of the intrahepatic bilirubin stones from 66% (31/47) to 95.5% (42/44). The noncontact irradiation used a power density of 70 W for 0.5 sec on calcium bilirubinate stones, and 80 W for 2 sec or more for cholesterol stones. The more efficient irradiation method with the contact laser endoprobe was shown for the cholesterol stones. This required an initial drilling of a hole which then resulted in the stones fracturing into multiple smaller pieces. This effect was obtained using a low power density of only 10-15 W for 1-2 sec. Cholangiolithotomy using the Nd:YAG laser with this new method makes the treatment of biliary stones much easier. Percutaneous choledocholithotomy for hepaticocholelithiasis will gain wider use and popularity as experience accumulates.

B. Pancreatic Surgery

Pancreatic surgery is associated with considerable morbidity and mortality resulting from blood loss, infection, and fistula formation. The trend in recent years for the treatment of pancreatic cancer, and occasionally for chronic pancreatitis, has been toward a total pancreatectomy. Furthermore, the pancreas is now being harvested for transplantation. Both Sutherland et al. (20) and Rossi et al. (21) with their recent clinical experience with living related pancreatic transplantation call for refining the technique of total pancreatectomy in cases of chronic pancreatitis. Any operative technique that can reduce operating time, blood loss, and associated morbidity and mortality would be of tremendous advantage. We are currently using the Nd:YAG laser in performing partial or near-total pancreatectomies.

The following two patients illustrate its use: A young male patient was diagnosed as having insulinoma at the junction of the body and neck of the pancreas and a young female patient was having protracted pancreaticobiliary problems caused by chronic pancreatitis. Both of the patients underwent a pancreatic resection (50 and 80%, respectively) with preservation of the head of the pancreas and duodenal loop. After lifting the pancreas from its inferior border, the dissection of the superior border was relatively easy and safe using Nd:YAG laser with the nontouch technique. The power density used was between 60 and 70 W for 1-3 sec. The time taken to do both procedures was about 1 hr. In the case of the male patient, no blood replacement was given (< 100 ml blood loss), and by dissecting the pancreas off the splenic vessels, the spleen and its vasculature were preserved. This would not have been possible without the laser.

The postoperative period was uneventful in both patients. The acute histological appearances of the resected human pancreas were similar in all respects to that of lasered rat pancreatic tissue (Fig. 9) with the exception of the depth of penetration of the acidophilic cells in zone 3, which reached a depth of 3.5 mm. The Nd:YAG laser using the contact probe can resect the head of the pancreas with preservation of the duodenal loop (Fig. 10).

The Nd:YAG laser offers a new therapeutic modality in the performance of tedious and often difficult pancreatic surgery. Contact laser surgery is expected to provide further advantages given its ability to use both coagulating and cutting properties.

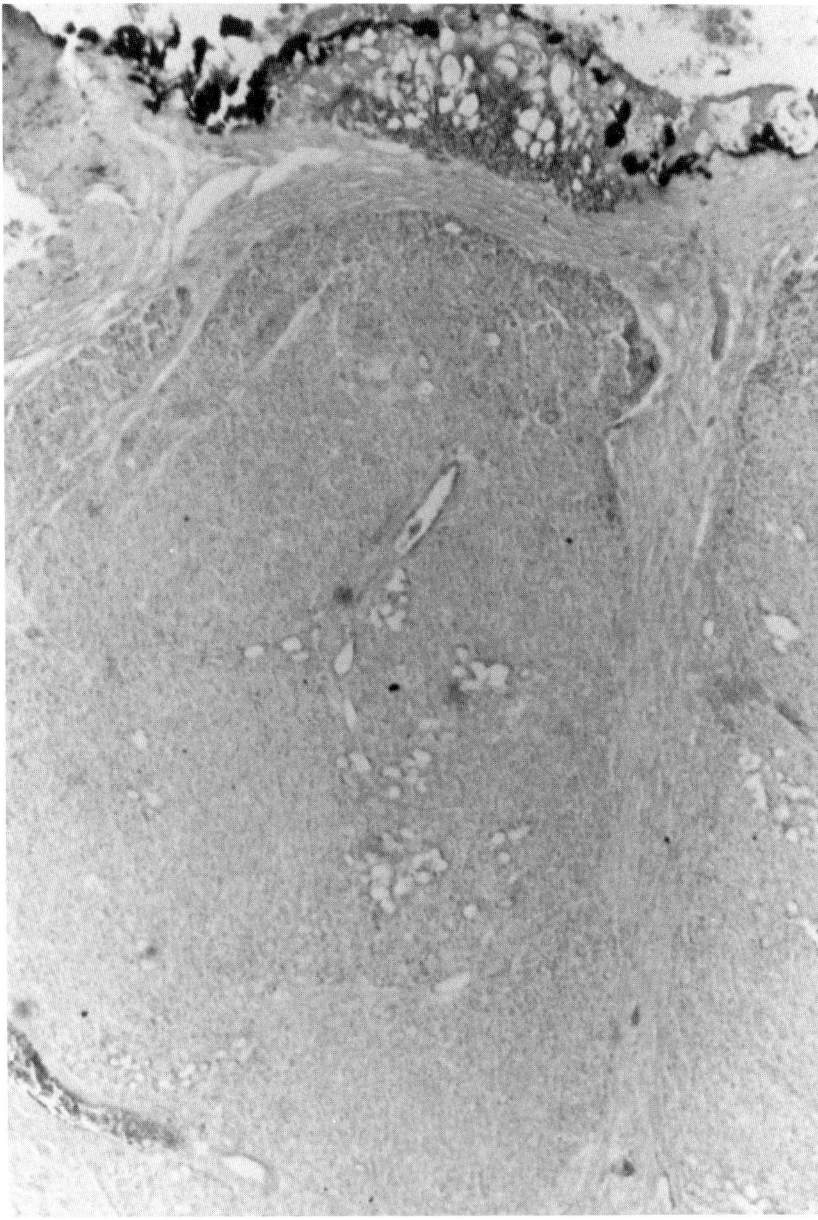

Figure 9 Acute histological appearance of resected human pancreas using Nd:YAG laser nontouch technique.

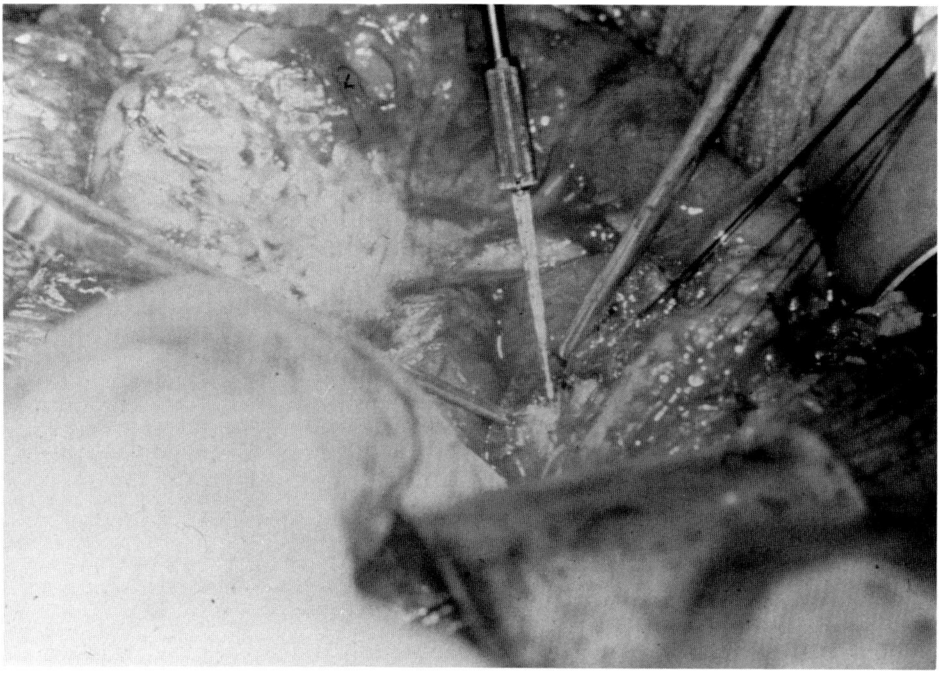

Figure 10 Laser pancreatectomy in clinical situation: resection of the head of the pancreas off the duodenal loop with preservation of the latter, using Nd:YAG laser (15 W/CW mode) and contact laser scalpel (Surgical Laser Technologies, Inc., Philadelphia). The head of the pancreas is being pulled aside by string sutures.

Other innovative procedures include drainage of pseudocyst of the pancreas by performing an endoscopic laser cystogastrostomy as performed by Dixon (22).

C. Laser Laparoscopy

Abdominal laparoscopy has gained wide popularity in gynecology. The role of laparoscopy as a diagnostic tool is becoming increasingly accepted because of the ease of use and availability of both rigid and flexible endoscopes and outpatient laparoscopy. If the laser could be coupled to the laparoscope, benefits would ensue in allowing definitive and therapeutic procedures to be performed that currently require a major abdominal laparotomy.

It is better to evaluate the laser laparoscopy as it is available and being used by the gynecologists. In treating endometriosis, both the argon and CO_2 lasers have been used. The latter is still favored by gynecologists, despite the cumbersome articulating arms and lack of flexible fibers. It is felt that although pelvic endometriosis (stage I and II) can be effectively treated by drugs such as danazol, the accompanying side effects of androgenicity, breakthrough bleeding, prolonged treatment time, and high cost make it essential to look for an alternative form of therapy. Kelly and Roberts (23) use a CO_2 laser adapted to the operating laparoscope in the treatment of pelvic endometriosis (stage I and II). This is performed at the time of the original diagnostic laparoscopy (23,24). The precision of use of the laser allows a safe treatment of the implants in the peritoneum, over the bladder and bowel, or near the ureter to be performed. In addition, ovarian endometriosis of < 1 cm in size and fibrous adhesions can be removed safely by tissue vaporization. Keye and Dixon (25) feel that the use of argon laser, because of its selective absorption by the hemoglobin of the endometriosis implants and the ease of photocoagulating such implants at laparoscopy, may eventually become an important treatment of endometriosis.

It has been suggested that the CO_2 laser, by virtue of its hypothetical capabilities for precise incision, minimizes tissue handling and bleeding and shortens the operating time, which may improve the success rate of gynecological infertility surgery. To assess this, a multicentric prospective study was undertaken to assess tubal patency and adhesion formation at early second-look laparoscopy after intraabdominal laser surgery. Procedures performed included salpingostomy, fimbrioplasty, lysis of adhesions, vaporization of endometriosis, and ovarian wedge resection. The results were compared with those of another multicenter prospective study that utilized nonlaser reconstructive pelvic surgery. Use of the CO_2 laser resulted in greater tubal patency at the time of the second-look procedure, and adhesions were reduced. However, nonlaser major open abdominal infertility surgery gave similar results regarding adhesion formation. The laparoscopic CO_2 laser is not necessarily better but avoids major abdominal surgery in the treatment of tuboperitoneal causes of infertility. Pregnancy rates following intraabdominal use of CO_2 laser remain to be established (26). Daniell and Herbert (27) cautiously state that operative laparoscopy with the CO_2 laser for distal salpingostomy may provide an acceptable alternative to laparotomy in selected patients.

Work is being done using both argon and Nd:YAG laser with the ability to have flexible fiberoptic coupling to a laparoscope. New areas are seen in gastroenterology as well.

V. FUTURE DEVELOPMENTS

The interesting possibilities of dividing abdominal adhesions using the contact laser endoprobe through the laparoscope will materialize in the near future. Gastroenterologists will vaporize metastatic tumor deposits in the liver or in the peritoneal cavity using the Nd:YAG laser through a fiberoptic laparoscope. Major procedures like cholecystectomy could well be carried out using the newly acquired armamentarium of laser-laparoscopy, which would not only reduce morbidity and mortality but be cost-effective in today's medical marketplace.

The innovative laser procedures are only expected to increase in number with the availability and improvement in the technology coupled to the constructive enthusiasm among trained medical personnel. The cautious implementation of these widening scientific frontiers of laser surgery offer new horizons in patient management.

REFERENCES

1. Fleischer D: Endoscopic laser therapy for gastrointestinal diseases. *Arch. Intern. Med. 144*:1225-1230, 1984.
2. Godlewski G, Miro L, Chevalier JM, and Bureau JP: Experimental comparative study on morphological effects of different lasers on the liver. *Res. Exp. Med. 180*:51-57, 1982.
3. Fidler JP, Moefler RW, Polanyi TG, Bredemeier MC, Siler VE, and Altemeir WA: Laser surgery in exsanguinating liver injury. *Ann. Surg. 181*:74-80, 1975.
4. Gonzales R, Edlich RF, Bredemeier MC, Polanyi TG, and Goodale, RL: Rapid control of massive hepatic hemorrhage by laser radiation. *Surg. Gynecol. Obstet. 131*:198-200, 1970.
5. Y. Nakajima. Experimental study on hepatic resection with CO_2 laser knife, in *Laser Tokyo 81* (K. Atsumi and N. Nimsakul, eds.). Intergroup Corp., Tokyo, 1981.
6. Tranberg KG, Rogotti P, Brackett KA, Bjornson HS, Fischer JE, and Joffe SN: Liver resection: A comparison using the Nd-YAG laser-an ultrasonic surgical aspirator, of blunt dissection. *Am J Surg. 151*:368-373, 1985.
7. Durtschi MB, Stothert JC Jr, Ashleman B, Auth DC, Lee MJ, Heimbach DM, and Carrico CJ: Laser scalpel for solid organ surgery. *Am. J. Surg. 139*: 665-668, 1980.
8. Godlewski G, Ginoves P, Chincholles JM, Viel E, Bureau JP, Rouy S, Mion H, Dubois A, and Fesquet J: Hepatic resection with an Nd:YAG laser in pig. *Laser Surg. Med. 3*:217-224, 1983.
9. T. Kouzu and H. Sato. *Endoscopic Laser Treatment of Intrahepatic Stones, Intrahepatic Calculi.* Alan R. Liss, Inc., New York, 1984, pp. 321-332.
10. Eloy R, Bouchet F, and Clendinnen G: New technique of total pancreatectomy without duodenectomy in the dog. *Am. J. Surg. 140*:400-412, 1980.
11. Munda R, Berlatzky Y, Jonung M, and Joffe SN: Studies on segmental pancreatic autotransplants in dogs. *Arch. Surg. 118*:1310-1315, 1983.
12. S. N. Joffe, K. Brackett, and M. Y. Sankar. The Nd:YAG laser in pancreatic

surgery, in *Laser and Infrared Light Applicators in the G. I. Tract and Abdominal Surgery* (Spinell, Bown, and Kiefhaber, eds.). Thieme, Stuttgart, Germany, 1986 (in press).

13. Nims TA and McCaughan JS Jr: Clinical experience with CO_2 laser vaporization of neoplasm. *Lasers Surg. Med. 3*:265-268, 1983.
14. Carpenter CM, Bowers JH, Luers PR, Dixon JA, and Miller FJ: Neodymium yttrium aluminum garnet laser treatment of hemobilia via a percutaneous biliary catheter tract. *Radiology 148*:853-854, 1983.
15. J. M. Brunetaud. Therapeutic applications of argon and Nd:YAG lasers, in *Laser Tokyo 81* (K. Atsumi and N. Nimsakul, eds.). Intergroup Corp., 1981, pp. 25-27.
16. Classen M: Endoscopic approach to papillary stenosis (PS). *Endoscopy 13*: 154-156, 1981.
17. Nakahara A, Takase Y, Tsukada K, and Orii K: The fundamental and clinical studies on the endoscopic laser therapy of the protruding lesions. *Gastroenterol. Endosc. 22*:1358-1370, 1980. (in Japanese)
18. Orii K, Ozaki A, Takase Y, and Iwasaki Y: Lithotomy of intrahepatic and choledochal stones with YAG laser. *Surg. Gynecol. Obstet. 156*:485-488, 1983.
19. Kouzu T: Cholangioscopic lithotomy using Nd:YAG laser by means of contact endoprobe (SLT) (Personal Communication).
20. Sutherland DER, Goetz FC, and Rynasiewicz JJ: Segmental pancreas transplantation from living related and cadaver donors: A clinical experience. *Surgery 90*:159-168, 1981.
21. Rossi RL, Breasch JW, O'Bryan EM, and Watkins E Jr: Segmental pancreatic autotransplantation for chronic pancreatitis. *Gastroenterology 84*:621-626, 1983.
22. J. A. Dixon. *Lasers in Surgery: Current Problems in Surgery*. Year Book Medical Publishers, Chicago, 1984, p. 32.
23. Kelly RW and Roberts DK: CO_2 laser laparoscopy—a potential alternative to danazol in the treatment of stage I and II endometriosis. *J. Reprod. Med. 28*:638-640, 1983.
24. Tadir Y, Kaplan I, Zuckerman Z, Edelstein T, and Ovadia J: New instrumentation and technique for laparoscopic CO_2 laser operations: A preliminary report. *Obstet. Gynecol. 63*:582-585, 1984.
25. Keye WR Jr and Dixon J: Photocoagulation of endometriosis by the argon laser through the laparoscope. *Obstet. Gynecol. 62*:383-386, 1983.
26. Diamond MP, Daniell, JF, Martin DC, Feste J, Vaughn WK, and McLaughlin DS: Tubal patency and pelvic adhesions at early second-look laparoscopy following intra-abdominal use of the CO_2 laser: Initial report of the intra-abdominal laser study group. *Fertil. Steril. 42*:717-723, 1984.
27. Daniell JF and Herbert CM: Laparoscopic salpingostomy utilizing CO_2 laser. *Fertil. Steril. 41*:558-563, 1984.

12

Laser Surgery for the Cervix and Vagina

V. C. WRIGHT
The University of Western Ontario, London, Ontario, Canada

I. LASER SURGERY FOR INTRAEPITHELIAL NEOPLASIA OF THE CERVIX

A. Introduction

The reasons for using the carbon dioxide laser for treatment of cervical intraepithelial neoplasia (CIN) over other methods include (1) microsurgical precision, (2) complete removal of diseased tissue to any depth or breadth required, (3) nontouch surgery with an unobstructed operating field, (4) minimal effect on adjacent normal tissue with rapid healing to normal or near normal volume with a new squamocolumnar junction at the level of the external os, (5) quick treatment with minimal postoperative complications and no apparent effect on subsequent fertility or cervical competence (1,2), (6) concomitant hemostasis, and (7) the highest success rate of all treatment modalities (2-4).

Conization, electrocautery, and cryosurgery have been employed with varying success and complication rates. Recently, laser surgery has been shown to be very effective for all grades of cervical intraepithelial neoplasia and has been observed to precipitate few complications (3,4).

1. The Susceptible Tissue

Cervical cancer does not appear to originate from the native or normal metaplastic squamous epithelium of the cervix (5). Squamous metaplasia usually results in normal squamous epithelium. This alteration appears irreversible and at no risk of development of squamous carcinoma (5). The area of susceptibility for the development of cervical cancer lies between the original squamo-

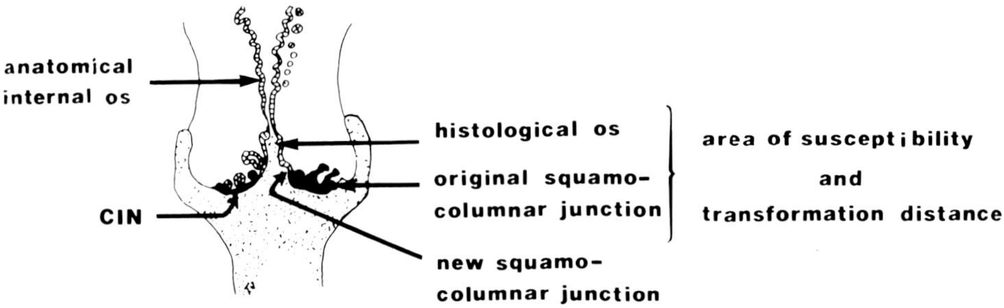

Figure 1 Cervical topography illustrating the familiar landmarks on the cervix and the area of cervical tissue susceptible to the development of neoplasia.

columnar junction and the histological os. This distance represents approximately 2-2.5 cm in length. This distance appears to be somewhat fixed for the life of the female. The transformation zone (the area from the original squamocolumnar junction to the new squamocolumnar junction) is only a part of the total area of susceptibility. The transformation zone containing the CIN is always located adjacent to the original squamocolumnar junction and expands upward toward the histological os (Fig. 1). The columnar epithelial cells in this area may incorporate the carcinogenic agent into their cell structure during the dynamic phases of metaplasia. Any atypical metaplastic pattern such as dysplasia provides the stage for future neoplastic development. The cells undergoing mutagenesis are expressed histologically, after a variable latency period, as cervical intraepithelial neoplasia (6).

Basically cellular patterns change with age (Fig. 2). As the female ages the area of susceptibility rises higher within the endocervical canal. In the elderly menopausal patient the histological os may be above the anatomical internal os. Usually, however, in the young female and early menopause, the histological os is below the anatomical internal os (7,8). These findings suggest that cervical intraepithelial neoplasia occupying the transformation zone can be destroyed without interfering with the anatomical internal os.

The length of cervical lesions involving the "transformation zone" varies from patient to patient. Findings indicate that this range can extend from 2 to 22 mm in length. The average linear involvement is usually less than 10 mm (9).

The distribution of cervical glands follows the shift of columnar epithelium. Cervical neoplasia coincides frequently with the distribution of the cervical glands (8). It is therefore of utmost importance to recognize the location of the columnar epithelium and glands and to consider this when planning the removal of tissue. This effectively individualizes the defect for each patient, depending

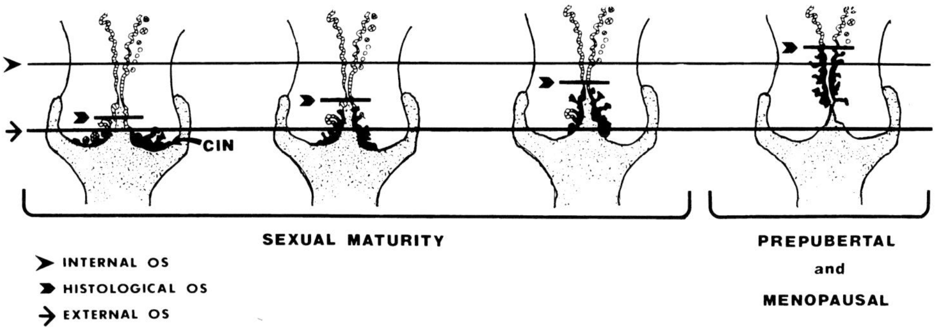

Figure 2 The changing cellular patterns on the cervix as a function of age. The internal anatomical os remains fixed as does the external os, but the histological os, with advancing age, recedes into the canal. The histological os is the uppermost limit of potential disease.

upon the location of the tissue at risk. Rarely does diseased columnar epithelium containing CIN extend higher than 1 cm up the canal above the external os and almost never more than 2 cm (8). Therefore when planning the extent of the disease to be removed, the surgeon should recall that the more exposed the columnar epithelium, the less likely is diseased tissue to be located high within the endocervical canal. The more endocervical the columnar epithelium, the more likely is involvement in the upper canal. This is very important to remember when one is planning the depth of an excisional procedure.

2. Cervical Colposcopy

Using the standard colposcopic techniques, the atypical cervical epithelium is graded, based upon the epithelium's degree of whiteness after the application of acetic acid; the fineness or coarseness of the epithelial blood vessels; the intracapillary distances; and the presence of regular or irregular punctation, and a mosaic or irregular blood vessel pattern. This colposcopic grading pattern indicates the site of the most severe lesion from which a biopsy specimen is taken and submitted for histological diagnosis. At times more than one biopsy is necessary. Frequently, the observations made with the colposcope accurately predict the histological diagnosis. Accurate colposcopic assessment indicates whether the lesion is on the ectocervix, extends into the endocervical canal, or is entirely within the canal. Endocervical currettage sometimes can be of value for proper evaluation of the endocervical canal.

Colposcopy evaluates the surface extent of the lesion. Despite directed biopsies, the presence of gland extension and the exact depth are not determined with

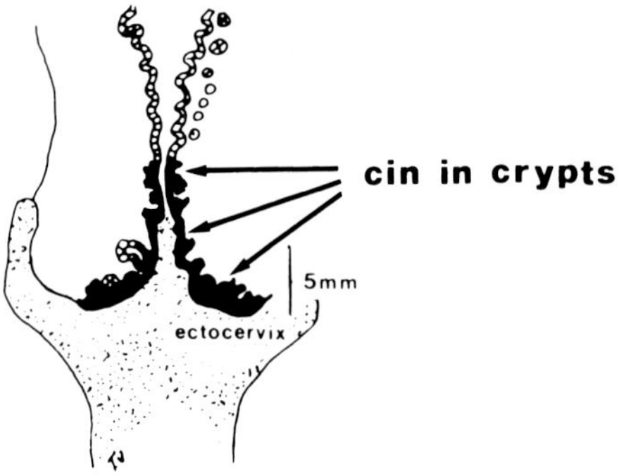

Figure 3 The maximal depth of extension of disease in the cervical crypts is 5 mm or less in 99.7% of cases. The higher the grade of disease the more likely is crypt involvement and the greater the depth.

certainty. Any observation of gland extension is important, because this potentential site of malignant invasion must be destroyed. Cervical disease left behind after treatment is a potential site for malignant invasion because intraepithelial neoplasia occurs in the area of cervical glands or crypts and often extends into them. The maximal depth of extension is the same whether the glands are on the portio, at the external os, or within the endocervical canal (Fig. 3). Maximal depth of extension of carcinoma in situ into the cervical crypts is less than 5 mm in 99.7% of cases (10,11). Therefore destroying cervical tissue to at least this depth along the transformation zone, including the area of disease, should eradicate the cervical intraepithelial neoplasia in virtually all cases. Dysplasia, with the more superficial surface involvement, has frequently less than 1 mm of glandular extension. But errors in diagnosis can occur so every case should be treated as CIN III disease (severe dysplasia and carcinoma in situ).

With recognition of the location and depth of obviously diseased and potentially involved tissue, a defect can be planned to remove this tissue. The need for a histological specimen for further evaluation must also be taken into consideration. Laser surgery can be employed for vaporization of tissue, excisional procedures, or a combination of the two. Each type of operation depends upon certain diagnostic prerequisites, cellular distribution, and the shape of the intended defect. These procedures will be discussed separately.

B. Laser Vaporization of Ectocervical Intraepithelial Neoplasia

1. Principles

Vaporization operations do not produce tissue for pathological interpretation. Therefore the principles of conservative laser vaporization are: (1) the colposco-

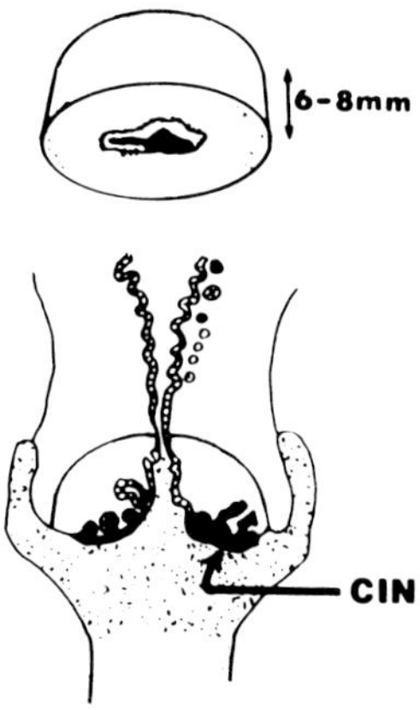

Figure 4 The ideal configuration of the vaporization defect for lesions on the portio: a cylindrical base with a domed top to incorporate all disease-bearing tissue.

pist must be certain from the qualitative assessment of the transformation zone that no invasive cancer is present; (2) the entire atypical transformation zone must be colposcopically defined; (3) correlation between cytology, colposcopy, and histopathology must indicate only CIN; and (4) the diseased lesion must occupy the ectocervix at or below the level of the external os with no extension into the endocervical canal (Fig. 4).

2. Operative Technique

A large, black, dulled or plastic, nonflammable speculum is inserted into the vagina exposing the cervix. These materials and surfaces reduce the potential for inadvertently deflecting the laser beam. The transformation zone and the area of pathology are again identified by washing the cervix with a 2-3% solution of acetic acid. A smoke evacuation system is necessary to remove the vapor plume. A hand-held suction tip can be positioned in the vagina by an assistant; an alternative method is to attach the suction to the upper blade of the speculum. Such suction devices are commercially available. The entire transformation zone including the CIN lesion is outlined with the laser beam. A power density range of 750-1500 W/cm^2 can be easily used and controlled. The marking with the laser beam must extend 3-4 mm beyond the area to be ablated. Employing a laser beam with a Gaussian power density profile, designated as TEM_{00} (the abbreviated formula is: *power density* = $W \times 100/d^2$, where W is the reading from the in-line power meter and d is the effective laser beam diameter measured in millimeters), the designated area is vaporizated.

Beginning at the 6 o'clock position, the laser beam is moved very quickly over the tissue to be evaporated. It is important to note that thermal damage is time related and independent of power density. Rapid beam movement reduces heat buildup in cervical tissue and minimizes cramping. Also, beginning at the posterior area prevents blood and/or ooze from dripping down from the anterior areas after laser application and obstructing the beam. It is best to use an effective laser beam diameter of 1.5-2 mm for vaporization. The beam is moved in multiple directions to prevent furrowing (horizontally, vertically, diagonally). For large surface areas, dividing the cervix into smaller units may facilitate the procedure.

Vaporization of the entire transformation zone including the area of pathology must be performed to a minimal depth of 6-8 mm. This is to ensure eradication of any CIN extension into the cervical crypts. The extra millimeters beyond the potential gland involvement compensate for any inadvertent technical error when vaporizing or measuring. A microruler can be employed to determine the depth of ablation. The end result should be a cylindrical-based, dome-topped defect with straight sides and a curved apex (Fig. 4). Destroying the cervical tissue to a measured minimum depth of 6-8 mm from every point along the lesion and transformation zone, particularly near the os, is essential. Frequently, the maximum depth of the defect from the most external plane of the ectocervix to the top of the dome, extends 8-10 mm. This yields a sizeable defect but does not appear to affect the volume of tissue that regenerates, the quality or duration of healing, or the postoperative complication rate.

Persistent disease after laser vaporization of the cervix occurs because (1) the initial marking of the defect's periphery by the laser beam did not provide a 3-4 mm normal tissue border or (2) a uniform measured minimum vaporization depth of 6-8 mm along the entire topography of the defect was not achieved (4). The surgeon usually obtains the proper depth at the lateral margin but fails to achieve this at the external os. It is therefore essential to dome the defect to match the anatomical topography of the cervix and ensure adequate depth near the external os.

C. Laser Excisional Procedure for Pure Endocervical Lesions

1. Principles

Please note that this procedure provides tissue for pathological interpretation (5). The principles for use of a laser excisional procedure should incorporate the following circumstances: (1) when discrepancies between cytology, colposcopy, and histology occur, (2) when lesions located within the endocervical canal require tissue for histological interpretation, or (3) when cytology or colposcopy suggests a possible invasive carcinoma that has not been proved by colposcopically directed biopsy.

2. Operative Technique

The operation is performed under general anesthesia in a properly equipped operating room. The patient is frequently operated on as a day patient. Hospital admission may be required for medical reasons or should the patient be traveling a long distance. The patient is placed in the lithotomy position, and the surgeon preferably should have an assistant. Surgical preparation including the vulva and vagina is accomplished with solutions that do not contain alcohol. Preparative solutions containing alcohol will be ignited by the laser light. Catheterization of the bladder is not necessary. Bimanual examination is undertaken, and the size and position of the uterus are noted. The length of the cervical canal is measured with a uterine sound. A weighted self-retaining speculum is placed in the proper position. A single-toothed tenaculum is placed at the 12 o'clock position, such that the area of traction is outside the area to be excised. This aids in exposing the cervix by downward and upward traction. Hemostatic sutures are placed deeply into the cervical tissue at the 3 and 9 o'clock positions. A 1.5% solution of Pitressin (0.15 ml of vasopressin mixed in 9.85 ml of normal saline) is injected into the cervix after placing the cervical angle sutures. The laser, coupled with the microscope, is placed into position. Matching lenses with focal lengths of 300 mm provide adequate working distance for the laser surgeon. It is best to use a laser beam diameter of 0.5-0.8 mm for excising the specimen. (Smaller beam diameters 0.1-0.4 mm increase troublesome bleeding.) The effective spot diameter can be measured before each procedure on a tongue depressor using the intended wattage. The cervix is washed with a 2-3% solution of acetic acid. The transformation zone and area of pathology are identified using the operating microscope. In many cases the ectocervix will appear normal because the diseased area is located within the endocervical canal. A smoke evacuation system is necessary. A hand-held plastic suction tip is held in the area of surgery by an assistant.

Power densities from 1000 to 16,000 W/cm^2 cause rapid incisional vaporization. The highest power density which the surgeon can control is used. Appropriate power is selected to correspond with the spot size. To decrease thermal conduction it is important to move the laser beam quickly along the incisional line. Higher power densities will create deeper incisions and vaporize more rapidly than lower power densities. Hemostasis is reduced, however. The ectocervix is circumscribed with the laser beam, creating a radius of at least 6 mm measured from the lateral border of the external os (Fig. 5). This radius may be increased if necessary to incorporate any atypical transformation zone noted on the ectocervix. The width and length of the specimen will be modified according to the obvious and suspected distribution of the disease as previously described.

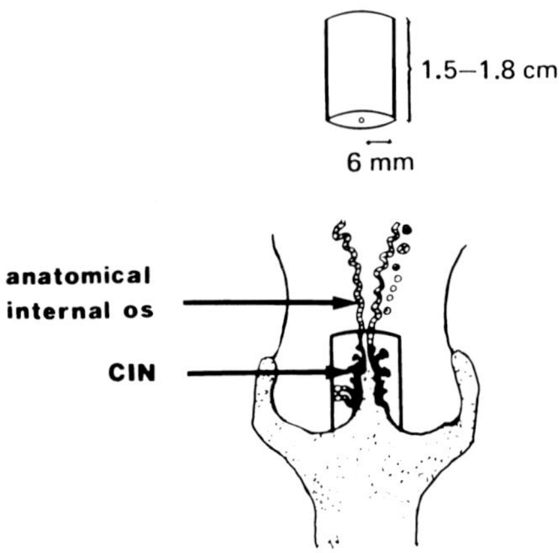

Figure 5 The ideal configuration of the excision procedure for endocervical disease: a tall cylinder with a radius of at least 6 mm and a height that varies with the cellular distribution of the individual patient but is usually 1.5-1.8 cm.

To increase exposure for developing the incision, forceps are placed at the 12 o'clock position of the developing specimen and downward traction is maintained. Forceps are similarly placed at the 3, 6, and 9 o'clock positions. A cylindrical specimen is developed to a length of 1.5-1.8 cm (see Fig. 5) for pure endocervical lesions. The total length, however, will be governed by the length of the previously measured endocervical canal and the anatomical location of the transformation zone. The uppermost limit must remain below the level of the internal cervical os in women of reproductive age. In the menopausal group the disease can extend above the internal os as noted in Fig. 2. Using a scalpel, the specimen is cut transversely at the upper pole. Because of the vasopressin, bleeding resulting from the scalpel cut will be negligible.

The surgeon holds the suction in one hand and clears the area free of any blood. The power density is now decreased to 650-1000 W/cm^2. The beam diameter is increased to 1.5-2 mm. The apex of the defect is flashed with the enlarged beam to seal the blood vessels severed by the scalpel, and bleeding is curtailed. During this particular step, vaporization will increase the length of the cylindrical defect by 1-2 mm.

A total defect of 1.6-2.0 cm in depth results. Again, the total defect size will be governed by the level of the internal cervical os and the anatomical distribution of disease. If slight oozing is still present from the apex of the defect, materials such as Oxycel or Avitene may be placed against the apex and held in place by a pack. The latter is removed 2 hr later, before the patient is discharged home from the outpatient surgical ward.

The cylindrical specimen produced has almost parallel sides. The only thermally damaged surface is the outward margin. This should not exceed 100 μm if the laser has been properly employed. For pathological orientation, the cylindrical specimen can be opened at the 12 o'clock position before being placed in the fixative.

Persistent disease occurs following the laser cylinder procedure presumably because (1) the initial marking of the defect did not have a radius of at least 6 mm, (2) adequate depth was not achieved, leaving disease high in the endocervical canal, (3) the shape of the defect resembled a cone with sloping sides. Such a cone-shaped defect, if produced, will cut across cervical crypts, leaving disease in situ especially high in the endocervical canal.

D. The Combination Procedure

1. Principles

This procedure produces tissue for pathological analysis. In some patients, extensive disease occupies the transformation zone on the ectocervix and also extends beyond colposcopic vision or evaluation into the endocervical canal (Fig. 6). Therefore diagnostic and therapeutic biopsy of the canal is necessary to completely evaluate this area and to eliminate the possibility that malignant disease is

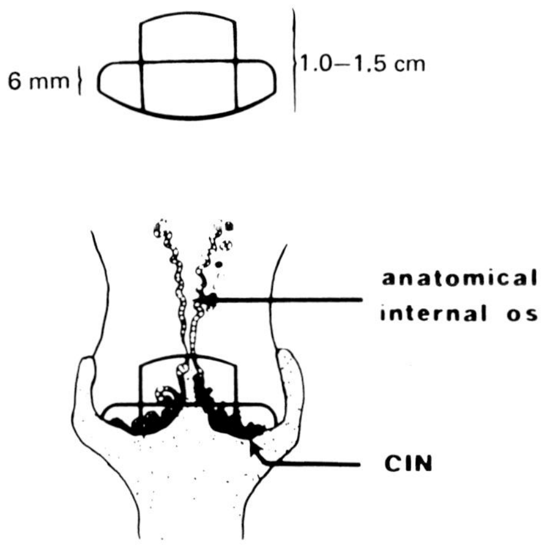

Figure 6 The configuration of the combination procedure, including central excision with peripheral vaporization to the depths indicated. This procedure is employed for lesions wide out on the portio but also extending beyond colposcopic view into the lower canal.

present. In treating such a patient, scalpel conization would frequently destroy the cervix because a very large volume of tissue would be removed. This could result in considerable immediate and delayed bleeding, infection and stenosis, and jeopardization of fertility. Because of these factors, laser incisional biopsy coupled with peripheral vaporization is preferable. This technique leaves a good surface and an outer border that enables organ regeneration. Colposcopically directed biopsies can be obtained from the peripheral ectocervix to verify only CIN.

2. Operative Technique

This operative technique is performed under general anesthesia. It employs both the laser cylinder procedure and the vaporization procedure. The cylindrical excision is first performed using the recommended techniques as previously described for the first two procedures. A shallow laser cylinder incision of the canal is first performed, and subsequently peripheral laser vaporization of the ectocervix is accomplished to remove any remaining lesion in the transformation zone. The total height of the laser cylinder will frequently be from 1 to 1.5 cm because as previously discussed, the more exposed the columnar epithelium, the less likely for disease to be located high within the endocervical canal. The peripheral zone is vaporized to the standard minimum depth of 6-8 mm. Therefore the actual configuration of the defect will vary in relative proportions of the top and flange, depending upon the location of disease. The resulting defect resembles a "cowboy" hat (Fig. 7). Should the cylindrical specimen verify only CIN disease, then this therapy will likely suffice. Should invasive disease be identified within the endocervical canal, the patient is managed accordingly.

The cervix heals from both the incised and vaporized surfaces following the combination procedure, and much of the removed volume is restored. The squamocolumnar junction is almost always located at the level of the external os making follow-up colposcopy and cytology satisfactory; persistent disease is unusual; and fibrosis and/or stricture are unlikely. In fact, an excellent looking cervix results.

E. Follow-Up, Complications, Therapeutic Effectiveness, and Retreatment

After laser surgery, the cervix heals rapidly with the new squamocolumnar junction forming at the level of the external os in 90% of the cases. This facilitates colposcopic evaluation and allows early detection of any persistent disease. Employing the colposcope will help identify any atypical area, and an adequate scraping to obtain excellent cytology is possible. The first follow-up examination with colposcopy and cytology is extremely accurate in predicting patients with residual disease (6,12).

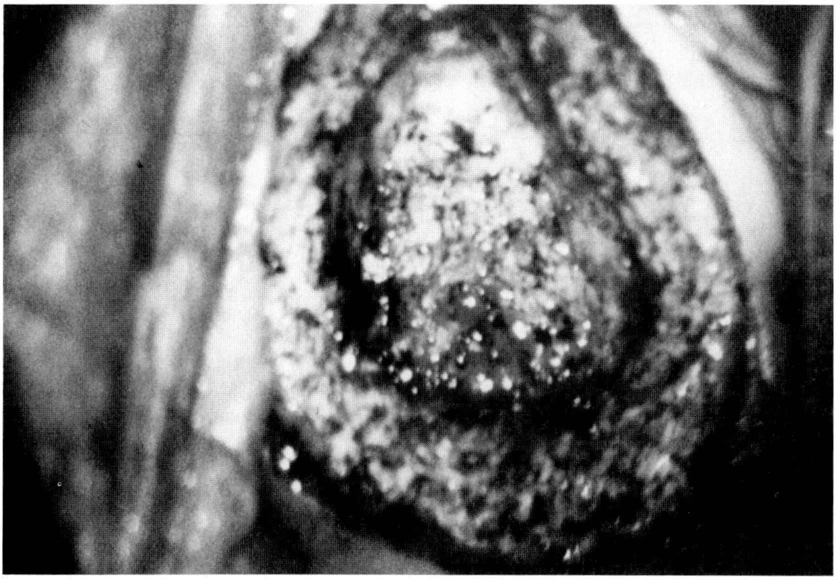

Figure 7 A photograph of the cervix after the combination procedure showing the central excision and peripheral vaporization, which resembles a cowboy hat.

It is recommended that the first follow-up examination be done at 3 months, the second at 6 months, and the third at 9 months after surgery. Patients can be considered cured if the results of three examinations, including colposcopy and cytology, are satisfactory and normal (6,17). The overall success of one laser surgery, regardless of the disease severity or location, will be greater than 95% (Tables 1 and 2). Any CIN lesion discovered by cytology or colposcopy must be verified by histology and considered a treatment failure. Patients who are treatment failures can be easily retreated with the laser using the same operative techniques, depending upon the location of the lesion (6,12). Success of one or two procedures approaches 100% in experienced hands (4,6,12). Less than 1% of patients with CIN should require a hysterectomy to eliminate their disease, assuming that there are no other indications for surgery.

The complication rate is low. Postoperative bleeding requiring hospital admission in this prospective study was less than 1%. A total of 144 pregnancies occurred in these patients. Two premature deliveries occurred in the vaporization group, one at 34 weeks and one at 36 weeks (Table 3). One pregnancy was lost at 20 weeks in the cylindrical excision group. One fetal death at term in the vaporization group resulted from a sudden abruptio placenta. Twenty patients

Table 1 Success After One Laser Surgery (Vaporization or Excision and Combination) in 1368 Consecutive Patients

Disease	Vaporization			Cylindrical excision/combination		
	Cases	Distribution (%)	Cured (%)	Cases	Distribution (%)	Cured (%)
CIN I	195	24.1	97.4	130	23.3	98.5
CIN II	270	33.3	96.3	110	19.7	93.6
CIN III	345	42.6	95.9	318	57.0	95.6
Total	810	100.0	96.4	558	100.0	95.7

Table 2 Success After One Laser Surgery (All Procedures Combined) by Disease Grade in 1368 Consecutive Patients

Disease	Cases	Cured	Cured (%)
CIN I	325	318	97.8
CIN II	380	363	95.5
CIN III	663	634	95.6
Total	1368	1315	96.1

are continuing with their pregnancies and are 16 weeks gestation or longer. These results are comparable to those of others (1) that indicate that laser surgery for CIN does not promote unexpected premature loss or increase fetal mortality.

F. Summary and Conclusions

These three procedures, laser vaporization, laser excision of a tall cylindrical specimen, and laser excision of a shallow cylinder coupled with peripheral vaporization, appear to be effective for eliminating all grades of cervical intraepithelial neo-

Table 3 Outcome of Pregnancy in 144 Cases of Cervical Intraepithelial Neoplasia Treated by Laser Surgery (Vaporization or Excision/Combination)

	Vaporization procedure		Excision and combination procedure		All three procedures	
Pregnancy outcome	No.	%	No.	%	No.	%
Induced abortion	2	1.9	2	5.4	4	2.8
Ectopic	3	2.8	2	5.4	5	3.5
Spontaneous abortion	9	8.4	1	2.7	10	6.9
Still pregnant, gestation 16 weeks or longer	13	12.1	7	18.9	20	13.9
Vaginal delivery at term	70	65.4	21	56.8	91	63.2
Cesarean section at term	8	7.5	3	8.1	11	7.6
Vaginal delivery less than 37 weeks gestation	2	1.9	1	2.7	3	2.1
Total	107	100.0	37	100.0	144	100.0

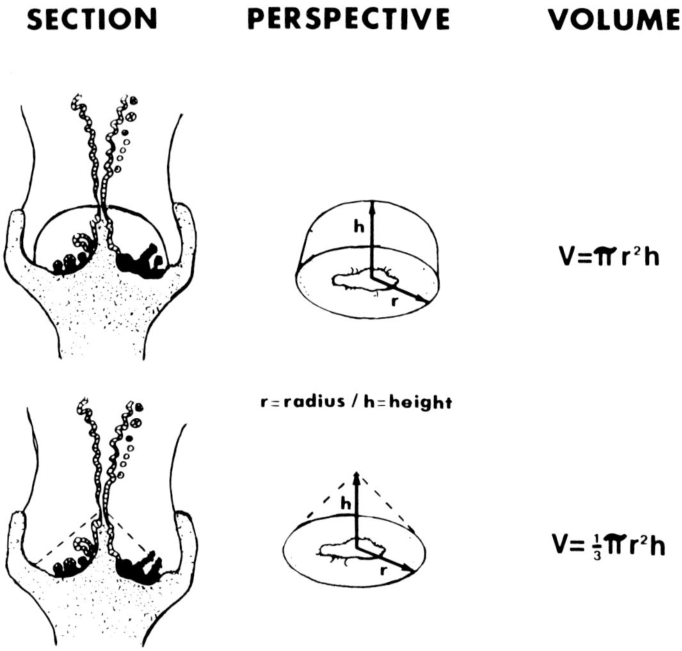

Figure 8 A comparison of the tissue volumes destroyed by cylindrical and conical destruction or excision. With equal radius and height a cylindrical specimen contains three times the volume of a conical one.

plasia, regardless of its severity, location, or lesion size. This is probably because the surgical approaches are designed to remove sufficient volumes of tissue (three times the volumes of conical specimens with equal radius and height, Fig. 8) appropriate for the individual patient's cellular distribution and the likely location of the disease. Presumably failures result from insufficient depth or radius of the defects.

These procedures also appear to preserve organ function and promote regeneration of a normal or near-normal tissue volume after healing. An ideal cellular distribution is produced in almost all cases with the squamocolumnar junction at the level of the external os. This yields early warning of persistent disease, most often at the first follow-up examination (4,6,12).

The success described here must be attributed to an expert cooperative unit that includes pathologists, operating room staff, and clinic nurses as well as to surgical technique and the laser. Knowledge and application of laser physics, instrumentation, optics, and surgical principles are essential for the laser surgeon.

Constant practice coupled with a large volume of patients contribute to expertise.

II. VAGINAL LASER SURGERY

A. Introduction—Anatomy and Structure of the Vagina

The vagina is covered by stratified squamous epithelium that also extends onto and covers the vaginal cervix to the squamocolumnar junction. The epithelium rests on a basement membrane with an underlying lamina propria consisting of collagenous and elastic connective tissue that reaches into the epithelial basal layer in the form of papillae similar to the dermal papillae of skin. The thickness of the vaginal epithelium is approximately 270-500 μm (0.27-0.5 mm) under normal conditions (Fig. 9). Normally the surface is devoid of keratin. The vaginal wall is composed of bundles of smooth muscle arranged in an interlacing circular and longitudinal fashion. Inferiorly, these interlace with the stratified fibers of the levator ani muscles. Lymphatics and an extensive blood supply from the branches of the vaginal artery lie external to the muscular coat and supply smaller branches into the lamina propria.

It is important that there are no glands normally found in the vaginal mucosa. This means that disease cannot occur in excessive depths, as it can in the cervix and vulva, unless it penetrates the basement membrane by invasion. The vaginal epithelium will vary in thickness depending upon different physiological states and pathological conditions.

B. Laser Surgery for Vaginal Condylomata

1. Introduction

The human papillomavirus (HPV) causing condyloma in the genital area appears to be sexually transmitted (12). The sexually transmitted DNA-containing human papillomavirus types responsible for genital condyloma are thought to represent types 6 and 11, a subgroup of the Papovaviridae family (12). This is opposed to types 16 and 18 that are found exclusively in high-grade cervical intraepithelial neoplasia (13). The virus invades the tissue and resides within the nuclei of the cells of the stratified squamous epithelium—the "prickle cells" of the stratum malpighii (14). The superficial involvement of the vaginal epithelium makes this disease amenable to superficial laser vaporization with the CO_2 laser. All lesions that are detected must be completely eliminated. But also, because the virus most likely resides in normal-appearing vaginal mucosa between and surrounding obviously infected areas, this region must be treated as well to prevent emergence of lesions in other areas (Fig. 10).

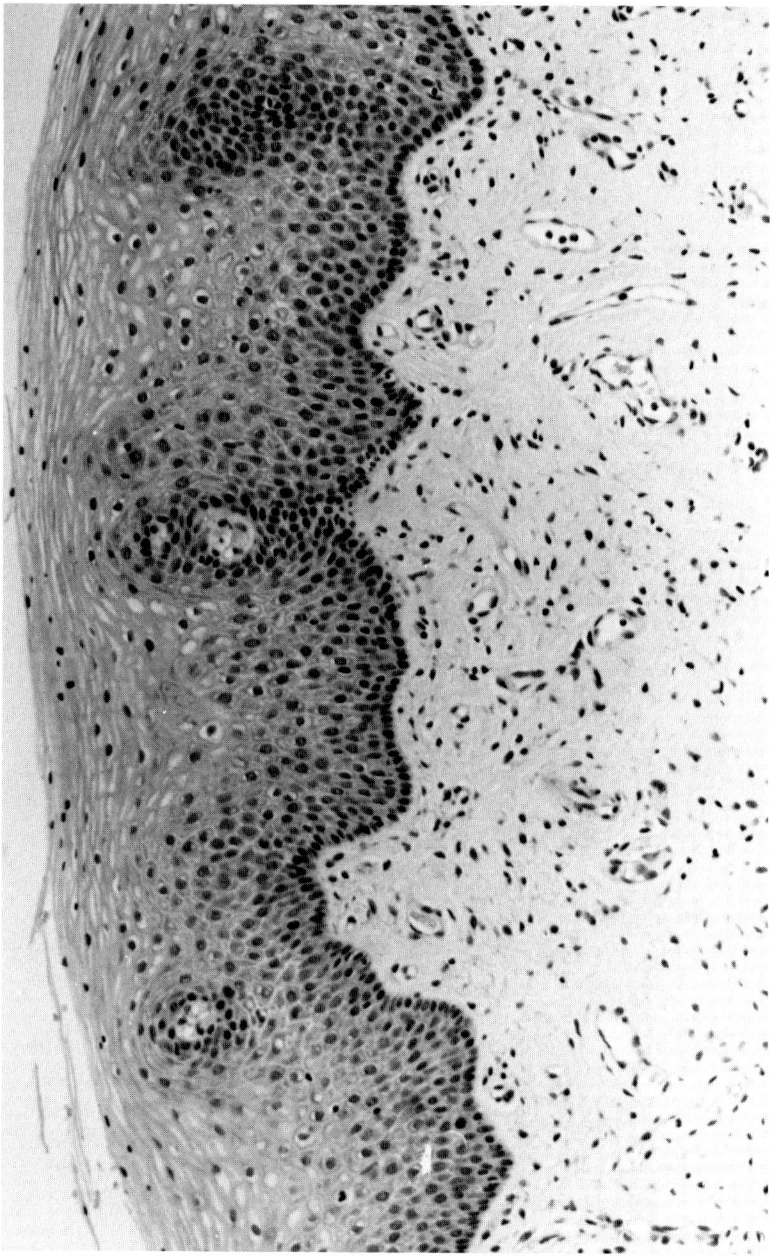

Figure 9 The normal histology of the vaginal mucosa. The epithelium measures 0.27 mm. Note the absence of keratin. The underlying lamina propria is seen extending into the basal layer in the form of papillae.

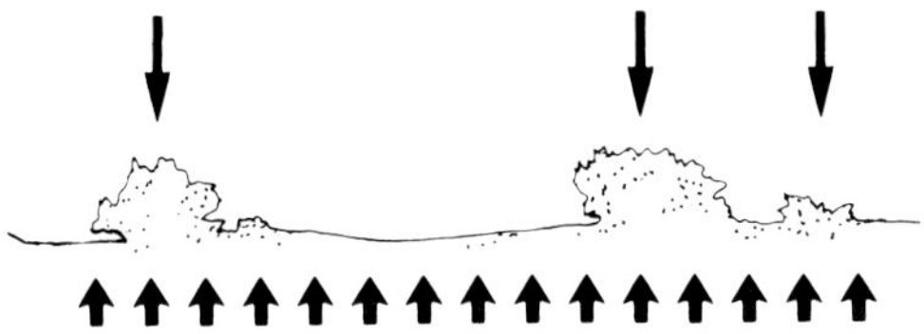

Figure 10 The HPV virus lies within the normal-appearing epithelium adjacent to the obvious condyloma. This is the rationale for flashing the laser beam over the adjacent normal tissue. This brief heat effect kills the virus and helps prevent new disease.

Patients identified with vaginal condyloma frequently have other sites of involvement such as the cervix, vulva, anus, and anal canal. These areas must be colposcopically scrutinized during the initial patient examination. Different clinical appearances of HPV infection, florid and flat condyloma, have been identified, the latter being more common (14). Each has its characteristic appearance. Frequently both types are recognized by the naked eye, and after the application of acetic acid a coarse leukoplakic pattern becomes evident. Vaginal and cervical HPV infections are associated with abnormal Papanicolaou (Pap) smears (15). Because the HPV is associated with genital neoplasia, selected biopsies may be necessary when an abnormal colposcopic pattern or an abnormal Papanicolaou smear is identified.

Most recent therapies recommended for vaginal condylomata include topical 5-fluorouracil (5-FU) (16), laser vaporization, and topical interferon (17). The response to topical 5-FU varies depending upon the type of condyloma present. Ferenczy (18) found the CO_2 laser more effective in eradicating flat condyloma as opposed to topical 5-FU. Both methods appeared equally effective for the florid type. Although topical interferon had a therapeutic effect on flat condylomata in most patients, it did not alter the Papanicolaou smears which, after treatment, still exhibited the HPV effect.

2. Technique of Laser Surgery for Vaginal Condyloma Acuminatum

Laser vaporization of all the obvious lesions is necessary. A nonreflecting vaginal speculum is placed in the vagina, and the laser is attached to the operating microscope. A working distance of 300 mm is preferred. This usually corresponds to the focal length of the microscope's objective lens. An effective laser beam diameter of 1.5-2 mm is mandatory. A simple evacuation system is necessary to remove the vapor plume. This is usually attached to the anterior blade of the speculum. Because of the thinness of the vaginal wall, low power densities of 750-1200 W/cm^2 are recommended to avoid any deep inadvertent laser beam penetration. It is important to remember that a great vascularity surrounds the vagina. Penetration into this plexus of vessels could present very troublesome bleeding.

Each individual vaginal condyloma is vaporized without penetrating into the subcutaneous tissue. Therefore, the total depth of laser destruction should be 150-250 μm. To destroy any potential virus in what microscopically appears to be normal vagina, the laser beam can be quickly flashed over the remaining adjacent vaginal epithelium. The purpose of flashing is to destroy the virus within the nuclei of the cell of the stratum malpighii by the thermal effect of the laser beam.

Frequently the cervix is involved with the HPV infection. This may be recognized as koilcytotic changes in the Pap smear, a visible lesion of flat or florid condyloma, or both. Colposcopic examination and directed biopsy are advised. The association of genital warts and cervical cancer has been previously described (13,19). Sould cervical HPV infection or obvious condyloma be identified, the cervix is treated with the laser similarly to the technique employed for ectocervical intraepithelial neoplasia.

3. Therapeutic Effectiveness and Retreatment

Therapeutic effectiveness is related to sizes and sites of lesions. The more sites that are involved, the more difficult it is to cure the patient with one surgical intervention. Recurrent condylomata occur and these must be managed. Ferenczy (18) noted that vulvar warts appeared within 3 weeks of initial laser therapy. Similar findings regarding the vagina have been observed. If no recurrent vaginal warts are observed by 6 weeks, then the chance of new condylomata is unlikely, unless the patient is reexposed to the virus. New lesions can be treated by a second laser application, topical 5-FU, or a combination of both. Topical 5-FU is available in a 5% cream (Efudex). One vaginal applicator-full inserted high into the vagina daily at night for 3 days upon retiring is usually effective. The 5-FU produces a sluffing of the epithelium. A protective layer of petrolatum should be applied at the introitus to prevent a symptomatic irritation of the mucosa. Should this complication occur, a mixture of 5% tannic acid and 95% petrolatum, applied three times a day, will relieve the pain. Tea sitz baths also produce symptomatic relief. Following both laser and/or 5-FU, the patient is advised against intercourse for a minimum of 6 weeks. Longer periods may be necessary if the condyloma requires additional therapy. To help prevent new disease, the current male partner(s) should be examined. Condoms are suggested during intercourse for 6 months following treatment.

Coaption of the vagina is rare; however, in extensively treated patients a speculum examination is advised at 3 weeks and weekly thereafter until healing is complete.

4. Summary and Conclusions

The CO_2 laser is effective in eradicating vaginal HPV infections. However, to assist in cure, the use of topical 5-FU is frequently warranted.

C. Laser Surgery for Carcinoma In Situ of the Vagina

1. Introduction

Vaginal epithelial neoplasia (VAIN) is uncommon. This disease is diagnosed in approximately 1% of patients who have had a previous diagnosis of squamous cell neoplasia of the cervix (20). Vaginal cancer has been observed in patients who have had a hysterectomy for benign disease (21). The actual incidence of this occurrence is difficult to determine. All cases of VAIN do not progress to invasive cancer, and some cases have been noted to undergo spontaneous remission (22). The contemporary histological grading scheme for cervical intraepithelial neoplasia is also utilized for the vagina (23): VAIN I, mild dysplasia; VAIN II, moderate dysplasia; VAIN III, severe dysplasia and/or carcinoma in situ.

Patients most often present with abnormal Pap smears and usually without symptoms. Careful colposcopic examination using acetic acid and Lugol's solution usually reveals the suspicious area(s) for biopsy. Lower grade lesions are seen as white epithelium as opposed to higher grades (VAIN II and III), which frequently show an associated punctate pattern. An accompanying mosaic pattern is not seen (21,24). The disease is usually multifocal, and most lesions are found in the upper one-third of the vagina (24,25). Because VAIN occurs in association with other genital neoplasia, careful colposcopic examination of the cervix (if present) and vulva should be part of the original examination.

The different modalities previously employed for VAIN therapy include radiation, local resection, total or partial vaginectomy with or without skin graft, electrocautery, and topical 5-FU (24,25) Most patients with VAIN can be treated successfully with 5-FU (24,26). Because of the nonaggressive nature of VAIN and the difficulty and sequelae of conventional therapy, surgery with the CO_2 laser has become another excellent method of treating VAIN (21). Excellent hemostasis can be achieved, and scarring usually is not a problem with laser vaporization. Careful beam control is essential, however, because of the underlying vascular plexus and associated near structures, the bladder and bowel. Because the thickness of the vaginal epithelium is usually less than 500 μm (0.5 mm) (Fig. 11) and because the vagina does not normally contain glands, only superficial laser vaporization is necessary.

2. Laser Operative Technique

Most patients require a general anesthetic for best therapeutic success. Small areas may be injected with local anesthetic and performed in a clinic or office

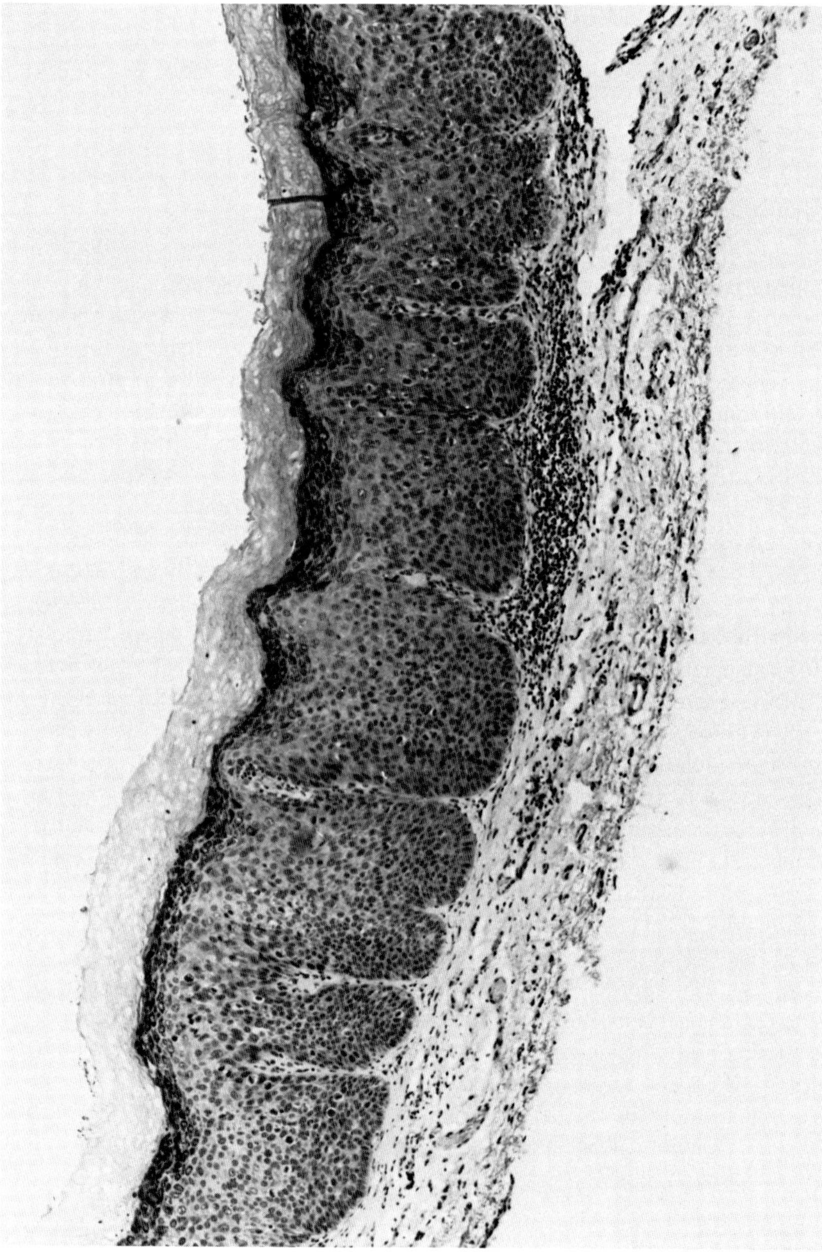

Figure 11 This histological section depicts carcinoma in situ of the vagina. The thickness without the superficial keratin measures 0.23-0.5 mm. With the keratin layer, the thickness ranges from 0.46-0.6 mm.

Table 4 Results of Laser Surgery For Vaginal Intraepithelial Neoplasia in Nine Cases

			Other lesions				Negative after one laser		Negative after one or two lasers		
Disease grade	No.	Location upper one-third of vagina	Vulvar in situ	CIN	Invasive cervical cancer	None	No.	%	No.	%	Length of follow-up (mo.)
VAIN II	3	3	1[a]	1[a]	0	2	3	66.7	3	100.0	12-36
VAIN III	6	6	0	1	1	4	5	83.3	6	100.0	18-36
Total	9	9	1	2	1	6	7	77.7	9	100.0	12-36

[a]Same patient.

setting. With the patient in the usual dorsal lithotomy position a nonreflective and nonflammable speculum exposes the vagina. A smoke evacuation system is necessary. All lesions are again identified with acetic acid and Lugol's solution. Low power densities of 650-1200 W/cm^2 are suggested, and an effective laser beam diameter of 1.5-2 mm is mandatory. Each lesion is encircled to a minimum of 3 mm beyond its periphery. Then, cautiously, the lesions are ablated by painting the entire encircled area in multiple directions with the beam. Vaporizing to a depth of 2 mm should account for the disease distribution. Depth should be carefully monitored, especially in postmenopausal patients with an atrophic vagina.

3. Results of Laser Surgery

Table 4 illustrates the results of CO_2 laser surgery for VAIN in nine patients. It is important that VAIN is associated with other sites of previous intraepithelial or invasive disease. The cure rates with laser for VAIN seem promising, although there is no evidence that laser vaporization is superior to topical 5-FU (21). However, in difficult cases, a combination of laser surgery and topical 5-FU should be considered.

REFERENCES

1. Anderson MC, Howell DH, and Broby Z: Outcome of pregnancy after laser vaporization conization. *Colpos. Gynecol. Laser Surg. 1*:35-39, 1984.
2. Wright VC: Carbon dioxide laser surgery for cervical intraepithelial neoplasia. *Lasers Surg. Med. 4*:145-152, 1984.
3. Wright VC, Davies E, and Riopelle MA: Laser surgery for cervical intraepithelial neoplasia: Principles and results *145*:181-185, 1983.
4. Evans AS and Monaghan JM: The treatment of cervical intraepithelial neoplasia using the carbon dioxide laser. *Br. J. Obstet. Gynecol. 90*:553-557, 1983.
5. Coppleson M: The origin and nature of premalignant lesions of the cervix uteri. *Int. J. Gynecol. Obstet. 8*:539-545, 1970.
6. Wright VC and Riopelle MA: Age at beginning of coitus versus chronologic age as a basis for Papanicolaou screening: An analysis of 747 cases of preinvasive disease. *Am. J. Obstet. Gynecol. 149*:824-830, 1984.
7. Hamperel H and Kaufman C: The cervix uteri at different ages. *Obstet. Gynecol. 14*:621-631, 1959.
8. C. Kaufman and K. G. Ober. In *Ciba Foundation Study Group No. 3, Cancer of the Cervix*. J. A. Churchill, London, 1959.
9. Adelman HC and Hajdu SI: Role of conization in the treatment of cervical carcinoma in situ. *Am. J. Obstet. Gynecol. 98*:173-179, 1967.
10. Przybora LA and Plutowa A: Histological topography of carcinoma in situ of the cervix uteri. *Cancer 12*:263-277, 1959.

11. Anderson MC and Hartley RB: Cervical crypt involvement by cervical intraepithelial neoplasia. *Obstet. Gynecol. 55*:546-549, 1980.
12. Gissman L, Wolnik L, Ikenberg H, Holdovsky U, Schnurch HG, and Zur Hanser H. Human papillomavirus types 6 and 11 DNA consequences in genital and laryngeal papillomas and in some cervical cancers. *Proc. Natl. Acad. Sci. U.S.A. 80*:560-564, 1983.
13. Crum CP, Ikenberg H, Richard R, and Gissman L: Human papillomavirus type 16 and early cervical neoplasia. *N. Engl. J. Med. 310*:880-883, 1984.
14. Meisels A, Fortier M, and Roy M: Human papillomavirus: The atypical condyloma. *Acta Cytol. 21*:379-390, 1977.
15. Purola E, Halila H, and Vesterinen E: Condyloma and cervical epithelial atypias in young women. *Gynecol. Oncol. 16*:34-34, 1983.
16. Ferenczy A: Laser therapy of genital condylomata acuminata. *Obstet. Gynecol. 63*:703-708, 1984.
17. Vesterinen E, Meyer B, Cantell K, and Purola E: Topical treatment of flat vaginal condyloma with human leukocyte interferon. *Obstet. Gynecol. 64*:535-538, 1984.
18. Ferenczy A: Comparison of 5-fluorouracil and CO_2 laser for treatment of vaginal condyloma. *Obstet. Gynecol. 64*:773-778, 1984.
19. Reid R, Stanhope RC, Herschmna BR, Booth E, Phibbs G, and Smith J: Genital warts and cervical cancer. I. Evidence of an association between subclinical papillomavirus infection and cervical malignancy. *Cancer 50*: 377-387, 1982.
20. Bell J, Bernd-Uwe S, Averette H, Mehradad N: Vaginal cancer after hysterectomy for benign disease: Value of cytologic screening. *Obstet. Gynecol. 64*:699-702, 1984.
21. Petrilli E, Townsend DE, Morrow CP, and Nakao CB: Vaginal intraepithelial neoplasia: Biologic aspects and treatment with topical 5-fluorouracil and the carbon dioxide laser. *Am. J. Obstet. Gynecol. 138*:321-328, 1980.
22. Guthrie D and Way S: Immunotherapy of non-clinical vaginal cancer. *Lancet 2*:1242-1243, 1975.
23. Richart RM: Natural history of cervical intraepithelial neoplasia. *Clin. Obstet. Gynecol. 10*:723-730, 1967.
24. Woodruff JD, Parmley TH, and Julian CG: Topical 5-FU in the treatment of vaginal CIS. *Gynecol. Oncol. 3*:124-127, 1975.
25. Bolten KA: Practical colposcopy in cervical and vaginal cancer. *Clin. Obstet. Gynecol. 10*:808-810, 1967.
26. Hernandez-Linares W, Puthawale A, Nolan JF, Jernstrom PH, and Morrow P: Carcinoma in situ of the vagina: Past and present management. *Obstet. Gynecol. 56*:356-360, 1980.

13

Endoscopic Laser Surgery in Gynecology

DAVID S. McLAUGHLIN
Humana Women's Hospital, Indianapolis Fertility Center, Inc., Indianapolis, Indiana

I. INTRODUCTION

The CO_2 laser has been used intraabdominally in gynecological surgery (to preserve reproductive potential) for the past 6 years. Several publications describe its applications when coupled to an operating microscope (microlaser surgery) as well as on freehand laser laparotomy in animals and in humans (1-23). As surgical techniques have evolved with time, so have improvements in surgical laser delivery systems. In particular, the development of a closed-cube micromanipulator enables the articulating arm of the laser to be coupled to the laser laparoscope. This preservation of the pneumoperitoneum has enabled the gynecological laser surgeon to perform conservative pelvic surgery on an outpatient basis, which obviates the need for laparotomy (24-37). This new modality, laser laparoscopy, is useful in treating pelvic endometriosis, adhesions, selected cases of distal tubal occlusion, and small uterine myomas. The first use of the CO_2 laser, aimed through the operating channel of the laparoscope, was reported by Bruhat in France in 1979 (25). In Israel, Tadir et al. (35) reported the use of the CO_2 laser through a second-puncture laparoscope in 1981. Both Bruhat (25) and Tadir (35) investigated the use of the CO_2 laser for tubal sterilization. They now both recommend against its use because of early tubal recanalization. Daniell and Brown (26) first reported the use of the single-puncture CO_2 laparoscope for animal and human surgery in 1982, which was followed by the use of a second-puncture laser laparoscope in treating endometriosis and pelvic adhesions in humans (27). Recently, Daniell and Herbert (28) reported the following results after laser laparoscopic terminal salpingostomy: 75% patency rate, 24% term pregnancy rate—for those women who had undergone at least one previous tuboplasty. Term pregnancy rates have varied from 14-76% (24,25,32-36) depending upon the underlying disease treated with the CO_2 laser laparoscope.

Two systems for this new modality are currently being used. First, the single-puncture laser laparoscope is used. It is coupled to a 300-mm zinc arcenide lens in a closed-cube micromanipulator, which centers the beam down a 5-mm operating channel while preserving the pneumoperitoneum. A disadvantage of this system is the diminished view of the helium-neon aiming beam of the laser when seen through the single-puncture laparoscope. Also, the second-puncture laser laparoscope is being used. It consists of an 8-mm hollow channel attached by a double ring to the zinc arcenide focusing lens in the closed-cube micromanipulator. This focuses a 0.5-mm spot at 2 cm from the distal end of the instrument. The power may be varied from 5 to 10 W to obtain low power densities (2000-5000 W/cm^2) as the laser is used in minimal continuous exposure or rapid repeat sequences to limit the depth of impact to less than 1 mm. By increasing the power setting (30-35 W), the surgeon is able to increase the power density to make a bloodless incision. It is imperative to mentally view the pelvic anatomy before activating the laser to avoid penetration of vital structures by the laser's beam. The advantages for using the precise CO_2 laser's energy are reduction of risk from electrical injury as well as reduction of thermal damage to surrounding normal structures. The surgeon is able to operate in a bloodless field because of the hemostatic properties of the CO_2 laser and to preserve more of the normal pelvic structures. This technique appears to be ideally suited for reducing the risk of recurrent adhesions. More, data however, are required for accurate assessment of the laser's long-term advantages in those women who wish to promote fertility or preserve their reproductive function.

This technique, while beneficial, is not without problems. These include an accumulation of smoke from the laser plume, difficulty in maintaining a pneumoperitoneum, fogging of the laser lens, and difficulty in beam alignment as it is aimed through the operating channels. This surgical procedure can also be cumbersome, often requiring multiple punctures. This can lead to a "clashing of swords" from the distal laser probes placed intraperitoneally. Moreover, as yet the laser beam cannot be delivered through a fiberoptic bundle.

Keye first reported the use of the argon laser delivered through a fiberoptic cable (38,39) for the photocoagulation of endometriosis as viewed through the laser laparoscope. This technique seems to be advantageous for endometriosis because the reddish-blue color of the endometrial implants selectively absorb the argon laser's energy. Limitations of this system include delayed tissue damage, which may not be apparent at the time of the procedure, as well as lack of fine control for depth of tissue destruction. The argon laser also appears to have reduced ability to incise pelvic adhesions and small myomas. Eye protection also is required, either over the laparoscope eyepiece or on the operator, to avoid retinal damage.

The neodymium:yttrium-aluminum-garnet (Nd:YAG) laser has also been used for outpatient ablation of peritoneal endometriosis as reported by Lomano (40). This Nd:YAG laser can also be delivered through fiberoptic bundles. Its coagulation depth appears to be 4-5 mm with significant scattering of the beam from the impact site. The absorption of the beam is also color-dependent, and again, special safety glasses must be used to protect the retina. The advantage of the Nd:YAG laser appears to be hemostasis, but as with the argon laser, it appears to be limited when incising tissue and vaporizing larger masses.

The Nd:YAG laser has also been used hysteroscopically to ablate the endometrium in women with menometrorrhagia who wish to avoid hysterectomy (41). The technique of Goldrath et al. (41) involves passing 0.4 mm fiberoptic cable through the operating channel of the hysteroscope. The danazol (Danocrine)-prepared endometrial lining is viewed transcervically as the uterine cavity is distended with saline solution (which is used to control temperature as well as remove debris). Using 40-60 W of power, the tip of the fiber is held a few millimeters from the uterine surface and slowly moved under hysteroscopic control to treat the entire endometrium. Again, the reddish-colored endometrium rapidly absorbs the Nd:YAG beam, with the beam's penetration limited to 3-4 mm by proper movement of the beam. The uterus is thicker than the depth of the desired tissue effect. Therefore the risk of damage to the exterior surface of the uterus, and thus other pelvic structures, appears unlikely. By using this laser, submucosal fibroids, intrauterine adhesions, and uterine septae may also be treated. The laser likewise may be passed through the cornua to effect sterilization. Although the argon laser has not been reported to be delivered hysteroscopically, it appears that this could be similarly used in treating the same uterine pathologies. A reduction in depth of penetration and cost of the argon systems may increase the availability of this technique in the future.

A CO_2 hysteroscope has been recently reported by Tadir (42). A rigid hysteroscope, using CO_2 as the distending medium, has been coupled by an endoscopic coupler to the articulating arm of the CO_2 laser. Again, airtight attachments are required to precisely vaporize any intrauterine adhesions, septae, or submucous and endometrial polyps. Bleeding that tends to obscure the operator's view when applying CO_2 hysteroscopy may limit the use of this technique, but the CO_2 laser appears to be advantageous for instantaneous vaporization compared with the Nd:YAG and the argon lasers.

II. OPERATING ROOM SET-UP FOR CARBON DIOXIDE LASER LAPAROSCOPY

To obtain an adequate view of the pelvis, a 10-mm laparoscope is inserted subumbilically with a second-probe puncture placed in the midline at the superior portion of the pubic hairline (Fig. 1). The pelvic pathology is photographed (Fig. 2) using an Olympus OM-2 camera coupled to the CLE-F flash unit (Olympus Corporation, Lake Success, New York). A high-flow CO_2 insufflator (3 L/min) is required to maintain the pneumoperitoneum, particularly when several

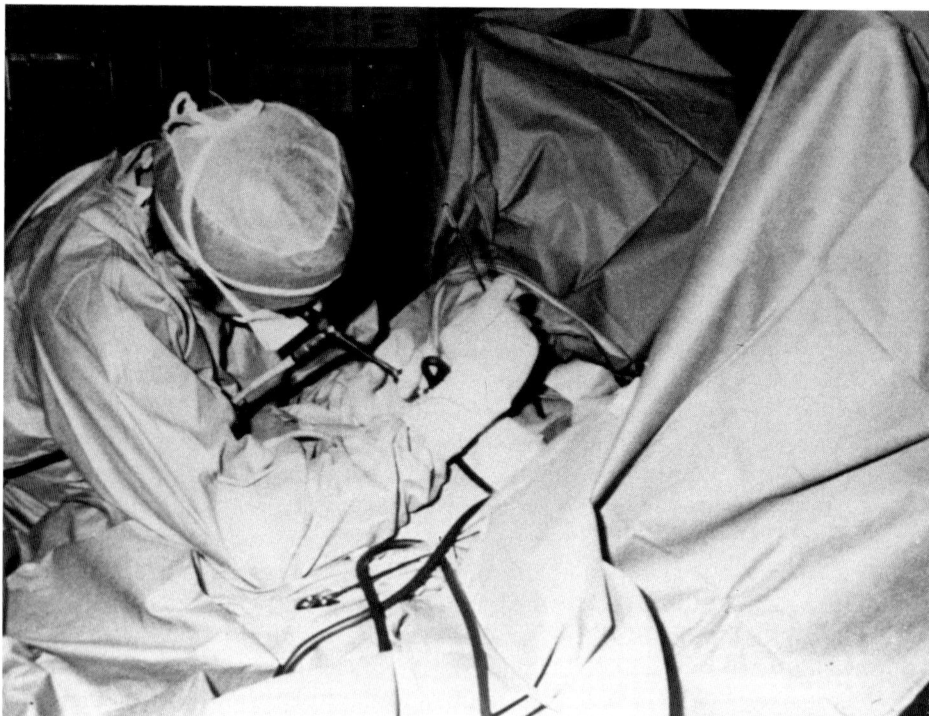

Figure 1 Laparoscopy: Two or three punctures may be required to view the pelvic pathology before beginning laser vaporization.

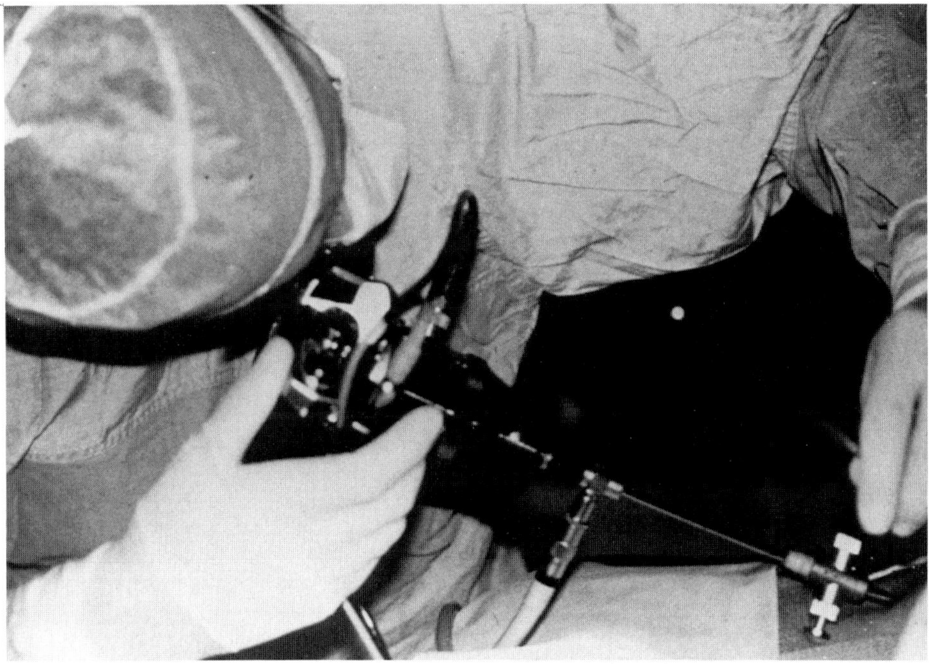

Figure 2 Laser photography: The Olympus OM-2 is coupled to the eyepiece and connected to the external flash CLE-F unit for reliable photodocumentation.

punctures are required. The laser plume is exhausted using suction or release of the pneumoperitoneum through the laparoscope trocar, thus rapid filling is required to continue the laser surgery. The closed-cube micromanipulator is coupled to either the single-puncture laparoscope to center the beam through the 5-mm operating channel, or to the second-puncture laser laparoscope (which has an 8-mm channel) (Fig. 3). Although many surgeons prefer to use the second-puncture laser laparoscope (Fig. 4), which is inserted through the right-lower quadrant, the single-puncture laser laparoscope may be advantageous when ablating disease in the posterior cul-de-sac or when performing distal tubal surgery for repeat tubal occlusion. An assistant or experienced scrub nurse is of help in steadying the pelvic structures before laser ablation is begun.

It is imperative to check the coaxial alignment of the helium-neon and CO_2 beams. The surgeon carefully centers the beams in the operating channel to avoid reflection off the interior of the channel. This can best be accomplished by firing the laser externally onto a tongue blade before applying the laser's energy intra-

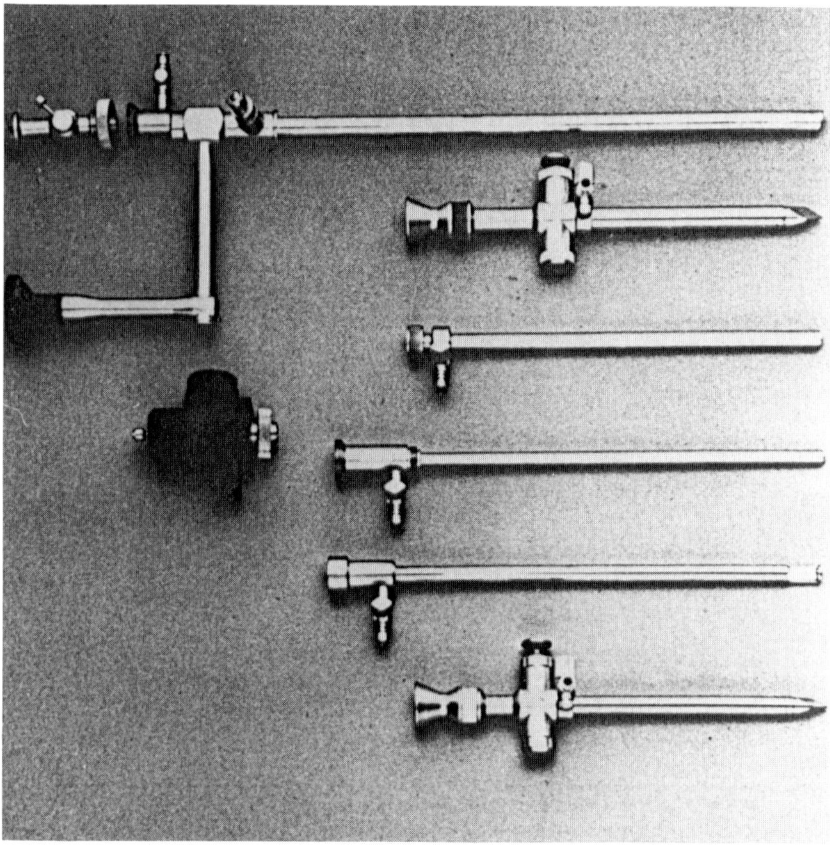

Figure 3 Single- and second-puncture laser laparoscopes: Aiming cube centers the CO_2 laser beam for aiming through single-puncture or double-puncture laparoscopes (*Source*: courtesy of Richard Wolf Medical Instruments Corp., Rosemont, Illinois).

peritoneally. After the laser laparoscope is placed intraabdominally, no eye protection is required. As a general rule, lower power densities (2000-5000 W/cm^2 – a power setting of 5-10 W for most lasers) are used to ablate superficial endometrial implants and adhesions. Higher power densities (10,000-20,000 W/cm^2, 20-30 W power) are needed to ablate deeper endometrial implants and small myomas, as well as to perform distal tubal surgery. An ebonized stainless steel

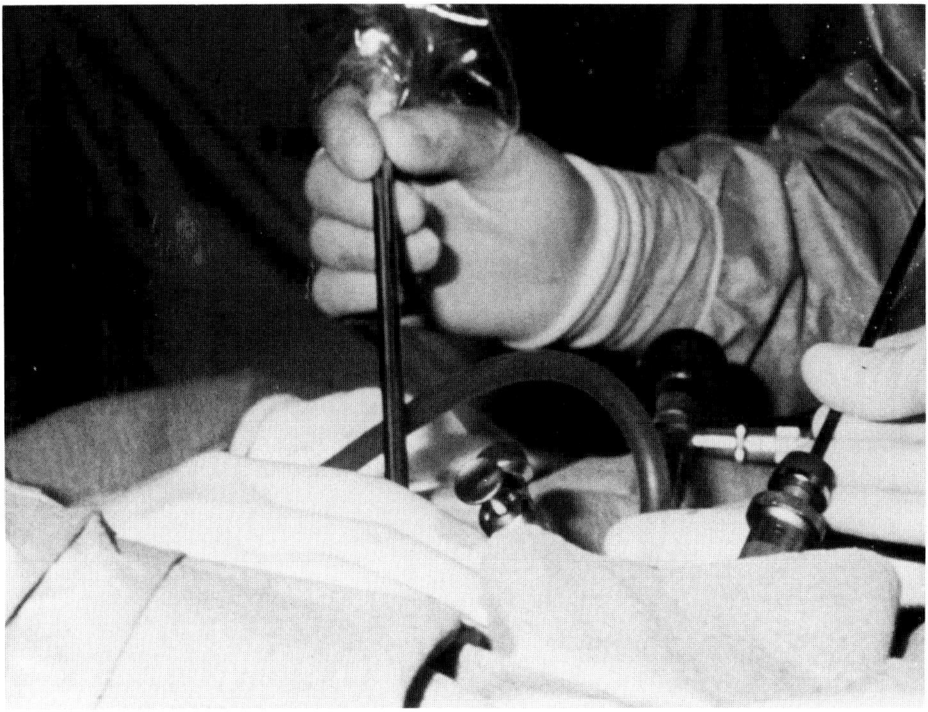

Figure 4 Second-puncture laser laparoscope in use: Often the scrub nurse is needed to steady one of the three laparoscopic instruments as well as the pelvic structures before laser vaporization.

probe has been most helpful as a backstop, although some laser laparoscopic instruments have a built-in backstop with a second-puncture probe. Daniell has devised a titanium suction-irrigating probe that may be used as a backstop for CO_2 laparoscopy, as well (Fig. 5). The laser plume may be evacuated by a wall suction controlled by opening the valve on the probe. A solution of heparinized Ringer's lactate also may be irrigated through the probe to remove carbon debris after laser vaporization. It is most important to be cognizant of the surrounding viscera that may be accidentally vaporized with the CO_2 laser beam. Common sense dictates that the surgeon should not activate the CO_2 laser without adequate visualization of the helium-neon beam throughout its course.

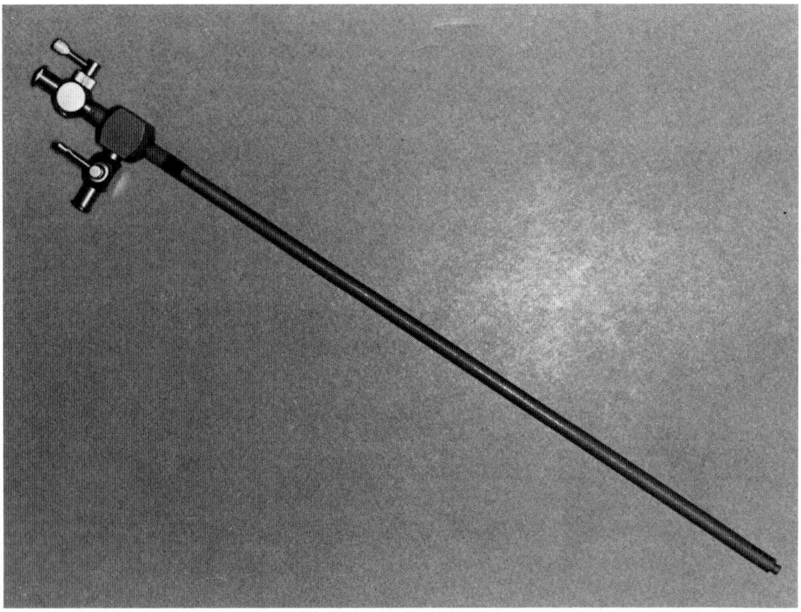

Figure 5 Suction irrigation probe for laparoscopy: Made of titanium, this probe designed by Daniell helps evacuate the laser plume, irrigate the remaining carbon debris, and serves as a backstop for the CO_2 laser's energy (*Source*: courtesy of James Daniell, Nashville, Tennessee).

III. CARBON DIOXIDE LASER LAPAROSCOPY

The advantages of the carbon dioxide laser coupled to the operating microscope include instantaneous vaporization, precision, and hemostasis, which, although the laser is cumbersome, override the apparent disadvantages that include lack of a presently available fiberoptic bundle to contain the laser's energy and occasionally decreased visibility because of accumulation of the laser plume inside the closed intraperitoneal space. These two disadvantages should be minimized as further improvements in laser technology occur.

A. Endometriosis

Using a laser power setting of 5-10 W, most superficial peritoneal implants of endometriosis (stage I) may be safely vaporized directly (Fig. 6a,b) or indirectly by reflecting the beam off a highly polished stainless steel mirror distally placed at

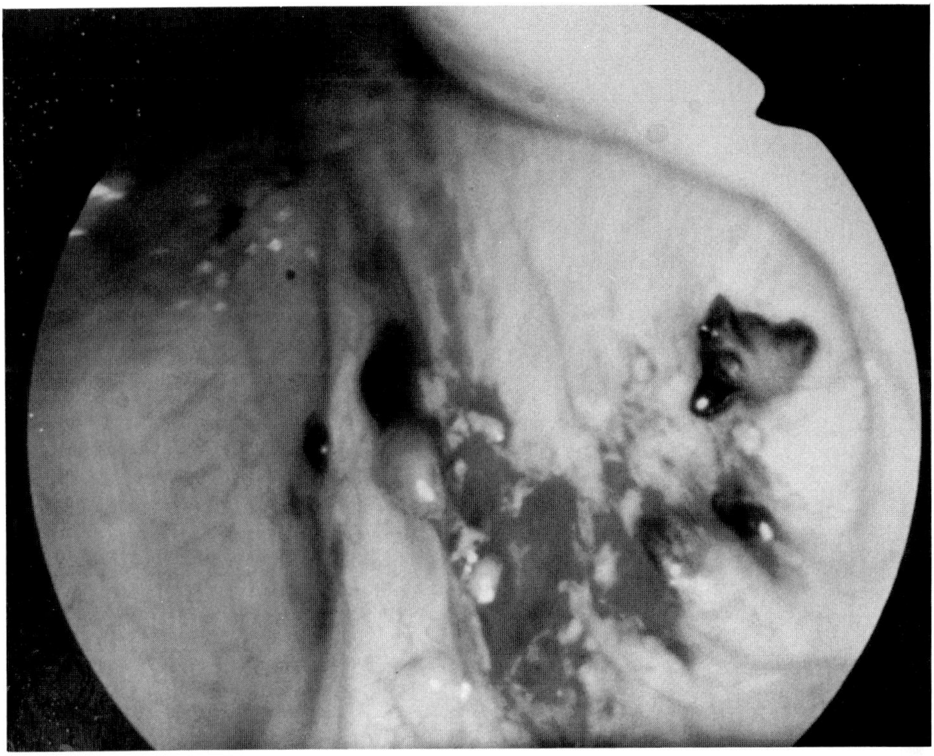

Figure 6a Endometriosis, stage I: Laparoscopic view of aggressive endometriosis on the left lateral pelvic wall.

the end of the second-puncture laser laparoscope (Fig. 7). Superficial ovarian endometriosis (stage II-A) may be vaporized at the same low-power density (2000-5000 W/cm^2) (Figs. 8a,b and 9). Deep ovarian implants (stage II-B) require higher power settings of 20-30 W (10,000-20,000 W/cm^2) to evaporate the aberrant endometrial glands (Fig. 10a,b). Oftentimes the chocolate fluid exuding from the ovarian endometriomas requires irrigation with heparinized Ringer's lactate solution to enable the laser laparoscopist to completely evaporate the endometrioma at its base. Superficial tubal endometriosis (stage III) may be vaporized safely at a lower setting, taking care to avoid unnecessary bleeding from the vascular mesosalpinx (Fig. 11a,b). Some implants on the bowel and ureter (stage IV) may be approached with the laser laparoscope if the implants have not pene-

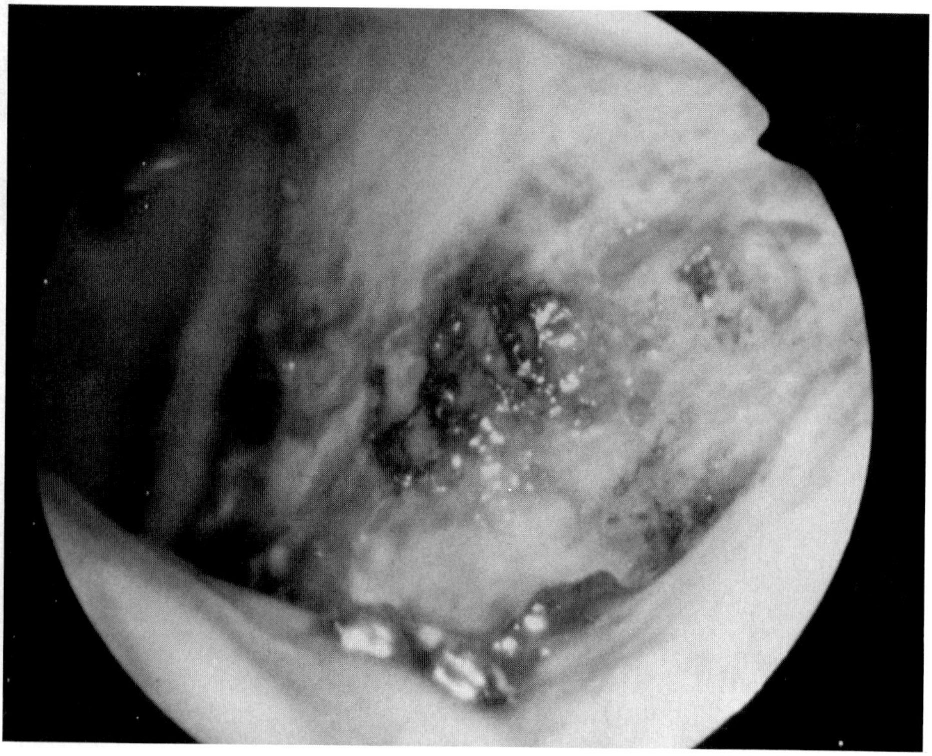

Figure 6b Endometrial implants partially vaporized: second-puncture laser laparoscope used to evaporate endometrial implants.

trated too deeply through the serosa (Figs. 12a,b and 13). Often by using the repeat pulse mode at 5-10 W, the depth of penetration can be safely controlled. Magnification through endoscopic optics helps compensate for the loss of perspective resulting from the monocular view obtained through the eyepiece. However, should the laparoscopist feel uncomfortable using laser laparoscopy for the more advanced stages of endometriosis, microlaser laparotomy is an alternative.

B. Pelvic and Peritubal Adhesions

Filmy adnexal adhesions may be directly vaporized with a low power setting (5-10 W) if a satisfactory backstop is positioned correctly to shield other vital intra-abdominal structures. By using the laparoscope probe (and sometimes a second

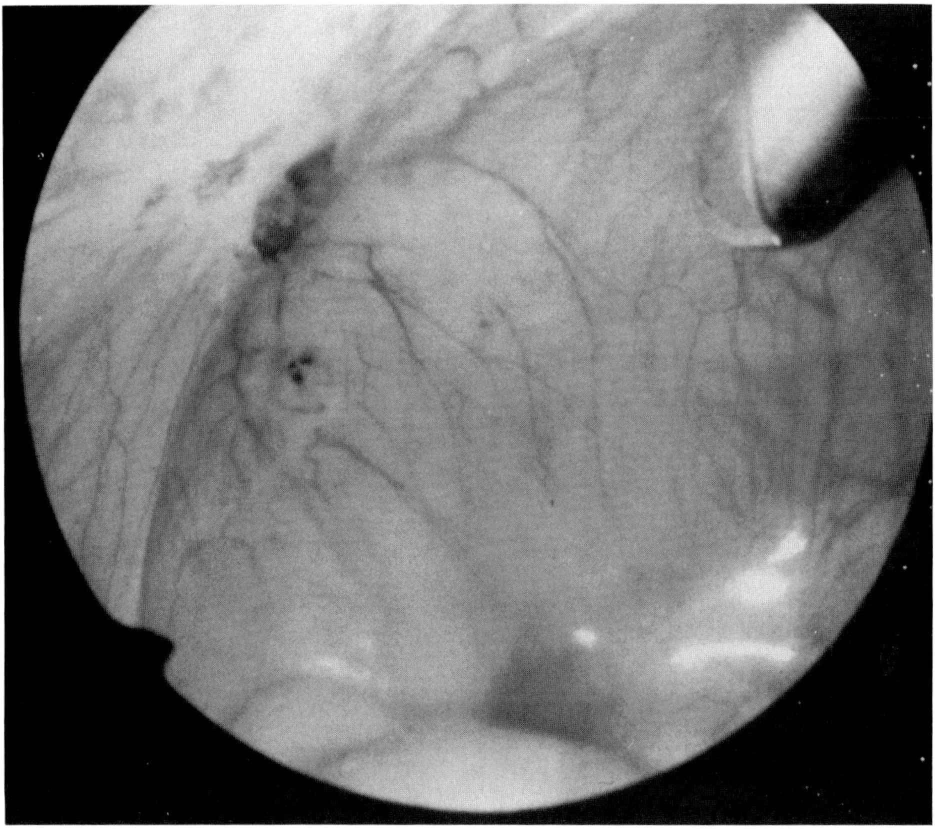

Figure 7 A 120°-angled laser laparoscope mirror: Carbon dioxide beam is reflected off a highly-polished stainless steel mirror at distal end of the second-puncture laser laparoscope to ablate endometrial implants.

probe and grasper), traction and countertraction will speed up vaporization of the adhesions (Fig. 14a-e). If meticulous technique is cautiously applied, an electrocautery for hemostasis will rarely be required, but one should be available because further application of the CO_2 laser laparoscope will rarely control severe bleeding.

C. Laser Laparoscopic Neosalpingostomy

Laser laparoscopy may be useful to reopen a hydrosalpinx that has recurred following previous microsurgery. Some women may elect this outpatient procedure

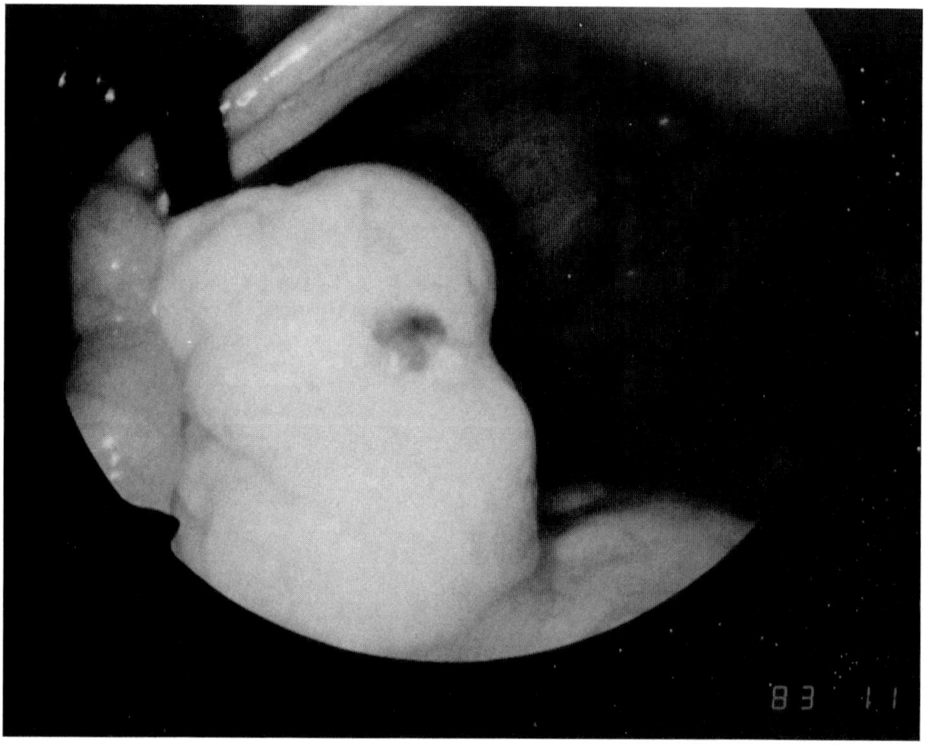

Figure 8a Superficial ovarian endometriosis, stage II-A: Laparoscopic view before CO_2 vaporization.

to avoid the expense and recuperation time needed for microlaser laparotomy. Certainly not all tubal surgery may be approached with this technique, particularly the thick-walled hydrosalpinges. With experience, the laser laparoscopist will develop his/her comfort level for appropriate case selection.

When performing this technique, the tube is mobilized and then distended by transcervical hydrotubation with methylene blue dye. High power densities (with the laser set at 30 W) are required to make radial incisions from the central tubal dimple. Penetration through the mucosa should be avoided until the stellate incisions have been outlined. Using traction and countertraction (often with an additional probe or grasper) the mucosa is penetrated posteriorly to anteriorly to ensure that blood or methylene blue dye does not interfere with the

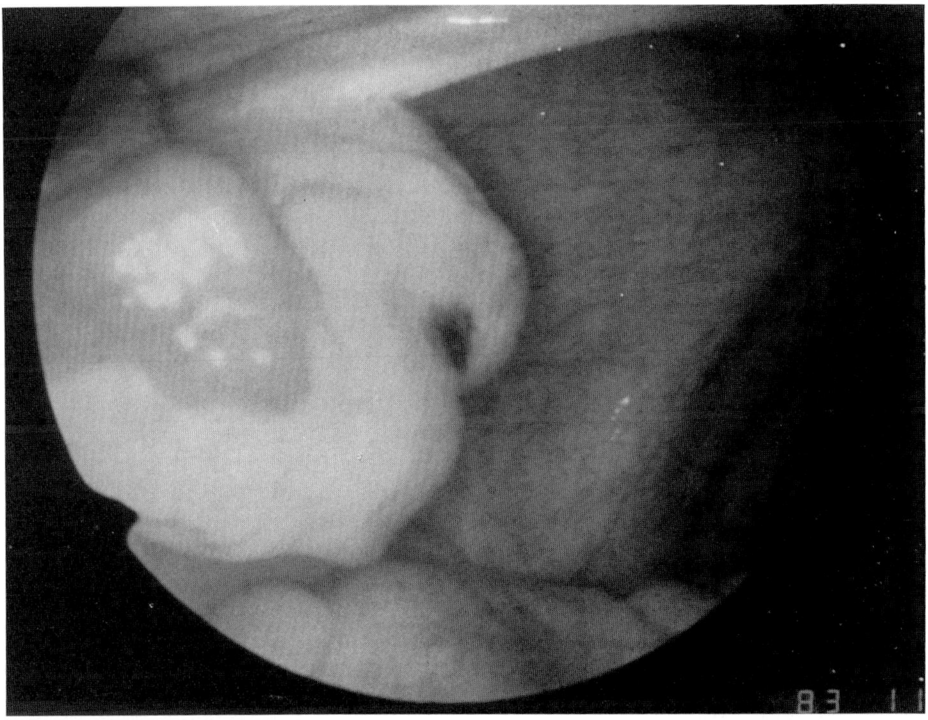

Figure 8b Ovarian endometriosis after carbon dioxide laser vaporization: Note the view is slightly decreased by a laser plume, which is contained intraperitoneally.

absorption of the laser's energy beam. The laser power setting should then be decreased to approximately 5 W to shrink the distal serosa adjacent to the fimbria. This flowering technique will help evert the fimbria of a thin-walled hydrosalpinx (Fig. 15a,b), thus obviating the need for suturing. It remains to be seen whether or not the pregnancy rates from CO_2 laser laparoscopic neosalpingostomy will surpass those of conventional microsurgical and microlaser neosalpingostomies. An apparent advantage of using the laser laparoscope is a seeming reduction in the risk of subsequent adhesion reformation, because there are less tissue trauma and fewer drying effects than seen during open laparotomy. A possible disadvantage is the reduced ability to evert fimbria as completely as can be done for most hydrosalpinges via a microlaser laparotomy.

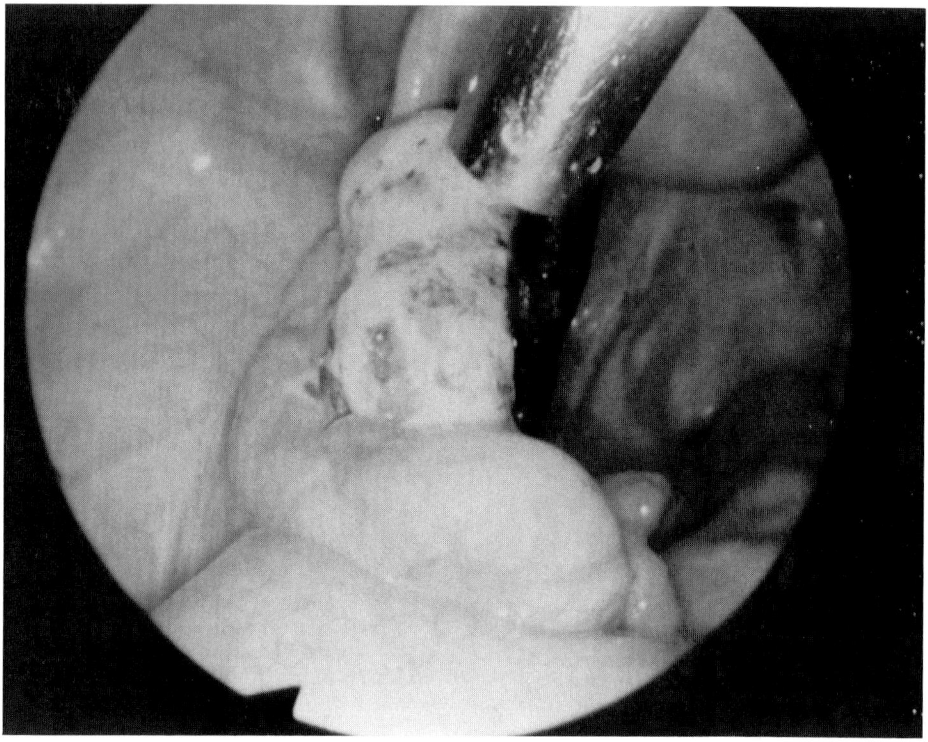

Figure 9 Multiple ovarian implants directly vaporized by second-puncture laser laparoscope: Note minimal carbon debris after superficial vaporization at low power densities.

Figure 10 (a) Stage II-B ovarian endometriosis before vaporization: Deeper implants require longer duration at higher power densities. (b) Impact site after vaporization: Healing within 3 months with minimal adhesion reformation.

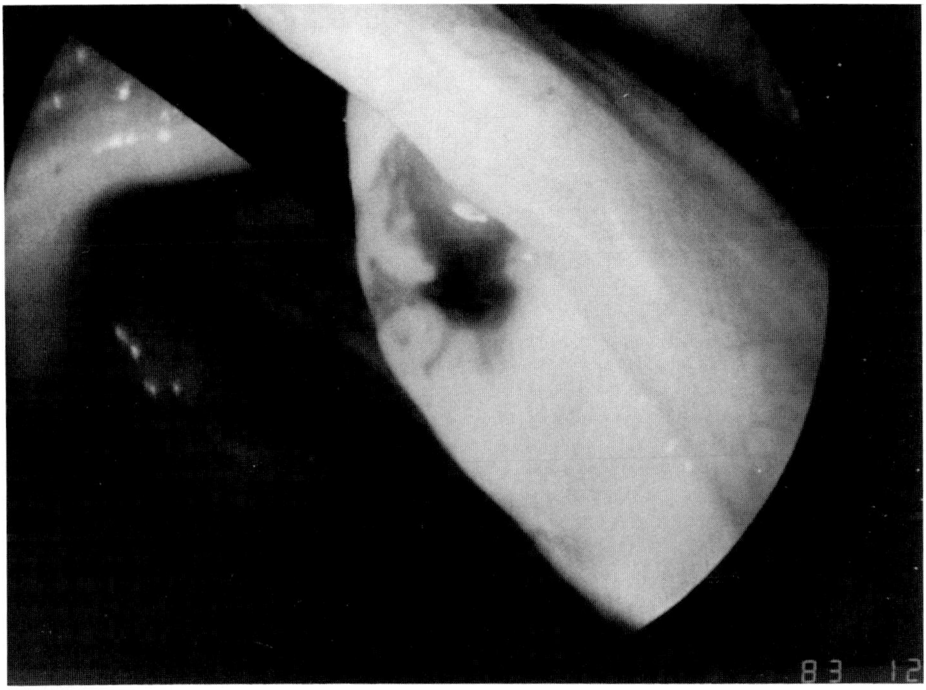

(a)

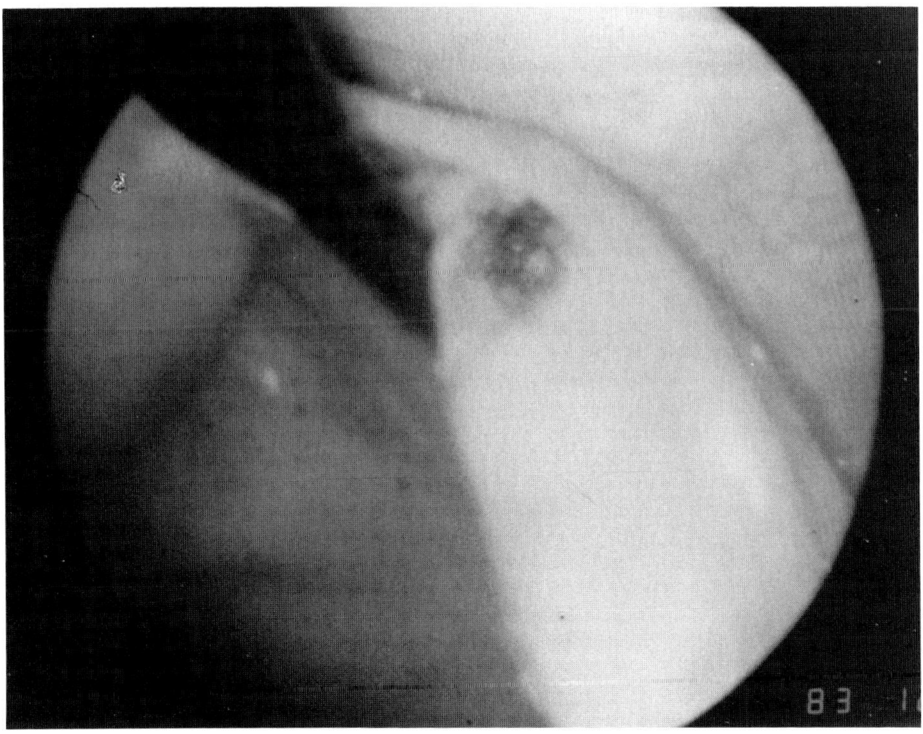

(b)

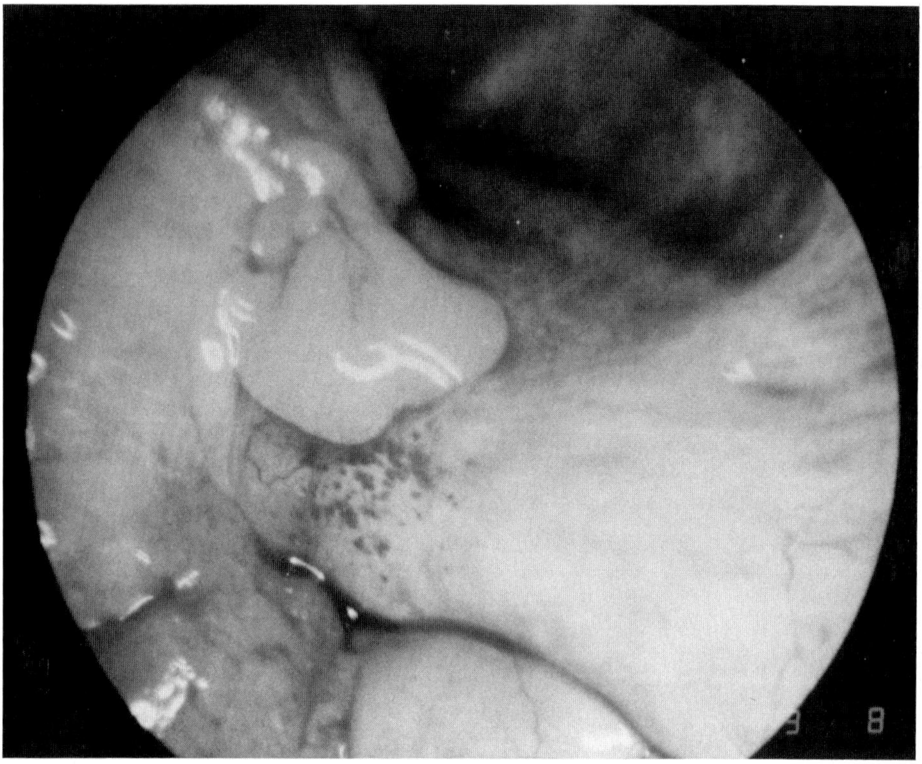

Figure 11a Tubal endometriosis, stage III: Although endometriosis rarely affects the fallopian tube, the implants may cause adhesions between the tube and surrounding structures.

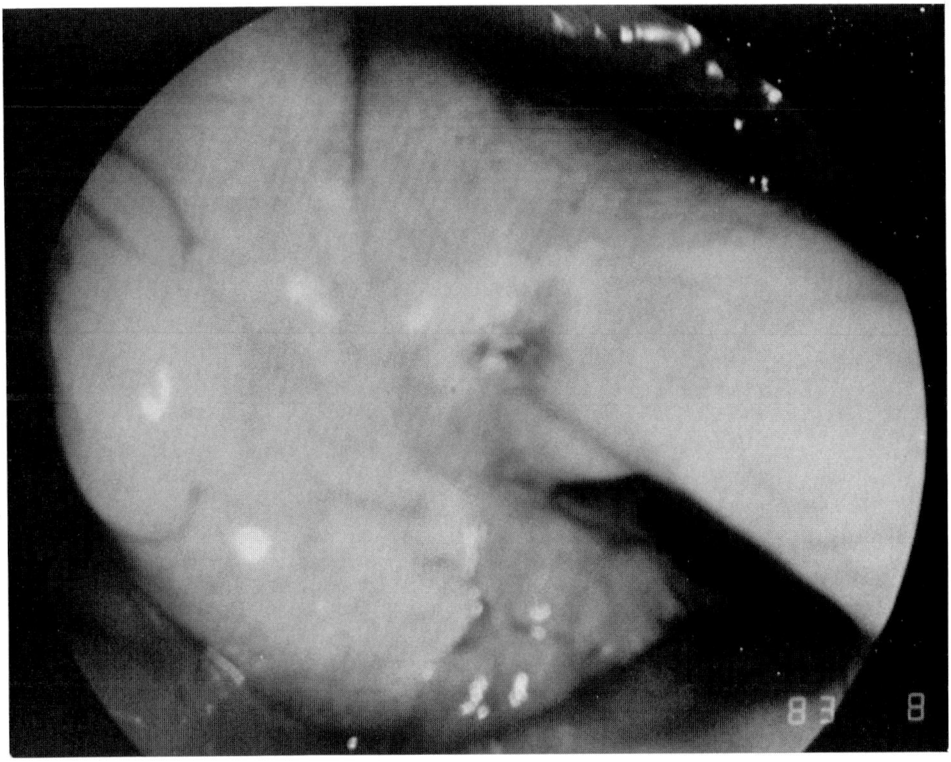

Figure 11b Stage III endometriosis after laser ablation: Superficial vaporization, using low power densities, can be safely accomplished.

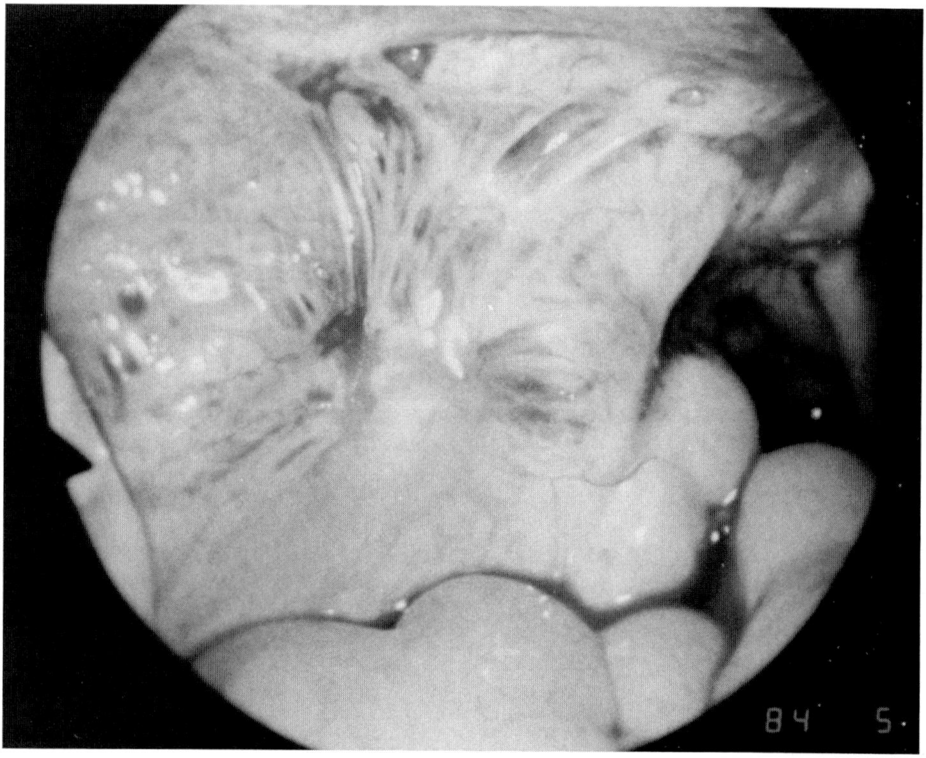

Figure 12a Stage IV endometriosis: Bowel adhesions have formed to the uterus, obliterating the posterior cul-de-sac.

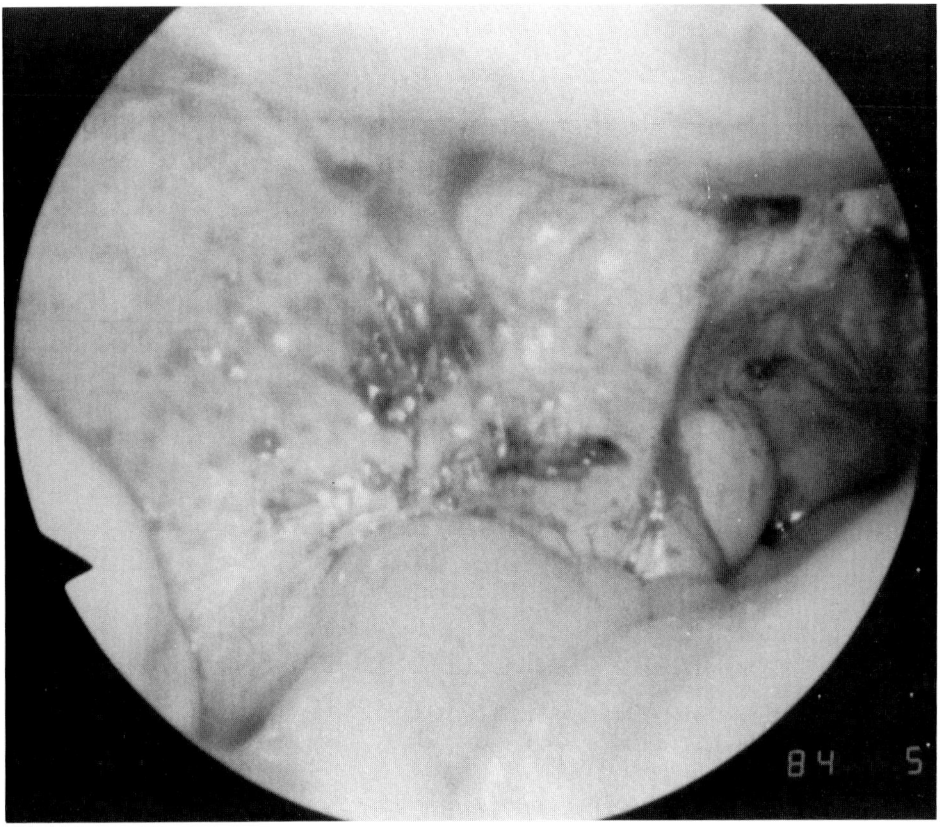

Figure 12b Ablated stage IV endometriosis: Low power densities with careful continuous or pulsed laser energy is used to avoid bowel perforations.

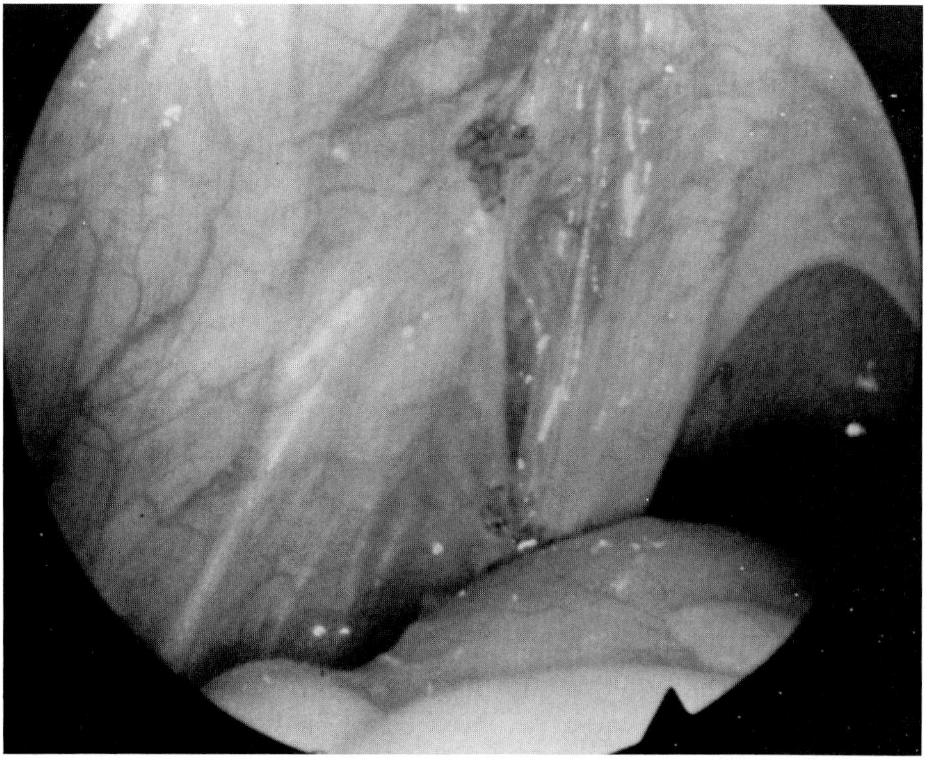

Figure 13 Ablated stage IV ureteral endometriosis: Again, low power density, with careful, meticulous technique, is required to ablate the implants safely using the CO_2 laparoscope.

D. Vaporization of Small Myomas

Incidental small subserosal fibroids may be vaporized directly (Fig. 16a,b). Usually 20-30 W of power are required, and efforts should be made to aim the laser beam such that it strikes the fibroid at a 90° angle. Larger myomas require laparotomy for excision as opposed to vaporization.

E. Laser Hysteroscopy

The CO_2 laser fired through the hysteroscope using carbon dioxide as a distending medium may be helpful in the transcervical dissection of uterine septae, intrauterine adhesions, and submucous myomas. Current pitfalls include dif-

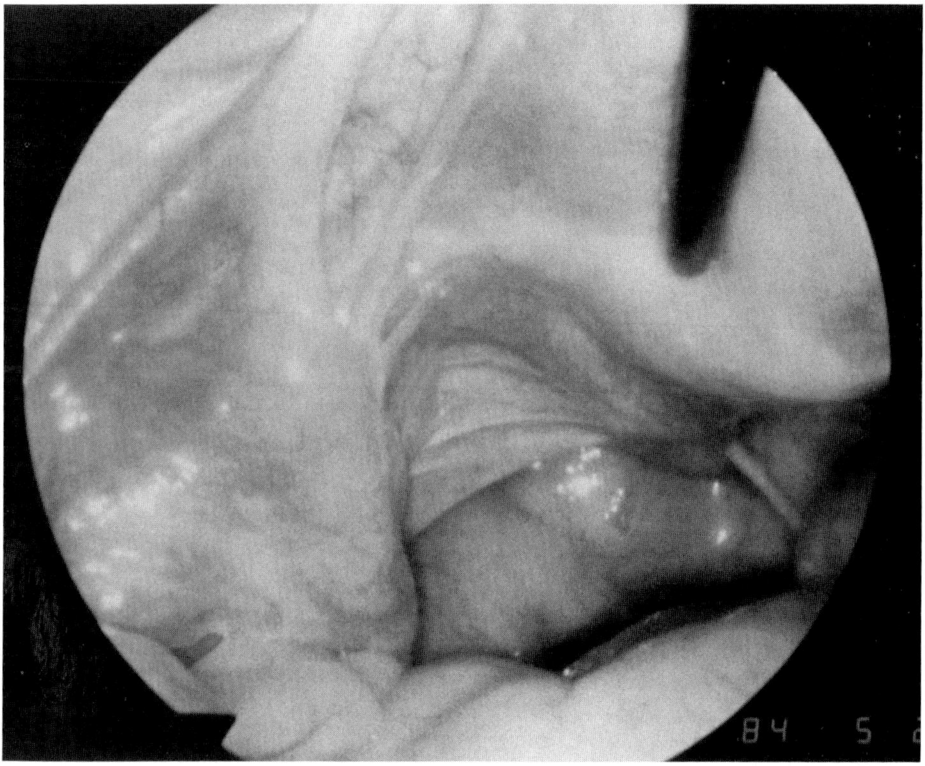

(a)

Figure 14 (a) Adnexal adhesions: The left fallopian tube is bound to the left posterior broad ligament, completely encompassing the left ovary. (b) Adhesions partially vaporized: A second-puncture CO_2 laser laparoscope is used to slowly evaporate the adhesions. (c) Tubal traction by laparoscope probe: Traction and countertraction, stretching the adhesion, helps release the adnexa as the laser's energy is applied continuously. (d) Base of adhesion vaporized: Using continuous technique, the remaining attachments of the adhesion are transformed into the laser plume. (e) Adnexa completely mobilized: Laser laparoscopy successfully avoids laparotomy, thus reducing the likelihood of additional adhesion reformation.

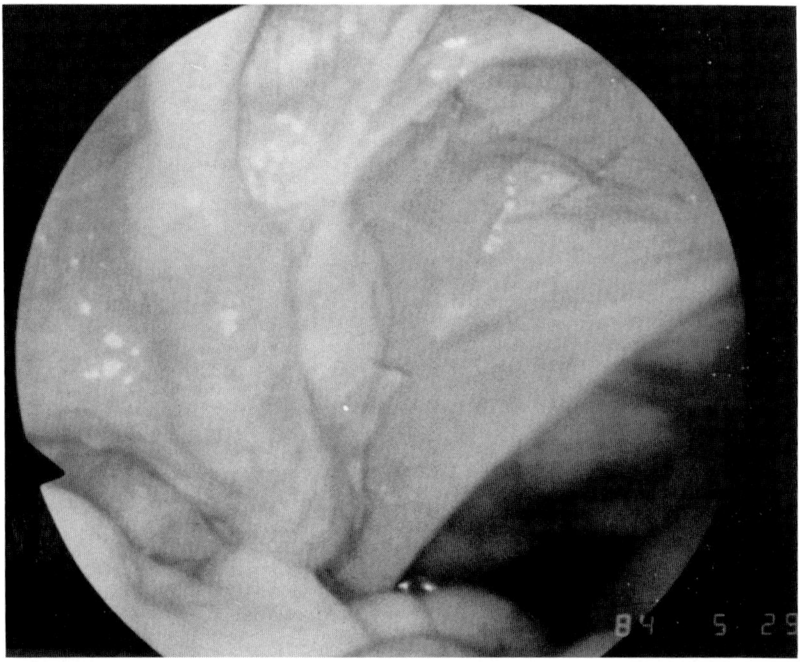

(b)

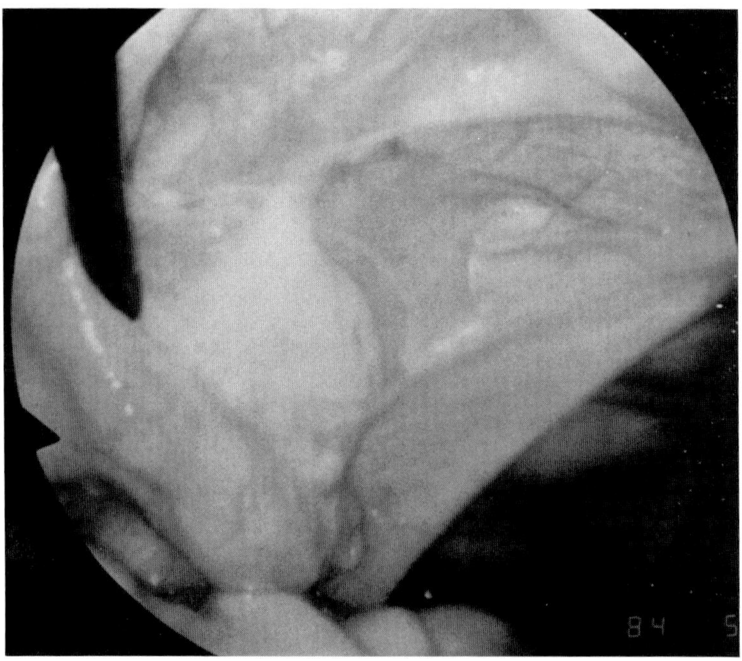

(c)

Figure 14 (continued)

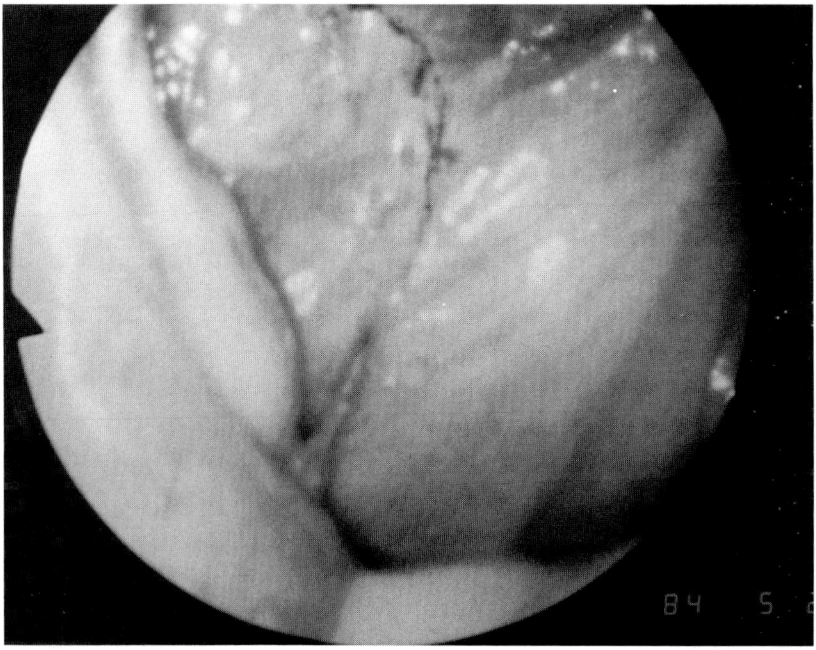

(d)

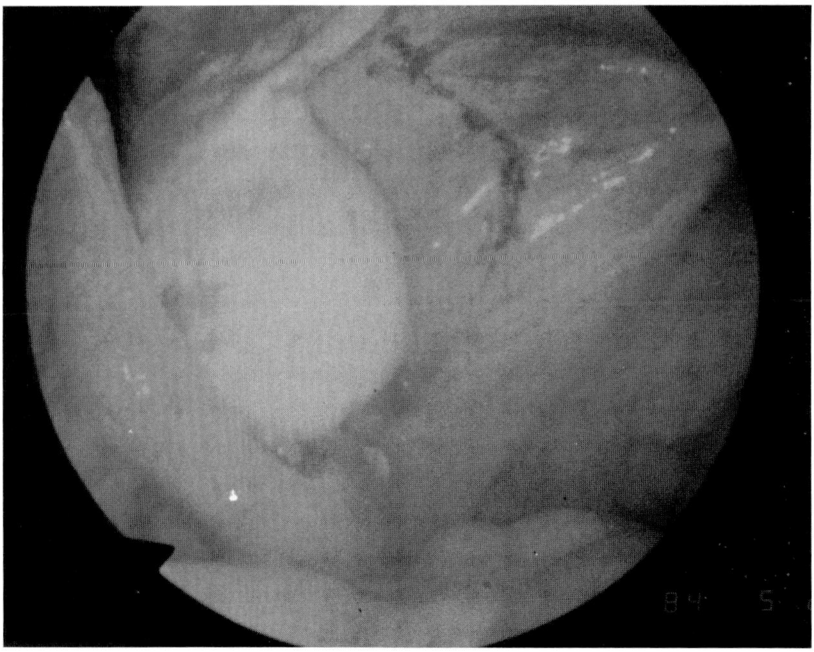

(e)

Figure 14 (continued)

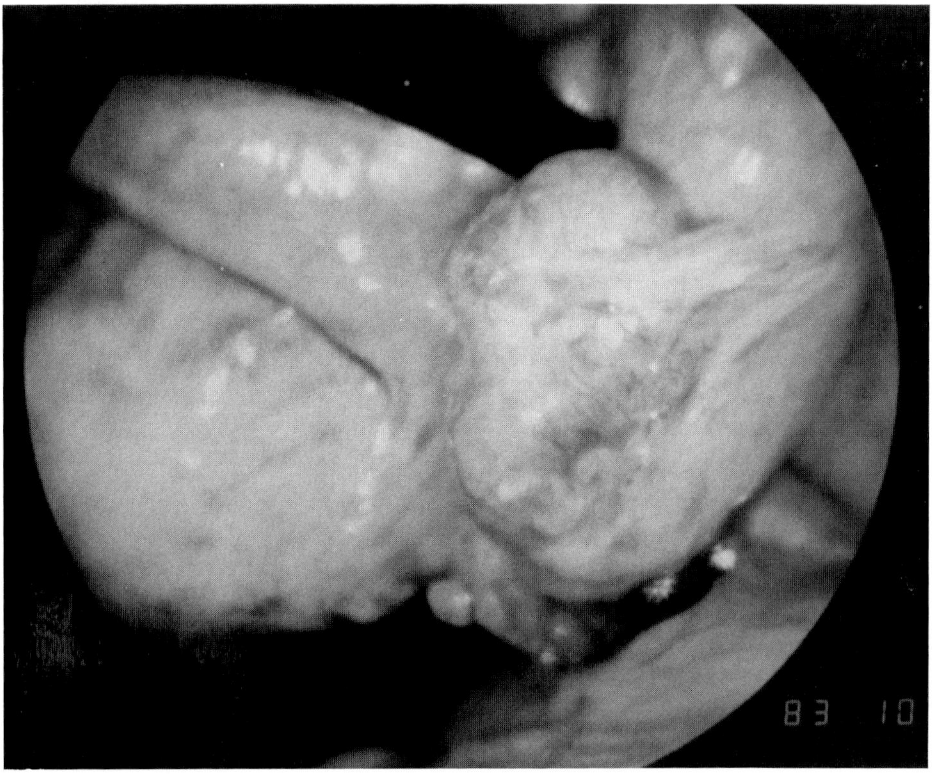

Figure 15a Recurrent tubal occlusion following previous microsurgical neosalpingostomy: Mild hydrosalpinx is present secondary to tubal reocclusion following previous microsurgical laparotomy.

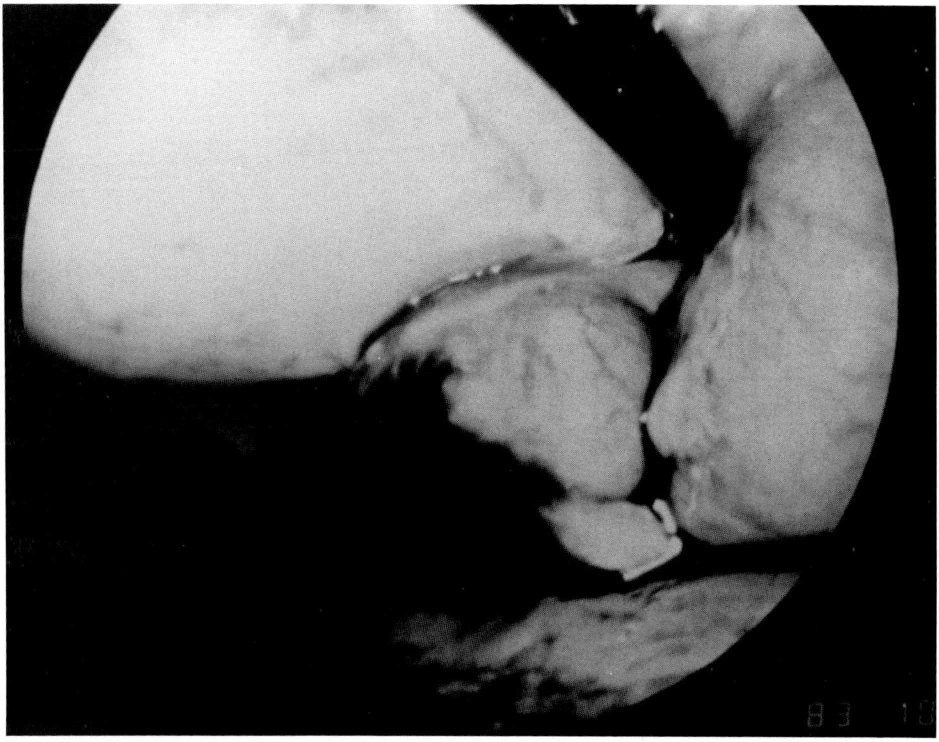

Figure 15b A CO_2 laparoscopic neosalpingostomy: Using high power densities, a stellate incision is made from the central dimple, and fimbria are everted by superficially vaporizing the serosa at low power densities, thus avoiding the need for suturing.

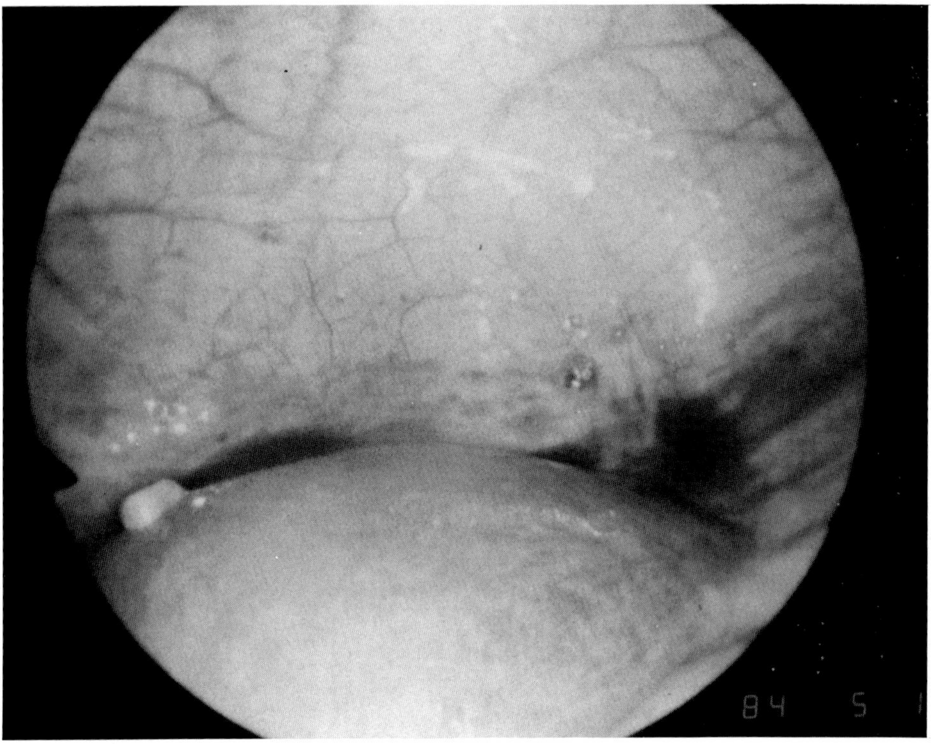

Figure 16a Superficial subserosal myoma: Small incidental fibroids may be safely vaporized using the CO_2 laser laparoscope.

ficulty in delivering the beam directly at the pathological site (particularly if the site is away from the midline of the uterus) and obscured vision should bleeding occur during hysteroscopy. The Nd:YAG laser has been used more extensively because it may be delivered through a fiberoptic bundle in a fluid-filled medium, which ia also useful to remove the blood and debris that would otherwise obscure the operator's view. This method has been used successfully to ablate the endometrium in women who wish to avoid hysterectomy for refractory menometrorrhagia.

IV. SUMMARY

Gynecological laser endoscopy has evolved as a more convenient way to treat mild to moderate pathology in women who wish to avoid pelvic laparotomy.

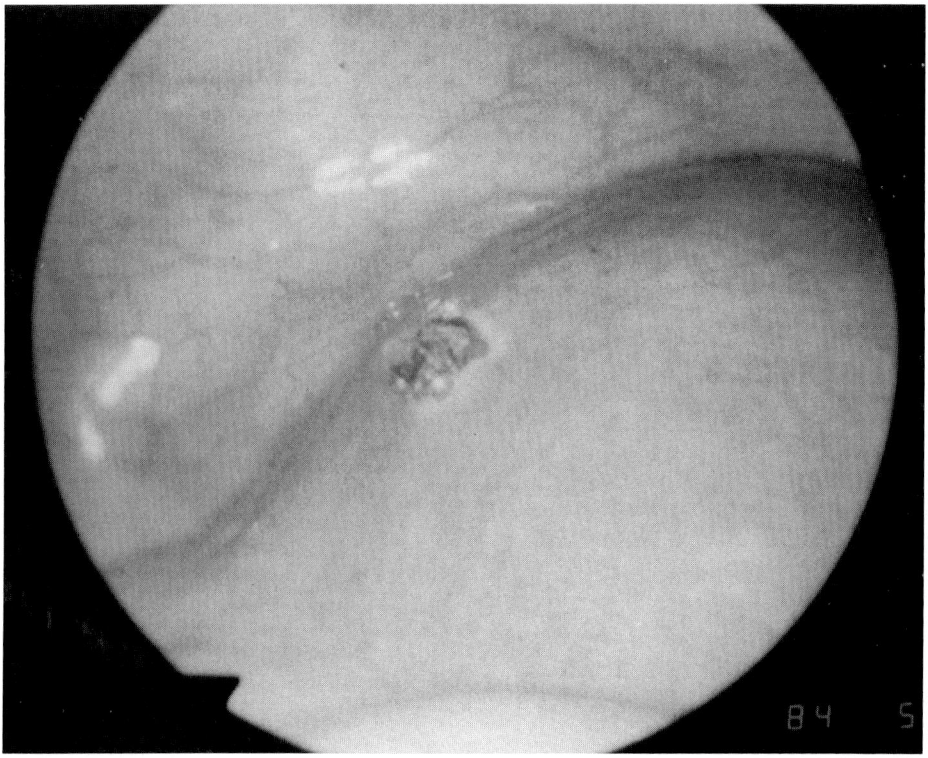

Figure 16b Myoma partially vaporized: Higher power densities are required to vaporize myoma because care is needed to apply the laser's energy at a 90° angle as myoma is vaporized.

It is likely that this new outpatient treatment modality will reduce the chance of subsequent adhesions and will be preferred for patient convenience. Although not for the timid, laser laparoscopy and hysteroscopy appear to be new treatment modalities for those gynecological surgeons who have a thorough understanding of the tissue effects of each surgical laser they wish to use and extensive experience in diagnostic laparoscopy and hysteroscopy. Moreover, a readily available CO_2 laser fiberoptic cable will further revolutionize laser laparoscopy by simplifying this new surgical technique and thus reduce operator frustration.

REFERENCES

1. Baggish MS: Status of the carbon dioxide laser for infertility surgery. *Fertil. Steril. 40*:442-445, 1983.

2. Baggish MS and Chong AP: CO_2 laser microsurgery of the uterine tube. *Obstet. Gynecol. 58*:111-116, 1981.
3. Baggish MS and Chong AP: Intraabdominal surgery with the CO_2 laser. *J. Reprod. Med. 28*:269-279, 1983.
4. Bellina JH: Microsurgery of the fallopian tube with the CO_2 laser: Eighty-two cases with follow-up. *Lasers Surg. Med. 2*:129-136, 1982.
5. Bellina JH, Hemmings R, Voros JI, and Ross LF: Carbon dioxide laser and electrosurgical wound study with an animal model: A comparison of tissue damage and healing patterns in peritoneal tissue. *Am. J. Obstet. Gynecol. 148*:327-334, 1984.
6. Bellina JH: Microsurgery of the fallopian tube with the carbon dioxide laser: Analysis of 230 cases with a two year follow-up. *Lasers Surg. Med. 3*:255-260, 1983.
7. Bellina JH: Reconstructive microsurgery of the fallopian tube with the carbon dioxide laser–Procedures and preliminary results. *Reproduction 5*:1-17, 1981.
8. Choe JK, Dawood MY, and Andrews AH: Conventional versus laser reanastomosis of rabbit ligated uterine horns. *Obstet. Gynecol. 61*:689-694, 1983.
9. Chong AP and Baggish MS: Management of pelvic endometriosis by means of intraabdominal carbon dioxide laser. *Fertil. Steril. 41*:14-19, 1984.
10. Daniell JF: The CO_2 laser in infertility surgery. *J. Reprod. Med. 28*:265-268, 1983.
11. Daniell JF: The role of lasers in infertility surgery. *Fertil. Steril. 42*:815-823, 1984.
12. Diamond MP, Daniell JF, Martin DC, Feste J, Vaughn WK, and McLaughlin DS: Tubal patency and pelvic adhesions at early second-look laparoscopy following intraabdominal use of the carbon dioxide laser: Initial report of the intraabdominal laser study group. *Fertil. Steril. 42*:717-723, 1984.
13. Fayex JA, McComb JS, and Harper MA: Comparison of tubal surgery with the CO_2 laser and the unipolar microelectrode. *Fertil. Steril. 40*:476-480, 1983.
14. Kelly RW and Roberts DK: Experience with the carbon dioxide laser in gynecologic microsurgery. *Am. J. Obstet. Gynecol. 146*:585-588, 1983.
15. Klink F, Grosspietzsch R, von Klitzing L, Endell W, Husstedt W, and Oberheuser F: Animal in vivo studies and in vitro experiments with human tubes for end-to-end anastomotic operations by a CO_2-laser technique. *Fertil. Steril. 30*:100-102, 1978.
16. Mage G and Bruhat M-A: Pregnancy following salpingostomy: Comparison between CO_2 laser and electrosurgery procedures. *Fertil. Steril. 40*:472-475, 1983.
17. McLaughlin DS: Micro-laser myomectomy technique to enhance reproductive potential: A preliminary report. *Lasers Surg. Med. 2*:107-127, 1982.
18. McLaughlin DS: Advanced surgical instrumentation needed for intra-abdominal application of the carbon dioxide laser in reproductive biology. *Lasers Surg. Med. 2*:241-254, 1983.

19. McLaughlin DS: Evaluation of adhesion reformation by early second-look laparoscopy following microlaser ovarian wedge resection. *Fertil. Steril. 42*: 531-537, 1984.
20. McLaughlin DS: Successful pregnancy outcome following removal of bilateral cornual polyps by microsurgical linear salpingotomy with the aid of the CO_2 laser. *Fertil. Steril. 42*:939-941, 1984.
21. McLaughlin DS: Metroplasty and myomectomy with the CO_2 laser for maximizing the preservation of normal tissue and minimizing blood loss. *J. Reprod. Med. 30*:1-9, 1985.
22. Pittaway DE, Maxson WS, and Daniell FJ: A comparison of the CO_2 laser and electrocautery on postoperative intraperitoneal adhesion formation in rabbits. *Fertil. Steril. 40*:366-368, 1983.
23. Tulandi, T: Hydrosalpinx: Comparison of electrosurgery and laser surgery (abstr.). *Fertil. Steril. 41*:73S, 1984.
24. Baggish MS: Laser endoscopy in obstetrics and gynecology. *Clin. Obstet. Gynecol. 26*:219-241, 1983.
25. M. Bruhat, G. Mage, and M. Manhes. Use of the CO_2 laser in laparoscopy, in *Proceedings of the Third International Society for Laser Surgery,* September 24-28, 1979, Tel Aviv, (I. Kaplin, ed.), pp. 235-238.
26. Daniell JF and Brown DH: Carbon dioxide laser laparoscopy: Initial experience in experimental animals and humans. *Obstet. Gynecol. 59*:761-764, 1982.
27. Daniell JF and Pittaway DE: Use of the CO_2 laser in laparoscopic surgery: Initial experience with the second-puncture technique. *Infertility 5*:15-1982.
28. Daniell JF and Herbert CM: Laparoscopic salpingosotomy utilizing the CO_2 laser. *Fertil. Steril. 41*:558-563, 1984.
29. J. F. Daniell and J. R. Feste. Laser laparoscopy, in *Laser Surgery in Gynecology and Obstetrics,* (W. R. Keye, Jr., ed.). G. K. Hall Medical Publishers, Boston, 1985, pp. 147-163.
30. Feste J: Utilization of the CO_2 laser laparoscopy. (in press).
31. Feste JR: Laser laparoscopy: A new modality (abstr.). *Fertil. Steril. 41*: 74S, 1984.
32. Feste JR: Laser laparoscopy: A new modality. *J. Reprod. Med. 30*:413-417, 1985.
33. Kelly RW and Roberts DK: CO_2 laser laparoscopy: A potential alternative to danazol in the treatment of stage II and III endometriosis. *J. Reprod. Med. 28*:638-640, 1983.
34. Martin DC: Interval use of the laser laparoscopy for endometriosis following danazol therapy (abstr.). *Fertil. Steril. 41*:74S-75S, 1984.
35. Y. Tadir et al. A second-puncture probe for laser laparoscopy, in *Proceedings of the Fourth Congress of the International Society for Laser Surgery,* September 1981, Tokyo, (K. Atsumi and N. Nimsakul, eds.), pp. 25-26.
36. Tadir Y, Kaplan I, Zuckerman Z, Edelstein T, and Ovadia J: New instrumen-

tation and technique for laparoscopic carbon dioxide laser operations: A preliminary report. *Obstet. Gynecol. 63*:582-585, 1984.
37. Y. Tadir, I. Kaplan, Z. Zuckerman, and J. Ovadia. Laparoscopic CO_2 laser sterilization, in *Human Reproduction: Proceedings of III World Congress.* (K. Semm and L. Mettler, eds.). Elsevier, New York, 1981, pp. 429-431.
38. Keye WR, Matson GA, and Dixon J: The use of the argon laser in the treatment of experimental endometriosis. *Fertil. Steril. 39*:26-29, 1983.
39. Keye WR Jr and Dixon J: Photocoagulation of endometriosis by the argon laser through the laparoscope. *Obstet. Gynecol. 62*:383-386, 1983.
40. Lomano JM: Photocoagulation of early pelvic endometriosis with the Nd: YAG laser through the laparoscope. *J. Reprod. Med. 30*:77-81, 1985.
41. Goldrath M, Fuller TA, and Segal S: Laser photovaporization of endometrium for the treatment of menorrhagia. *Am. J. Obstet. Gynecol. 140*:14-19, 1981.
42. Tadir Y, Raif J, Dagan J, Kaplan I, Zuckerman Z, and Ovadia J: Hysteroscope for CO_2 laser application. *Lasers Surg. Med. 4*:153-156, 1984.

14

Clinical Applications of Neodymium:Yttrium-Aluminum-Garnet Laser Surgery in Urology

JOHN A. LIBERTINO
Lahey Clinic Medical Center, Burlington, Massachusetts

JOHN F. CYR
Wakefield, Massachusetts

I. INTRODUCTION

Interest in the surgical application of lasers has increased greatly since 1978 when only 12 titles were listed under laser treatment in the *Cumulated Index Medicus.* Only 4 years later, the 1982 *Cumulated Index Medicus* listed 212 articles about laser therapy, and recently a journal entitled *Lasers in Surgery and Medicine* was established. Interest in the urological application of laser surgery continues to grow at a rapid rate and is currently being incorporated into the urological armamentarium.

In the early 1970s, consideration was given to the use of the laser to treat tumors of the bladder. Frank (1-3) of Munich, West Germany, worked out the endoscopic application of the neodymium:yttrium-aluminum-garnet (Nd:YAG) laser. His research led to an understanding of the biophysical principles by which the Nd:YAG laser functions endoscopically, and he developed the instrumentation necessary for laser application. Frank, together with his clinical colleagues Hofstetter et al. (1-4), were the first to apply and to evaluate the use of the Nd: YAG laser for destroying bladder tumors.

Initially, the argon laser was used (5) in a gas-filled bladder, but use of the Nd: YAG laser was adopted because depth of penetration with the argon laser was inadequate and the Nd:YAG laser can be used in a water-filled bladder in a fashion similar to other urological endoscopic procedures. The Nd:YAG laser also has a 2- to 5-mm depth of penetration, making it potentially useful for the treatment of invasive bladder tumors. In addition, transmission is possible through a flexible fiber that permits easy access to most portions of the urinary bladder.

II. LASER SYSTEM

The system employed at the Lahey Clinic for endoscopic laser treatment consists of a Nd:YAG laser power source (MediLas 2 MBB-AT, ENDO-LASE, Inc., New York, New York) similar to that used in gastroenterology, neurosurgery, and other specialties, and a quartz rod fiber 0.6 nm in diameter. The laser system consists of a mobile desk casing with an operating console that houses the power supply, the control unit, and the cooling system (Fig. 1). The console enables a digital selection of the pulse power (watts) and duration (seconds). A printer records the indicated data for each laser pulse administered. This device is simple to operate, compact, and portable.

The laser head accommodates all of the optical components (i.e., the crystal, pump lamps, resonator mirror, and the lamp housing) together with a 1-mw pilot laser (helium-neon laser) that emits light in the visible range for targeting (Fig. 2). The pilot light beam is coincident with the therapeutic Nd:YAG laser beam. Both radiations are transmitted simultaneously to the operating field by a flexible fiberoptic light guide. In this way, the pilot light marks the target point of the therapeutic Nd:YAG laser beam.

A standard Storz urethroscope (Karl Storz, Inc., Culver City, California) with variable viewing lenses and a special laser channel attachment (Fig. 3a) with a specially devised Storz Albarran bridge (Fig. 3b) permits the flexible fiber to be moved through an 80° arc, thus permitting access to most sections of the bladder. In addition, the fiber can be advanced distally to obtain the proper distance from the bladder wall area to be irradiated. The quartz fiber is usually held 2-3 mm away from the section of the bladder requiring irradiation. A constant flow of irrigating fluid into the bladder permits clear visualization and cooling of the bladder wall, decreasing the likelihood of injury to adjacent viscus from forward scatter of the laser beam.

With minor modifications, this system can be used in the urethra to treat urethral strictures and in the ureter and the renal pelvis through a ureterorenoscope to treat low-grade papillary tumors. A hand-held adapter (Fig. 4) is used to treat external lesions of the genitalia, such as condyloma acuminatum and cancer of the penis. Presently, we are evaluating the role of the hand-held device for partial nephrectomy and the role of a CO_2 milliwatt laser for vasovasostomy and for small vascular anastomosis. The wearing of wavelength-appropriate protective glasses (Fig. 5) is required to avoid retinal damage.

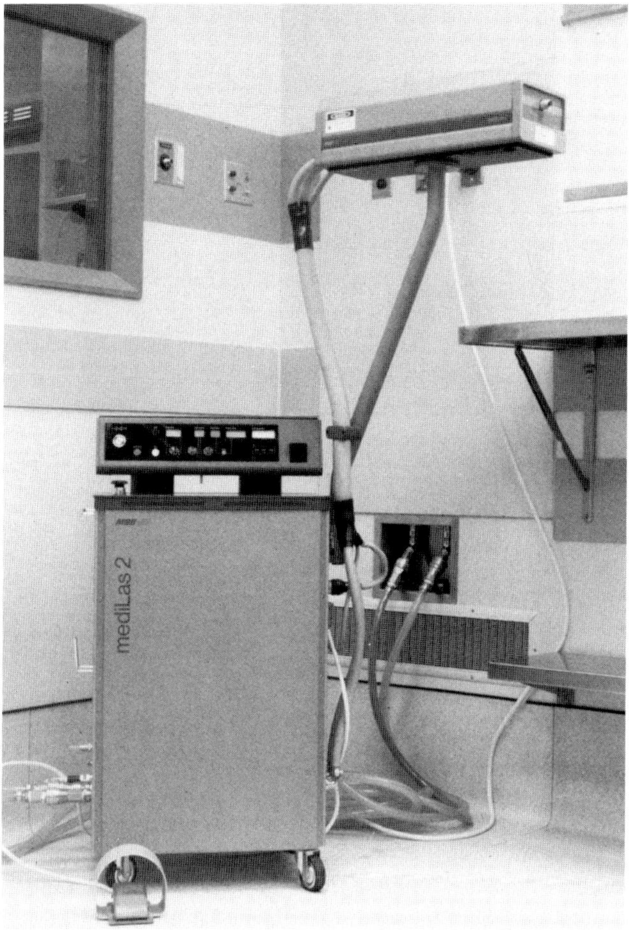

Figure 1 Operating console of the Nd:YAG laser (MediLas 2, MBB-AT ENDOLASE, Inc., New York, New York).

Figure 2 Laser head containing the optical components into which the quartz fiber is inserted.

III. INDICATIONS AND ADVANTAGES

The Nd:YAG laser is the most versatile and preferred laser for use in urology because the source of the fiberoptic laser is easily adapted to our cytoscopes, ureteroscopes, and nephroscopes, and the equipment can be used in a medium of water or urine. It provides the highest degree of tissue penetration of the presently available surgical lasers, and therefore it permits transmural laser irradiation and destruction of lesions by thermal coagulation necrosis rather than by cutting, thus making it a superior hemostatic device for endoscopic use. The use of the Nd:YAG laser in patients with bladder cancer (3) has been extensively reported and to a lesser degree in patients with transitional tumors of the ureter and renal pelvis (Böwering R, unpublished data, 1984), urethral strictures (6), bladder neck contractures, and carcinoma of the penis (7).

Practical and theoretical advantages and disadvantages exist with the use of the Nd:YAG laser in urological surgery. Some of the practical advantages are that it is a procedure technically similar to electrocautery, making it easy to incorporate into the urologist's armamentarium; it is usually performed on an ambulatory

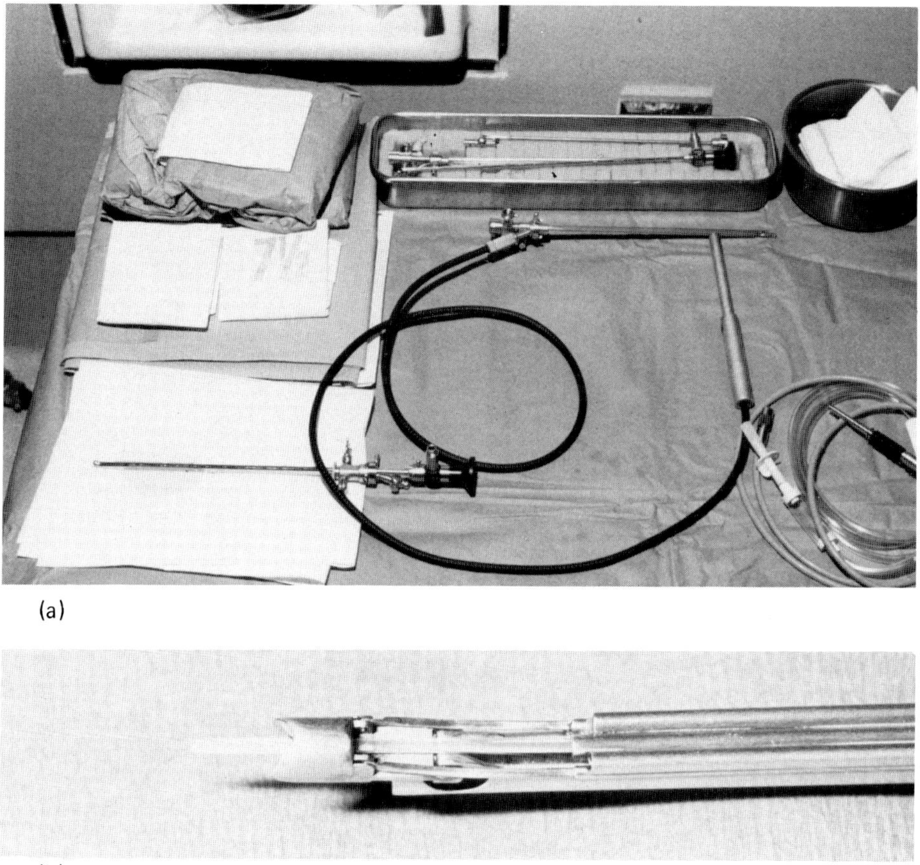

(a)

(b)

Figure 3 (a) Laser endoscopy equipment with Storz urethroscope (Karl Storz, Inc., Culver City, California) and flexible quartz fiber. (b) Close-up of quartz laser fiber and specially designed Storz Albarran bridge.

basis, and in high-risk patients it may also be carried out with the patient under local or regional anesthesia. Because hemostasis is excellent, an indwelling Foley catheter is usually not needed thus reducing the incidence of infection and troublesome bladder spasm. Hospital stays are shorter, and therefore the procedure is cost-effective, an especially important point in the era of cost-containment and diagnosis-related groups.

Pathologically, architectural destruction of the bladder wall is minimal. Even

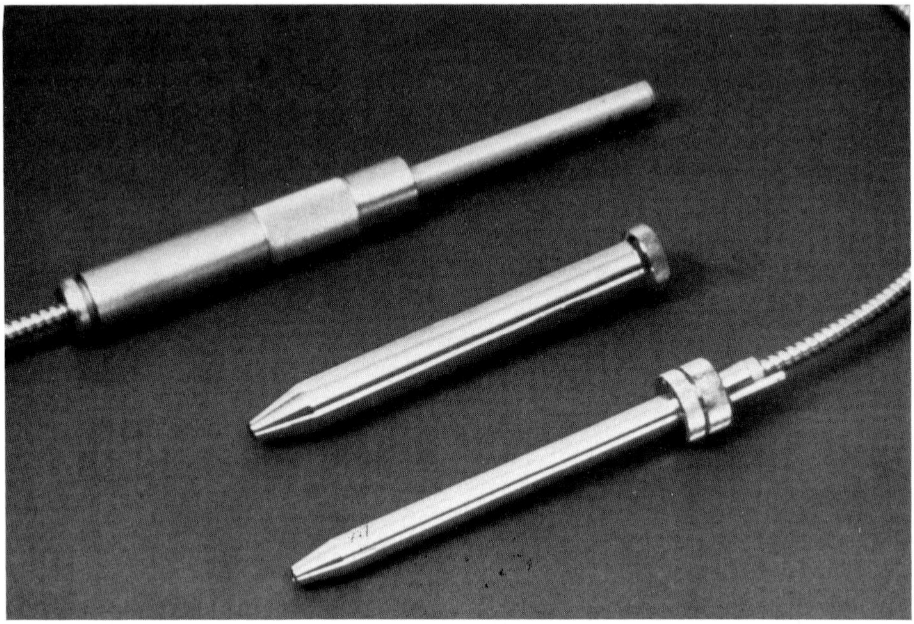

Figure 4 Hand-held laser adapter for use on the external genitalia.

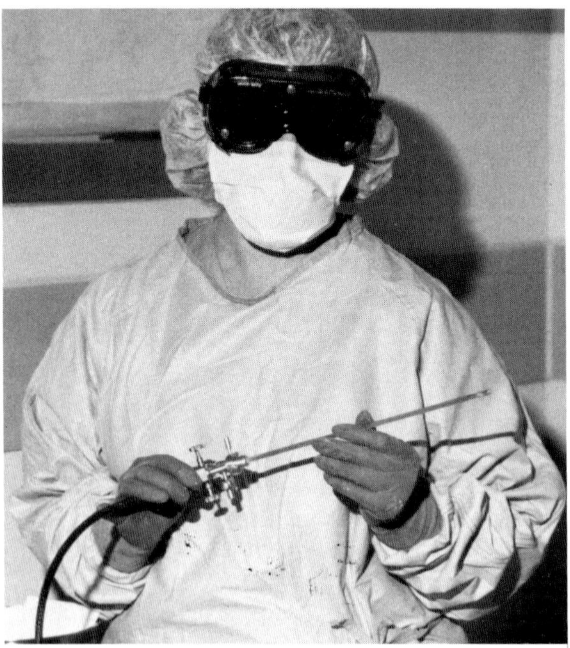

Figure 5 LASER-GARD goggles (Glendale Optical Co., Woodbury, New York) must be worn to protect the eyes.

with transmural necrosis, the elastic elements of the bladder remain intact and form the supporting system that permits regrowth or regeneration of detrusor muscle. This property also reduces the frequency of bladder perforation, so common with a resectoscope, and the potentially devastating complication of extravesical seeding of malignant cells.

In addition, this no-touch technique, without contact, has the theoretical advantage of preventing tumor seeding after electroresection, as suggested by Soloway (8). Other theoretical advantages include early lymphatic sealing around the tumor, decreased incidence of local recurrence, and a deep coagulation effect found to be curative in some patients with T_1 and T_2 lesions.

Certain disadvantages associated with use of the Nd:YAG laser should be mentioned. Several instances of perforation of the bladder and bowel were reported in the early experience of Pensel et al. (9). If the laser beam is focused on one area for a prolonged period it may cause necrosis of the bladder wall and adjacent bowel. Hofstetter and Böwering (personal communication, 1983) had one instance of perforation of the bowel that required resection. Another problem that occasionally arises is that some areas of the bladder are not accessible to the laser beam, such as the section behind the bladder neck at the 12 o'clock position and lesions near the bladder neck in a patient with a large obscuring prostate gland.

IV. BLADDER TUMORS

Bladder tumors are staged according to their depth or penetration into the bladder wall (Fig. 6). Bladder tumors up to 3 cm in size have been treated effectively by the Nd:YAG laser. This laser produces a homogenous bandlike necrosis from the mucosa of the bladder to the serosa with little appreciable effect on the stability of the connective tissue of the bladder wall and little likelihood of perforation of the bladder. Typical dosimetry is 40 W of power applied in 4- to 5-sec pulses in a paintbrushlike fashion such that one area is not overly irradiated. The laser is directed so that the tumor area is treated horizontally and then vertically in a continuous motion, dispersing the power applied in a uniform rather than concentrated manner in one precise area for a prolonged period. This produces a whitish conical area of necrosis 3-5 mm in depth (Fig. 7). In tumors larger than 3 cm, electroresection removes the bulk of the superficial portion of the tumor, permitting deep muscle biopsy of the base with the cold-cut biopsy forceps and subsequent laser irradiation of the base to prevent recurrent tumor and permit treatment of the deeper muscularis.

The most extensive use of the Nd:YAG laser for treatment of bladder tumors is from the pioneering group of Hofstetter and associates (3). Their series includes 302 bladder tumors treated between 1976 and 1979. An additional 196 tumors were treated by a combination of electroresection and laser irradiation of

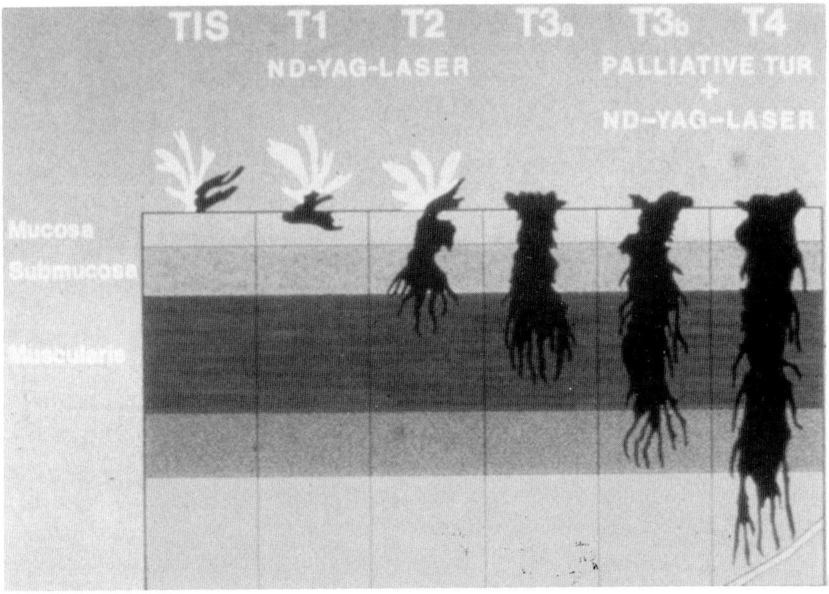

Figure 6 Staging of bladder tumors: TIS, carcinoma in situ; T_1, invasion of lamina propria; T_2, invasion of superficial muscle; T_{3a}; invasion of deep muscle; T_{3b}, invasion of serosa; T_4, distant metastasis (*Source*: Alfons Hofstetter and Frank Frank, Urologischen Abteilung, University of Munich, Munich, West Germany).

the base of the tumor. In a follow-up period of 2 months, the local recurrence rate was 9% in patients treated with a rigid endoscopic laser and 1% with the flexible fiberoptic light guide and the Albarran bridge now in use. Unfortunately, no other results are available. Okada et al. (10) have also reported satisfactory results treating bladder tumors with the Nd:YAG laser.

From September 1983 to December 1984 we have treated 86 patients with more than 200 discrete bladder tumors with the Nd:YAG laser, and we have used the laser to treat neoplasms of the bladder, condyloma acuminatum, stricture of the urethra, contracture of the bladder neck, erythroplasia, and bleeding from interstitial and radiation cystitis. Of the 86 patients treated, 64 had various forms of bladder cancer. Patients with carcinoma in situ, T_a (papillomas), T_1, T_2, and T_3 bladder cancers whose disease failed to respond to electroresection and intravesical chemotherapy were treated with the Nd:YAG laser. Of the 64 patients, 55 have had a foll v-up period longer than 3 months, and 33 of the patients have been followed up for 12 months or more. The local recurrence rate of carcinoma

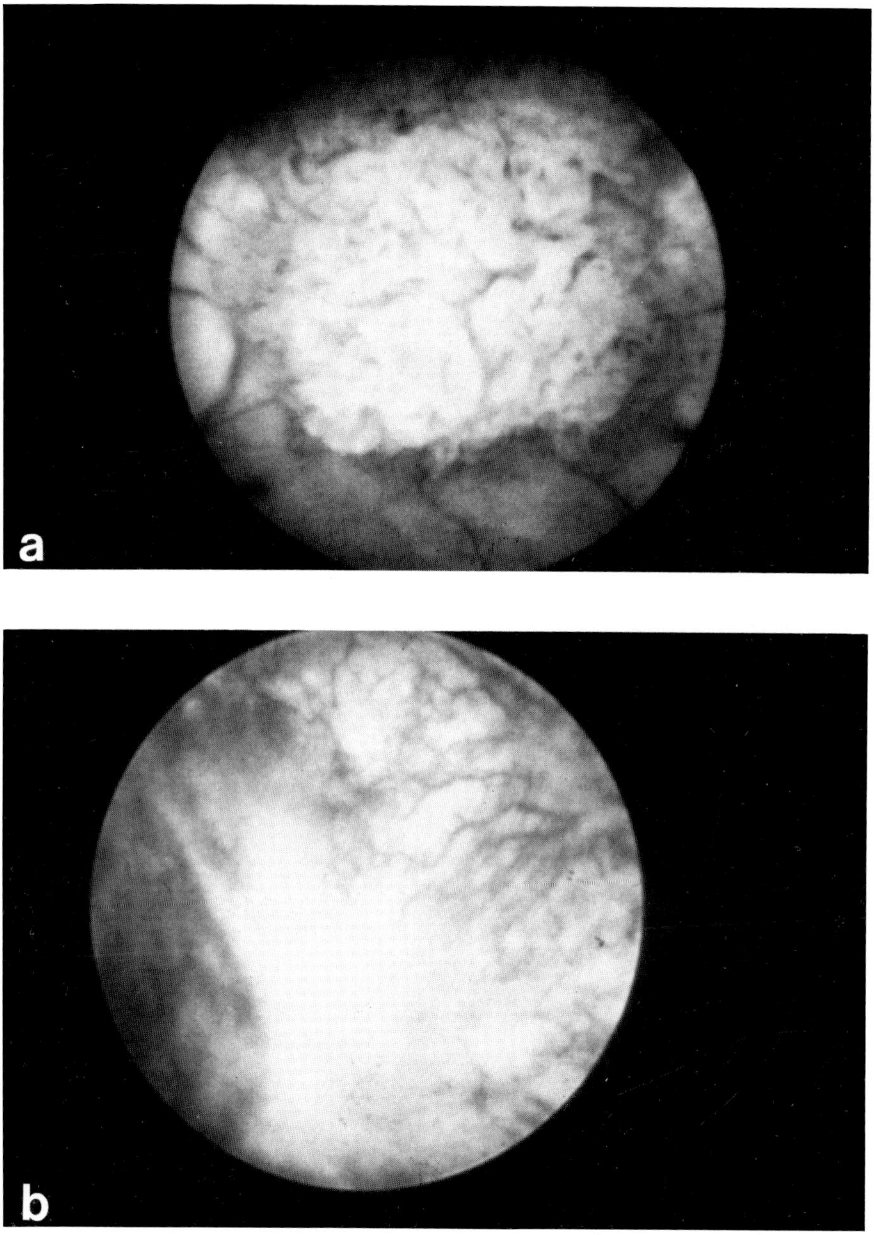

Figure 7 (a) Bladder tumor before laser treatment. (b) Coagulation necrosis and scarring 6 weeks after laser irradiation of bladder tumor (endoscopic views).

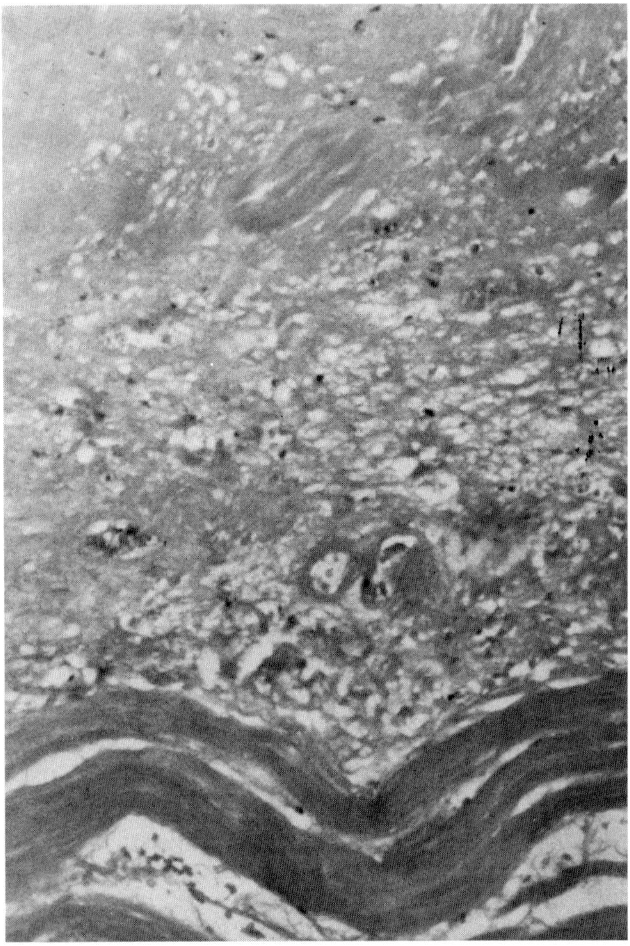

Figure 8 Microscopic section of invasive T_2 transitional cell carcinoma of the bladder demonstrating effect of laser irradiation (transmural necrosis) without evidence of residual tumor (hematoxylin and eosin; original magnification, 100X).

in situ is 0%; T_a (papillomas), 21%; T_1 invasive bladder cancer, 36%; and T_2 invasive bladder cancer, 0%. Patients with T_2 and T_3 cancer of the bladder were treated as an unusual subset.

Ten patients with T_2 and T_3 cancer of the bladder underwent electroresection followed by laser irradiation of the base of the tumor. Four to six weeks

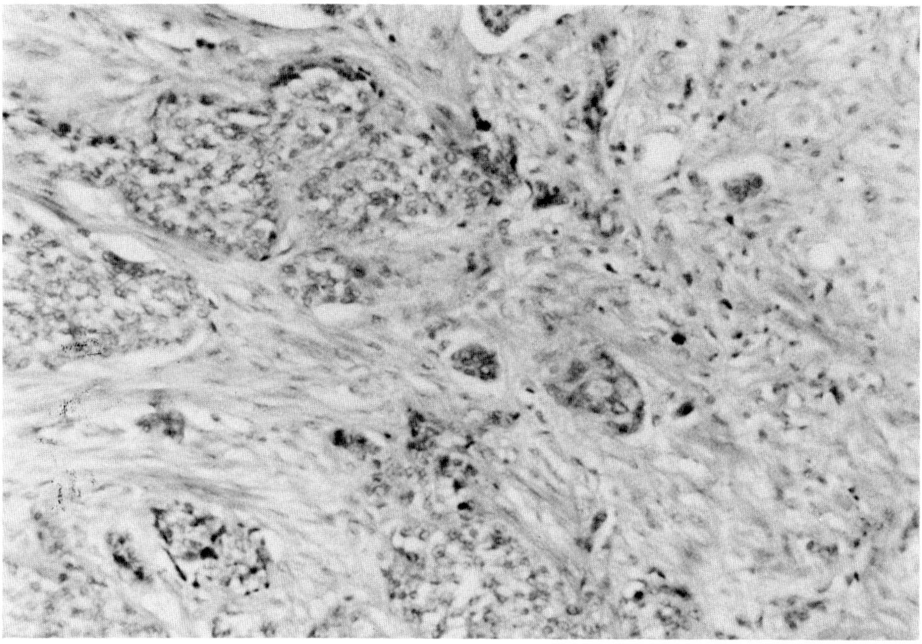

Figure 9 Microscopic section of invasive T_3 transitional cell carcinoma of the bladder demonstrating effect of laser irradiation on deep residual tumors in areas adjacent to irradiated region (hematoxylin and eosin; original magnification, 100X).

after this therapy, these patients underwent standard cystectomy and ileal conduit. In five of the 10 patients, disease was downstaged to P_0, and they were free of residual tumor as a result of the laser irradiation therapy (Fig. 8). Five patients with T_{3a} and T_{3b} cancers still had evidence of residual tumor in the region irradiated by the Nd:YAG laser (Fig. 9). These findings suggest that a combination of electroresection and laser irradiation of the base of the bladder tumor may have an important role in the future management of patients with invasive cancer of the bladder, especially if we can improve our ability to detect submucosal extension of bladder tumors so that unseen cancer is not left undetected and untreated by the laser beam. Laser holography, ultrasonography, or computed tomography may possibly be refined to detect unseen malignant extensions into the detrusor muscle.

V. TRANSITIONAL CELL TUMORS OF THE UPPER TRACTS

The Nd:YAG laser with its fiberoptic fiber can be used to treat transitional cell tumors of the ureter and the renal pelvis through a ureteroscope or nephroscope. Böwering (unpublished data, 1984) has reported eight cases of ureteral tumors treated through a ureteroscope, and Malloy (unpublished data, 1985) has presented two patients with localized transitional cell carcinoma in a solitary kidney who were treated with the Nd:YAG laser at the time of partial nephrectomy.

We have treated one patient with transitional cell carcinoma in a solitary kidney by partial nephrectomy. Three months later ureteral and bladder papillomas were treated satisfactorily with the Nd:YAG laser through the ureterorenoscope with good results up to 10 months. Laser treatment is obviously not meant to replace standard radical surgery at this juncture, but in patients with a solitary kidney and low-grade ureteral lesions, laser surgery may be a preferable and viable alternative to chronic dialysis if follow-up studies in these patients show limited recurrence rates.

VI. URETHRAL STRICTURES

The Nd:YAG laser has been employed in the treatment of patients with urethral strictures with satisfactory short-term results. The strictures are vaporized by the laser beam rather than cut as with the conventional cold-knife urethrotomy. Our initial results in five patients with a urethral stricture and contracture of the bladder neck are disappointing, and therefore we cannot recommend laser treatment for these entities at the present time. Whether or not results of other investigations will prove the laser to be superior to optical cold-knife urethrotomy awaits long-term follow-up. Although results with the laser technique in the management of patients with urethral strictures are unsatisfactory, our results have been good in patients with intraurethral condyloma acuminatum. This has been especially helpful in immunosuppressed transplant recipients who have a propensity for the development of urethral condylomas.

VII. EXTERNAL APPLICATION

External uses of the Nd:YAG laser have been effective in patients with penile carcinoma, malignant melanoma, and condyloma acuminatum. The hand-held device has an adapter that enables blood to be blown away from the operative field. These procedures can be performed on an ambulatory basis and usually with the patient under local (penile block) or general anesthesia.

Hofstetter (7) reported 17 cases of squamous cell carcinoma of the penis in T_1 and T_2. Seven patients who had no evidence of metastatic disease were followed up from 10 to 24 months and were free of recurrent disease. The Nd:

YAG laser may provide curative or palliative treatment for patients with penile carcinoma. Patients with early T_1, N_0, or M_0 lesions are candidates for curative laser surgery; none of Hofstetter's patients in this group experienced local relapse.

Of 30 patients with condyloma acuminatum treated by laser surgery, only two had recurrent lesions. The hand-held device can also be used to treat urethral caruncles and condyloma of the vulva.

VIII. PARTIAL NEPHRECTOMY

In our laser laboratory, we have found that the Nd:YAG laser may be helpful in performing partial nephrectomy. Clinically this may become relevant in patients with renal cell carcinoma in a solitary kidney. The laser would irradiate tumor cells in the area of enucleation, seal lymphatics, and enhance hemostasis.

IX. SUMMARY

At this early juncture, the Nd:YAG laser has already demonstrated its clinical efficacy and safety and is thus far the most efficient laser for overall use in urological surgery. We have not encountered any complications as a result of its use, and our encouraging early results are the motivating force for our continued investigation of the Nd:YAG laser. Prospective randomized studies now under way in Germany and the United States will determine its future role in urological surgery.

REFERENCES

1. Staehler G, Hofstetter A, Frank F, et al.: Cited by Hofstetter and Frank (Ref. 7).
2. A. Hofstetter and F. Frank. *The Neodymium-YAG Laser in Urology*. Editiones Roche, Basel, Switzerland, 1980.
3. Hofstetter A, Frank F, Keiditsch E, and Böwering R: Endoscopic neodymium-YAG laser application for destroying bladder tumors. *Eur. Urol.* *7*:278-282, 1981.
4. Hofstetter A, Staehler G, and Keiditsch E: Experimentelle Erzeugung von malignen Tumoren an der Kaninchen-Harnblase. *Fortschr. Med.* *95*:346-347, 1977. (Eng. abstr.)
5. Staehler G, Hofstetter A, Gorisch W, Keiditsch E, and Müssiggang M: Endoscopy in experimental urology using an argon-laser beam. *Endoscopy* *8*:1-4, 1976.
6. Bülow H, Bülow U, Levine S, Wurster H, and Frohmüller H: Zum gegenwärtigen Stand der transurethralen Laser-Technik in der Behandlung der Harnröhrenstriktur. *Urologe* [Ausg A] *20* (suppl):328-332, 1981. (Eng. abstr.)

7. A. Hofstetter and F. Frank. Laser use in urology, in *Surgical Application of Lasers* (J. A. Dixon, ed.). Year Book Medical Publishers, Chicago, 1983, pp. 146-162.
8. Soloway MS: Rationale for intensive intravesical chemotherapy for superficial bladder cancer. *J. Urol. 123*:461-466, 1980.
9. Pensel J, Hofstetter A, Frank F, Keiditsch E, and Rothenberger K: Temporal and spatial temperature profile of the bladder serosa in intravesical neodymium-YAG-laser irradiation: An investigation to determine the therapeutic range of a new process for the treatment of bladder tumors. *Eur. Urol. 7*:298-330, 1981.
10. Okada K, Asoaoka H, Amagai T, Onoe Y, and Kishimoto T: Transurethral neodymium:YAG laser surgery for bladder tumors. *Urology 20*:404-407, 1982.

15

Laser Surgery in Cardiovascular Disease

JEFFREY M. ISNER
Tufts Medical School and New England Medical Center, Boston, Massachusetts

RICHARD H. CLARKE
Boston University and Tufts Medical School, Boston, Massachusetts

I. INTRODUCTION

The heart, coronary arteries, and peripheral arteries represent the ultimate development of a complex anastomotic network that originated embryologically as a hollow tube. In this respect, the cardiovascular system lends itself well to the concept of laser endoscopic surgery (Fig. 1). The cardiovascular system is sufficiently unique, however, to require major modifications of previously developed laser endoscopic surgical techniques. In particular, the diminutive luminal diameter of coronary and peripheral arteries, the fact that these vessels are blood filled and that the coronary arteries, as well as the heart, are moving 60-100 times per minute have forced investigators to consider a variety of technological innovations to establish cardiovascular laser endoscopic surgery as a viable treatment modality. Accordingly, this chapter will discuss the manner in which laser technology has thus far been tailored to accomplish in vitro, and in selected cases in vivo, cardiovascular endoscopic laser surgery.

II. LASER ANGIOPLASTY

A. Background

Expiation of atherosclerotic plaque from the coronary or peripheral arterial lumen represents the cardiovascular application of laser therapeutics that best exploits the unique properties of laser light. Cardiovascular disease is the most widespread cause of death in this country annually (1); almost all of these deaths are due to coronary heart disease. Although the pharmaceutical armamentarium available to patients with coronary disease has expanded greatly over the past 10

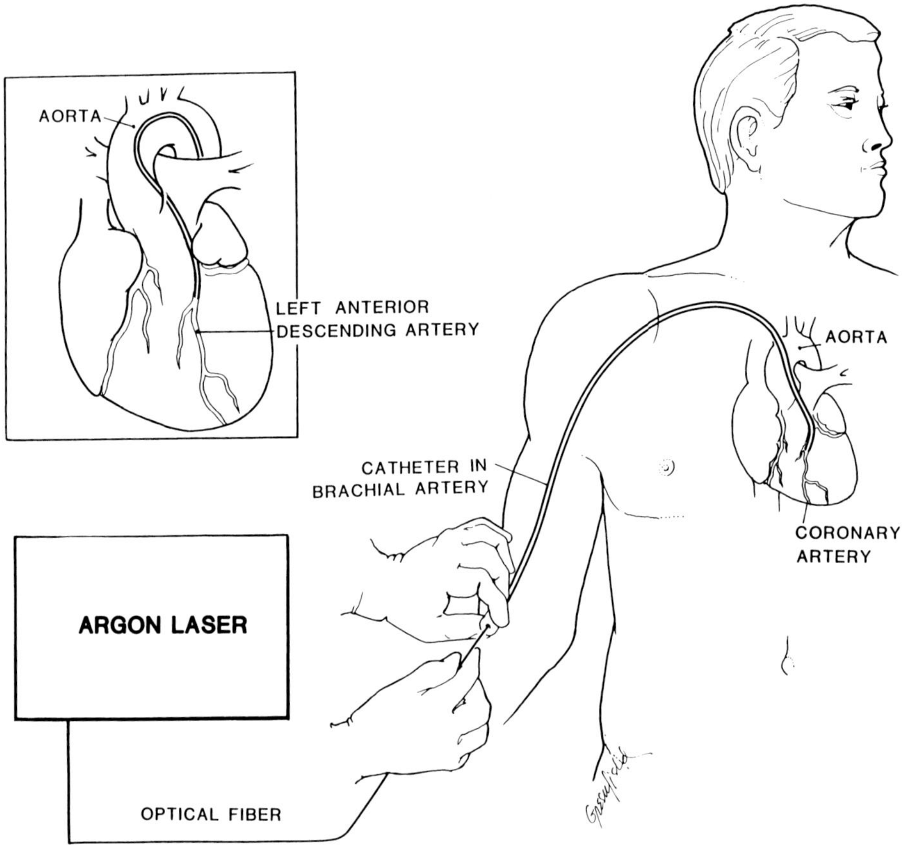

Figure 1 Schematic diagram of percutaneous manner in which laser coronary angioplasty might be ultimately performed. Optical fiber, coupled to, in this particular case, an argon ion gas laser; fiber is delivered to appropriate coronary artery via specially designed catheter delivery system.

years, nearly 200,000 patients each year continue to require definitive intervention. A few patients may be treated successfully with balloon angioplasty; for most patients, however, definitive intervention means coronary artery bypass surgery. If laser coronary angioplasty could be suitably developed as a percutaneous alternative to bypass surgery, the savings to patients would be considerable. At the very least, the patient would be spared the adverse effects of a thoracotomy. The cost of such a procedure would be substantially less than the current cost (20,000-30,000 dollars) of bypass surgery; additional savings would be expected as the result of reduced hospital stay, use of catheterization rather than operative facilities, and possibly, earlier return to work.

Even if the technical difficulties involved in developing a percutaneous catheter delivery system were to limit laser coronary angioplasty to intraoperative application, the technique would still represent a major improvement in definitive intervention. It could be argued, in fact, that laser coronary angioplasty constitutes the only intervention yet developed that is in any sense definitive. Bypass surgery literally "bypasses" the offending stenosis; long-term sequelae include premature occlusion of the bypassed stenosis (2) and the predictable attrition of bypass graft patency (3). Balloon angioplasty simply redistributes the atherosclerotic stenosis (4); consequently, the frequency of restenosis in the first year after angioplasty may be as high as 30% (5) and increases further both as a function of the number of lesions treated and the time after angioplasty (6). Debridement of atherosclerotic plaque by laser would presumably result in improved long-term sequelae.

Finally, two important subsets of patients exist for whom definitive therapeutic intervention is not an option: those patients in whom atherosclerotic narrowing of the coronary arteries is so diffuse that bypass surgery is precluded, and those who have undergone extensive revascularization, but who now represent with both a diffusely diseased native circulation and occluded or diffusely diseased bypass grafts, or both. For these patients, intraoperative application of laser coronary angioplasty would constitute a new and unique therapeutic option.

B. Results of In Vitro Studies

The demonstration in vitro that laser irradiation could vaporize atherosclerotic plaque represents an important hallmark in the history of attempts to treat coronary arterial narrowing because *no other* potential percutaneously transmissible instrument has even been shown to be capable of physically eliminating atherosclerotic plaque.

Macruz et al. (7), Lee et al. (8), Abela et al. (9), and Choy et al. (10) used atherosclerotic coronary arterial segments removed from necropsy hearts to demonstrate that atherosclerotic plaque could be vaporized by any one of three continuous wave (CW) lasers–argon, neodymium:yttrium-aluminum-garnet (Nd:YAG), or carbon dioxide (CO_2). Similar results were demonstrated using pulsed lasers, including the excimer laser, in our (11-13) and in others' (14) laboratories. Atherolysis in these in vitro experiments was accomplished with relatively low energy (typically less than 4.0 W and less than 20 sec cumulative exposure) and without regard for the qualitative composition of the atherosclerotic plaque.

As with most lasers previously employed for biomedical purposes (15), the mechanism by which plaque vaporization is achieved with CW laser irradiation is thermal. This is supported both by the consistent finding of gross, light micro-

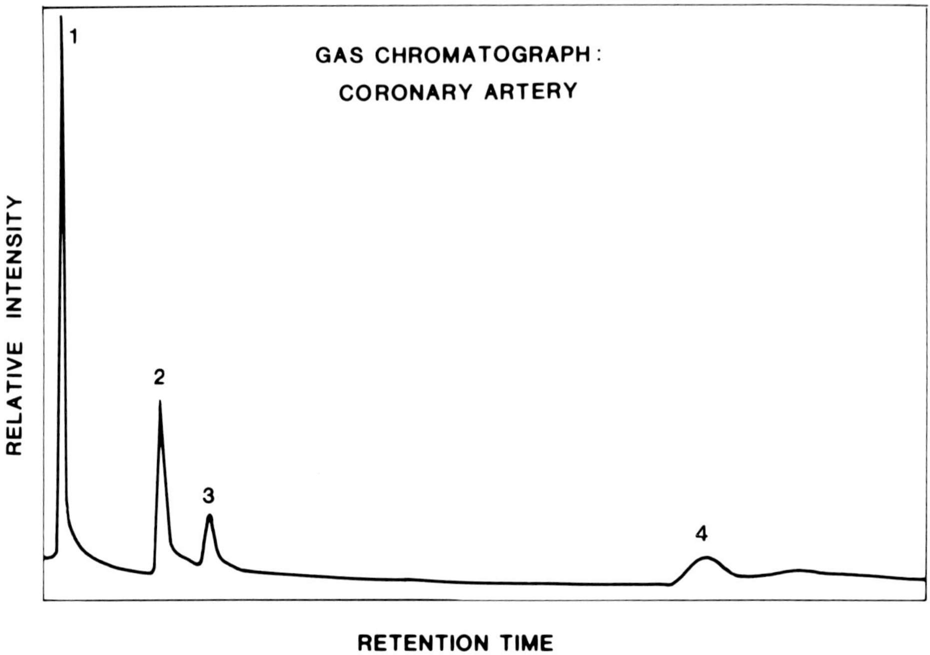

(a)

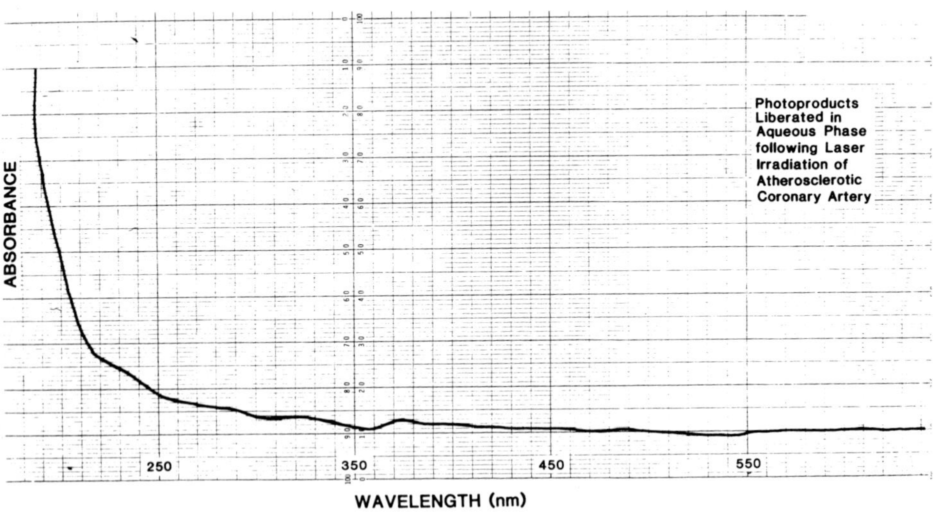

(b)

Figure 2 Photoproducts liberated following CW argon laser irradiation of atherosclerotic coronary arterial segment. (a) Representative results of gas chromatography (GC)/mass spectroscopy include hydrogen (1), ethylene (2), ethane (3), and propylene (4). Water vapor and trace amounts of carbon monoxide were detected on a supplemental GC column. (Retention time is in minutes). (b) Representative results of absorbance spectroscopy include distinct absorbance only in the ultraviolet region, typical of small protein fragments.

scopic, and ultrastructural signs of thermal injury of the residual laser-irradiated surface and analysis of photoproducts (Fig. 2) generated by CW argon laser irradiation of calcified and noncalcified human plaque and of foam-cell-type atherosclerotic rabbit plaque (16).

In our laboratory, we have attempted to extend the in vitro findings just described to a simulated intraoperative model of laser coronary angioplasty, using intact hearts obtained in a fresh, unfixed state at the time of postmortem examination (17). Following selective postmortem (prelaser) arteriography, simulated intraoperative laser coronary angioplasty was performed using a 4-F guiding catheter and 240 μm (200 μm core) quartz optical fiber. The cladding was stripped from the distal 2 cm of the optical fiber. A Tohey-Borst adapter at the proximal end of the catheter was used to position the optical fiber tip in relation to the tip of the guiding catheter. The fiberoptic/guiding catheter was then advanced into the extramural coronary artery as illustrated in Fig. 3. The laser used was in every case a 20-W argon ion gas laser (Innova Model, Coherent, Inc., Palo Alto, California). Its energy profile consisted of 1.0-3.0 W power, delivered intermittently in 2-5 sec bursts, with cumulative exposure times of 15 sec to 3.5 min. The coronary artery was continuously perfused with saline through the side-port of the Tohey-Borst adapter during both antegrade transmission of the fiberoptic/guiding catheter and administration of argon laser irradiation. This saline perfusate accomplished two objectives. First, it distended the coronary arteries during manual antegrade catheter movement, obviating spurious mechanical perforation of the vessel wall caused by vessel wall collapse. Second, it prevented accumulation of debris at the tip of the fiberoptic/guiding catheter during laser irradiation.

After completion of the simulated intraoperative laser coronary angioplasty, postmortem, postlaser arteriograms were obtained, and the coronary arteries were prepared for histological examination. Simulated intraoperative laser coronary angioplasty was performed according to this protocol at a total of 53 sites in the 17 specimens.

In 18 of 53 attempts, the right coronary artery was the site of attempted laser angioplasty. Perforation (Fig. 4) complicated the attempted angioplasty in 13/18 (72%) cases. Of the 13 perforations, 11 were thermal and two mechanical. Of the 11 thermal perforations, eight occurred at sites of extensive calcific deposits (Fig. 5), and two involved branch points. One of two mechanical perforations involved a branch point. Extensive tortuosity of the proximal right coronary artery contributed to mechanical perforation in one case, thermal perforation in one case, and prevented laser angioplasty altogether in one case (Fig. 6).

In 23 cases, the left anterior descending coronary artery was the site of attempted laser angioplasty. Perforation occurred in 11 (48%) of the 23; all were thermal perforations. In nine cases, the perforation site was heavily calcified,

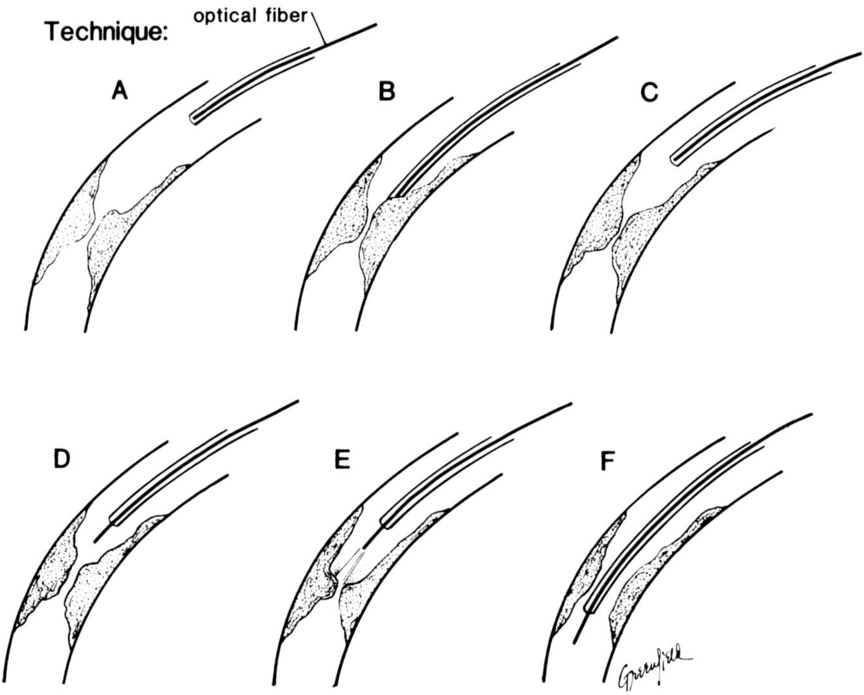

Figure 3 Technique used for simulated intraoperative laser coronary angioplasty. (A) The tip of the optical fiber (OF) is flush with the tip of the guiding catheter (GC) as both are advanced antegrade from the coronary ostia through the coronary lumen; (B) GC/OF advancement is discontinued at the site of an atherosclerotic lesion; (C) the GC/OF is withdrawn slightly so that it is now proximal to the lesion; (D) the tip of the OF is advanced out of the GC, and then locked in place. (E) administration of laser irradiation is initiated and as vaporization of the atherosclerotic lesion is accomplished, the GC/OF traverses the previously stenotic site (F).

and in nine cases, the perforation involved a branch point (Fig. 7). In one case, laser angioplasty was attempted in the diagonal branch and resulted in a thermal perforation unrelated to calcific deposits or anatomical peculiarities.

In 11 cases, the left circumflex or major left obtuse marginal branch of the circumflex was the site of attempted laser angioplasty. Perforations occurred in eight (73%) of 11 cases, including six thermal and two mechanical perforations. Calcific deposits were observed at the site of four of the six thermal perforations. Two of the thermal perforations involved branch points, as did one of the two mechanical perforations. In two cases, excessive tortuosity contri-

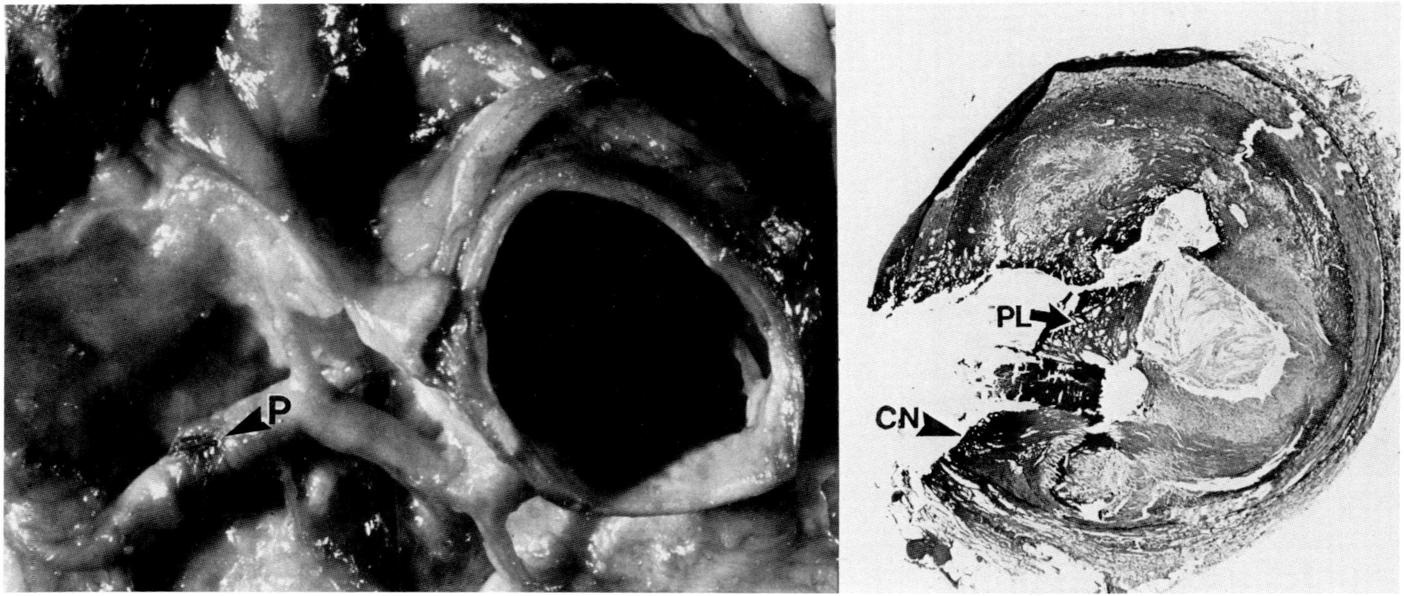

Figure 4 Gross and light microscopic appearance of thermal perforation (P). Grossly apparent charring is observed at P site; light microscopy discloses the histological hallmarks of thermal injury, coagulation necrosis (CN) and polymorphous lacunae (PL). (Verhoeff's elastic tissue stain, 4X).

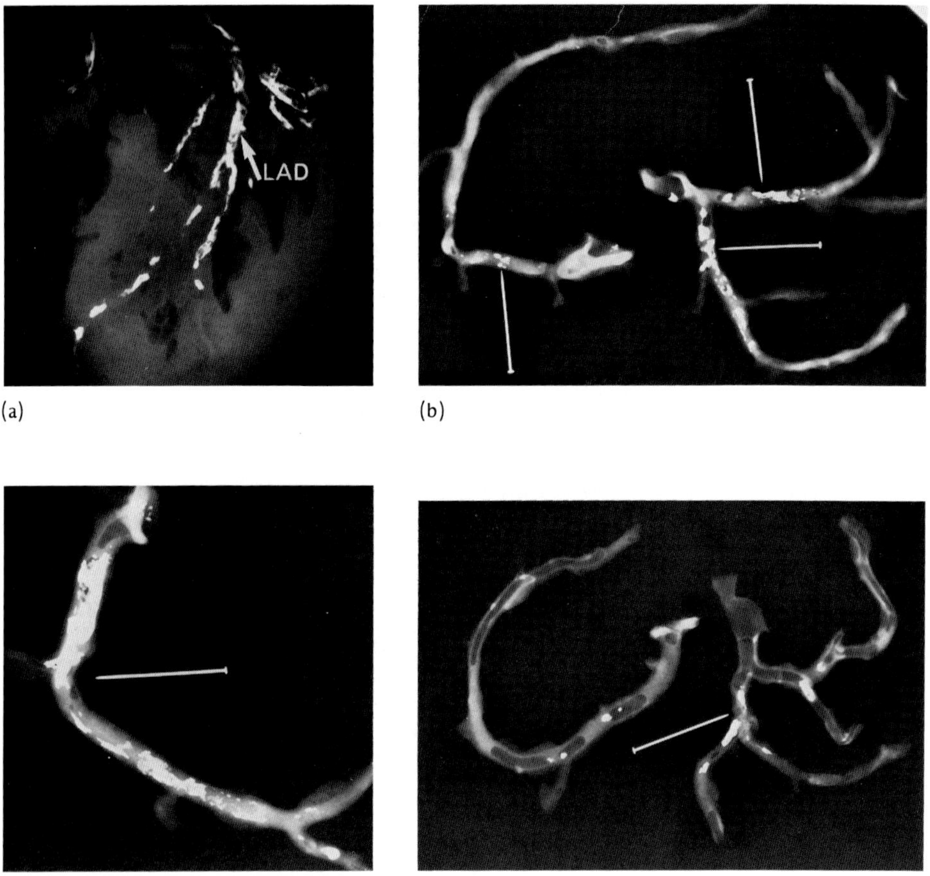

(a) (b) (c) (d)

Figure 5 Postmortem radiographs of coronary arteries in which perforation complicated simulated intraoperative laser coronary angioplasty. Shown in top left is postmortem, in situ (a), *pre*laser radiograph performed *without* injection of contrast media: extensive calcific deposits in the left anterior descending coronary artery (LAD) predisposed to thermal perforation. Remaining three photographs show excised coronary arteries following laser angioplasty complicated by perforation (pin). In each, the site of perforation was heavily calcified.

buted to mechanical perforation, and in one case, excessive tortuosity prevented attempts at laser coronary angioplasty.

In all, perforation complicated laser coronary angioplasty in 33 (62%) of 53 cases. Most perforations (29) were thermal; four were purely mechanical. Perforation sites were characterized by extensive calcific deposits in 21 (64%) of 33 cases and the origin of a side branch in 13 (39%) of 33. Excessive tortuosity of the extramural coronary artery contributed to arterial perforation in four cases, and in two other cases precluded attempts to perform laser angioplasty.

Experience with the postmortem model has indicated that even when access problems associated with percutaneous laser angioplasty are obviated by simulated intraoperative laser coronary angioplasty, perforation of the underlying coronary arterial wall continues to represent the "rate-limiting" complication of laser angioplasty. In this particular model of laser coronary angioplasty, most perforations occurred in relation to calcific deposits, branch points, and to a lesser extent, tortuous segments of the extramural coronary arteries.

Our interpretation of the manner in which these items result in perforations during attempted laser coronary angioplasty is illustrated in Figs. 8-11. Branch points and tortuous coronary segments, present in this series in over 40% of perforation sites, lead to mechanical perforations when they prevent the fiberoptic/guiding catheter from being coaxial with the artery. When a branch point is the site of an atherosclerotic lesion, the lesion may cause the fiberoptic/guiding catheter to be deflected away from the coronary lumen and into the coronary arterial wall (see Fig. 8). Branch-point atherosclerotic lesions may also predispose to thermal perforations: if the fiberoptic penetrates the lesion via a "track" that preferentially leads to the branch point vertex rather than either of the two intersecting coronary lumens, thermal perforation may result.

Plaque composition also appears to represent an important determinant of the potential for perforation. Nearly one-third of all the perforations observed in the present study occurred at sites of heavy calcific deposits. Analysis of both the postmortem radiographs and histological sections of laser-perforated coronary arteries in this series suggested that during laser irradiation, the fiberscope preferentially "tracks away" from the calcified, severely fibrotic portions of the plaque, toward the "softer" portions of the plaque, such as foci of yellow pultaceous debris (see Fig. 8). These sites constitute a potential path of least resistance, which not infrequently promotes eccentric fiber penetration through plaque into the thin, outer adventitia, resulting in perforation. This problem is compounded by the fact that the principal component of the "normal" arterial wall—the medial smooth muscle—is characteristically depleted or attenuated in segments of atherosclerotic coronary arteries (see Fig. 9) (18). Thus, a limited margin for error exists between the target of vaporization (i.e., plaque) and the underlying arterial wall. In fact, extensive plaque vaporization of an athero-

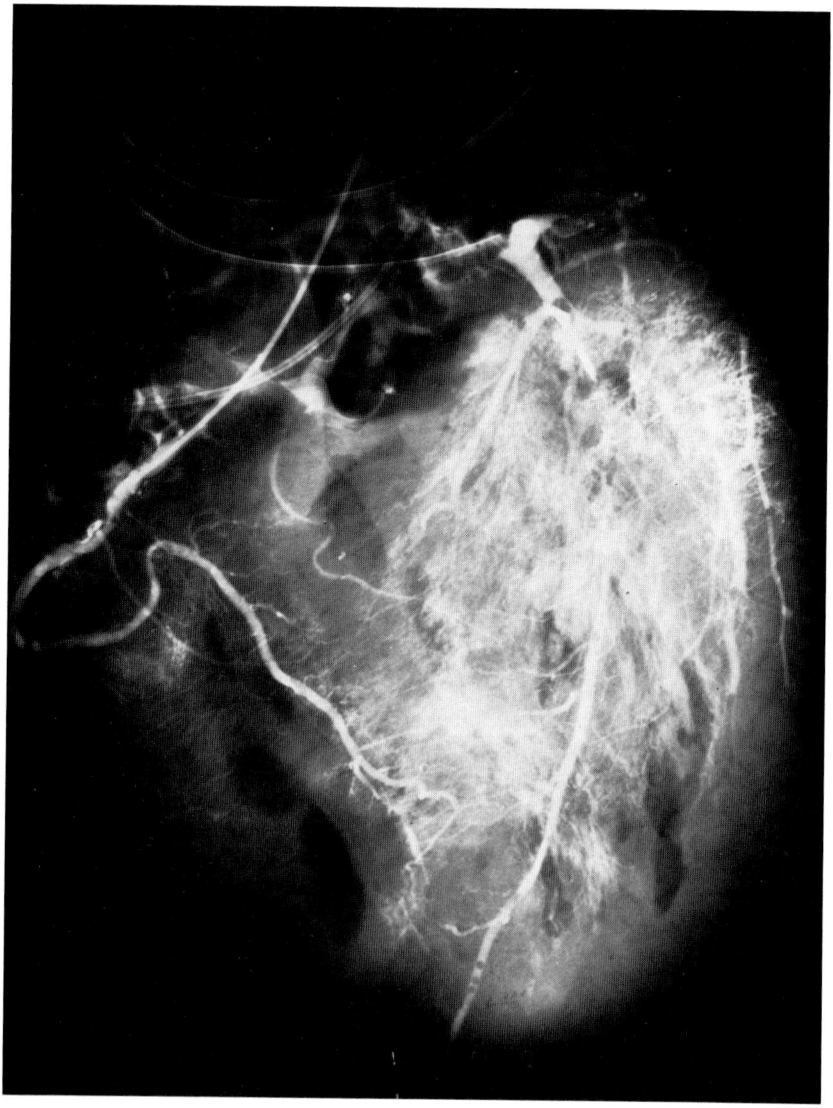

(a)

Figure 6 Excessive tortuosity of right coronary circulation in postmortem. (a) Prelaser arteriogram. (b) Left obtuse marginal artery in two different specimens precluded attempts to align guiding catheter/optical fiber in a coaxial manner.

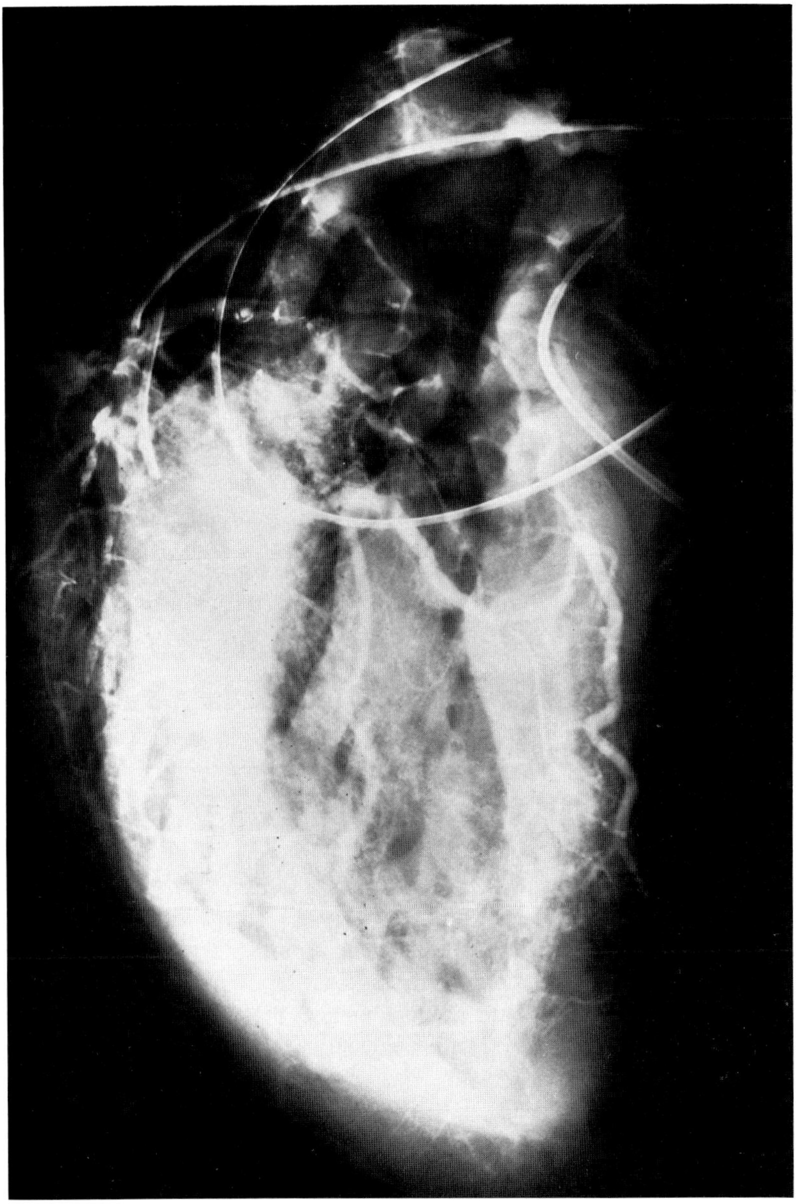

(b)

Figure 6 (continued)

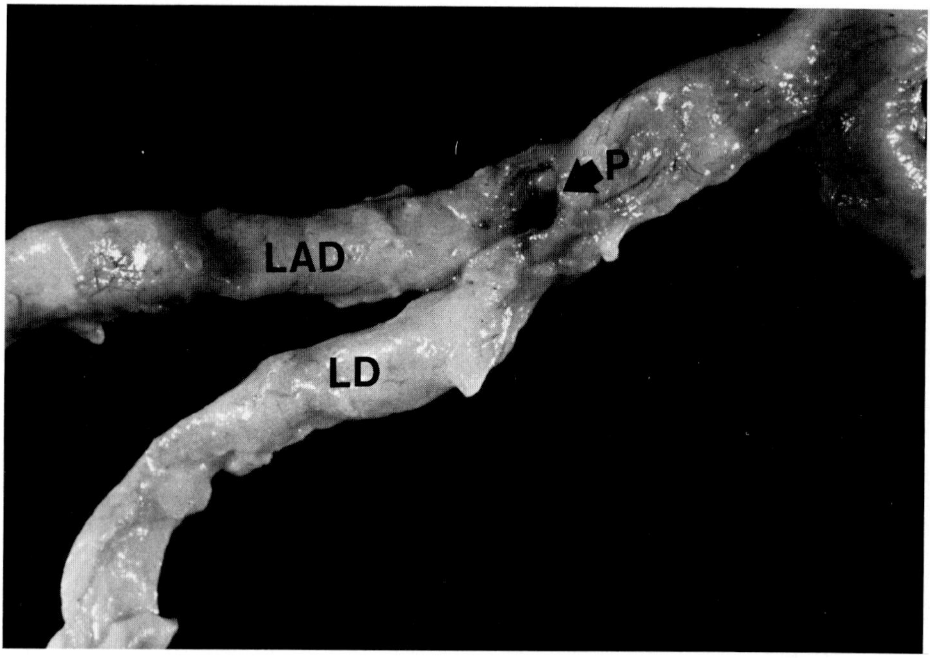

Figure 7 Thermal perforation (P) during attempted laser coronary angioplasty of stenotic lesion involving the left anterior descending (LAD) coronary artery. Perforation occurred at a branch point, the origin of the left diagonal (LD) coronary artery.

sclerotic segment devoid of smooth muscle may result in a de facto pseudoaneurysm (see Fig. 10).

Eccentric lumens represent a third factor predisposing to both mechanical and thermal perforations. When atherosclerotic lesions result in greater than 75% reduction in cross-sectional area, the residual lumen is most often eccentric in location. Eccentric lumens, particularly those that are slitlike and adjacent to nonatherosclerotic, nonattenuated, highly absorbent (red) media, appear to contribute to perforations by offering a path of "no resistance" (see Fig. 11).

C. Results of In Vivo Studies

In vivo studies performed both in experimental animal models, and to a more limited extent in humans, have confirmed the two principal findings demonstrated in vitro: the results of laser atherolysis may be accomplished in vivo with an energy requirement only slightly greater than that required in vitro; and perforation remains the rate-limiting complication.

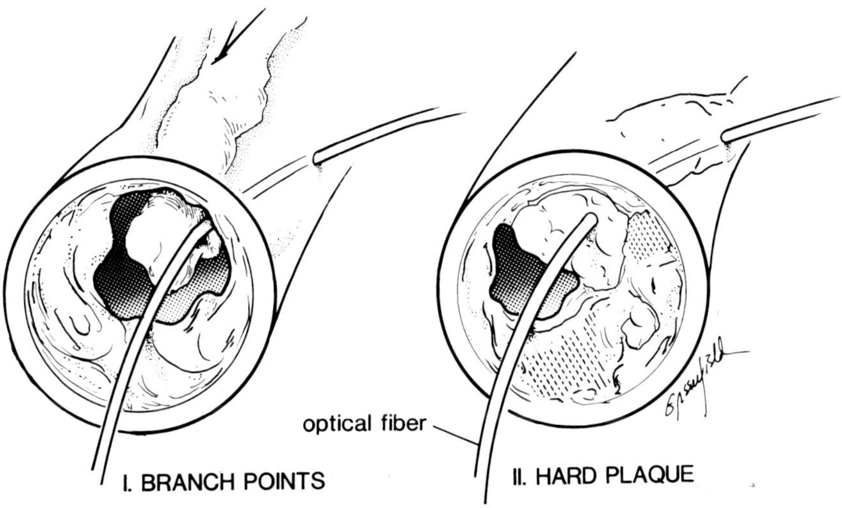

Figure 8 Perforations resulting from laser coronary angioplasty. I. Branch-point lesions predispose to perforations if the fiber penetrates a "track" that preferentially leads to the branch point vertex rather than to either of the two intersecting coronary lumens. II. During laser irradiation of plaque, the fiber appears to preferentially "track away" from the calcified, severely fibrotic portions of the plaque, and burrow through sites of pultaceous debris. The latter, if adjacent to the adventitia, may predispose to perforation.

Abela et al. (19), for example, used CW argon (2-35 J) and Nd:YAG (36-75 J) lasers to irradiate atherosclerotic rabbit arteries. The severity of iliofemoral stenoses in eight of 15 arteries was diminished from 78 ± 18% to 32 ± 11% (mean ± SD). In one additional artery in which stenosis was diminished from 45 to 25%, arterial perforation occurred. Perforation complicated attempts to perform laser angioplasty in six other arteries as well without obvious improvement in severity of stenosis.

Gerrity et al. (20) used a nonfocused beam from a CW CO_2 laser to irradiate segments of a swine atherosclerotic descending thoracic aorta isolated with tandem cross-clamps. The energy employed was always less than 10 J, consisting of exposure times of 0.1-0.5 sec and powers of 1-20 W. At 2 days, 2 weeks, or 8 weeks, postoperatively, the animals were sacrificed and the sites of laser irradiation examined by light and ultrastructural microscopy. Two days postoperatively, platelet fibrin thrombi lined the laser crater but did not protrude above the level of adjacent endothelium. Two weeks postoperatively, the depressed crater surface was mostly reendothelialized. At 8 weeks postoperatively, the crater remained depressed (and therefore less stenotic) and a fibrous cap had formed over the necrot-

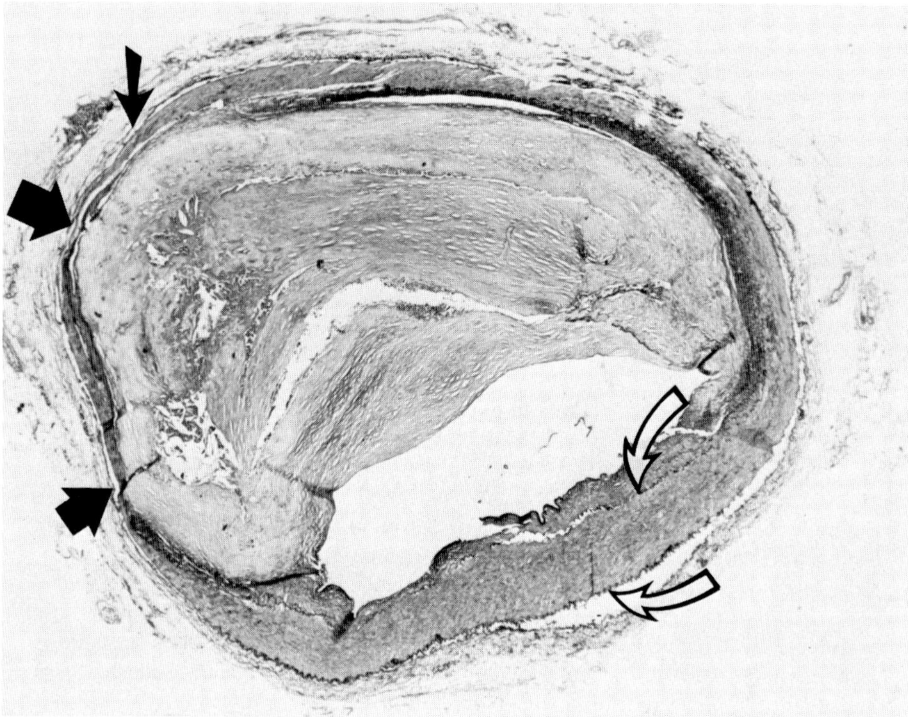

Figure 9 Depletion of smooth muscle from the wall of an atherosclerotic segment of coronary artery. There is severe attenuation of medial smooth muscle (black arrows) in portion of artery underlying atherosclerotic stenosis; media is of normal thickness (white arrows) where atherosclerotic plaque is minimal (Verhoeff's elastic tissue stain, 4X).

ic area. Minimal thrombogenesis and no enhanced accumulation of lipid material were observed despite diet-induced hyperlipidemia.

Carbon dioxide laser irradiation of atherosclerotic plaque was also performed in vivo by Eldar and associates (21). In contrast to the focused-beam irradiation utilized by Gerrity et al. (20), Eldar et al. employed a reportedly nontoxic, flexible optical fiber fabricated by the Department of Applied Physics of the Tel Aviv University. These fibers, 0.1-1.0 mm in diameter and up to 1 meter in length, were produced by extrusion of silver chloride-silver bromide crystals. In three rabbit aortas, atherolysis of varying degrees was accomplished via these fibers, cladded by a loose Teflon tube, and inserted through the right iliac artery. Energy requirements consisted of a combination of 15 W of power applied intermittently in 0.2 sec bursts.

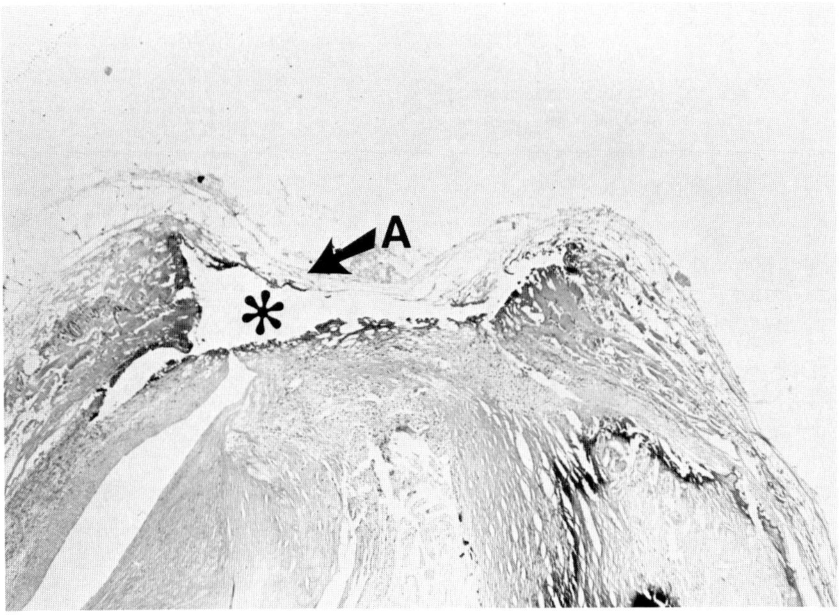

Figure 10 Pseudoaneurysm formation following laser coronary angioplasty. A CW argon ion gas laser was used to perform simulated intraoperative laser coronary angioplasty. The absence of medial smooth muscle at the site of recanalization (asterisk), however, has resulted in a de facto psuedoaneurysm, as the outer wall of the artery at this site now consists solely of a thin layer of adventitia (A) (hematoxylin and eosin, 10X).

Therapeutic attempts to vaporize atherosclerotic plaque in humans have thus far been limited to percutaneous administration of argon and Nd:YAG irradiation to atherosclerotic peripheral arteries and intraoperative irradiation of coronary arteries. Ginsburg and coworkers (22) reported using a continuous wave argon laser, coupled to a 200-μm quartz fiber, delivered through a balloon angioplasty guiding catheter. To minimize the risk of perforation, the optical fiber was advanced across the target stenosis using only 2 W power; higher output (7 W) was administered only during "pullback" through the stenosis.

Geschwind et al. (23) reported similar results in three patients with obstructed femoral or popliteal arteries. A CW Nd:YAG laser, coupled to a 0.2 cm diameter optical fiber, was advanced toward the stenotic lesion using an inflated balloon catheter designed to ensure coaxial placement of the fiberoptic unit within the vessel. During laser irradiation, the fiber tip was advanced beyond the inflated balloon catheter while diluted, cooled blood perfusate was circulated distal to the inflated balloon. Three doses of 12-20 W administered for

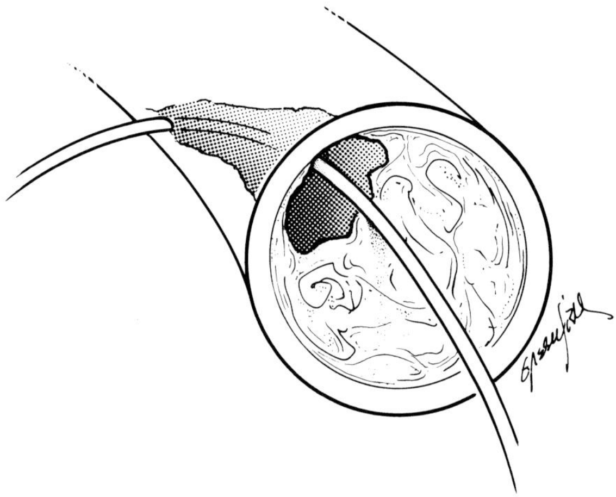

Figure 11 Eccentric lumens, particularly those that are slitlike and adjacent to nonatherosclerotic, nonattenuated, and highly absorbent (red) coronary arterial media, appear to contribute to perforations during argon laser angioplasty by offering a path of "no resistance."

50 sec were required to recanalize two totally occluded arteries. A third stenosis was reduced from 80 to 40% diameter narrowing using three doses at 30 sec. While no perforation resulted, "only a narrow tunnel could be obtained in the totally occluded arteries . . . ."

Finally, intraoperative application of CW argon laser irradiation has been used experimentally in conjunction with coronary arterial bypass surgery. Choy and coworkers (24) employed a CW argon laser (60-1231 J) coupled to a specially designed fiberoptic laser catheter and 85-μm quartz fiber with continuous saline flush. Laser irradiation was applied to a proximal coronary stenosis in conjunction with construction of a conventional bypass graft distal to the same stenosis. In one patient, improved luminal patency was documented 25 days postoperatively. In the remaining four patients, no angiographic improvement in luminal patency was described, including one patient in whom the procedure was complicated by a mechanical perforation.

III. ALTERNATIVE APPROACHES

The results of the previously described in vitro and in vivo investigations have underscored two points. First, laser irradiation represents a potentially unique therapy for the debridement of atherosclerotic stenoses, whether applied intraoperatively or percutaneously. Second, the principal factor limiting routine

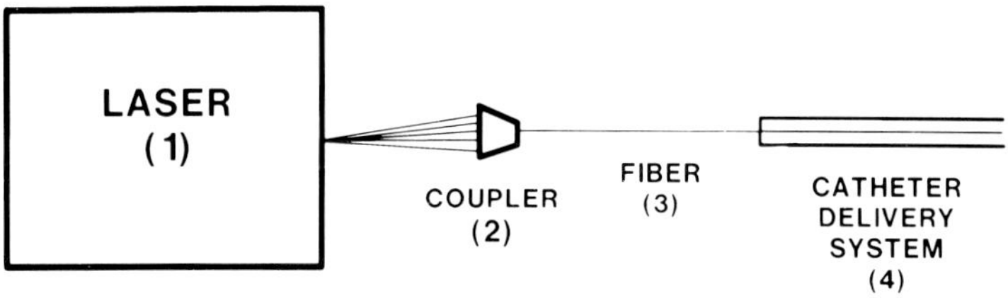

Figure 12 Attempts to improve the safety and efficacy of laser coronary angioplasty include consideration of alternative laser sources (1); development of necessary coupling devices (2) and novel fiberoptic systems (3) for wavelengths that cannot presently be effectively delivered via conventional fiberoptic systems; and modifications of the catheter delivery system (4) that will permit more precise control of emitted laser light.

clinical use of laser angioplasty in humans is the risk of mechanical or thermal perforation, or both. Efforts to eliminate the problem of vessel perforation encompass a variety of alternative approaches; some have been realized, others are presently under active investigation, and still others have simply been proposed. These alternatives may be considered according to the individual components of the system illustrated in Fig. 12.

A. Alternative Laser Sources

Most biomedical applications of laser technology employed to date have utilized CW lasers. If tissue vaporization could be accomplished without heating nontarget, "normal tissue" (e.g., arterial wall), then perforation of the underlying wall might be more readily avoided. Consequently, we have investigated a variety of alternative laser sources in an attempt to identify those capable of vaporizing pathologic cardiovascular tissue without producing thermal injury (11-13).

Grossly visible charring was observed at all laser wavelengths when the laser temporal energy profile was continuous or pulsed at high repetition rates (greater than 2 kHz) with low pulse energies (less than 3 mJ) (Fig. 13). Charring was absent (Fig. 14) when laser irradiation was delivered in the pulsed mode at low repetition rates (less than 200 Hz) with high pulse energies (greater than 10 mJ), independent of the wavelengths employed. The tissue charring, observed on gross examination, was consistently associated with light microscopic findings of coagulation necrosis and polymorphous lacunae. In the absence of gross tissue charring, these light microscopic findings were not observed (Fig. 15). Likewise, scanning electron microscopy consistently confirmed the histological findings associated with gross charring and, in the absence of charring, disclosed only "wrinkling" or "pruning" of the tissue bordering the laser crater (Fig. 16). The

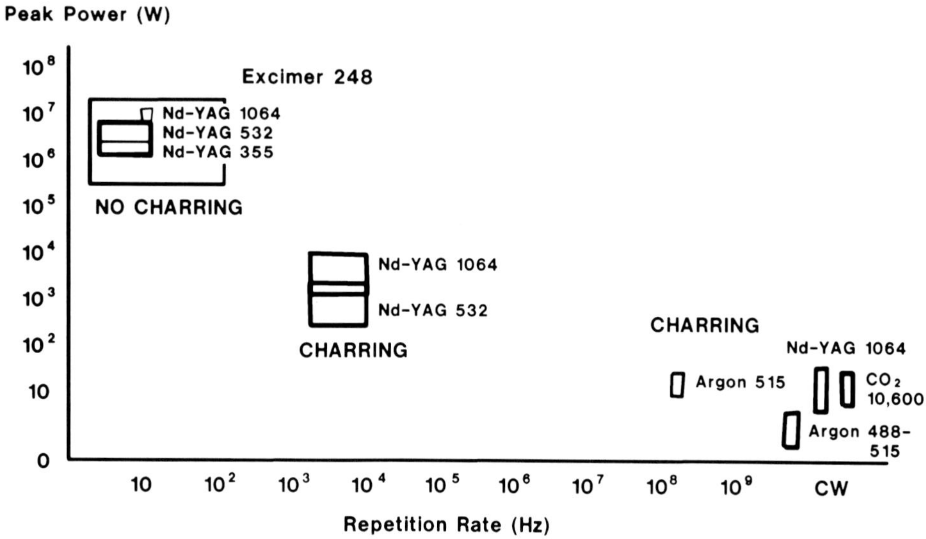

Figure 13 Continuous wave lasers, or pulsed lasers operating at relatively low peak power and relatively high repetition rate, consistently result in thermal injury. In contrast, when pulsed lasers operate at relatively high peak power (greater than 700 kw) and relatively low repetition rate (less than 200 Hz), tissue vaporization or ablation is accomplished without gross or microscopic evidence of thermal injury.

extent of ultrastructural alteration adjacent to the laser crater was, in the worst case, limited to less than 650 μm in each specimen irradiated with a pulsed laser at low repetition rate.

Although it is clear that pulsed delivery at low repetition rates eliminates gross microscopic evidence of thermal injury, it remains uncertain whether this superior histological appearance is due to enhanced dissipation of heat during a thermal process or is a representative result of a nonthermal (photochemical or photodisruptive) process. Experiments by Hayes and Wolbarsht (25), Welch (26), and Anderson and Parrish (27) have suggested that thermal injury of adjacent, nontarget tissues can be avoided when exposure time utilized for each pulse is shorter than the thermal relaxation time (characteristic time for first-order temperature dissipation) of the irradiated tissue. If the pulse duration and cumulative exposure are sufficiently brief, and if the interval between each pulse is sufficiently long, then the heat generated by each pulse may be sufficiently dissipated before delivery of the subsequent pulse. Consequently, thermal injury of tissue outside the path of the laser beam can be avoided.

Alternatively, it is possible that the photon density resulting from the mega-

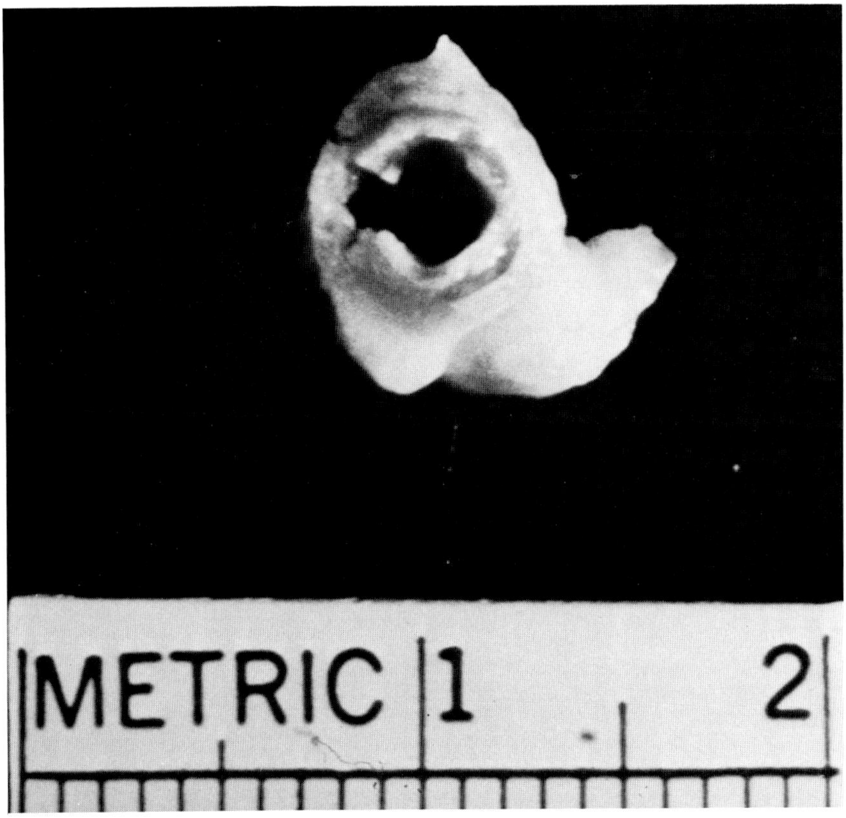

Figure 14 A segment of coronary artery, originally occluded by atherosclerotic plaque. Recanalization has been accomplished (in vitro) using a krypton fluoride excimer laser (248 nm). Note absence of gross charring.

watt peak power characteristics of such thermal injury-free pulsed systems may ablate tissue by a nonthermal (photochemical) (28,29) or photodisruptive (30) process. Interestingly, it was possible to produce charring using an energy profile that otherwise did not produce charring by irradiating the tissue *above* the focal spot of the lens used to focus the laser beam. Such defocusing results in a lower energy density or photon flux at the tissue surface. The resulting pathological injury may reflect this lower energy density; alternatively, the longer exposure time (11-fold increase) required to bore through the tissue slice may have overwhelmed the heat-dissipative capabilities of the tissue adjacent to the target site.

In addition to the potential advantage of diminishing arterial wall perforation,

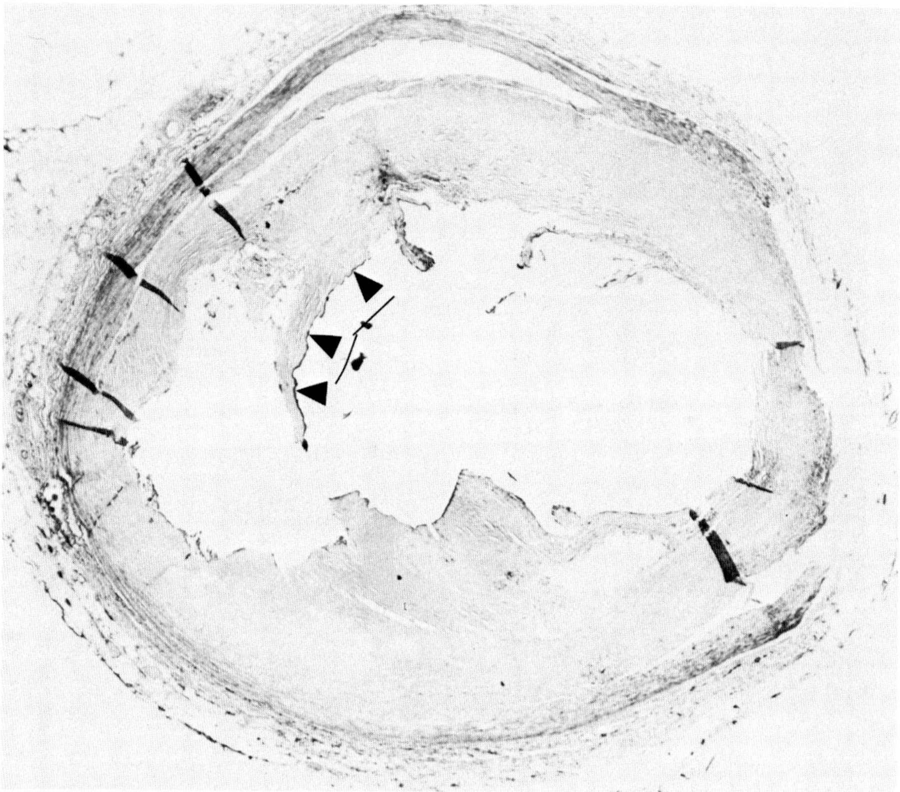

Figure 15 Light microscopic findings following krypton fluoride excimer laser irradiation (248 nm). There is a superficial zone of linear discoloration (arrows) around the perimeter of the neolumen; otherwise, no remarkable alterations are observed (left 4X; right 10X; top, trichrome stain; bottom, Verhoeff's elastic tissue stain).

elimination of pathological alterations, such as coagulation necrosis and polymorphic lacunae in residual laser-irradiated tissue, is potentially advantageous for other reasons as well. A more benign, less thrombogenic healing process, might be reasonably anticipated. At present, however, these remain only potential advantages because no long-term, in vivo data are available.

B. Alterations in Laser Fiberoptic Couplers and Alternative Optical Fibers

Because the high peak powers characteristic of the pulsed energy profiles employed to avoid pathological tissue injury cannot presently be coupled to com-

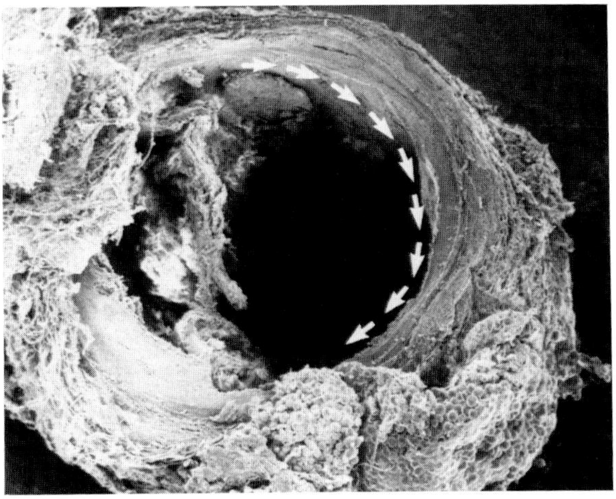

Figure 16 Scanning electron photomicrograph (22X) of atherosclerotic coronary arterial segment following excimer laser recanalization (248 nm, 330 mJ, 25 Hz, 10 sec); note precise edging along arrows.

mercially available optical fibers, alterations in coupling design and/or development of novel optical fibers are presently under investigation for this purpose. Percutaneous application of excimer laser irradiation in particular, but also megawatt-range pulsed Nd:YAG laser irradiation, are critically dependent upon transmission through optical fibers that can be advanced by guiding catheters. In contrast to noncardiac applications of laser endoscopic surgery, the value of hollow wave guide transmission, with or without a series of reflecting mirrors, is limited to direct surface (e.g., intraoperative) irradiation. Given the tortuous anatomy of the coronary arterial bed, the nonavailability of suitable optical fibers would seriously compromise the full potential of laser coronary angioplasty.

Likewise, clinical application of CO_2 laser irradiation treatment of cardiovascular disease will no doubt be limited pending the advent of commercially available infrared optical fibers similar to the Israeli prototypes described earlier. Even without available fibers, certain intraoperative therapies may be performed. In our own institution, using a focused beam, two patients have thus far undergone left ventricular CO_2 laser-assisted endocardiectomy (31) for refractory ventricular tachyarrhythmia. More extensive use of CO_2 laser irradiation, however, awaits completion of optical fiber testing. Use of CO_2 laser irradiation for transarterial cardiovascular applications is further compromised by the poor transmission of 10,600 nm through a saline or blood-filled system. Although certain strategies have been proposed to solve this problem (e.g., tandem balloon occlusion of the arterial target, or placement of a balloon at the distal tip of the guiding catheter), superimposition of this problem on the catheter delivery problems common to all laser angioplasty systems may make the CO_2 laser a suboptimal instrument for laser endarterectomy.

C. Alterations in Catheter Delivery System

Routine clinical application of laser angioplasty, particularly laser coronary angioplasty, will almost certainly require a guiding catheter delivery system of an order of magnitude more sophisticated than any previously manufactured. Simpler alterations of existing catheter systems have been advocated. For example, as mentioned previously, Ginsberg and associates (22) have advocated first crossing the target stenosis with the optical fiber (but with the laser turned off) and then applying laser irradiation to the stenotic lesions during pullback through the stenosis. We have not found this approach to be useful in the simulated intraoperative model. First, mechanical perforations are not eliminated by this technique (even a 100-μm optical fiber is more inflexible than a 0.014 or 0.016 wireguide). More importantly, laser irradiation applied during pullback of the optical fiber implies optimal use of the power along the divergent wings of the laser light beam. Because most of the power is directed straight ahead, increasing power to achieve adequate power in the "wings" may result in excessive power in the straight-ahead portion of the beam. If the augmented straight-ahead laser emission strikes a tortuous or curved distal portion of the artery, perforation at the distal site may be the result.

For intraoperative laser coronary angioplasty, we have explored the use of a high-frequency (12 MHz) two-dimensional echocardiographic probe, applied directly to the extramural coronary arteries, to direct the guiding catheter/fiber-optic bundle and image the target plaque (Fig. 17) before, during, and after laser coronary angioplasty (32). Our preliminary experience with this imaging device established the following. First, the probe is not damaged by thermal effects of intravascular argon laser irradiation, including up to 14 W of power for up to 30 sec continuous exposure. Second, the probe improved the manual antegrade manipulation of the guiding catheter/fiberoptic bundle by optimally directing it away from branch points and toward target sites. As a result only one mechanical perforation (in an extremely tortuous circumflex system) occurred when ultrasonic imaging was used. Third, although arterial tortuosities, in our experience, did not represent an important source of false-positive stenoses, as reported with the earlier generation device (33), use of the probe did not guarantee that the most excessive tortuosities could be successfully negotiated: in two cases, we were unable to advance the catheter through tortuous segments of the right and circumflex coronary arteries, respectively. Fourth, resolution with this probe was consistently adequate to discriminate the guiding catheter from the optical fiber from the normal wall on either side of the coronary lumen; in most

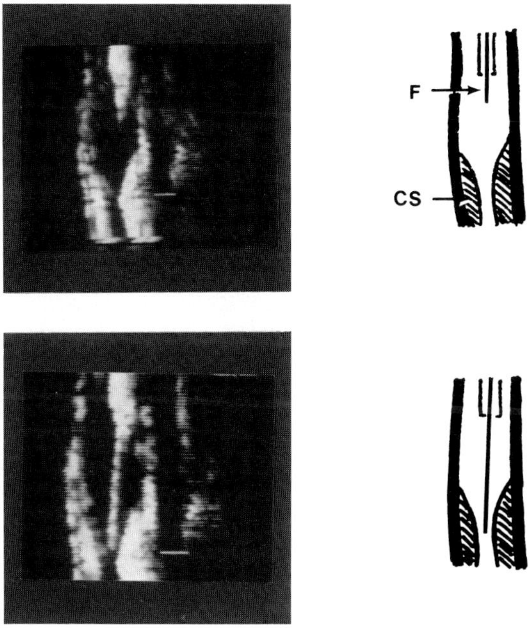

Figure 17 Still frame (left) and schematic (right) sequential images recorded during simulated intraoperative laser coronary angioplasty, using 12 MHz ultrasonic probe placed directly onto the extramural coronary artery harboring the target stenosis. Fiber (F) is observed as it is advanced out of guiding catheter toward atherosclerotic coronary stenosis (CS).

cases, atherosclerotic plaque, particularly when densely calcified, was easily recognized, including during laser irradiation. Extravascular ultrasonic imaging thus appears to represent a promising means of assisting intraoperative laser coronary angioplasty. Ultimately, it may be useful for both establishing the end point of laser irradiation and allowing the operator to recognize, and thereby avoid, impending thermal perforation.

Specific minimum requirements of a guiding catheter/fiberoptic delivery system for percutaneous, as well as intraoperative, use include a diminutive profile (preferably less than 3.5 F), maximum flexibility, at least two lumens (one for the optical fiber, one for flushing), and a heat-resistant tip. Incorporation of a focusing lens at the distal fiber tip, protection by a transmissive, heat-resistant balloon at the distal tip, and steerability of the distal tip are all features that are currently under investigation by multiple groups. Another proposed modification designed to diminish the frequency of arterial perforation is incorporation of a Doppler probe as part of the guiding catheter; use of the Doppler probe to detect high-velocity flow near the sites of coronary stenoses (34) could theoretically identify the true coronary lumen, thus keeping the guiding catheter/fiberoptic bundle coaxial with the coronary lumen.

An approach designed to obviate many of the problems inherent in attempting to precisely control energy delivery at the distal tip of the guiding catheter/fiberoptic bundle consists of administering hematoporphyrin derivative (HpD) to "stain" the atherosclerotic plaque. This approach, pioneered by Spears and coworkers (35), is based upon the use of HpD in oncology: HpD is selectively taken up by certain neoplasms, and fluoresces after exposure to the appropriate wavelength of light, resulting in selective tumor kill. Spears et al. (35) have demonstrated that HpD, administered systemically, may be taken up by atherosclerotic plaques of experimental animal models; exposure of the HpD-positive plaques to an appropriate wavelength of laser light might then be expected to induce atherolysis. Fundamental to this line of investigation are ongoing experiments designed to determine if the complex atherosclerotic plaque of man, which is typically much less cellular and much more fibrotic than that found in experimental animal models, can actively incorporate systemically administered HpD. Nevertheless, the virtue of this approach is that it minimizes problems of catheter design required for precisely directed laser irradiation.

Finally, and perhaps most analogous to noncardiac laser endoscopic surgery, the use of intravascular angioscopy in guiding catheter delivery systems is currently under investigation for controlling beam emission at the distal end of the optical fiber (36-38). By borrowing heavily from gastrointestinal and otolaryngological techniques, intravascular angioscopy could theoretically allow the operator to distinguish plaque from normal wall and thus improve the safety associated with delivery of laser light. It is in this area of technology that the distinguishing features of the cardiovascular system are most problematic: the diminutive size of the relevant vessels requires an extreme degree of miniaturization of the angioscopic working channels and optical fibers while blood flow through these vessels severely compromises the quality of images obtained. Necessary design features of an intravascular imaging system must include, as a minimum, an illumination optical fiber, an imaging optical fiber, and a separate working channel. The peripheral arterial system will likely accommodate an outer diameter as large as 3.5 mm, while for intracoronary work, the maximum outer diameter ideally should be smaller than 2.0 mm. Moreover, the instrument must be flexible, particularly for intracoronary applications. Finally, incorporation of an occluding balloon to transiently occlude antegrade blood flow, a feature of at least one commercially available device, might be expected to improve image quality.

IV. SUMMARY

Thus, while considerable potential for lasers in the treatment of cardiovascular disease exists, the biomedical application of this tool to clinical medicine will require further fundamental, innovative investigations. Noncoronary cardiovascular applications of laser endoscopic surgery have already been employed successfully in humans. Laser myoplasty is currently the preferred operative treatment for hypertrophic cardiomyopathy at our institution; preliminary experience in a canine model has indicated that percutaneous laser myoplasty may soon be possible (39). Laser endocardiectomy is now a routine adjunct in antiarrhythmia surgery at our institution. However, laser angioplasty, whether intraoperative or percutaneous, will demand substantially more technological modifications for this approach to be established as a safe and efficacious mode of laser endoscopic surgery.

ACKNOWLEDGMENTS

Much of the work described in this chapter was supported by a grant from the GenRad Foundation, GenRad Corp., Concord, Massachusetts.

REFERENCES

1. American Heart Association: *Heart Facts, 1985.* Dallas, 1984.
2. Cashin WL, Sanmarco ME, Nessim SA, and Blankenhorn DH: Accelerated progression of atherosclerosis in coronary vessels with minimal lesions that are bypassed. *N. Engl. J. Med. 311*:824-828, 1984.
3. Campeau L, Lespérance J, and Bourassa MG: Natural history of saphenous vein aortocoronary bypass grafts. *Mod. Concepts Cardiovasc. Dis. 53*:59-63, 1984.
4. Isner JM and Salem DN: The persistent enigma of percutaneous angioplasty. *Int. J. Cardiol. 6*:391-400, 1984.
5. Holmes DR Jr, Vlietstra RE, Smith HC, Vetrovec GW, Kent KM, Cowley MJ, Faxon DP, Gruentzig AR, Kelsey SF, Detre KM, Van Raden MJ, and Mock MB: Restenosis after percutaneous transluminal coronary angioplasty (PTCA): A report from the TCA Registry of the National Heart, Lung, and Blood Institute. *Am. J. Cardiol. 53*:77C-81C, 1984.
6. Zaidi AR, Hollman JL, and Galan K: Multivessel angioplasty, procedure, and follow-up trends (abstr.). *Circulation 70*:II-266, 1984.
7. Macruz R, Martins JR, Tupinamba A da S, Lopes EA, Vargas H, Pena AF, de Carvalho VB, Armelin E, and Decourt LV: Possibilidades terapeuticas do raio laser em atermoas (Eng. abstr.). *Arq. Bras. Cardiol. 34*:9-12, 1980.
8. Lee G, Ikeda RM, Kozina J, and Mason DT: Laser-dissolution of coronary atherosclerotic obstruction. *Am. Heart J. 102*:1074-1075, 1981.

9. Abela GS, Normann S, Cohen D, Feldman RL, Geiser EA, and Conti CR: Effects of carbon dioxide, Nd-YAG, and argon laser radiation on coronary atheromatous plaques. *Am. J. Cardiol. 50*:1199-1205, 1982.
10. Choy DSJ, Stertzer SH, Rotterdam HZ, and Bruno MS: Laser coronary angioplasty: Experience with 9 cadaver hearts. *Am. J. Cardiol. 50*:1209-1211, 1982.
11. Isner JM, Clarke RH, Donaldson RF, Muller DF, Foxall TL, Laliberte SM, Libby P, Alroy J, and Ucci AA: The excimer laser: Gross, light microscopic, and ultrastructural analysis of potential advantages for use in laser therapy of cardiovascular disease (abstr.). *Circulation 70*:II-35, 1984.
12. Isner JM and Clarke RH: The current status of lasers in the treatment of cardiovascular disease. *IEEE J. Quantum Electron 20*:1406-1420, 1984.
13. Deckelbaum LI, Isner JM, Donaldson RF, Clarke RH, Laliberte S, Aharon AS, and Bernstein JS: Reduction of laser-induced pathologic tissue injury using pulsed energy delivery. *Am. J. Cardiol. 56*:662-667, 1985.
14. Grundfest W, Litvack F, Forrester J, Fishbein M, Morgenstern L, McDermid S, Pacala T, Rider D, and Laudenslager J: Pulsed ultraviolet lasers provide precise control of atheroma ablation (abstr.). *Circulation 70*:II-35, 1984.
15. F. Hillenkamp, R. Pratesi, and C. A. Sacchi (eds.). *Lasers in Biology and Medicine,* NATO ASI Series A, Life Sciences: Vol. 34. Plenum Press, New York, 1981.
16. Isner JM, Clarke RH, Donaldson RF, and Aharon A: Identification of photoproducts liberated by in vitro argon laser irradiation of atherosclerotic plaque, calcified cardiac valves and myocardium. *Am. J. Cardiol. 55*:1192-1196, 1985.
17. Isner JM, Donaldson RF, Funai JT, Deckelbaum LI, Pandian NG, Clarke RH, Konstam MA, Salem DN, and Bernstein JS: Factors contributing to perforations from laser coronary angioplasty: Observations in an intact human postmortem preparation of intraoperative laser coronary angioplasty. *Circulation 72*:II-191, 1985.
18. Isner JM, Donaldson RF, Fortin AH, and Clarke RH: Attenuation of the medial layer of the coronary arterial wall in segments with advanced atherosclerosis: A potential factor predisposing to perforation during laser coronary angioplasty (abstr.). *J. Am. Coll. Cardiol. 5*:409, 1985.
19. Abela GS, Normann SJ, Cohen DM, Franzini D, Feldman RL, Crea F, Fenech A, Pepine CJ, and Conti CR: Laser recanalization of occluded atherosclerotic arteries in vivo and in vitro. *Circulation 71*:403-411, 1985.
20. Gerrity RG, Loop FD, Golding LAR, Ehrhart LA, and Argenyi ZB: Arterial response to laser operation for removal of atherosclerotic plaques. *J. Thorac. Cardiovasc. Surg. 85*:409-421, 1983.
21. Eldar M, Battler A, Neufeld HN, Gaton E, Arieli R, Akselrod S, Levite A, and Katzir A: Transluminal carbon dioxide-laser catheter angioplasty for dissolution of atherosclerotic plaques. *J. Am. Coll. Cardiol. 3*:135-137, 1984.
22. Ginsburg R, Kim DS, Guthaner D, Toth J, and Mitchell RS: Salvage of an ischemic limb by laser angioplasty: Description of a new technique. *Clin. Cardiol. 7*:54-58, 1984.

23. Geschwind H, Boussignac G, Teisseire B, Vieilledent C, Gaston A, Becquemin JP, and Mayiolini P: Percutaneous transluminal laser angioplasty in man (letter). *Lancet 1*:844, 1984.
24. Choy DS, Stertzer SH, Myler RK, Marco J, and Fournial G: Human coronary laser recanalization. *Clin. Cardiol. 7*:377-378, 1984.
25. Hayes JR and Wolbarsht ML: Thermal model for retinal damage induced by pulsed lasers. *Aerospace Med. 39*:474-480, 1968.
26. Welch AJ: The thermal response of laser irradiated tissue. *IEEE J. Quantum Electron 20*:1471-1481, 1984.
27. Anderson RR and Parrish JA: Selective photothermolysis: Precise microsurgery by selective absorption of pulsed radiation. *Science 220*:524-526, 1983.
28. Srinivasan R and Mayne-Banton V: Self-developing photoetching of poly-(ethylene terephthalate) films by far-ultraviolet excimer laser radiation. *Appl. Phys. Lett. 41*:576-578, 1982.
29. Trokel SL, Srinivasan R, and Braren B: Excimer laser surgery of the cornea. *Am. J. Ophthalmol. 96*:710-715, 1983.
30. Puliafito CA and Steinert RF: Short-pulsed Nd:YAG laser microsurgery of the eye: Biophysical considerations. *IEEE J. Quantum Electron 20*:1442-1448, 1984.
31. Isner JM, Michlewitz H, Clarke RH, Estes NA III, Donaldson RF, Salem DN, Bhan I, Payne DD, and Cleveland RJ: Laser photoablation of pathological endocardium: In vitro findings suggesting a new approach to the surgical treatment of refractory arrhythmias and restrictive cardiomyopathy. *Ann. Thorac. Surg. 39*:201-206, 1985.
32. Funai JT, Pandian NG, Isner JM, Clarke RH, Lojeski EW, Donaldson RF, Konstam MA, and Salem DN: Utility of high-frequency two-dimensional echocardiography in the performance of laser coronary angioplasty: Experimental studies (abstr.). *Clin. Res. 32*:672A, 1984.
33. Sahn DJ, Barratt-Boyes BG, Graham K, Kerr A, Roche A, Hill D, Brandt PWT, Copeland JG, Mammana R, Temkin LP, and Glenn W: Ultrasonic imaging of the coronary arteries in open-chest humans: Evaluation of coronary atherosclerotic lesions during cardiac surgery. *Circulation 66*:1034-1044, 1982.
34. Tomonaga G, Mitake H, Hoki N, and Kajiya F: Measurement of point velocity in the canine coronary artery by laser Doppler velocimeter with optical fiber. *Jpn. J. Surg. 11*:226-231, 1981.
35. Spears JR, Serur J, Shropshire D, and Paulin S: Fluorescence of experimental atheromatous plaques with hematoporphyrin derivative. *J. Clin. Invest. 71*:395-399, 1983.
36. Lee G, Ikeda RM, Stobbe D, Ogata C, Theis J, Hussein H, and Mason DT: Laser irradiation of human atherosclerotic obstructive disease: Simultaneous visualization and vaporization achieved by a dual fiberoptic catheter. *Am. Heart J. 105*:163-164, 1983.
37. Spears JR, Marais HJ, Serur J, Pomerantzeff O, Geyer RP, Sipzener RS, Weintraub R, Thurer R, Paulin S, Gerstin R, and Grossman W: In vivo coronary angioscopy. *J. Am. Coll. Cardiol. 1*:1311-1314, 1983.

38. Litvack F, Grundfest WS, Lee ME, Carroll RM, Foran R, Chaux A, Berci G, Rose HB, Matloff JM, and Forrester JS: Angioscopic visualization of blood vessel interior in animals and humans. *Clin. Cardiol. 8*:65-70, 1985.
39. Isner JM, Clarke RH, Pandian NG, Donaldson RF, Salem DN, Konstam MA, Payne DD, and Cleveland RJ: Laser myoplasty for hypertrophic cardiomyopathy: In vitro experience with human postmortem hearts and in vivo experience in a canine model (transarterial) and human patient (intraoperative). *Am. J. Cardiol. 53*:1620-1625, 1984.

16

Orthopedic Surgery

TERRY L. WHIPPLE
Orthopedic Research of Virginia; Medical College of Virginia; University of Virginia School of Medicine; and Humana Hospital-St. Lukes Center for Orthopedics, Richmond, Virginia

I. LASER ENERGY IN ORTHOPEDIC SURGERY

A. Mesenchymal Tissues

In contrast to most areas of surgical specialization, orthopedics addresses a wide variety of tissues, mostly of mesenchymal origin.

The multipotential primitive mesenchymal cell under various influences may develop into bone, cartilage, muscle, ligament, tendon, or fascial divisions. These tissues have differing consistencies and density, varying vascular supply, and a wide range of function from distribution of compressive loads to transfer of tensile forces. In the extremities and spinal column, orthopedics also encompasses disorders of vascular and neurological tissue.

B. Orthopedic Subspecialization

There is a growing tendency within the specialty of orthopedic surgery toward subspecialization based on tissue type, tissue size, or classification of disorders. In certain circumstances surgeons may limit their practice and expertise to fracture management, cartilage injury or degeneration, or muscle pathophysiology. Others may practice microsurgery exclusively, repairing or reconstructing vascular and nerve supply in the extremities.

Still other orthopedists concentrate on disorders of common etiology such as athletic injuries, inflammatory disorders, tumors and metabolic bone disease, or congenital anomalies.

C. Surgical Manipulation

With such a variety of tissue type, tissue size, tissue function, and nature of disorders, orthopedic surgery as a specialty has become a very instrument-dependent field. Repair and reconstruction of tissues less than 2 mm in size requires extreme precision, adequate magnification devices, and miniaturized, necessarily delicate instrumentation. The orthopedic oncologist or traumatologist must be able to handle the most sclerotic bone immediately adjacent to major vessels, as well as the most friable necrotic tissue which may envelop critical neurological elements.

The surgeon's objectives may include any combination of the following functions: removal of tissue, anatomical repair of injured structures, reconstruction of unrepairable tissue by transfer or rearrangement of structures into nonanatomical relationships, or the implantation of foreign materials to augment or replace deficient structures.

D. Conventional Instrumentation

Historically, only mechanical instruments have been available to accomplish these surgical objectives. The notable exception is the electrocautery for coagulation or hot-knife incision of vascular tissues. Occasionally cryosurgery or chemosurgery is employed for tissue ablation, as in resection of intraosseous giant cell tumors or herniated intervertebral disks. Otherwise, cutting functions are accomplished with scalpels, drills, chisels, saws, curettes of various sorts, and scissors. Fixation of tissue is achieved with sutures, staples, screws, pins, rods, and grouting substances such as polymethyl methacrylate.

While these instruments may vary in size or material, these few devices constitute the orthopedist's instrument armamentarium.

E. Disadvantages of Conventional Orthopedic Techniques

While conventional instruments are extremely effective for most surgical procedures and have served the orthopedist well for decades if not centuries, purely mechanical devices do frequently present disadvantages in the face of particular surgical circumstances. All mechanical devices require force application to the point of tissue yielding. This applies to all of the previously noted devices for cutting or fixation, except for grouting substances. Force is applied through the instrument to the tissue until the tissue separates. Accuracy and precision may depend solely upon an understanding of natural tissue cleavage planes and the surgeon's ability to modulate the mechanical force he applies to the instrument. Minor errors in application of force may result in unintended tissue laceration or disruption of tissue apposition. The potential for these misapplications of mechanical energy is exemplified by the difficulty in positioning and pinning small osteochondral fragments when treating displaced finger fractures, or in wrinkling the anastomosis of a small artery.

Furthermore, with the use of mechanical instrumentation, the tissue to be manipulated must always be accessible. Access may require greater surgical exposure than would otherwise be desired. The trend toward minimally invasive surgery, in fact, is requiring not only improved methods of visualizing intracorporeal structures but also less bulky and less cumbersome means of manipulating, incising, and reattaching tissues. This is a formidable challenge when mechanical force is required for tissues as dense as bone and fibrocartilage.

Finally, fixation of musculoskeletal tissues has most commonly required the implantation of foreign substances. Whether rods, plates, screws, or suture, these implants carry certain inherent risks of infection or foreign-body reaction. While most such implants are well tolerated, their undesirability is evidenced by the increasing efforts to produce biodegradable or resorbable fixation materials.

The ideal tissue glue has not yet been developed, but it still would represent the introduction of a foreign substance into the body with associated inflammatory reactions.

Newer developments are advantageous and therefore commendable to the extent that they minimize any disadvantages of mechanical instruments and fixation devices.

II. THEORETICAL ADVANTAGES OF LASER ENERGY IN ORTHOPEDIC SURGERY

A. Target Effect

The primary effect of any laser energy on target tissue is the conversion of light energy to heat. Regardless of the laser-generating material or the wavelength of energy produced, when that energy is absorbed by biological tissue, thermal injury results. The heat produced is based on the absorption coefficient of the target tissue for the particular wavelength of laser energy employed.

B. Incision, Excision

Heat alters the tissue structurally and possibly chemically. No mechanical force is involved whatever, thus no pressure is applied to the tissue treated. As mesenchymal tissue absorbs coherent light energy, the heat produces a permanent change in tissue architecture. Depending upon the specific wavelength, this change may be focal or regional. The carbon dioxide (CO_2) laser, for example, generating light with a wavelength of 10.6 μm, is very efficiently absorbed. The thermal effect therefore is essentially confined to the surface of the target tissue, whereas neodymium:yttrium-aluminum-garnet (Nd:YAG) laser energy of 1.06 μm wavelength is less well absorbed and the thermal effect extends radially from the point of impact. The Nd:YAG laser effects on mesenchymal tissue therefore can be measured at some depth below the surface as compared with those of the CO_2 laser.

As most tissues of mesenchymal origin are relatively hypocellular, the most significant thermal effect is on the structure of the collagen molecule's triple helix. As collagen fibers are heated, they contract, becoming less flexible and more brittle. As the brittle fibers break, the tissue separates.

This is in contradistinction to the laser effect on cellular tissues wherein the intracellular fluid is boiled, rupturing the cell membranes and causing tissue separation. Thermal regulation best determines the effect on target tissue irrespective of the type of laser energy employed.

1. Thermal Energy

Thermal regulation can be modulated by local cooling, adjustment of the laser output wattage, the focal spot size which further determines power density, and the duration of exposure. Higher power, smaller focal spot, and longer periods of exposure increase the heat production. All heat is dissipated radially from the site of exposure. Protein begins to denature at approximately 45°C. Coagulation with tissue necrosis occurs between 45 and 100°C. Tissue water boils at 100°C and carbonization occurs between 300 and 400°C. At temperatures in excess of 530°C, tissue combustion or incineration will occur. Excessive accumulation of carbonaceous material may have an adverse effect on tissue healing, much as an eschar impedes epithelialization of cutaneous lesions. The appropriate temperature must therefore be achieved but not exceeded to produce the desired target effect.

2. Mechanical Energy

The avoidance of mechanical energy would be very advantageous in several surgical situations. If small fragments of tissue are precariously apposed to one another, their relationship will not be disturbed by the resection or reconstruction of tissue in the immediate vicinity. Fragile or friable tissues will not be torn or otherwise unintentionally disrupted. Because the energy applied to the target tissue is electronically regulated, it can be more uniformly and accurately applied to achieve the intended effect. This introduces an additional element of precision to the surgical manipulation of delicate or precariously positioned anatomical structures. Without vibration or mechanical pressure, adjacent tissues are minimally affected.

C. Fixation

As noted, conventional fixation for stabilization of musculoskeletal, vascular, and neurological structures requires the introduction of foreign material. Inevitably this provokes some degree of foreign-body reaction. Although fixation devices are fabricated of relatively inert materials, the biological immune mechanism is sufficiently sensitive to recognize that they are of exogenous origin. As in any surgical procedure, the introduction of foreign materials carries a certain risk of contamination of surgical wounds via the implantation of bacteria. The potential for infection is especially great when fixation devices such as Kirschner wires or Steinmann pins are left protruding through the skin, as frequently happens with small bone fragment fixation. This greatly increases the risk of septic seeding of bones or joints.

As noted, collagen begins to denature and will coagulate when heated to between 45 and 100°C. If this temperature can be well controlled and focally applied, the concept of tissue welding becomes a practical means of tissue fixation or stabilization. With the ability to precisely focus a laser beam to fractions of a millimeter in diameter, and regulate the power in milliwatts, tissue can be heated so as not to exceed 100°C—as evidenced by the boiling of surface moisture or cellular water—producing a collagen amalgam. Any devitalized material will be debrided by physiological means. The inflammatory reaction is short-lived and self-limiting. The devitalized tissue is merely altered host tissue and does not represent foreign substances. This concept of laser abrazement for tissue fixation is especially attractive in consideration of the speed with which it can be performed. Once appropriately positioned, tissues may be laser welded in fractions of a second, compared with the time required for the placement of sutures or pins.

D. Photocoagulation

Intraoperative or postoperative bleeding is a hindrance to any surgical procedure. Securing hemostasis in the surgical field where extremity tourniquets cannot be employed may be laborious and time consuming. The argon laser energy with a wavelength of 488-514 nm is readily absorbed by red substances and is particularly effective for photocoagulation. Similarly, the Nd:YAG and CO_2 lasers will coagulate blood or seal vessels when the beam is defocused to cause generalized heating of the tissue rather than incision or vaporization. Photocoagulation is virtually instantaneous and is convenient since defocusing the beam requires no change of instruments. The evolution of combined lasers, which allow alternation between argon and other energies at will, further increases the appeal and convenience of photocoagulation mechanisms for hemostasis. In orthopedic surgery, these concepts are proving useful in shoulder, hip, and spine procedures where tourniquets cannot be employed.

E. Sterilization

As laser energy can rupture host cells by boiling intracellular water, so too can it disrupt microbial organisms. Thus it has particular appeal in the sterilization of contaminated wounds. The bacteriostatic or sporicidal effects of laser energy through alteration of organelles has not been extensively investigated.

F. Remote Access

The ability to project a powerful cutting or fixation means across distances without hampering visualization presents numerous advantages for minimally invasive surgery. For procedures in remote or difficult access areas, the effective reduction in the size of the surgical device reduces the necessary surgical exposure. The advantages of minimally invasive surgery and less extensive dissection are well documented in terms of reductions in surgical morbidity including pain, complication risks, and cosmetic scarring, and with regard to earlier mobilization and rehabilitation.

III. LASER ENERGY AS A CUTTING DEVICE

A. Arthroscopy

Perhaps the most significant progress in orthopedic surgery toward minimally invasive techniques has been the development of intraarticular surgery under arthroscopic control. Meniscectomy, synovectomy, retinacular release, and debridement of degenerative articular lesions can be performed arthroscopically at least as well, if not better than, through traditional arthrotomy. These procedures have been developed with mechanical instrumentation which, although adequate, have some notable disadvantages. Resection of dense cartilaginous or mineralized tissue requires sharp rigid instruments. However, the confining space between the articular surfaces of joints, such as the knee or shoulder, provides little room for the manipulation of instruments strong enough to deal with such tissues without risk of breakage or of damage to the delicate articular cartilage.

Meniscus pathology occurs most commonly in the posterior region. Arthroscopic mechanical instrumentation generally ranges in size from 3.4 to 5.0 mm in diameter, with the smaller instruments being considerably slower and less effective. Laser energy capable of vaporizing or excising torn fragments of meniscus fibrocartilage can be transmitted in a beam diameter of 1 mm or less. This represents a 65-80% reduction in "instrument" size. Preliminary experience with laboratory and clinical laser meniscectomy has documented the effectiveness of a 1 mm CO_2 laser beam (10.6 μm wavelength) for the resection of human meniscus

(1,2). The procedure preserves the advantages of minimally invasive surgery under arthroscopic control for common disorders of the knee and has the potential for simplification of these procedures with effective cutting through direct frontal or tangential approaches to the meniscus. Loose bodies and synovium can also be vaporized easily.

The CO_2 laser produces the most effective energy wavelength for tissues as dense as cartilage or bone. These tissues absorb 10.6 μm energy efficiently, confining the thermal effect to the surface of the target. Thus there is no penetration of energy to deeper elements of the tissue. Whipple and coworkers (1) have demonstrated no alteration of collagen fibers or cell structure in meniscus fibrocartilage beyond 70 μm from the target site using CO_2 energy of 30-35 W at shutter speeds of 0.2-0.5 sec and examining the tissue with light microscopy (Fig. 1). Electron microscopic sections indicate some loss of collagen striation and organelle changes within 300 μm of impact (3) (Fig. 2). Thus the effects of 10.6 μm energy on meniscus tissue can be well controlled with minimal side effects.

Whipple et al. (4) demonstrated, in 1983, that the healing potential of the remaining peripheral rim of meniscus following laser meniscectomy was unimpaired and proceeded to a smooth sealed margin of fibrocartilage with normal architecture within 10 weeks of surgery in animal models (Fig. 3).

Glick (5) reported preliminary experience with meniscectomy using the Nd:YAG laser (wavelength 1.06 μm) with suboptimal results. While Nd:YAG laser energy has the advantage of fiberoptic transmission, it is poorly absorbed by white cartilaginous tissue, and the thermal alteration of collagen fibers can be found at considerable distances from the site of impact. Thus the remaining peripheral rim following Nd:YAG laser meniscectomy is distorted and brittle, which is unacceptable.

Whipple described the technique for CO_2 laser meniscectomy (1) under arthroscopic control using a cannula attached to the articulated arm for the introduction of the CO_2 laser beam. Power settings were recommended in the range of 30-35 W with a focal spot size of 1 mm and shutter speeds of 0.2-0.5 sec. The procedure is performed with gaseous joint distention using nitrogen or carbon dioxide delivered through the laser cannula by an insufflation pump at pressures of 60-80 mmHg (6) (Fig. 4). Smoke is evacuated through the arthroscope sheath and filtered externally. Magnetization of the smoke particles produces gradual opacification of the arthroscope lens interfering with visualization.

Smith and Nance (2) undertook a clinical series of 109 CO_2 laser meniscectomies from 1982 to 1984 using a laser cannula with a mirror at the tip to angle the beam to greater advantage (Fig. 5). All meniscectomies were successful (Fig. 6). Using nitrogen for joint distention, Smith reported an occasional problem with subcutaneous emphysema with spontaneous, although occasionally

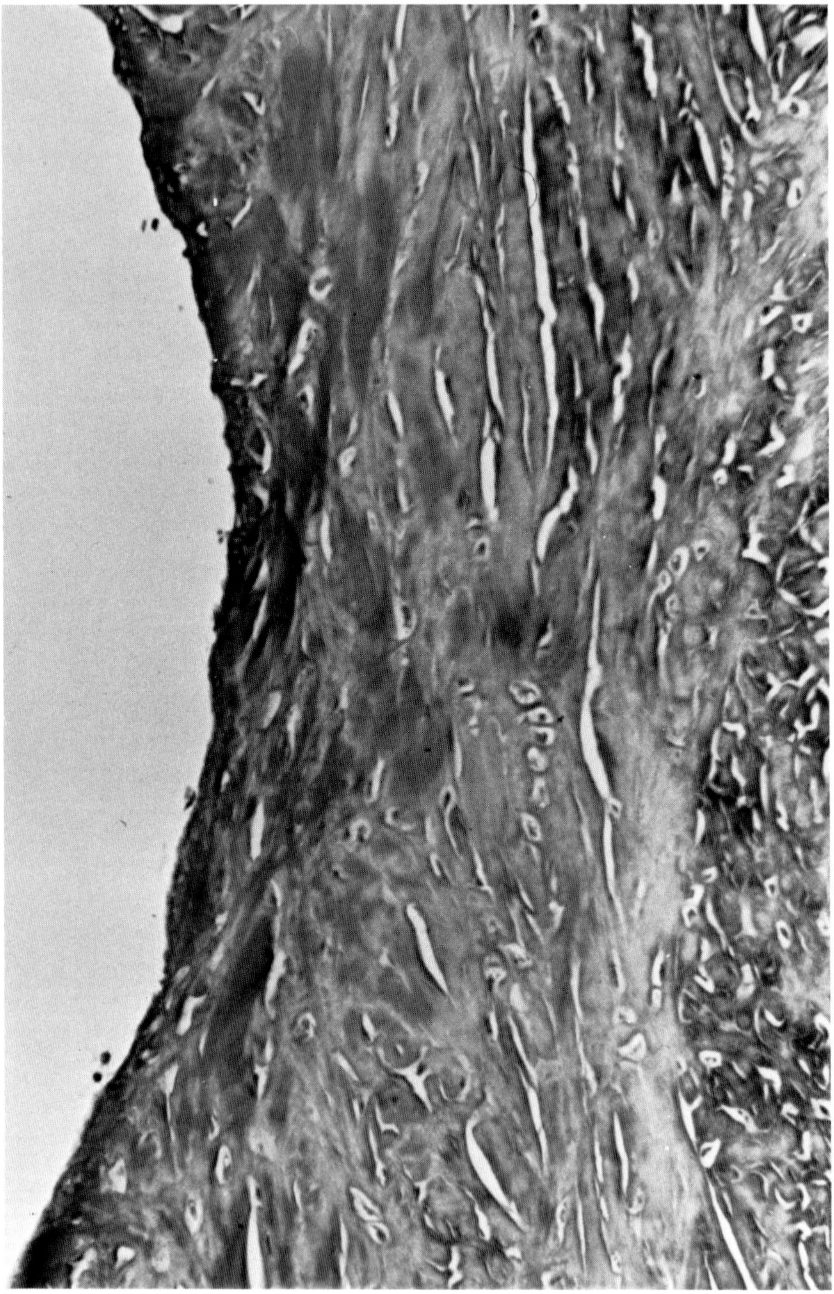

Figure 1 Human meniscus remnant after laser resection of free margin. Note viable cells and preservation of collagen fiber architecture near resection edge (hematoxylin and eosin X100).

delayed, resolution. This complication did not occur when the joint was insufflated with carbon dioxide.

Smith and Nance (2) recommended saline rinses periodically during the procedure to purge the joint of carbon ash residue. Saline was introduced through a needle placed in the posterior compartment, filling the joint from below and displacing the gas anteriorly through the arthroscope sheath. This avoids the inconvenience of bubble formation in the joint, which impairs vision (Fig. 7). While clearing the carbon residue from the joint seems desirable, Whipple et al. (7) demonstrated, with no rinsing, complete resorption of all carbon residue by the synovium within 6 weeks of laser meniscectomy in animal models.

Philandrianos (8), in France, reported in 1984 a series of CO_2 laser meniscectomies under arthroscopic control with results similar to Smith's. Smith and Nance (2) noted a 15% incidence of tense bloody effusions forming within 48 hr of the procedure with spontaneous resolution. This has been attributed to postoperative synovial response to remaining carbon residue and possibly smoke particles.

Whipple has designed a laser-competent arthroscope sheath similar to that used in bronchoscopy for laser surgery (Fig. 8). This device provides the advantage of visual field aiming of the beam rather than the conventional helium-neon laser-aiming spot (Precision Surgical Instruments, Richmond, Virginia). A ventilation mechanism has also been designed to allow automated gaseous joint distention with low pressure and high flow to evacuate smoke and obviate the problem of lens clouding and possible synovial irritation (Snyder Labs, Dover, Ohio). These devices have not been employed clinically as of this writing.

While the potential for arthroscopic laser meniscectomy appears in many respects advantageous over mechanical means, it remains in the developmental stages at this time. Other arthroscopic procedures, such as synovectomy, may well employ Nd:YAG laser energy more effectively. Synovium is neither as thick nor as dense as meniscus fibrocartilage, and the advantage of fiberoptic delivery, as well as transmission through a fluid medium, can be employed.

B. Cartilage

Laser energy has been used very little on articular cartilage. Because of its density and white color, the most appropriate laser for articular cartilage absorption is the CO_2.

Schultz and colleagues (9) have reported a single experiment when the knees of guinea pigs were fused after laser debridement of the articular surfaces. They noted a 100% incidence of solid arthrodesis, although with some tendency toward self-mutilation by the animals.

Articular cartilage is difficult to excise completely in an efficient manner with laser energy. The tissue tends to coagulate on the surface of subchondral

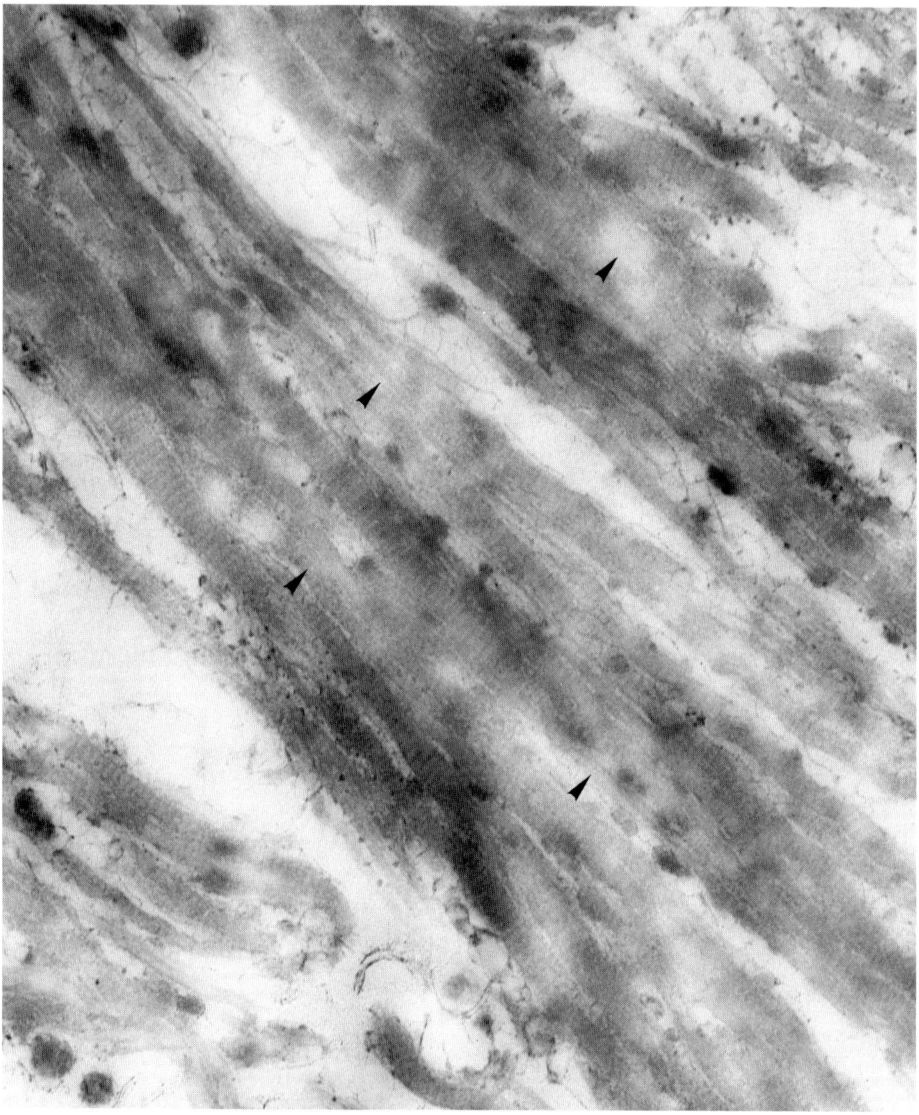

(a)

Figure 2 Electron microscopy of human meniscus following laser resection of free margin. There is some loss of collagen striation (large arrow) and cellular organelle changes indicated by smudging (small arrow) within 300 μm of resection edge [(a) ×42,500; (b) ×11,500].

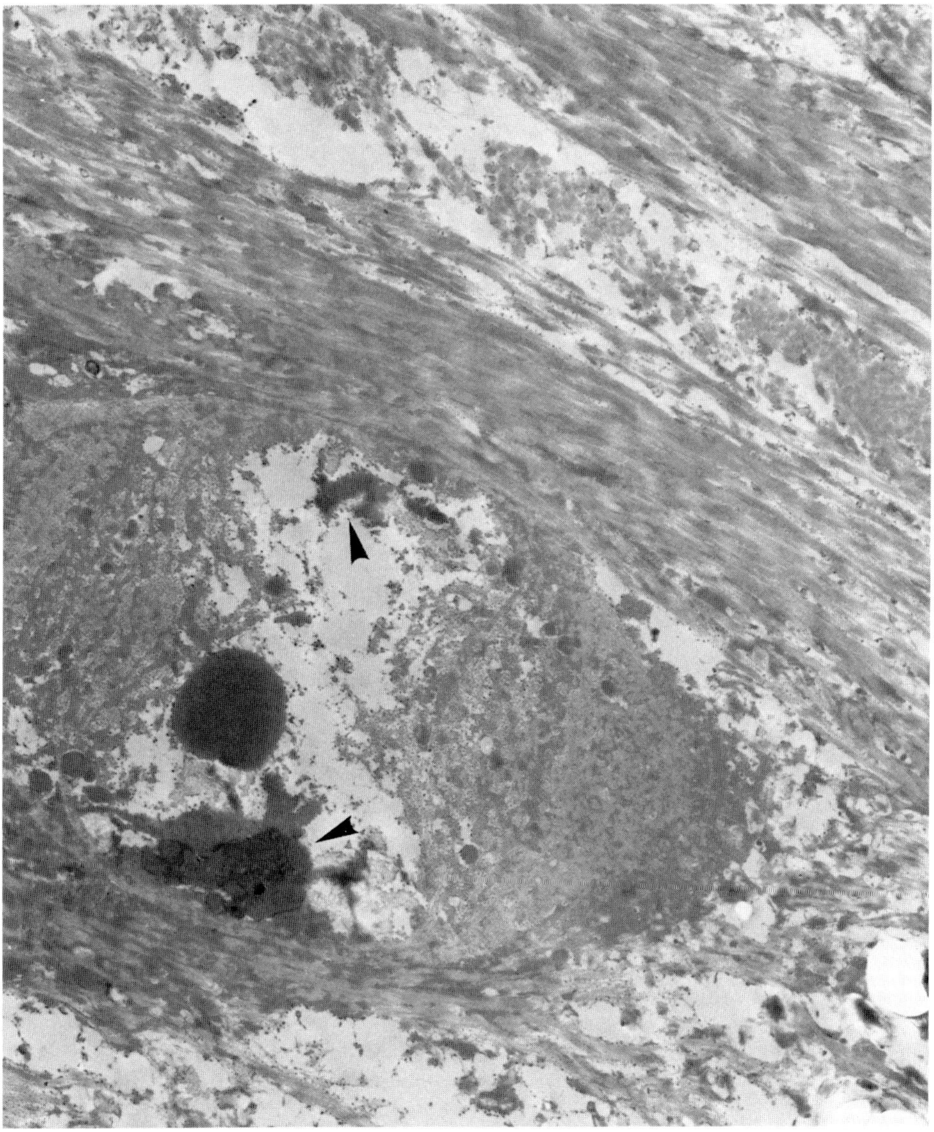

(b)

Figure 2 (continued)

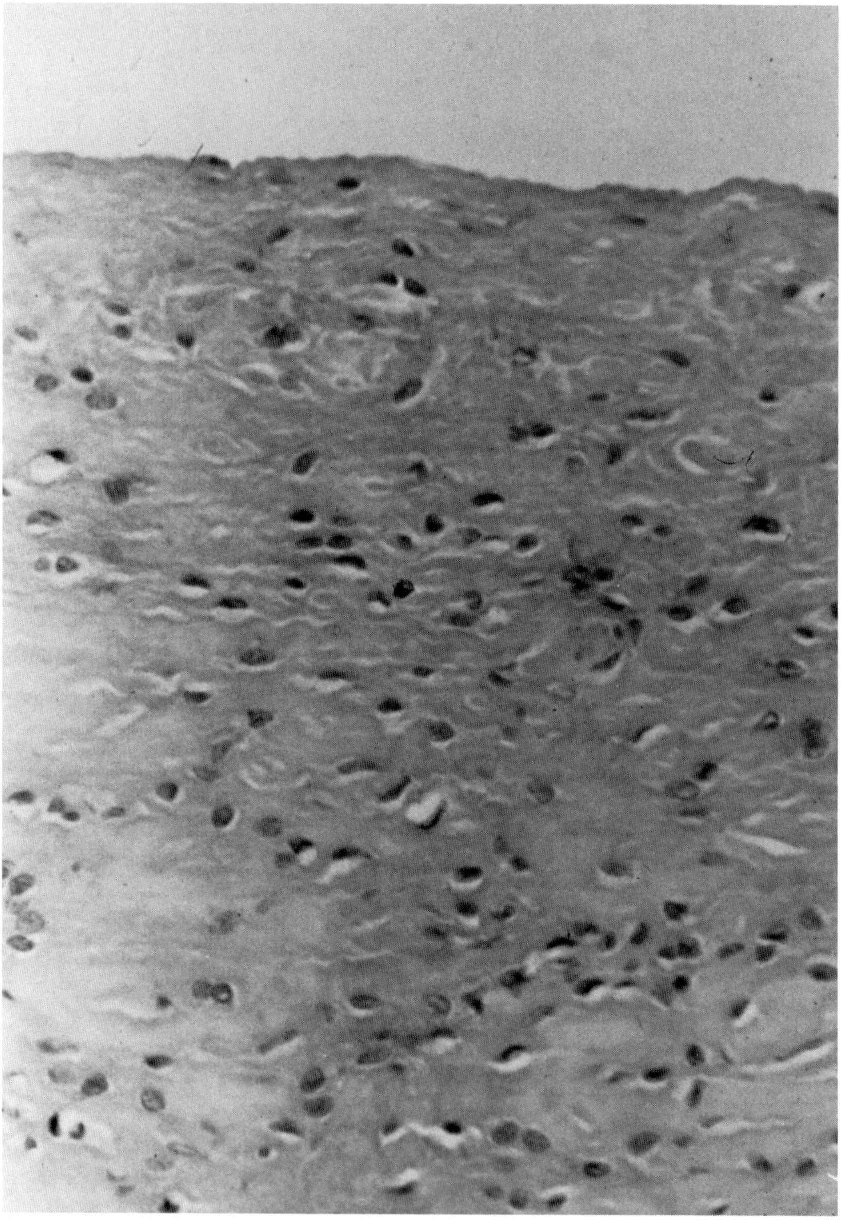

Figure 3 Rabbit meniscus 10 weeks postlaser subtotal meniscectomy. Free margin is completely healed with reorganization of logically oriented collagen fibers and essentially normal cell population (hematoxylin and eosin).

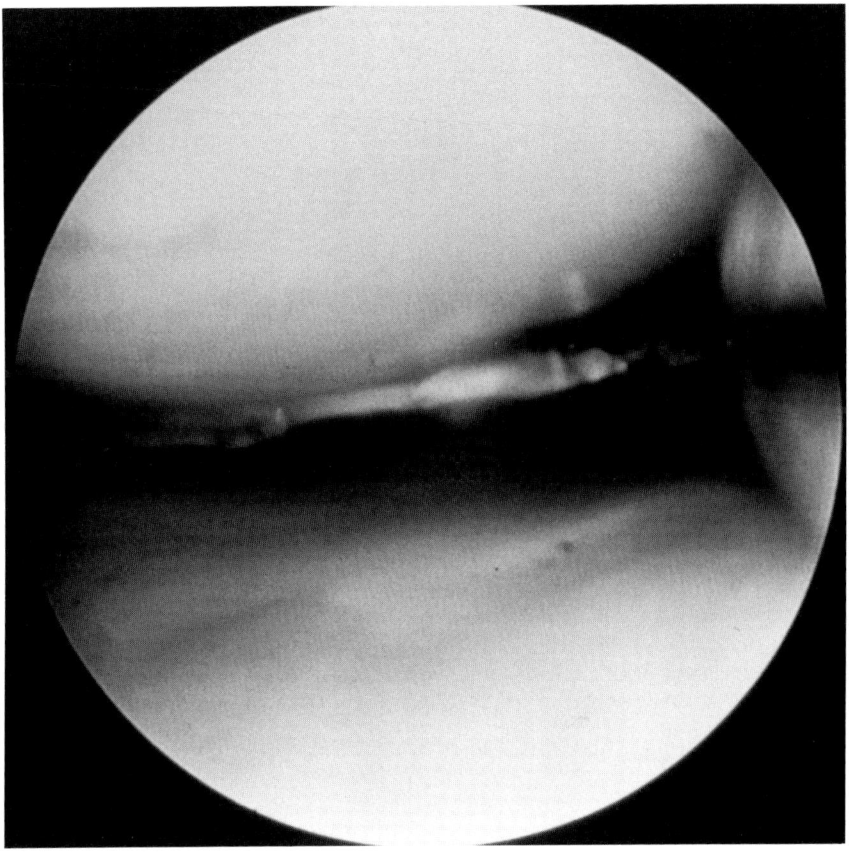

Figure 4 Projection of CO_2 laser beam from intraarticular cannula (right) to edge of torn meniscus seen between femoral condyle above and tibial plateau below (*Source*: courtesy J. B. Smith).

bone and must be scraped away to remove the gummy residue and carbon ash. Removal of small areas of articular cartilage or localized partial thickness degenerative fronds, however, may be practical.

C. Skin

Skin can be readily incised with the CO_2 laser. Used extensively in dermatology, the laser has been used sparingly in open surgery. Again, the CO_2 laser seems most appropriate as a laser knife for skin

I have performed 40 shoulder and hand procedures incising the skin with a

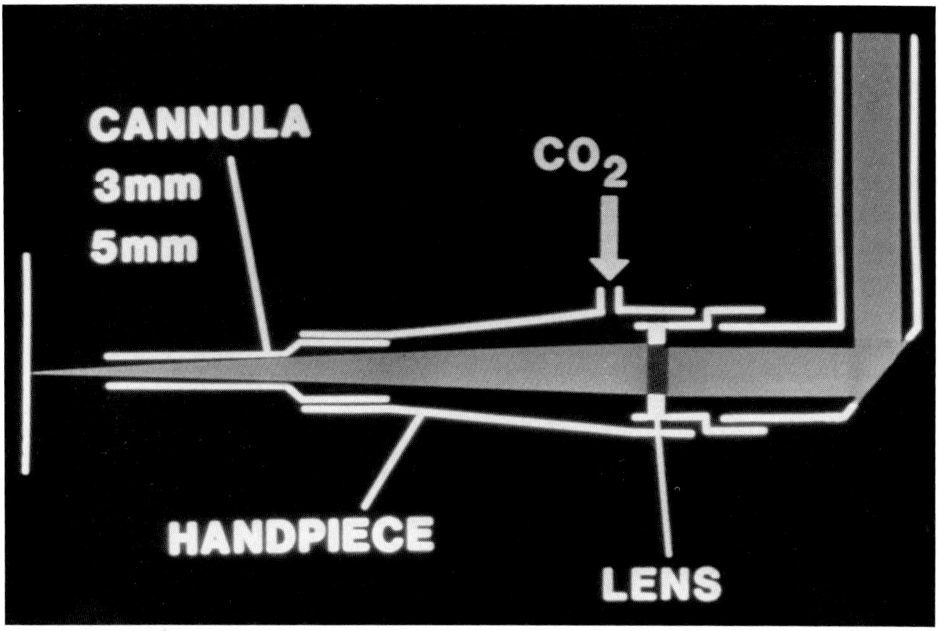

Figure 5 Diagram of CO_2 laser cannula used for arthroscopic meniscectomy. The tip can be fitted with a mirror to angle the intraarticular beam (*Source*: courtesy J. B. Smith).

CO_2 beam of 20 W with a focal spot size of 0.6 mm. Hemostasis is instantaneous, and larger subcutaneous bleeders can be coagulated by lifting the handpiece slightly to defocus the beam. Tension must be maintained on the skin edges in a direction perpendicular to the line of incision to part the tissues and to deepen the incision.

No difference is noted in the postoperative discomfort, inflammatory reaction, or healing rate of these incisions compared with cold-knife techniques. The incisional scars were consistently narrow and cosmetic and sealed rapidly with conventional suture closure. Thus in vascular regions such as the hand, face, and neck, laser energy may be employed for skin incision with slight advantage because of photocoagulation.

D. Bone

There is little advantage to incising bone with laser energy. For appropriate surgical speed, power output in excess of 100 W is required. Still the laser is no

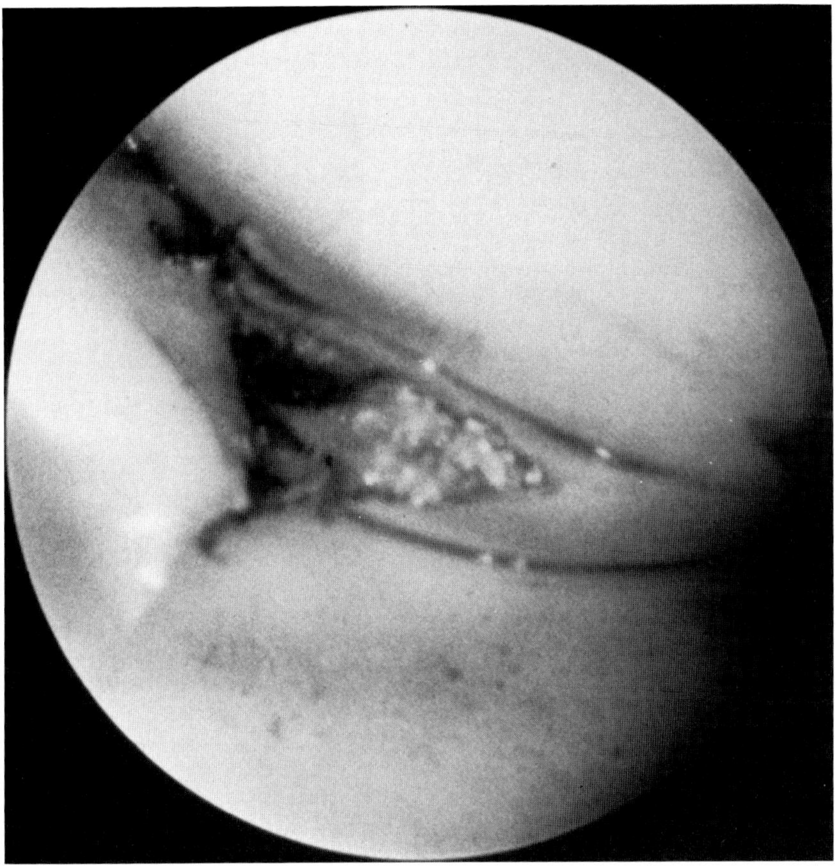

Figure 6 Cleavage cut and torn medial meniscus during laser partial menisccc-tomy. Note carbonization of resection margin (*Source*: courtesy J. B. Smith).

more effective nor precise for orthopedic purposes than modern surgical saws or high-speed surgical sculpting instruments. For more delicate bone work, as in the middle ear, perhaps there may be some advantage.

E. Inflammatory Tissue

Rheumatoid arthritis is a common disorder that is becoming effectively palliated with arthroscopic techniques. Mechanical cutting devices are now small enough to fit into the crevices and recesses of most major joints to mechanically excise hypertrophic rheumatoid pannus and synovium under arthroscopic control.

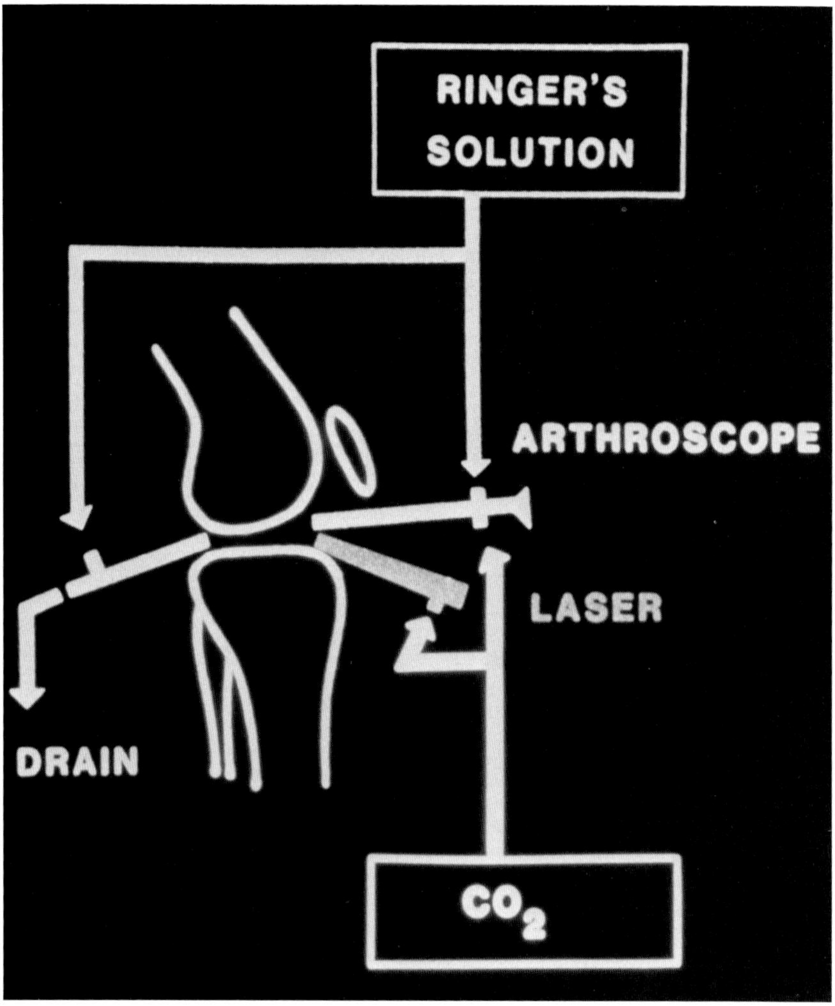

(a)

Figure 7 Arrangement of posterior drain for saline rinses during arthroscopic laser meniscectomy in a CO_2 medium (*Source*: courtesy J. B. Smith).

While this is a vast improvement over surgical synovectomies through conventional arthrotomy, postoperative bleeding remains a significant problem. Even when precautions are taken to minimize hemarthrosis following synovectomy by using indwelling suction drains, pressure bandages, and the instillation of

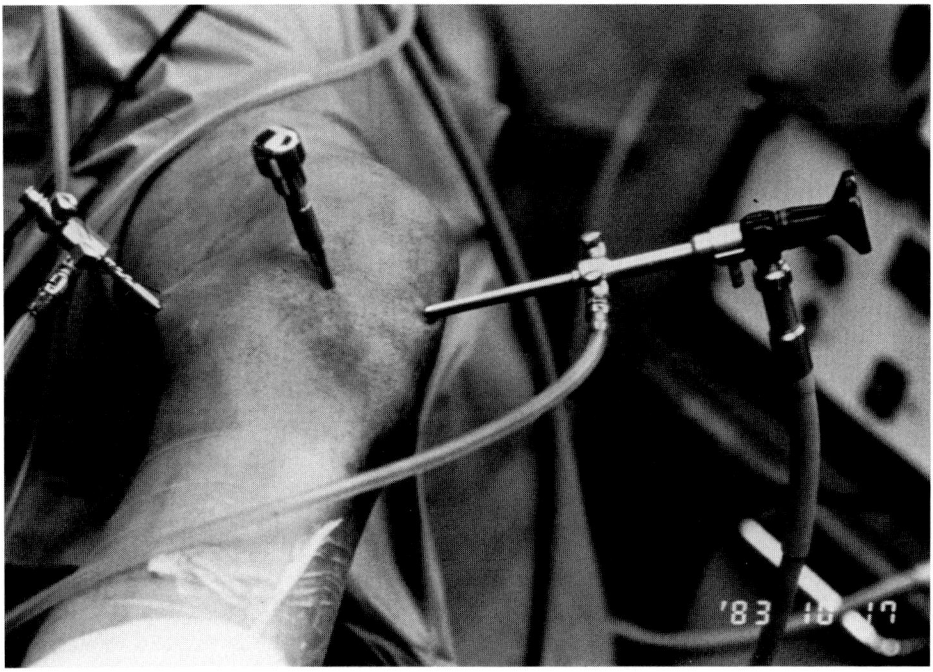

(b)

Figure 7 (continued)

dilute epinephrine solutions, intraarticular bleeding remains the major obstacle to postoperative recuperation.

As a photocoagulator or photocautery device, the laser offers some potential for arthroscopic synovectomy with decreased postoperative bleeding in rheumatoid patients. Rheumatoid pannus can be readily vaporized with CO_2 energy with no harm to the underlying subsynovial structures. The CO_2 beam can be delivered in a focused or defocused mode either through a parallel channel in the arthroscope sheath, or through a separate cannula.

Argon lasers are not powerful enough to remove large volumes of inflammatory synovium. The Nd:YAG lasers are more likely alternatives to CO_2 for inflammatory tissue; however, they risk deeper penetration of thermal injury to subsynovial and capsular tissues.

Pigmented villonodular synovitis can also be resected under arthroscopic control with CO_2 or Nd:YAG laser energy with the same advantage of instant hemostasis.

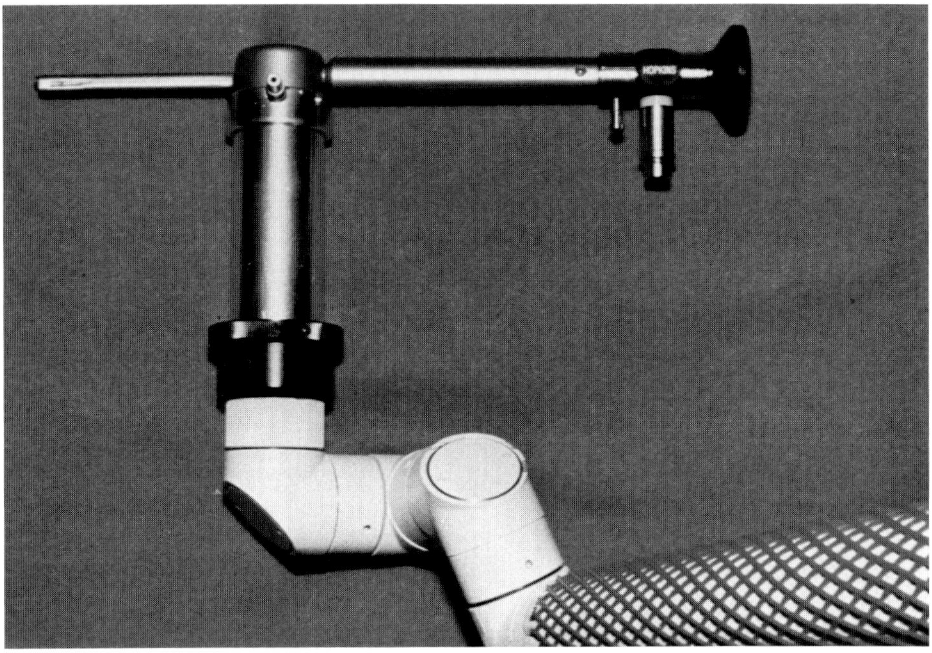

Figure 8 Laser competent arthroscope uses fiberoptic illumination and CO_2 laser beam introduced via articulated arm. Intraarticular tip dimensions: 4.0 × 2.7 mm.

IV. LASER ENERGY AS A FIXATION DEVICE

A. Microsurgery

Microvascular anastomosis with laser energy is a promising experimental technique in orthopedics. Replantation of severed digits has become a successful means of restoring hand function after traumatic amputation. Digits can also be reconstructed effectively in certain cases of congenital anomaly. These are complex and lengthy procedures that are largely dependent upon the successful anastomosis of vessels and their continued patency. Similarly, the transfer of vascularized skin, muscle, and bone graft entails tedious microvascular anastomoses. Irregular suture lines, which precipitate thromboses or stricture of the vessel lumen, jeopardize these procedures.

1. Vessels

There is good evidence that laser anastomoses of small-caliber vessels may be at least as successful as, and considerably faster than, suture techniques. The procedure is accomplished by denaturing the collagen of opposing vessel ends into a confluent bridge that "welds" the tissues together. Laser energy focused to extremely small spot size and with properly regulated thermal effects on the vessel walls can seal vessels together with a patent lumen and can do so without disturbing the precarious apposition of the vessel ends or tearing the fragile tissue with suture material.

Yahr and Strully (10,11) were the first to employ laser energy in vascular anastomoses by gluing the tissues together side to side, then incising their common wall with a Nd:YAG laser. This technique proved unsuccessful, because the Eastman 910 monomer glue (2-methyl-cyanoacrylate) did not maintain the anastomosis.

Jain (12-15) reported the first successful in vivo laser arterial anastomosis in a series of experiments on vessels up to 1.0 mm diameter. Rat carotid arteries were approximated over a stent or balloon catheter and the vessel edges were welded together with a Nd:YAG laser focused to 0.2 mm spot size. The patency rate in four series ranged from 85 to 92.5% up to 6 months, as demonstrated by angiographic studies and scanning electron microscopy examination. With less trauma to the vessel wall, the laser anastomoses required only 5 min compared with 15 min for suture technique.

In a subsequent report (16), Jain noted that similar techniques using an argon laser for arterial repair caused significant damage to the vessel wall and weak anastomoses.

Carbon dioxide laser energy has also been successfully used for vascular anastomoses and it may offer some advantages. Morris et al. (17) used 10-0 nylon stay sutures and 75 mW of CO_2 laser energy focused to 30-40 μm with exposure times of 0.5 sec for arterial repairs. Whether the laser welds added significant tensile strength to the repair or only sealed the vessel between stay sutures is not known.

Neblett (18), collaborating with Morris, modified the CO_2 laser technique somewhat in attempt at more precise thermal control. A film of moisture was placed over the anastomosis site, and while using a spot size of 150 μm and 60-80 mW of power, the tissue was exposed until the moisture boiled into a wisp of steam. This limited the heating to 100°C to avoid overexposure and tissue necrosis with carbonization.

Dew et al. (19) introduced a technique of bridging small gaps or leaks in CO_2 laser arterial anastomoses with a muscle patch. A few muscle fibers were placed over the gap in a dry field and were heated with a relatively large (3-mm) spot size to diffuse the heat. This fused the muscle patch to the vessel wall to seal the anastomosis. Dew has obtained successful CO_2 laser anastomoses of rat arteries and veins with a spot size of 0.5 mm and brief exposures to limit heat production without the steam indicator technique employed by Jain (19,20) (Fig. 9).

Hayashi and coworkers (21-24) have reported extensively on the use of CO_2

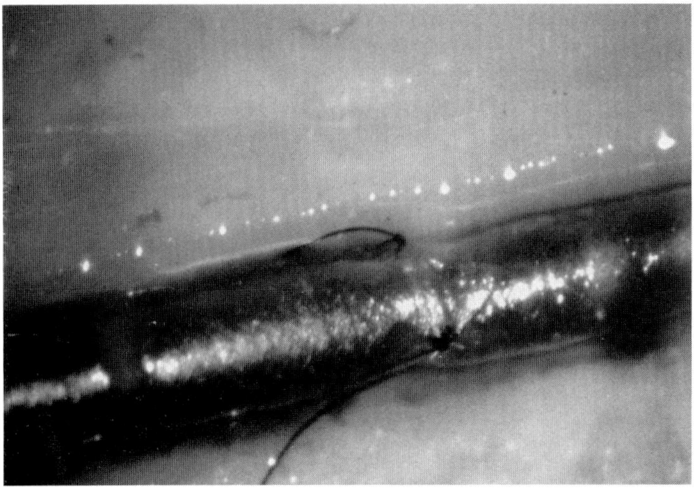

Figure 9 A 1-mm rat femoral artery following CO_2 laser anastomosis (*Source*: courtesy D. K. Dew).

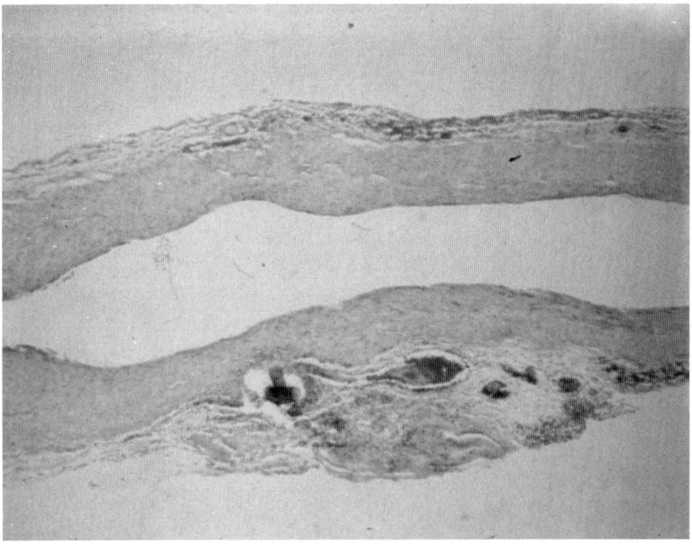

Figure 10 A 1-mm rat femoral artery 90 days post CO_2 anastomosis. Note preservation of vessel endothelium and patency of lumen (hemotoxylin and eosin) (*Source*: courtesy D. K. Dew).

laser energy for vascular anastomosis. They have concluded the ideal spot size should be less than 120 μm with an energy application of 1.5-2.0 J/mm^2. The analogous difference in the techniques of Hayashi and Dew is that Hayashi sprinkles the anastomosis with energy and Dew allows it to flow onto the area. The latter is obviously easier and probably more uniform. Both techniques produce less trauma to the vessel endothelium than the Nd:YAG laser anastomosis because of the surface absorption of 10.6 μm energy, but both have the disadvantage of requiring stay sutures (Fig. 10).

As the use of laser reduces the time and foreign materials required for microvascular anastomosis, its advantage and application in reconstructive hand surgery and vascularized graft transfer will become more widespread.

2. Nerves

Whereas there are vast circulatory collaterals and arterial loops in the extremities, there is minimal duplication of nerve supply to muscles and cutaneous zones. Injury to a single peripheral nerve is therefore a substantially greater loss than most peripheral vascular injuries, and nerve repair or grafting is more frequently necessary. In addition, nerve repair is as critical to the success of replantation as vascular anastomosis. Because of the common difficulties with suture techniques, there is widespread interest in the use of laser energy for peripheral nerve repair. Sutures are tedious, traumatic, and sometimes bulky for epineurial repairs, and more so for fascicular repairs. Suture material evokes a chronic inflammatory response and may promote neuroma formation.

Epineurial repair with CO_2 laser energy has been effectively accomplished in rat sciatic nerves using milliwatt power, a defocused beam of 2-3 mm diameter, and 0.05 sec exposure (25). Stay sutures are initially used for approximation of the nerve ends, then are removed after the laser neurorrhaphy. The sealed epineurium appears advantageous in promoting axonal regeneration into neural sheaths.

The surface absorption of 10.6 μm energy is commendable in avoiding injury to underlying axons (26). With the laser mounted on the operating microscope, laser neurorrhaphy is also considerably faster than suture techniques.

Using a different technique and argon laser energy, Almquist et al. (27,28) performed vascular repairs by coagulating hemoglobin around axon ends to produce a miniature adherent cuff. The white nerve tissue is protected by virtue of its poor absorption of argon laser energy.

Numerous problems remain in the development of microneurorrhaphy and microvascular techniques. The most advantageous energy source has not been determined, nor have the specific absorption properties of various tissues and tissue components been defined. Finally, practical application of laser energy for anastomoses will demand an automated mechanism for regulating power output and exposure time of the respective tissues treated.

B. Cartilage Grafts

Whipple is investigating the use of CO_2 and Nd:YAG laser energy for fixation of articular cartilage allografts and the stabilization of small osteochondral fracture fragments. Mechanical fixation means are entirely too traumatic and disruptive. Defocused Nd:YAG energy appears most promising because of its penetration, thereby producing a larger weld than the CO_2 laser.

C. Menisci

Similar techniques are being attempted for the repair of certain tears in knee menisci. This fibrocartilage structure has very limited circulation where tears usually occur, is slow to heal and remodel, and is subjected to extraordinary stresses. This places greater demands on any repair technique, but if sufficient durability and tensile strength can be obtained with laser tissue "melting" techniques, arthroscopic meniscus repair may be possible without penetrating sutures or jeopardy to extracapsular neurovascular structures.

V. TISSUE STERILIZATION

The application of laser energy in orthopedics for photocoagulation and surgical field sterilization is beyond the scope of this chapter on endoscopy. However, it should be noted that laser energy can be used effectively for orthopedic procedures in vascular regions where tourniquet control is impossible, such as the spine, shoulder, and hip. The CO_2 laser will also effectively debride contaminated wounds such as decubiti and pyogenic granulomas without risk of injury to deeper tissues or adjacent neurological structures.

VI. ORTHOPEDIC APPLICATIONS OF PROJECTED ENERGY

Whenever surgical intervention represents as much insult as the pathology it addresses, its justification or indication is questionable. Recent trends toward minimally invasive surgery have produced significant advances in the treatment of numerous disorders. The ability to project laser energy as an effective cutting means to relatively remote anatomical regions potentially represents a major surgical advantage. Arthroscopic laser surgery has been discussed previously. The removal of intervertebral disk material under microscopic control using CO_2 laser energy of 20-40 W can be accomplished with considerably less surgical exposure and mechanical trauma than diskectomy with rongeurs and curettes. Moreover, anterior cervical diskectomy can be more delicately performed by using the CO_2 laser to remove foraminal bone spurs and posterior bony ridges that may impinge or encroach on the dural sac. Carbonization is a minor problem that can be overcome with gentle mechanical debridement.

Endoscopic intramedullary surgery for the removal of malignant metastases in cases of pathological fracture of long bones, or for the removal of methyl methacrylate from the femoral canal in revision hip surgery, can be accomplished with Nd:YAG or CO_2 lasers introduced through instruments similar to the laser bronchoscope or Whipple's laser arthroscope. While there has been no report of such applications of intraosseous laser endoscopy, the theoretical applications are obvious.

VII. FUTURE DEVELOPMENTS

A. Delivery System

The application of laser energy in orthopedic surgery is in its infancy. Many deficiencies must be addressed and techniques developed.

A major obstable to the endoscopic use of CO_2 laser energy is the development of an ideal delivery system. This problem is not peculiar to orthopedics, and a suitable fiberoptic cable for 10.6 μm energy will undoubtedly spur the development of techniques for endoscopic laser surgery for musculoskeletal pathology. Mechanisms for rapid change of focal length of the projected beam are also needed for applications other than in the field of microsurgery. Finally, computer-assisted automation for the regulation of power output and exposure time to produce the optimal power density for respective tissues to be treated is also needed.

B. Environmental Safety

Major attention is needed also in the field of environmental safety. Flameproof drapes, towels, and absorbent sponges are imperative, as well as nonreflective accessory instruments.

C. Education

Finally, a major eductional effort is needed for surgeons and operating room personnel alike. A fundamental understanding of the capabilities and hazards of projected radiation in a surgical environment must be familiar to all who participate in its application.

ACKNOWLEDGMENTS

The author recognizes with appreciation the contribution to the microsurgery section of this chapter by D. K. Dew, M.D.

REFERENCES

1. Whipple TL, Caspari RB, and Meyers JF: Laser energy in arthroscopic meniscectomy. *Orthopedics 6*:1165-1169, 1983.
2. Smith JB and Nance TA: Arthroscopic laser surgery. Presented in part at the Annual Meeting of the Arthroscopy Association of North America. Coronado, California, January 1983.
3. Whipple TL: Electron microscopic evidence of thermal effects of CO_2 laser energy on human meniscus. 1984 (unpublished).
4. Whipple TL, Caspari RB, and Meyers JF: Laser subtotal meniscectomy in rabbits. *Lasers Surg. Med. 3*:297-304, 1984.
5. Glick J: YAG laser meniscectomy. Presented at the Triannual Meeting of the International Arthroscopy Association. American Arthroscopy Association of North America. Rio de Janeiro, Brazil, August 1981.
6. Whipple TL: Applications of the CO_2 laser to arthroscopic meniscectomy in a gas medium. Presented at the Triannual Meeting of the International Arthroscopy Association. American Arthroscopy Association of North America. Rio de Janeiro, Brazil, August 1981.
7. Whipple TL, Caspari RB, Meyers JF: Synovial response to laser induced carbon ash residue. *Lasers Surg. Med. 3*:291-295, 1984.
8. Philandrianos G: Early comparative results of CO_2 laser surgery in knee pathology. Presented at the Fifth International Congress of Laser Medicine and Surgery. Detroit, Michigan, October 1983.
9. Schultz RJ, Krishnamurthy S, Cicatelli M, and Delbello D: Laser arthrodesis of guinea pig knee joints. Presented at the Fifth International Congress of Laser Medicine and Surgery. Detroit, Michigan, October, 1983.
10. Yahr WZ, Strully KJ, and Hurwitt ES: Nonocclusive small arterial anastomosis with a neodymium laser. *Surg. Forum 15*:224, 1964.
11. Yahr WZ and Strully KJ: Blood vessel anastomosis by laser and other biomedical applications. *J. Assoc. Adv. Med. Instrum. 1*:28, 1966.
12. Jain KK: Suture-less microvascular anastomosis using a neodymium:YAG laser. *J. Microsurg. 1*:436, 1980.
13. Jain KK: Lasers in neurosurgery: A review. *Lasers Surg. Med. 2*:217, 1983.
14. K. K. Jain. *Handbook of Laser Neurosurgery*. Charles C Thomas, Springfield, Illinois, 1983.
15. Jain KK: Neurosurgical forum, To the Editor. *J. Neurosurg. 60*:872, 1984.
16. Jain KK: Laser in microvascular surgery—comparative studies. *Trans. Am. Soc. Laser Med. Surg.* Third Annual Meeting, New Orleans, Louisiana, January 10-12, 1983.
17. Morris J, Carter M, and Thomson D: Microsurgical anastomosis of arteries using a laser. *Trans. Orthop. Res. Soc. 6*:142, 1981.
18. Neblett CR: Laser reconstructive neurosurgery. *Trans. Int. Cong. Appl. Lasers Electro-Optics.* ICALEO 83, Los Angeles, California, November 14-17, 1983.

19. Dew DK, Hart WS, and Evans TA: Carbon dioxide laser microsurgical vessel anastomosis techniques. *Dig. Conf. Lasers Electro-Optics.* CLEO 83, Optical Society of America, Washington, DC, 1983.
20. Dew DK, Serbent R, and Hart WS: Laser assisted microsurgical vessel anastomosis techniques: The use of the argon and carbon dioxide lasers. *Lasers Surg. Med. 3*:135, 1983.
21. Hayashi NT and Green BA: Microlaser surgery in anastomosis of small vessels and the effect of laser on blood vessels. *J. Jpn. Soc. Laser Med. 3*: 587, 1982.
22. Hayashi NT and Green BA: Laser technique for the anastomosis for small vessels. *J. Nihon Univ. Med. 42*:135, 1983.
23. Hayashi NT and Tsuji Y: Small vessel anastomosis by laser microsurgery. *Noshinkei Geka 11*:17, 1983.
24. Hayashi NT and Kumano K: Surgical techniques of the laser vascular anastomosis and thermal effects in the vascular wall by carbon dioxide laser irradiation. *J. Jpn. Soc. Laser Med. 4*:181, 1984.
25. Dew DK, Verdeja JC, and Shebert R: Carbon dioxide laser is successful in the repair of transected rat sciatic nerves. *Neurology 32*:132, 1982.
26. Dew DK and Burkhalter WE: The carbon dioxide laser is successful in the repair of rat sciatic nerve grafts. Presented before the 96th Annual Meeting of the American Orthopedic Association, Hot Springs, Virginia, June 27-30, 1983.
27. Almquist EE, Nachemson A, and Auth D: Evaluation of the use of the argon laser in repairing rat and primate nerves. *Trans. Am. Soc. Surg. Hand* 1982.
28. Almquist EE, Nachemson A, and Auth D: Evaluation of the use of the argon laser in repairing rat and primate nerves. *J. Hand Surg.* 9(A):792, 1984.

17

Photodynamic Therapy

THOMAS J. DOUGHERTY
Roswell Park Memorial Institute, Buffalo, New York

I. HISTORICAL OVERVIEW AND INTRODUCTION

The use of light-absorbing chemicals to cause photoreactions in biological systems dates back to at least 1900 when Raab (1) reported a lethal effect on paramecia exposed to light and an acridine dye. Since that time, numerous examples of the *photodynamic effect* have been reported for a wide range of photosensitizers both in vitro and in vivo (2). Oxygen is usually required in addition to a photosensitizer and light. The mechanism generally accepted for inactivation of biological systems is an energy-transfer process from the excited triplet state of the sensitizer to oxygen, producing singlet oxygen, which causes irreversible oxidation of some essential cellular component.

Although a number of photosensitizers can be considered for localization and possible therapy of tumors, the porphyrins have received the most attention since the observation of Policard in 1924 (3) that certain malignant tumors in animals and man demonstrated a reddish fluorescence attributed to accumulation of endogenous porphyrins resulting from secondary infection by hemolytic bacteria.

In 1966 Lipson, Gray, and Baldes (4) reported the first use of hematoporphyrin derivative (HPD) to treat cancer. While the primary purpose of their study was tumor detection, Lipson realized the potential for selective destruction of tumors containing HPD by making use of its photodynamic properties. In this report a patient with a large ulcerating recurrent breast tumor was treated by multiple HPD injections and local exposure of the tumor to filtered light from a xenon arc lamp (spectrum not specified). While the lesion remained after several weeks of repeated treatment, objective evidence of response was found. No further therapy using this method was reported by this group or any others until the 1970s.

For a photosensitizer to be widely clinically useful, it should be nontoxic in clinical dosages, be selectively taken up or retained or both in malignant tissue, be activated by penetrating light (> 600 nm), and be relatively photochemically efficient. The active fraction of hematoporphyrin derivative, which appears to be dihematoporphyrin ether (DHE), meets these criteria reasonably well (5).

In 1975 Dougherty and colleagues (6) reported that systemic HPD, activated by red light (600-700 nm) from a xenon arc lamp, could cause complete eradication of a transplanted mouse mammary tumor without excessive damage to any surrounding skin that was also included in the light field. Neither the light nor HPD alone, as applied, resulted in any gross effects on tumor or skin. After toxicological and preclinical studies were completed, a clinical trial was begun by this group in 1976.

Also in 1976, Kelly and Snell (7) reported specific HPD uptake in malignant and premalignant lesions of the bladder and reported treatment of a single case of bladder cancer using HPD and light from a mercury lamp directed into the bladder via a glass light guide. Tumor destruction was found only in the illuminated tumor areas. No further work has been reported by this group. Since 1976, and especially since 1980, numerous studies of HPD photoradiation applied to a wide variety of tumors have been initiated and reported.

II. HEMATOPORPHYRIN DERIVATIVE AND ITS ACTIVE COMPONENT

Numerous porphyrins have been tested as in vivo photosensitizers of experimental SMT-F animal tumors, e.g., hematoporphyrin, protoporphyrin, uroporphyrin (8). Among these, only tetraphenylporphine sulfonate (TPPS), hematoporphyrin derivative (HPD), and its active component DHE have been found to be active in this system (this is not necessarily true if activity is measured in vitro; discussed later). Hematoporphyrin (HP) itself has been reported to be active in vivo by some investigators (9) and inactive by others (10). It is now known that various preparations of hematoporphyrin may contain material similar to the active material in HPD, and when this is removed, in vivo activity disappears (10).

Although TPPS was at least as active as HPD, its low rate of serum clearance (D. A. Bellnier, Ph.D. thesis, 1982) and reported neurotoxicity (J. Kennedy, unpublished data, 1983) precluded clinical testing.

When HP is acetylated, in the first step to preparing HPD, it yields monoacetate (20-30%), diacetate (50-60%), and unchanged HP (5-20%). In addition, the mono- and di-dehydration products of HP are formed: hydroxyethylvinyldeuteroporphyrin (HVD) and protoporphyrin (proto), respectively. None of the components of this acetate mixture are active in photosensitizing tumors (11). However, when dissolved for injection in dilute sodium hydroxide solution or slightly acidic saline, the acetate groups are rapidly hydrolyzed producing a mixture of porphyrins that is very active as a tumor localizer and photosensitizer and termed here *hematoporphyrin derivative* (HPD), or by some authors, alkali-treated HPD. Because none of the known components of HPD, e.g., HP, HVD, and proto, were active tumor photosensitizers, it was clear that the activity was a result of an unidentified component. In 1983, Dougherty et al. (12) separated HPD by gel exclusion chromatography and identified a new fraction representing about 45% of the mixture. This material was found to be responsible for the tumor photosensitizing ability of HPD (12). In addition, it appeared (in animal testing) to provide a higher therapeutic ratio (tumor/skin) than the HPD mixture. Using a combination of nuclear magnetic resonance (NMR) spectroscopy, fast-atom bombardment mass spectroscopy, and various chemical reactions, this material tentatively has been identified as bis-1-[3(1-hydroxyethyl)deuteroporphyrin-8-yl] ethyl ether and/or other isomers of this structure. This ether apparently is formed during hydrolysis of the acetate(s) provided the hydrolysis is carried out under conditions such that self-aggregation of the acetate occurs. Under dilute conditions in which self-aggregation does not occur, hydrolysis produces none of this ether, which can be considered conceptually as the ether derived from hematoporphyrin by elimination of water from hydroxyethyl groups on two separate HP molecules. A common name suggested for this ether is dihematoporphyrin ether (DHE).

This material has a strong tendency to self-associate in aqueous solutions, even in the presence of albumin (12a) at a concentration similar to that found in serum (approximately 30 mg/ml). L. Grossweiner, personal communication, has estimated that 15-20% of this material remains unbound to serum protein in patients. In fact, spectroscopy of sera taken from patients 0.5-3 hr after receiving DHE indicated considerable self-association of the porphyrin, whereas at longer times it appeared to be mainly dissociated (probably bound to serum protein). It is this strong tendency for self-association that may be responsible for long periods of retention of HPD and DHE in tumors. Phagocytosis of DHE by macrophages, Kupffer cells, mast cells, and cells of the reticuloendothelial system has been demonstrated (13). There is preliminary evidence that tumor endothelial cells may also take up DHE and may be an important site of tumor activation by PDT (unpublished data).

Dihematoporphyrin ether in aqueous solutions demonstrates no fluorescence. However, in tetrahydrofuran, a solvent that causes disaggregation, the typical red porphyrin fluorescence was seen. Further, bright fluorescence, similar to that found in HPD, was seen in tumors in animals or humans. The actual fluorescent yield of DHE has not yet been determined; for HPD it is 2-3% (14).

Also, when aggregated, DHE does not generate singlet oxygen; it is less than one-quarter as efficient as HP at similar concentrations in water (12). Therefore, it is reasonable to assume that the DHE aggregates, although they may be important in tumor uptake and retention, are not responsible as such for the fluorescent or photosensitizing effect of DHE that to be effective must apparently first disaggregate (and/or be metabolized) within the tissue.

The uptake of DHE and HPD in DBA/2 Ha mouse tissue indicates that when compared with HPD, only half of the injected amount of DHE is necessary to achieve similar tissue levels, e.g., 3-5 μg/g tumor (SMT-F) 3-24 hr after intraperitoneal administration of 10 mg/kg HPD of 5.0 mg/kg DHE (15).

The distribution of HPD in various organs and tissues of the mouse as delineated by visual fluorescence is not consistent with the distribution determined using radioactively tagged prophyrins. For example, even though little fluorescence is apparent in liver, kidney, and spleen after systemic HPD injection, these organs take up more [^{3}H]- or [^{14}C] HPD than any other tissues (16). Although in principle, it could be argued that porphyrin metabolites are responsible for this discrepancy, animals that received HPD and illumination of the peritoneal cavity (red light) died of liver damage (see later discussion). It is more likely that fluorescence in these organs cannot be observed because of the high absorptivity of the blood for both the activating light (usually blue) and the fluorescent light (630, 690 nm). Thus, when tumors are treated in areas near or involving these organs, great care should be exercised to prevent their exposure to the therapeutic light.

Because of the tendency of HPD and DHE to be retained in skin, patients receiving these materials are cautioned to remain out of bright sunlight for a minimum of 30 days.

The HPD uptake, in at least some spontaneous tumors in animals and man, appears to be considerably higher than that measured in rodent tumors. For example, a spontaneous, primary adamantinoma tumor removed from a pet dog 72 hr after administration of 5.0 mg/kg HPD demonstrated considerably brighter fluorescence than that seen in a transplanted SMT-F tumor in DBA/2 mice 24 hr after receiving the same dosage of HPD. A sample of this primary tumor, when assayed for porphyrin, had a tissue level of porphyrin of approximately 50 μg/g tissue, more than ten times that of the SMT-F tumor (Dougherty TJ, unpublished data, 1983). A study reported by Wharen et al. (17) that included a transplanted rat glioblastoma tumor and three glioblastoma tumors in humans, used superficial tissue fluorescence of frozen biopsy samples to determine porphyrin levels in normal brain tissue and tumors. An astrooligodendroglioma (grade III) removed from a patient had a porphyrin level of 2.5 μg/g 24-hr postinjection of 5.0 mg/kg HPD. Interestingly, the porphyrin level in a patient with a similar tumor, examined 48-hr after injection of 5.0 mg/kg HPD, was only 0.1 μg/g, indicating a possibly rapid clearance of HPD from these types of tumors. This is quite unlike the transplanted mammary tumors reported by Gomer and Dougherty (16), which showed little clearance over many days.

III. PRECLINICAL STUDIES

Before initiating clinical studies, our group at Roswell Park Memorial Institute undertook extensive studies in experimental rodent tumors to demonstrate the efficacy and therapeutic benefit of PDT. The ability of HPD, activated in vivo by red light, to completely eradicate transplanted tumors and to spare normal skin was demonstrated (6). More recently, similar studies using DHE indicate an increased efficacy and a greater therapeutic ratio for this material when compared with HPD (see previous discussion).

IV. TOXICOLOGY

Direct drug toxicity was determined (Dougherty TJ, unpublished data, 1983) in albino Swiss mice (Ha/ICR) by injecting the porphyrin intraperitoneally and maintaining the animals in the dark. The LD_{50} dose, 24 hr after injection of HPD in male or female mice, was 275 mg/kg. At 14 days the LD_{50} values were 230 mg/kg and 180 mg/kg for male and female mice, respectively. Multiple doses (50 mg/kg every day for 5 days) or 25 mg/kg every other day for 10 days yielded no deaths. On a weight basis, DHE is approximately twice as toxic as HPD with LD_{50} values near 130 mg/kg. Death in all cases was attributed to liver and kidney necrosis. Necrosis of splenic and thymic lymphocytes was also observed. Bone marrow was unaffected.

Both porphyrins were considerably more toxic when animals were exposed to light. In these experiments, after shaving the entire back, the animals were exposed for 5 hr to full-spectrum xenon light (6 mw/cm^2) with the IR filtered out, immediately following systemic injection of various doses of HPD or DHE. Immediately after exposure, skin irritation and erythema on ears and feet were apparent at the higher drug doses. Moribund animals exhibited apparent hypoxia. The LD_{50} was 7.5 mg/kg for 24 hr and 14 day toxicity tests. The values for DHE were approximately 4 mg/kg. Death of animals in this group could not be attributed to any specific organ damage and was considered to be caused by a shock syndrome. Similar studies were done with rats (receiving 20 mg/kg HPD) but no deaths occurred in these animals, indicating a possible relationship to size or to the relative area of the exposed surface.

Because some applications of PDT include intraabdominal exposure, the toxicity of HPD plus light was studied in a variety of surgically exposed rabbit organs. In some cases the Brown-Pearce tumor was implanted into the organ 1 or 2 weeks before treatment. The animals received 5.0 mg/kg HPD 48-72 hr before surgical exposure and treatment. Five days later the animals were sacrificed and all exposed organs and tumors were examined histologically. In general, the stomach, small intestine, liver, pancreas, kidney, and bladder demonstrated slight to no effects at light doses of 70-100 J/cm^2 (630 ± 5 nm) whereas higher light doses produced more extensive damage. Also at 72 hr or longer after injection, light exposure was less damaging than at earlier intervals. The tumor underwent necrosis at all doses (30 J/cm^2 or higher) and time intervals studied.

V. EFFICACY STUDIES IN EXPERIMENTAL ANIMALS

Hayata and colleagues have carried out a preclinical study of PDT for treating intrabronchial squamous cell carcinoma induced in dogs (18). They also determined that they could detect most of these tumors by observing the HPD fluorescence induced by violet light from a krypton laser. In these studies the exciting light was delivered via a single quartz fiber in the biopsy channel of the bronchoscope fitted with appropriate filters and an image intensifier to observe the fluorescing (red) light. Normal bronchial mucosa and tumors in dogs that did not receive HPD were not fluorescent. This method has also proved feasible in detecting occult lung tumors in humans (19-22) as well as in delineating tumors and dysplasia in the bladder (23).

For treatment of these tumors, the red light from an argon dye laser system (approximately 630 nm) was directed through the fiber in the biopsy channel of the bronchoscope. The tumors received 60-240 J/cm^2 48 hr following administration of 2.5-3.0 mg/kg HPD. Complete response was documented histologically in the three animals thus treated. Only minor normal tissue effects were seen. Tumors in control animals treated with light at 60 or 120 J/cm^2 but without HPD and tissues in normal animals receiving HPD but no light underwent no histological changes.

Gomer and coworkers (24) examined acute toxicity in the normal rabbit eye following application of PDT. A dye laser system (635 ± 5 nm) similar to that used by Hayata et al. (18) was the light source directed onto a 1-cm^2 area of the retina through the cornea of pigmented rabbits. Rabbits received HPD (1-10 mg/kg) 48 hr before a 15-min exposure (40-400 mw/cm^2). Use of a variety of methods (fundus photography, fluorescein angiography, and histological examination) showed that light alone at power levels below 200 mw/cm^2 (180 J/cm^2 produced no damage, whereas higher doses (400 J/cm^2) led to damage ranging from edema to retinal detachment, even outside the treatment field. After administrating HPD at 2.5 mg/kg (the usual dose used in humans) treatment damage was minimal at light dosages of up to 45 J/cm^2 (40 mw/cm^2), whereas after administration of 10 mg/kg HPD, extensive damage was seen, even at the lowest light dose (45 J/cm^2). It was concluded that because the safe range for normal tissue damage occurred within the dose range usually used to treat tumors in humans, this method might be applicable to treatment of a variety of ocular tumors (see later discussion).

To determine the safety of PDT in the treatment of bladder tumors, Nseyo and coworkers exposed normal bladders in dogs to a variety of light doses (630 nm) 3 days after a systemic injection of DHE (2.5 mg/kg) (unpublished data, 1983). Using a 1-cm^2 spot size from the optical fiber within the bladder (fiber placed via a rigid cystoscope), doses as high as 360 J/cm^2 were used. At this dose, a lesion was produced that extended to the muscle layer (examined 5 days after treatment) but the dose was not sufficient to cause a fistula. At doses from 100-360 J/cm^2 only superficial lesions were produced. In some cases the bladder was first irritated chemically, or an ulceration was induced by biopsy before HPD was administered. No enhancement of damage was seen in these areas. In some cases, whole-bladder illumination (30 J/cm^2) was used without producing evidence of irreversible damage.

Jocham and colleagues (25) have used PDT to treat the Brown-Pearch carcinoma transplanted into the bladder of rabbits. They have demonstrated both a selective uptake of HPD in the bladder tumors and the ability to destroy the tumor without damage to normal urothelium. Occasionally the tumor was placed into the neck of the bladder and treated with an optical fiber that produced spherical illumination. In combination with a dispersing medium they demonstrated the ability to eradicate these tumors even in the neck of the bladder without damage to normal tissue. Light doses were 100-200 J/cm^2 (approximately 630 nm). (The dose of HPD was not stated, but presumably it was 5.0 mg/kg given 2 days before treatment, as in their detection studies.)

Because tunable dye lasers are most commonly used for PDT, it is important to determine the wavelength most effective in producing tumoricidal effects with HPD or DHE. Red light is used for activation rather than the shorter wavelengths (which actually are absorbed more strongly by these porphyrins) because the red is considerably more penetrating through tissue (discussed later). In solution the maximum wavelength for absorption of HPD is near 620 nm, whether or not serum albumin is added. However, when absorbed onto or into cells, the maximum red absorption is near 630 nm. The action spectrum for control of the transplanted SMT-F tumor in DBA/2 mice was determined by Dougherty and colleagues (Dougherty TJ, unpublished data, 1983). By using standard conditions (75 mw/cm^2), 30 min, one day after injection of 7.5 mg/kg HPD or 5.0 mg/kg DHE and by varying only the wavelength, it was determined that maximum response occurred at 630 nm. In an earlier study, this group had determined that maximum normal skin response of HPD in albino mice occurred at the same wavelength.

Shortly after PDT (immediate to 1 day, depending upon conditions) tumors in experimental animals became hemorrhagic (e.g., subcutaneous SMT-F tumor in the mouse), or in some cases totally blanched (e.g., green melanoma in the rabbit eye). When effectively treated, the tumor mass becomes nonpalpable or nonobservable, usually within a matter of a few days. Complete eradication is possible depending upon treatment conditions. It has been demonstrated that oxygen levels in experimental tumors drop rapidly during PDT treatment (26, 27). When examined histologically, the earliest changes occur in and around the tumor vasculature with red cell congestion and some enucleation of cells adjacent to the vessels. At later times, obvious red cell extravasation is seen, along with apparent total destruction of tumor cells (13).

It has been shown autoradiographically with the SMT-F tumor or the sarcoma 180 mouse tumor, that tritium-labeled HPD is located in the vascular stroma at a level five times that found in or around the tumor cells (13). A similar pattern can be observed by fluorescence microscopy of frozen sections of these tumors. However, a common characteristic of these experimental animal tumors is their poor vascularity. It is unclear whether similar distribution and vascular damage occur in tumors in humans.

VI. IN VITRO STUDIES

Several groups have studied the uptake of HPD in normal cells in culture for comparison with cells of malignant tissue origin. Moan et al. (28) examined four types of mouse embryo fibroblast cells of different oncogenic potential for their ability to take up HPD measured cytofluorometrically. There was no increase in HPD uptake or in photosensitivity with the increasing tendency for these cells to produce tumors in syngeneic immunosuppressed mice. Also, Chang and Dougherty (29) demonstrated no differences in uptake or clearance of HPD or TPPS between normal cells (CHO, L929) and cells of malignant character (HeLa, PC-1). Similar results for a variety of cultured cells were found by Henderson et al. (30). On the other hand, Christensen et al. (31) have demonstrated 25-50% more HPD uptake in transformed cells than in the normal counterparts (10T1/2). Also, Andreoni et al. (32) have demonstrated enhanced destruction of a transformed epithelial rat thyroid cell compared with the normal cell following exposure to HPD and light.

Berns et al. (33) examined a kangaroo epithelial cell line (PTK_2) after exposure to HPD and found a rapid uptake for the first 2 hr, followed by slower accumulation over the next 2 days. They used a fluorescence detection system for individual cells. Excretion was measured after HPD was removed from the medium, which demonstrated that a residual amount of HPD could not be removed, the residual increasing with time of initial exposure. Also in this study, the fluorescence in myocardial cells was seen to be primarily associated with mitochondria located near the nuclear membrane. This observation correlates with early damage to mitochondria observed by Coppola et al. (34) in lymphoma cells treated in vitro with hematoporphyrin and light.

Much of the data indicate the central role of cellular membrane as the site of damage resulting from porphyrins and light. This damage may be a result of cross-linking of membrane protein reported by Dubbelman et al. (35) and Girotti (36). However, Hilf et al. (37) have demonstrated inactivation of mitochondrial membrane enzymes as an early event in cells exposed to light in vitro after in vivo uptake of HPD.

Recently, several investigators have demonstrated a "tightly bound" porphyrin fraction in cells exposed to HPD or DHE. Its fraction of total porphyrin increases with time of incubation to a maximum of 50-60% at 24 hr. Henderson et al. (30) and Bellnier and Lin (38) have demonstrated increased photosensitivity of cells containing tightly bound porphyrin compared with those with "loosely bound" (i.e., removed by washing in media) porphyrin. It is expected that these in vitro conditions may be better correlated with the in vivo situation. Kessel and Chou (39) and Christensen et al. (31) have demonstrated an enhanced in vitro uptake of DHE compared with uptake of the other components of HPD. The DHE appears to correspond to the tightly bound fraction of HPD because it is the only HPD component that is not removed from cells by washing following even long-term exposure.

Although very little of the HPD in cells is taken up by the nucleus, effects of photoradiated porphyrins on DNA can be measured. Single-strand breaks have been demonstrated by Moan and Boye (40) in NHIK 3025 cells exposed to hematoporphyrin and light, but the frequency is low compared with that of ionizing radiation and does not correlate with cell survival. Also, sister chromatid exchange has been demonstrated by Moan et al. (41) and by Gomer et al. (42). On the other hand, HPD plus light does not produce mutants in CHO cells when light doses are used that are sufficient to produce survival similar to that produced by x-irradiation, which produced a high degree of mutation in these cells (42). It has also been reported (J. Kennedy, unpublished data, 1983) that HPD plus light is negative in the Ames test for mutagenicity in bacteria.

VII. DOSIMETRY OF LIGHT IN TISSUE

An important factor determining the extent of tumor necrosis is the penetrability of the necessary visible light through tissue. Svaasand et al. (43,44) have considered this question theoretically and made measurements of penetration of various wavelengths in various tissues in experimental animals and tissues. For a diffusion-dominant, one-dimensional case (surface illumination)

$$\Phi = \Phi_0 e^{-\alpha X}$$

where Φ = space irradiance

Φ_0 = irradiance at surface

α = total attenuation coefficient (β/ζ) 1/2 where β = absorption coefficient and ζ = diffusion coefficient
X = distance

Measured values for δ (mm) = I/α, where δ is the attenuation distance or the depth at which the incident intensity falls off to I/ϵ or approximately 37%, range from 1 to 4 mm. When fibers are inserted into the tumor for interstitial PDT, the equation for space irradiance in the diffusion-dominant case is

$$\phi = \phi_0 \frac{(a)}{(r)} e^{-\alpha(r-a)}$$

where r = radial distance in tissue
a = radius of fiber

The pathological appearance of tumors removed after treatment by PDT generally indicated an area of necrosis of 5-10 mm. Coagulation necrosis of a tumor obstructing the bronchi after PDT has been demonstrated by inserting special fibers that allow distribution of lateral light. Such fibers have been demonstrated to cause necrosis within a radius of 9-10 mm from the point of insertion (Dougherty TJ, unpublished data, 1983).

Although many of the early clinical studies were conducted using a filtered xenon arc lamp with a fairly broad spectral output of 600-700 nm, such systems are rarely used today, although they may have some application for superficial skin lesions. Most clinical studies use lasers to take advantage of the high coupling efficiency of optical fibers that allow light to be delivered conveniently to many areas.

A variety of delivery fibers are available: those with an optically flat end, with lenses to produce a spot of homogeneous intensity, usually used to treat cutaneous or subcutaneous tumors; those with various lenses on the ends to further expand the spot; and those with diffusers of various lengths that produce a cylindrical pattern along the diffuser length and are generally used within a lumen or for interstitial PDT. Fibers without lenses or diffusers should not be used because they produce inhomogeneous or hot spots when used interstitially.

Although PDT is a photochemical rather than a thermal ablation or coatulative process as with the CO_2 or Nd:YAG laser, considerable heat can be generated at the tips or broken surfaces of the fibers if they come in contact with strongly absorbing tissue or blood. One instance has been reported of a fire during intratracheal PDT under 100% oxygen when a hot spot on the fiber came in contact with tissue, or blood, or both. No instances have been reported for treatments performed in air, although it must be borne in mind that even though a fire is not likely in air, the fiber can nonetheless be destroyed if high power is applied when in contact with blood or other strongly absorbing tissue.

VIII. PHOTODYNAMIC THERAPY FOR TREATMENT OF CANCER IN HUMANS

Since the early clinical reports from Roswell Park Memorial Institute describing the efficacy of PDT for treatment of cutaneous and subcutaneous lesions (45, 46), considerable additional clinical data have been reported from several groups in the United States and abroad. The following is a compilation of the current state of these studies.

A. Cutaneous and Subcutaneous Cancer

Tumors in this category include basal and squamous cell carcinoma, malignant melanoma, metastatic breast carcinoma, mycosis fungoides, Kaposi's sarcoma, and Bowen's disease (Table 1). Note that the response percentages are approximated because not all investigators used the same criteria of response and not all reported their data as indicated here.

A number of points must be made concerning these data. Most patients had received multiple therapies before the PDT including ionizing radiation in most cases. The tumors ranged in size from a few millimeters to more than 10 cm and in pigmentation from slight color to black (highly pigmented melanoma). In most cases, the light source was a laser emitting at approximately 630 nm, but some patients were treated with other light sources making direct comparisons difficult. In general, the dose was 15-100 mw/cm^2 to total delivered light dose of 20-120 J/cm^2. Many patients, especially those with large tumors treated aggressively, experienced moderate to severe pain. This invariably occurred when overlying or adjacent skin became necrosed forming deep eschars. This did not occur at the lower-dosage range. The superficial tumors (1-3 mm) responded completely at the lower dose range, usually preserving the normal skin. Another side effect, besides the overall photosensitivity, was increased sensitivity to PDT in previously irradiated (ionizing) fields in patients who had received doxorubicin (Adriamycin) up to a few months before PDT. Also, patients with inflammatory breast cancer frequently were unusually oversensitive to treatment.

Patients with small, superficial tumors can benefit from PDT even if the lesions are widespread and located in areas previously receiving ionizing irradiation. Optimum conditions for treatment are administration of 3.0 mg/kg HPD, or 1.0-1.5 mg/kg DHE, 20-35 J/cm^2 (630 nm) for HPD and 30-72 J/cm^2 for DHE delivered 2-5 days before injection. Higher light doses destroy normal skin as well as the tumor. Patients with large tumors ($\geqslant$ 1 cm) usually benefit only if the tumors are relatively few and localized. These tumors are best treated aggressively by a combination of superficial treatment ($>$ 70 J/cm^2), confined to the tumor as much as possible, and by interstitial PDT. Some must be treated repeatedly after sloughing or debriding of dead tissue. As a general rule, a depth of tumor 0.5-1.0 cm from the light source will be destroyed by a given treatment whether applied

TABLE 1 Clinical Results from Various Centers: Skin Cancers

Type	Approx. no. pts./sites	Response (%)[a]			Longest follow-up without recurrence (to date) (yr)
		CR	PR	NR	
Metastatic breast cancer	120/>1000	60-70	20-30	10	4
Basal cell carcinoma	15/50	70-80	20-30	0	4
Squamous cell carcinoma	5/10	20	70-80	10	1
Malignant melanoma	50/>1000	50	30	20	1
Mycosis fungoides	5	20	60	20	1
Kaposi's sarcoma	5/>100	80	20	0	3
Bowen's disease	2/10	100	-	-	1.5

[a]CR = Complete response: disappearance of tumor or proved by biopsy. PR = Partial response: at least 50% reduction in tumor volume. NR = No minimal response: Less than 50% reduction in tumor volume.

Source: Data from Refs. 46-51.

superficially or interstitially. Therefore, multiple interstitial treatments with multiple fibers (500 mw/cm length of diffuser) are necessary to destroy tumors of large volume. Although an occasional patient may benefit from this aggressive approach, debulking of a tumor is not the best application of PDT, which optimally should be applied before the patient reaches this advanced stage, whether or not previous therapies have been tried. Photodynamic therapy can be beneficial to many of these patients if it is applied before the diesase is clearly out of control.

B. Lung Cancer

The first patient to receive PDT for lung cancer was treated by Hayata and associates at the Tokyo Medical College in early 1980 (52). Nearly 4 years after treatment this patient was cancer-free at autopsy following a heart attack.

The procedure used most often was to inject 2.5-3.0 mg/kg HPD 3 days before endoscopic PDT. Some of the patients were anesthetized, and some were mildly sedated. Light from the laser (about 630 nm) was delivered through 200 μm or 400 μm quartz fibers placed into the channel and externally directed at the lesion for smaller tumors or inserted into the mass for large or obstructing tumors. It has been recently discovered that the fibers with diffusing tips (described earlier), which allow lateral light distribution, are most convenient for these treatments or for others in which intraluminal tumors are treated. Although accurate determination of light dosimetry is difficult in these cases because of the wide variation in size and shape, doses of approximately 100-200 J/cm^2 have been effective for treatment of most lesions (e.g., 200 mw/cm^2 for 10-15 min).

Hayata and colleagues (53) have recently reported treatment results of 73 patients with lung cancer, 13 of whom had early-stage squamous cell carcinoma (mucosal involvement only) and eight had stage I tumors. Eight of the patients with early-stage cancer had inoperable tumors and five had subsequent surgery after PDT. Macroscopically complete tumor remission was obtained in the eight patients (13-41 months posttreatment). In the five patients who underwent resection following PDT, a complete histological response was seen in two, and there was > 60% response in the other three. Of the eight patients with stage I tumors, three had surgery after PDT. Macroscopically complete response was seen in two of the patients with nonresected tumor, and there was > 60% response in the other three patients. The three patients with stage I tumors who had surgery all showed some residual tumor. These authors recommend PDT for the early-stage tumor patients with superficial or intramural invasion but not for those in whom the tumor has invaded beyond normal muscular or cartilage layers. Viable tumor cells tend to remain behind these structures. In the latter types of patient, combination with other modalities, such as chemotherapy or immunotherapy, is recommended. In some cases, inadequate light delivery (diffusing fibers were not used) was apparent because the invading tumor was eradicated in some areas but not in other directly adjacent areas.

A major advance in the application of PDT to advanced, partially, or totally obstructing lung tumors has been reported by Balchum et al. (54). Their protocol consisted of injecting 3.0 mg/kg HPD and 3 days later illuminating the tumor via a quartz fiber with a special end made for light distribution. In most cases the fiber, with a lateral diffusing end of 1-2 cm, was inserted directly into the tumor. The power at 630 nm was 400 mw/cm of diffusing length. The treatment time was approximately 8.5 min, thus producing 200 J/cm^2 of diffuser. For smaller lesions, the same fiber was held within the remaining lumen adjacent to the tumor. Patients were rebronchoscoped 3 days after PDT and tumor debris and firm exudate were removed. If residual tumor was found (11/35 cases) PDT was repeated, again 3 days after a second injection of HPD (3.0 mg/kg). Cleanup bronchoscopy was performed 3 days after PDT. If indicated by dyspnea, cleanup bronchoscopy was done as early as 8 or 24 hr after PDT and whenever necessary thereafter. Thirty-three of the 35 patients with endobronchial cancer had total opening of the lumen with no remaining tumor visible, although biopsy results were frequently positive at the cleanup procedure. One other patient had a partially opened lumen, and one patient with a benign, fibrous mass showed no response. Twenty-one patients had squamous cell carcinoma, one had large cell carcinoma, and four had adenocarcinoma. Other patients had adenoid cystic carcinoma, malignant fibrous histiocytoma, and carcinoid. Metastatic tumors included those from thyroid, colon, and breast.

In an earlier study by Balchum (Balchum OJ, unpublished data, 1983) and others it had been noted that some patients were at considerable risk after PDT because of heavy secretions that could cause further obstruction or result in obstructive pneumonia. Balchum has obviated this problem by selecting patients who are nonterminal and who have reasonable respiration upon mild exertion. Also, he routinely rebronchoscopes the patients at 3 days after PDT to remove debris and secretions. An earlier rebronchoscopy was necessary because of respiratory distress in only a few patients.

Of the four patients with atelectasis of the lung, all experienced reexpansion and reaeration after PDT 1-3 days after cleanup bronchoscopy. Dyspnea in each patient became less and respiratory function improved. Complications included infection (3/35) which was controlled by antibiotics. Twenty-four of 31 patients with obstructing endobronchial tumors lived 3 or more months after PDT (one patient is still living after 14 months). These studies indicate that PDT can be carried out safely and effectively for palliation in this group of patients.

Hayata and colleagues (52) and Kato et al. (55) have applied PDT as an adjunct to surgery in patients in whom a surgical procedure had been precluded because of extension of the disease or where the extent of surgery could be reduced by prior

PDT. In four of five patients whose tumors were considered inoperable, surgery became possible following PDT to remove tumors in the trachea, carina, or main bronchi. In five other patients, a lobectomy was possible rather than a pneumonectomy after PDT. Two patients had two primary cancers. In one patient the lower lobe was resected and the upper lobe tumor treated with PDT, and in the other, both lesions were treated with PDT followed by middle and lower bilobectomy. Surgery was performed 3-4 weeks after PDT. In one patient the tumor recurred 11 months after treatment. Several other patients are currently surviving nearly 2 years after treatment. Overall results, to date, for 180 patients are summarized in Table 2 (54,56-59). Recent studies reported by Cortese and Kinsey (59) and Hayata et al. (57) demonstrate that early-stage lung cancers may be effectively treated with PDT and that this may be a useful technique to delay surgery in patients at high risk for new tumors. These investigators have recently reported results of PDT for treatment of carcinoma in situ (CIS) or early-stage endobronchial tumors. Seventeen of 32 patients had a complete response to treatment. Follow-up study ranged from 1 to 4 years.

C. Cancer of the Esophagus

The rationale for the use of PDT in esophageal cancer is similar to that for the treatment of lung cancer, i.e., early-stage disease can be treated locally with little risk, and advanced cases may receive palliation. It should be emphasized that PDT provides the clinician with another modality that should be appropriately worked into the overall management of the patient. The guiding principle is that PDT is more localized and selective than some of the other treatments but cannot be expected to entirely eliminate large, bulky tumor, especially outside the lumen or in lymph nodes.

To date, approximately 30 patients with esophageal cancer have received PDT. Because in most instances, esophageal cancer, like lung cancer, is not often detected in an early stage (except in China), most patients have had advanced disease and previously had received therapy (surgery, radiation therapy). A case described by Aida and Hirashima (60) indicates the benefit, however, even for some patients with advanced stage disease. A patient had undergone total gastrectomy, total resection of the thoracic esophagus, and right hemicolectomy. Two early-stage lesions were found in the esophagus as well as in the residual cervical esophagus for which he received two courses of radiation therapy (>5,000 rad). Further surgery had been ruled out. However, within a few months of radiation, the biopsy results remained positive. He received an injection of 2.5 mg/kg HPD and 2 days later endoscopic PDT was performed followed by a second treatment the next day. Two years later he remains free of disease.

Tian, of the Henan Tumor Institute (M. Tian, unpublished data, 1983) has treated 13 patients with early and stage I esophageal cancer which had been de-

Table 2 Photodynamic Therapy Clinical Results from Various Centers[a]: Lung Cancer

Institution	Patients (no.)	Disease category	Method[b]	Response CR	Response PR	Response NR	Complete opening	Adverse experiences
RPMI	17	Very advanced	S	-	13	2	-	Exudate, fever, abscess: 14
	34	Advanced	S	7	10	17	-	Exudate: 6 Hemoptysis: 5(?)
	5	Advanced	I	-	-	-	5	-
USC	70	Advanced	I	-	-	-	70	-
Tokyo Med. Coll.	107	Advanced	S	30	77	0	-	Exudate
	46	100% obstructed	S/I	-	-	-	28	-
	13	Early	S	10	3	0	-	-
Mayo Clinic	19	Early	S	7	8	c	-	Sunburn: 3; cough: 4; obstruction (temporary): 4
Other	50	-	-	-	-	-	-	-
Totals	311			54	111	19	103	

[a]S = Superficial treatment; I = Interstitial treatment.
[b]Maximum follow-up: RPMI, 1 yr; USC, 1 yr; Tokyo Med. Coll., 4 yr; Mayo Clinic, 2 yr.
[c]4 lost to follow-up.

Source: Data from Refs. 54, 56-59.

tected in a mass screening program. He used a procedure similar to that described by Hayata for treating early bronchial tumors with PDT; i.e., 630 nm light delivered endoscopically to the lesion 72 hr after injection of 2.5 mg/kg HPD. The dose was 100-200 J/cm^2 delivered from a fiber without any type of lens or diffuser. In four cases (stage I) the patients proceeded to surgical intervention. In three of these, residual tumor nests were found in the submucosal layers. This may have been a result of inadequate light delivery, a common problem with these types of fibers because they produce an inhomogeneous spot. Fibers with diffusing ends are recommended. The remaining patients have shown no recurrences of disease for up to 1 year after treatment. McCaughan et al. (61) has reported on seven patients with partial or total esophageal obstructions. Photodynamic therapy was used to relieve symptoms of dysphagia and to restore the patients' ability to eat solid foods. With these patients the fiber was inserted into the tumor or remaining lumen. Light delivery was 200-300 J/cm^2 of diffuser. All patients responded to treatment with massive sloughing of tumor within a few days of treatment. Dysphagia was relieved and normal diet was resumed in all patients. McCaughan points out that although residual tumor may remain after treatment, palliation has been achieved in one patient for up to 9 months. Similar results have been obtained by Aida and Hirashima (60) and H. Nava (unpublished data, 1983), although these authors noted that patients complained of severe mediastinal pain for several days after treatment. Also, one patient developed an esophageotracheal fistula presumably because of sloughing of the invading tumor. One patient in this group of five remained free of symptoms for more than 1 year after treatment.

D. Bladder Cancer

Photodynamic therapy may become the treatment of choice for multicentric superficial cancer of the bladder because it may provide a greater degree of selectivity than other treatments currently applied in these cases. Benson et al. (23) have demonstrated a high degree of selectivity of HPD for dysplasia and carcinoma in situ in the bladder. To date, approximately 62 patients with bladder cancer have been treated with PDT, ranging from CIS to invasive carcinoma (Table 3). Both sessile and pedunculated tumors have been treated. The HPD dose was 2.5 mg/kg, and the wavelength was 630 nm from dye laser systems delivered endoscopically. In these cases the light dose has ranged from 150 to 360 J/cm^2 for treatment of individual lesions. The nonresponding tumors were those receiving only 50 J/cm^2 regardless of size.

Benson (67) had treated patients 3 hr, as well as at 48 hr, after HPD injection with equivalent results, whereas all the other studies were done at 48-72 hr after HPD injection.

To date, only a few patients have been treated by whole-bladder illumination

Table 3 Photodynamic Therapy Clinical Results from Various Centers: Bladder Cancer

				Response (to 2 yr)			
Institution	No. of pts./(sites)	Disease category	Method[a]	CR	PR	NR	Adverse experiences
RPMI	4	Very advanced	WB	1	3	-	Bladder shrinkage: 2
Mayo Clinic	14	CIS/TCC[b]	WB/F	11	-	3	None
WILT (Santa Barbara)	6	CIS/TCC	WB/F	4	2	-	Temporary bladder shrinkage: 1; sunburn: 2
Kanazawa University	9 (46)	CIS/TCC	F	(28)	(8)	(10)	None
	18	CIS/TCC	WB/F	18	-	-	Temporary bladder shrinkage; 3-4 mo follow-up
Tokyo Medical College	11	CIS/TCC	F	8	3	-	None
Totals	62 (93)			(70)	(16)	(13)	

[a]WB = whole-bladder treatment; F = focal treatment.
[b]CIS = carcinoma in situ; TCC = transitional cell carcinoma.

Source: Data from Refs. 62-66.

after administration of HPD or DHE. All patients had previously received resection, radiation, and chemotherapy. Following light doses of 35-100 J/cm^2 (whole-bladder) bladder shrinkage was observed. In most patients some shrinkage had already occurred because of previous treatments. Dose ranges of 10-25 J/cm^2 appear to reduce this problem. Also, H. Hisazumi (personal communication) has found that the bladder shrinkage is related to the bladder volume before treatment and recommends a pretreatment volume of at least 150 ml. It is also likely that the extensive degree and the time of bladder distention are sufficient to produce fibrosis and shrinkage (68).

E. Gynecological Tumors

The application of PDT for this type of tumor has not been extensive, but it is impressive. In 1982, Ward et al. (69) reported treatment of five cases of recurrent vaginal cancer by PDT. Also in 1982, Soma et al. (70) reported the use of PDT in a single patient with primary vaginal carcinoma; more recently he has treated 13 patients. These patients received 2.5-5.0 mg/kg HPD, usually 2-3 days before PDT. Treatment was performed by surface illumination in most cases, at a dose rate of 25-100 mw/cm^2 and a total dose of approximately 200-300 J/cm^2. Many of the lesions reported by Ward were large ($>$ 3 cm in depth) and were treated by multiple interstitial placements of the fiber at doses of approximately 600 J per insertion where thermal effects are likely to be substantial. However, Soma and Nutahara (71) have successfully eradicated large lesions using only surface illumination of up to 500 J/cm^2 at a low dose rate of 25 mw/cm^2, where thermal effects would be of little consequence. In several patients, initial treatment produced only partial response. These patients were retreated within 1-3 weeks resulting in complete responses. Rettenmaier et al. (72) recently reported treatment of six patients with recurrent cervical, vaginal, and endometrial tumors, using total delivered light doses of only 30-40 J/cm^2. Two of these patients demonstrated a complete response to treatment. Aside from the usual photosensitivity to treatment, no side effects were noted. Excellent reepithelialization was seen after tumor necrosis and sloughing, which usually occurred over 3-4 weeks.

F. Cancers of the Head and Neck

Wile and coworkers (51,73) studied various head and neck tumors. Nineteen sites in 16 patients were treated for recurrent squamous cell carcinoma. Five of these demonstrated complete response to treatment (up to 1 year, currently) and 10 demonstrated partial responses. In some patients an endoscope was used to visualize the lesion, which was then treated through the scope. In most cases, patients were given 3.0 mg/kg HPD, 72 hr before treatment. Light doses were in the range of 25-90 J/cm^2. Among the patients who benefited from treatment was a patient with a tongue cancer that had persisted following radiation therapy but that responded completely to PDT with improvement in the patient's eating and speaking. A second patient with a recurrent tumor of the vocal cord that recurred after radiation therapy also responded completely to PDT. Wile is currently carrying out a more extensive study of these types of tumors. Also, Taketa and Imakiire (74) have reported on six cases of squamous cell carcinoma including laryngeal carcinoma, cancer of the nasopharynx, tongue, and vocal cords. In most patients, a reduction in the tumor mass was found, but the biopsy results usually remained positive in the deeper portion of the tumors involving areas beneath the mucosa.

G. Other Tumors

Photodynamic therapy is currently being investigated for treatment of a variety of other types of tumors, but results are still very preliminary. For example, Laws et al. (75) have reported a feasibility study for PDT treatment of brain tumors, and McCulloch et al. (76) have applied PDT in combination with surgery and ionizing radiation for the treatment of at least 17 patients with glioblastoma.

One of the most promising new areas for application of PDT is in treatment of intraocular tumors, such as malignant melanoma of the choroid and retinoblastoma. Early results of Murphree et al. (77) have shown that malignant melanoma of the choroid responds to PDT (3.0 mg/kg HPD) when the light is applied both through the cornea and through the sclera. Similar results have been obtained by Bruce (78). A long-term follow-up report of these patients is not yet available, although there is one recurrence, at approximately 1 year after PDT, of a choroidal melanoma treated by L'Esperance and Dougherty (F. L'Esperance and T. J. Dougherty, unpublished data, 1983). The proper light dose and means of delivery are still under investigation. In general, patients treated by Bruce (78) were treated with high power densities (up to 600 mw/cm^2), whereas Gomer et al. (24) have limited the dose rate to 200 mw/cm^2, at which dose little damage occurred to ocular structures in rabbits. The high degree of pigmentation of some of the melanoma lesions raises the possibility of a major contribution by thermal effects to the overall results.

IX. MECHANISM OF ACTION

Questions about the uptake and distribution of HPD and DHE in animal tissue, cellular target sites, tissue target sites, and mechanism of tissue distribution following PDT have been addressed for several years by several groups. However, until recently, little definitive data were available. As previously pointed out, effects of PDT on mitochondria, cellular membranes, and DNA have been found, although the initial target(s) is/(are) still uncertain.

In a study by Bugelski et al. (13), the distribution of $[^3H]$HPD matched the distribution of labeled albumin. At longer periods (over 6 hr) the porphyrin was cleared from most normal tissue, with certain exceptions. No change was seen in the tumor within a 3-hr to over a several-day period, although dilution occurred, presumably caused by cell proliferation. Within the tumor the porphyrin was primarily associated with vascular stroma (macrophages, mast cells, possibly fibroblasts). The relative density of porphyrin in stroma to that in tumor cells was approximately 5:1. In the liver, hepatocytes appeared to retain little if any prophyrin, but Kupffer cells retained high levels. The porphyrin appears to be generally associated with reticuloendothelial cells in all tissues. Fluorescent microscopy of experimental mouse tumors indicates that fluorescence occurs intracellularly in cells associated with the vascular stroma. Although some of these are probably macrophages, it is unclear if other cells, e.g., the endothelium of the tumor vasculature, also retain the material. It is also unclear if such cells in the tumor behave fundamentally differently from cells in normal tissues. Only recently has HPD fluorescence been demonstrated in endothelial cells of connective tissue surrounding a recurrent rectal carcinoma.

Little is known about HPD distribution within tissues and spontaneous tumors in man other than that most are seen to be fluorescent after HPD injection. In general, with equal doses per weight of injected drug, tumors in humans and spontaneous tumors in dogs are much more fluorescent than are the transplanted tumors that have been studied to date in rodents.

As indicated earlier, the porphyrin in the experimentally induced mouse tumors appears to be associated primarily with the vascular stroma. In fact, the earliest histological effects of HPD plus light occur in this area. Bugelski et al. (13) examined tumors by electron microscopy at various times following treatment. The earliest change that could be observed was blebbing of tumor cells adjacent to the microvasculature. The endothelium appeared to be intact at this point. After longer periods, tumor cells more remote from the vessels demonstrated blebbing, endothelial cells were retracted and damaged, and red cells had extravasated. Finally, massive extravasation and tumor cell destruction were observed.

Such data raised the question about the role of direct tumor cell destruction, such as that studied in vitro (discussed earlier), and in the destruction of the vasculature in the overall process of tissue destruction after PDT.

Henderson et al. (79) have studied this question by examining tumor cell survival after PDT in vivo, using the EMT-6 tumor system—an in vivo to in vitro colony-formation survival assay. Tumors were treated using DHE and 630 nm light with doses that produced destruction of tumor bulk in all animals and 52% long-term cures. No appreciable tumor cell inactivation was found following PDT in vivo when tumors were explanted immediately after completion of treatment. However, when tumors were allowed to remain in situ for various lengths of time (1-10 hr) following treatment, tumor cell death occurred rapidly and progressively. This delayed cell death was similar to that caused by shutdown of the tumor blood circulation by simply killing the mouse. Thus at least in this system, death of tumor cells seemed to be the result of vascular damage by PDT rather than of direct photodynamic damage. Although it is possible that sublethal damage to tumor cells combined with vascular destruction is necessary for cure of the tumor, this has not yet been demonstrated. Recently, Henderson et al. (79) have found that the subcutaneously implanted L1210 mouse leukemia is destroyed by PDT through what seems to be primary tumor cell destruction. The situation in patients is likely to be a combination of both vascular and tumor cell effect.

Singlet oxygen appears to be involved in the in vitro destruction of cells containing prophyrins (80) (as discussed earlier). While no comparable experiments have been carried out in vivo, it is likely that singlet oxygen may be involved at least as an initially formed cytotoxic species. It has been demonstrated (K. Weishaupt and T. J. Dougherty, unpublished data, 1983) that if tumors in the legs of mice are shut off from oxygen by clamping the legs the usual destructive effect of HPD plus light is prevented.

Recent results indicate a synergistic interaction between PDT and heat in experimental animal systems (81). A reasonable means of heat delivery may be to use the Nd:YAG laser concurrently with, or subsequent to, delivery of 630 nm light.

X. CONCLUSION

Table 4 summarizes current results for patients treated in the past 5 years by PDT at several medical centers. Most of these patients had failed or had had recurrence of the disease with multiple other therapies before PDT. In spite of this, most patients responded to treatment, many showing complete response. However, in some of the patients with more advanced disease, especially those with multiple skin metastases, some tumors responded completely and some did not, because in the early stage of development the emphasis was on determining the appropriate light dose for complete response without unacceptable normal tissue damage. In some areas, e.g., the bladder, this is still the intention. Even though DHE is retained in malignant tissue at higher levels than in most other tissues, in many cases the light dose is still limited by normal tissue reactions. Clearly, the optimum conditions need to be determined for each site because normal tissue reactions vary considerably. For example, normal skin can tolerate only 30-50 J/cm^2 without damage, whereas normal bronchial mucosa shows only slight edema and erythema at 100 J/cm^2. The differences reflect the variation in porphyrin content of normal tissue as well as the tissue's optical properties.

Photodynamic therapy has been effective in the treatment of tumors of the skin, bronchus, esophagus, bladder, head and neck, and in gynecological tumors, and it is also showing promise in many other areas. One of its major advantages is that it can be applied effectively and safely after other modalities, such as sur-

Table 4 Summary of Response of Various Tumors to Photodynamic Therapy (to 1984)

Tumor type	No. of patients or (sites)	Response[a] CR	PR	NR	Longest follow-up to date (yr)
Skin (met. breast, basal cell, squamous, melanoma)	(219)	(147)	(29)	(43)	4
Endobronchial: late stage	262	106[b]	136	20	1.5
Endobronchial: early stage	32	17	10	-	4.5
Bladder: superficial	(70)	(38)	(16)	(13)	2
Head/neck: recurrent	49	9	26	14	2
Gynecological: recurrent	11	4	5	2	2
Esophagus: advanced	17	0	11[c]	6	1

[a]CR = Complete response: disappearance of tumor or proved by biopsy. PR = Partial response: at least 50% reduction in volume of tumor. NR = No/minimal response: less than 50% reduction in volume of tumor.
[b]Complete opening to wall of lumen following physical removal after PDT.
[c]Partial relief of obstruction.

gery, chemotherapy, and radiation therapy, thus adding one more modality to those already available to the oncologist. Although it can be used to eradicate relatively large tumors (4-5 cm), it appears to be especially advantageous to the patient with early disease or early recurrence or, as clearly demonstrated by Balchum et al. (19), for relief of bronchial obstructions. In some patients it may be a viable alternative to debilitating surgery, which can be considered a later option, if necessary. For some patients with inoperable tumors (e.g., lung cancer patients with impaired respiratory function), it provides a reasonable alternative to other modalities, and in some cases may be the preferred treatment. While some of the studies are still considered to be at an early phase because of numbers of patients and short follow-up time, several phase III controlled trials are currently underway for treatment of superficial bladder tumors and endobronchial tumors.

XI. SUMMARY

Photodynamic therapy is a new, experimental method of treating malignant tumors by utilizing the relatively selective retention of the photosensitizer (hematoporphyrin derivative or dihematoporphyrin ether) and its ability to elicit an efficient photodynamic reaction upon activation with penetrating viable light. Application of this therapy to tumors in the bronchus, bladder, skin, and several other sites has demonstrated both safety and efficacy even in advanced cases. Eradication of early-stage tumors in the bronchus and bladder has been demonstrated. Selective retention of the photosensitizers in tumors is apparently related to the relatively large size of the aggregates of these materials that cause phagocytosis by reticuloendothelial cells and prevent rapid clearance from tumor interstitial fluid and subsequent uptake in lipophylic components of cells. Upon light activation, generally delivered from lasers via fiberoptics, the sensitizers generate singlet oxygen, the apparent cytotoxic agent, causing both vascular damage and injury to tumor cells.

REFERENCES

1. Rabb O: Üeber die Wirkung fluorescirender Stoffe auf Infusorien. *Z. Biol.* *39*:524-546, 1900.
2. Spikes JD and Straight R: Sensitized photochemical processes in biological systems. *Ann. Rev. Phys. Chem.* *18*:409-436, 1967.
3. Policard A: Experimental tumors studied by Wood's light. *Compt. Rend. Soc. Biol.* *91*:1423-1424, 1924.
4. Lipson RL, Gray MJ, and Baldes EJ: Hematoporphyrin derivative for detection and management of cancer. *Proc. 9th Int. Cancer Cong.,* Tokyo, Japan, 1966.
5. T. J. Dougherty, W. R. Potter, and K. R. Weishaupt. In *Porphyrins in Tumor Phototherapy* (A. Andreoni and R. Cubeddu, eds.). Plenum Press, New York, 1984, pp. 23-35.
6. Dougherty TJ, Grindey GB, Fiel R, Weishaupt KR, and Boyle DG: Photoradiation therapy. II. Cure of animal tumors with hematoporphyrin and light. *J. Natl. Cancer Inst.* *55*:115-121, 1975.
7. Kelly JF and Snell ME: Hematoporphyrin derivative: A possible aid in the diagnosis and therapy of carcinoma of the bladder. *J. Urol.* *115*:150-151, 1976.
8. Dougherty TJ: Photodynamic therapy (PDT) of malignant tumors. *CRC Crit. Rev. Oncol. Hematol.* *2*:83-116, 1984.
9. Tomio L, Zorat PL, Corti L, Calzavara F, Reddi E, and Jori G: Effect of hematoporphyrin and red light on AH-130 solid tumors in rats. *Acta Radiol.* *22*:49-53, 1983.
10. Dougherty TJ: Hematoporphyrin as a photosensitizer of tumors. *Photochem. Photobiol.* *38*:377-379, 1983.
11. Berenbaum MC, Bonnett R, and Scourides PA: In vivo biological activity of the components of haemotoporphyrin derivative. *Br. J. Cancer* *45*:571-581, 1982.

12. T. J. Dougherty, D. G. Boyle, K. R. Weishaupt et al. Photoradiation therapy –clinical and drug advances, in *Porphyrin Photosensitization* (D. Kessel and T. J. Dougherty, eds.). Plenum Press, New York, 1983, pp. 3-14.
12a. Grossweiner LI and Goyak GC: Binding of hematoporphyrin derivative to human serum albumin. *Photochem. Photobiol. 40*:1-4, 1984.
13. Bugelski PJ, Porter CW, and Dougherty TJ: Autoradiographic distribution of hematoporphyrin derivative in normal and tumor tissue of the mouse. *Cancer Res. 41*:4606-4612, 1981.
14. Doiron DR, Profio E, Vincent RG, and Dougherty TJ: Fluorescence bronchoscopy for detection of lung cancer. *Chest 76*:27-32, 1979.
15. T. J. Dougherty, W. R. Potter, and K. R. Weishaupt. The structure of the active component of hematoporphyrin derivative, in *Porphyrin Localization and Treatment of Tumors* (D. R. Doiron and C. J. Gomer, eds.). Alan R. Liss, Inc., New York, 1985, pp. 301-314.
16. Gomer CJ and Dougherty TJ: Determination of [^{3}H]- and [^{14}C] hematoporphyrin derivative distribution in malignant and normal tissue. *Cancer Res. 39*: 146-151, 1979.
17. Wharen RE Jr, Anderson RE, and Laws ER Jr: Quantitation of hematoporphyrin derivative in human gliomas, experimental central nervous system tumors, and normal tissues. *Neurosurgery 12*:446-450, 1983.
18. Hayata Y, Kato H, Konaka C, Hayashi N, Tahara M, Saito T, and Ono J: Fiberoptic bronchoscopic photoradiation in experimentally induced canine lung cancer. *Cancer 51*:50-56, 1983.
19. Balchum OJ, Doiron DR, Profio AE, and Huth GC: Fluorescence bronchoscopy for localizing early bronchial cancer and carcinoma in situ. *Recent Results Cancer Res. 82*:97-120, 1982.
20. Cortese DA, Kinsey JH, Woolner LB, Payne WS, Sanderson DR, and Fontana RS: Clinical application of new endoscopic technique for detection of in situ bronchial carcinoma. *Mayo Clin. Proc. 54*:635-641, 1979.
21. Hayata Y, Kato H, Ono J, Marsushima Y, Hayashi N, Saito T, and Kawate N: Fluorescence fiberoptic bronchoscopy in the diagnosis of early stage lung cancer. *Recent Results Cancer Res. 82*:121-130, 1982.
22. King EG, Doiron D, Man G, Profio AE, and Huth G: Hematoporphyrin derivative as a tumor marker in the detection and localization of pulmonary malignancy. *Recent Results Cancer Res. 82*:90-96, 1982.
23. Benson RC Jr, Farrow GM, Kinsey JH, Cortese DA, Zincke H, and Utz DC: Detection and localization of in situ carcinoma of the bladder with hematoporphyrin derivative. *Mayo Clin. Proc. 57*:548-555, 1982.
24. Gomer CJ, Doiron DR, Jester JV, Szirth BC, and Murphree AL: Hematoporphyrin derivative photoradiation therapy for the treatment of intraocular tumors: Examination of acute normal ocular tissue toxicity. *Cancer Res. 43*: 721-727, 1983.
25. Jocham D, Staehler G, Chaussy C, Hammer C, and Lehrs U: Laserbehandlung von Blasentumoren nach Photosensibilisierung mit Hämatoporphyrin-Derivat. Erste Experimentelle Erfahrungen *Urologe [Ausg A] 20*(suppl):340-343, 1981. (Ger).

26. Bicher HI, Hetzel FW, Vaupel P, and Sandhu TS: Microcirculation modifications by localized microwave hypothermia and hematoporphyrin phototherapy. *Bibl. Anat. 20*:628-632, 1981.
27. F. W. Hetzel, H. Farmer, J. Zdral, and B. Shumaker. Dose effects relationships in a mouse mammary tumor, in *Porphyrin Localization and Treatment of Tumors* (D. R. Doiron and C. J. Gomer, eds.). Alan R. Liss, Inc., New York, 1985, pp. 583-590.
28. Moan J, Steen HB, Feren K, and Christensen T: Uptake of hematoporphyrin derivative and sensitized photoinactivation of C3H cells with different oncogenic potential. *Cancer Lett. 14*:291-296, 1981.
29. Chang C and Dougherty TJ: Photoradiation therapy: Kinetics and thermodynamics of porphyrin uptake and loss in normal and malignant cells in culture (abst.). *Radiat. Res. 74*:498, 1978.
30. B. W. Henderson, D. A. Bellnier, B. Ziring, and T. J. Dougherty. Aspects of the cellular uptake and retention of hematoporphyrin derivative and their correlation with the biological response to PRT in vitro, in *Porphyrin Photosensitization* (D. Kessel and T. J. Dougherty, eds.). Plenum Press, New York, 1983, pp. 129-138.
31. Christensen T, Sandquist T, Feren K, Waksuik H, and Moan J: Retention and photodynamic effects of haematoporphyrin derivative in cells after prolonged cultivation in the presence of porphyrin. *Br. J. Cancer 48*:35-43, 1983.
32. Andreoni A, Cubeddu R, DeSilvestri S, Laporta P, Ambesi-Impiombato FS, Esposito M, Mastrocinque M, and Tramontano D: Effects of laser irradiation on hematoporphyrin-treated normal and transformed thyroid cells in culture. *Cancer Res. 43*:2076-2080, 1983.
33. Berns MW, Dahlman A, Johnson FM, Burns R, Sperling D, Guiltinan M, Siemens A, Walter R, Wright W, Hammer-Wilson M et al.: In vitro cellular effects of hematoporphyrin derivative. *Cancer Res. 42*:2325-2329, 1982.
34. Coppola A, Viggiani E, Salzarulo EL, and Rasile G: Ultrastructural changes in lymphoma cells treated with hematoporphyrin and light. *Am. J. Pathol. 99*:175-181, 1980.
35. Dubbelman TM, deGoeij AF, and van Steveninck J: Protoporphyrin-induced photodynamic effects on transport processes across the membrane of human erythrocytes. *Biochim. Biophys. Acta 595*:133-139, 1980.
36. Girotti AW: Photodynamic action of protoporphyrin IX on human erythrocytes: Cross-linking of membrane proteins. *Biochem. Biophys. Res. Commun. 72*:1367-1374, 1976.
37. Hilf R, Smail DB, Murant RS, Leakey PB, and Gibson SL: Hematoporphyrin derivative-induced photosensitivity of mitochondrial succinate dehydrogenase and selected cytosolic enzymes of R3230AC mammary adenocarcinomas of rats. *Cancer Res. 44*:1483-1488, 1984.
38. Bellnier DA and Lin C-W: Photodynamic destruction of cultured human bladder tumor cells by hematoporphyrin derivative: Effects of porphyrin molecular aggregation. *Photobiochem. Photobiophys. 6*:357-366, 1983.

39. Kessel D and Chou T: Tumor-localizing components of the porphyrin preparation hematoporphyrin derivative. *Cancer Res. 43*:1994-1999, 1983.
40. Moan J and Boye E: Photodynamic effect on DNA and cell survival of human cells sensitized by hematoporphyrin. *Photobiochem. Photobiophys. 2*: 301-307, 1981.
41. Moan J, Waksvik H, and Christensen T: DNA single-strand breaks and sister chromatid exchanges induced by treatment with hematoporphyrin and light or by x-rays in human NHIK 3025 cells. *Cancer Res. 40*:2915-2918, 1980.
42. Gomer CJ, Rucker N, Banerjee A, and Benedict WF: Comparison of mutagenicity and induction of sister chromatid exchange in Chinese hamster cells exposed to hematoporphyrin derivative photoradiation, ionizing radiation, or ultraviolet radiation. *Cancer Res. 43*:2622-2627, 1983.
43. Svaasand LO, Doiron DR, and Dougherty TJ: Temperature rise during photoradiation therapy of malignant tumors. *Med. Phys. 10*:10-17, 1983.
44. Svassand LO and Ellingsen R: Calibration of applications of nonionizing electromagnet radiation. Part 2. Experimental results. Report No. RB/ PE029, Division of Physical Electronics, University of Trondheim (Norway), Norwegian Institute of Technology, 1983.
45. Dougherty TJ, Kaufman JE, Goldfarb A, Weishaupt KR, Boyle D, and Mittleman A: Photoradiation therapy for the treatment of malignant tumors. *Cancer Res. 38*:2628-2635, 1978.
46. Dougherty TJ, Lawrence G, Kaufman JH, Boyle D, Weishaupt KR, and Goldfarb A: Photoradiation in the treatment of recurrent breast carcinoma. *J. Natl. Cancer Inst. 62*:231-237, 1979.
47. Forbes IJ, Cowled PA, Leong AS, Ward AD, Black RB, Blake AJ, and Jacka FJ: Phototherapy of human tumours using haematoporphyrin derivative. *Med. J. Aust. 2*:489-493, 1980.
48. J. Kennedy. HPD photoradiation therapy for cancer at Kingston and Hamilton, in *Porphyrin Photosensitization* (D. Kessel and T. J. Dougherty, eds.). Plenum Press, New York, 1983, pp. 53-62.
49. D. R. Doiron and C. J. Gomer (eds.) *Porphyrin Localization and Treatment of Tumors.* Alan R. Liss, Inc., New York, 1985.
50. Y. Tokuda. Primary skin cancer, in *Lasers and Hematoporphyrin Derivative in Cancer* (Y. Hayata and T. J. Dougherty, eds.). Igaku-Shoin Medical Publishers, Inc., Tokyo/New York, 1983, pp. 88-96.
51. Wile AG, Dahlman A, Burns RG, and Berns MW: Laser photoradiation therapy of cancer following hematoporphyrin sensitization. *Lasers Surg. Med. 2*:163-168, 1982.
52. Hayata Y, Kato H, Konaka C, Ono J, Matsushima Y, Yoneyama K, and Nishimiya K: Fiberoptic bronchoscopic laser photoradiation for tumor localization in lung cancer. *Chest 82*:10-14, 1982.
53. Y. Hayata, H. Kato, R. Amemiya et al. Indications of photoradiation therapy in early stage lung cancer on the basis of post-PRT histological findings, in *Porphyrin Localization and Treatment of Tumors* (D. R. Doiron and C. J. Gomer, eds.). Alan R. Liss, Inc., New York, 1985, pp. 747-758.

54. Balchum OJ, Doiron DR, and Huth GC: Photoradiation therapy of endobronchial lung cancers employing the photodynamic action of hematoporphryin derivative. *Lasers Surg. Med. 4*:13-30, 1984.
55. H. Kato, C. Konaka, J. Ono, Y. Matsushima, K. Nishimiya, J. Lai et al. Effectiveness of HPD and radiation therapy in lung cancer, in *Prophyrin Photosensitization* (D. Kessel and T. J. Dougherty, eds.). Plenum Press, New York, 1983, pp. 23-40.
56. Vincent RG, Dougherty TJ, Roa U, Boyle DG, and Potter WR: Photoradiation therapy in advanced carcinoma of the trachea and bronchus. *Chest 85*: 29-33, 1984.
57. Hayata Y, Kato H, Konaka C, Amemiya R, Ono J, Ogawa I, Kinoshita K, Sakai H, and Takahashi H: Photoradiation therapy with hematoporphyrin derivative in early and stage 1 lung cancer. *Chest 86*:169-177, 1984.
58. Y. Kato, et al. In *Lasers and Hematoporphyrin Derivative in Cancer* (Y. Hayata and T. J. Dougherty, eds.). Igaku-Shoin Medical Publishers, Inc., Tokyo/New York, 1983, pp. 39-56.
59. Cortese DA and Kinsey JH: Hematoporphyrin derivative phototherapy in the treatment of bronchogenic carcinoma. *Chest 86*:8-13, 1984.
60. M. Aida and T. Hirashima. Cancer of the esophagus, in *Lasers and Hematoporphyrin Derivative in Cancer* (Y. Hayata and T. J. Dougherty, eds.). Igaku-Shoin Medical Publishers, Inc., Tokyo/New York, 1983, pp. 57-64.
61. McCaughan JS Jr, Hicks W, Laufman L, May E, and Roach R: Palliation of esophageal malignancy with photoradiation therapy. *Cancer 54*:2905-2910, 1984.
62. Nseyo UO, Dougherty TJ, Boyle D, Potter W, Englander LS, Huben RP, and Pontes JE: Experimental photodynamic treatment of canine bladder. *J. Urol. 133*:311-315, 1985.
63. Benson RC Jr, Kinsey JH, Cortese DA, Farrow GM, and Utz DC: Treatment of transitional cell carcinoma of the bladder with hematoporphyrin derivative phototherapy. *J. Urol. 130*:1090-1095, 1983.
64. Hisazumi H, Misaki T, and Miyoshi N: Photoradiation therapy of bladder tumors. *J. Urol. 130*:685-687, 1983.
65. Hisazumi H, Miyoshi N, Naito K, and Misaki T: Whole bladder wall photoradiation therapy for carcinoma in situ of the bladder: A preliminary report. *J. Urol. 131*:884-887, 1984.
66. T. Ohi and A. Tsuchiya. Superficial bladder tumors, in *Lasers and Hematoprophyrin Derivative in Cancer* (Y. Hayata and T. J. Dougherty, eds.). Igaku-Shoin Medical Publishers, Inc., Tokyo/New York, 1983, pp. 79-84.
67. R. C. Benson. In *Porphyrin Localization and Treatment of Tumors* (D. R. Doiron and C. J. Gomer, eds.). Alan R. Liss, Inc., New York, 1985, pp. 795-804.
68. H. J. Reuter, R. J. Kohen, and M. A. Reuter. In *Atlas of Urologic Endoscopic Surgery* (H. J. Reuter, ed.), 2nd Ed. W. B. Saunders Co., Philadelphia, 1982, pp. 21-34.
69. Ward BG, Forbes IJ, Cowled PA, McEvoy MM, and Cox LW: The treatment

of vaginal recurrences of gynecologic malignancy with phototherapy following hematoporphyrin derivative pretreatment. *Am. J. Obstet. Gynecol. 142*: 356-357, 1982.

70. Soma H, Akiya K, Nutahara S, Kato H, and Hayata Y: Treatment of vaginal carcinoma with laser photoirradiation following administration of haematoporphyrin derivative: Report of a case. *Ann. Chir. Gynaecol. 71*:133-136, 1982.
71. H. Soma and S. Nutahara. Cancer of the female genitalia, in *Lasers and Hematoporphyrin Derivative in Cancer* (Y. Hayata and T. J. Dougherty, eds.). Igaku-Shoin Medical Publishers Inc., Tokyo/New York, 1982, pp. 97-109.
72. M. Rettenmaier, M. Berman, P. DiSaia, et al. In *Porphyrin Localization and Treatment of Tumors* (D. R. Doiron and C. J. Gomer, eds.). Alan R. Liss, Inc., New York, 1985, pp. 767-776.
73. A. G. Wile, J. Movotny, G. R. Mason, et al. Photoradiation therapy for head and neck cancer, in *Porphyrin Localization and Treatment of Tumors* (D. R. Doiron and C. J. Gomer, eds.). Alan R. Liss, Inc., New York, 1985, pp. 681-692.
74. C. Taketa and M. Imakiire. Cancer of the ear, nose and throat, in *Lasers and Hematoporphyrin Derivative in Cancer* (Y. Hayata and T. J. Dougherty, eds.). Igaku-Shoin Medical Publishers, Inc., Tokyo/New York, 1983, pp. 70-78.
75. Laws ER Jr, Cortese DA, Kinsey JH, Eagan RT, and Anderson RE: Photoradiation therapy in the treatment of malignant brain tumors: A phase I (feasibility) study. *Neurosurgery 9*:672-678, 1981.
76. G. A. J. McCulloch, I. J. Forbes, K. S. Lee, et al. Phototherapy in malignant brain tumors, in *Porphyrin Localization and Treatment of Tumors* (D. R. Doiron and C. J. Gomer, eds.). Alan R. Liss, Inc., New York, 1985, pp. 709-718.
77. A. L. Murphree, D. R. Doiron, C. J. Gomer, et al. In *Clayton Foundation International Symposium on Porphyrin Localization and Treatment of Tumors, Santa Barbara, California, April 24-28, 1983* (D. R. Doiron and C. J. Gomer, eds.). Alan R. Liss, Inc., New York, 1984.
78. R. A. Bruce. An approach to the treatment of ocular malignant melanomas, in *Porphyrin Localization and Treatment of Tumors* (D. R. Doiron and C. J. Gomer, eds.). Alan R. Liss, Inc., New York, 1985, pp. 774-784.
79. B. W. Henderson, T. J. Dougherty, and P. B. Malone. Studies on the mechanism of tumor destruction by photoradiation therapy, in *Porphyrin Localization and Treatment of Tumors* (D. R. Doiron and C. J. Gomer, eds.). Alan R. Liss, Inc., New York, 1985, pp. 601-612.
80. Weishaupt KR, Gomer CJ, and Dougherty TJ: Identification of singlet oxygen as the cytotoxic agent in photoinactivation of a murine tumor. *Cancer Res. 36*:2326-2329, 1976.
81. Waldow SM and Dougherty TJ: Interaction of hyperthermia and photoradiation therapy. *Radiat. Res. 97*:380-385, 1984.

18

Laser Surgery in the Pediatric Airway

GERALD B. HEALY
The Children's Hospital, Boston, Massachusetts

I. INTRODUCTION

Endoscopic surgery of the pediatric airway traditionally has been difficult because of the lack of appropriate equipment. Moreover, the imprecision of conventional cold-knife, electrocautery, and cryosurgical techniques often compounded the problem with inadvertent but unavoidable damage to healthy tissues, resulting in consequent edema and scarring. As a result, a postoperative tracheotomy was often necessary, despite its undesirability.

The development of fiberoptic technology increased the ability of surgeons to use more precise transoral and transnasal approaches to the airway. In addition, the union of the carbon dioxide (CO_2) laser to the operating microscope permitted the surgeon to excise or ablate lesions with minimal bleeding, postoperative edema, or obstructive scarring, without the need for tracheotomy.

Currently, the CO_2 laser appears to be the most useful in the pediatric population, while the neodymium:yttrium-aluminum-garnet (Nd:YAG) laser has had very limited application in the pediatric age group. The argon laser also has not proved effective in treatment of lesions of the pediatric airway. Although the laser has been used in all areas of the upper aerodigestive tract in children, its main usefulness has been in the larynx. Application in the tracheobronchial tree has been limited to infrequently occurring benign processes, such as recurrent respiratory papillomatosis.

The pediatric larynx has both similarities to and differences from the adult larynx. It is similar in its basic makeup and neurological components. It differs in four basic ways: location, configuration, size, and tissue consistency (1). The pediatric larynx is located somewhat higher in the neck than that of an adult.

At birth, it is located at approximately the level of the fourth cervical vertebra, and gradually descends to approximately the level of the seventh cervical vertebra in the adult.

The larynx of a child is usually somewhat narrower than that of the adult. The angle between the epiglottis and the glottis is more acute.

The infant larynx is approximately one-third the size of the adult larynx. The vocal cords average about 7 mm in length, with the posterior two-thirds being the vocal process of the arytenoid. As in the adult, the subglottic space is the narrowest portion of the airway, measuring approximately 6 mm in diameter in the normal infant.

The soft tissue and supporting structures of the larynx are much softer in the child. Submucosal planes are somewhat looser, thus increasing the chance of edema during acute inflammatory processes.

Because of these differences, the pediatric larynx has always been somewhat more difficult to evaluate and treat than that of the adult. Congenital and acquired lesions of the pediatric larynx will frequently necessitate multiple examinations to assess treatment outcome and future therapeutic planning. A method of examination and treatment that permits precise and consistent observation and a precise means of planning future therapy is paramount.

The advent of fiberoptic binocular laryngoscopes and the use of the surgical microscope for evaluating the larynx opened new avenues of evaluation and treatment never before possible (2). This modality, although first used in adults, is also extremely useful in pediatrics. In addition, the introduction of fiberoptic bronchoscopes with diagnostic and operative telescopes has been a major step forward in the treatment of pediatric airway disease.

With the introduction of the CO_2 laser, it soon became apparent that still another major advance had emerged that would eliminate the need for many external procedures of the airway, thus markedly diminishing morbidity and even mortality.

The advent of modern anesthesia techniques, including the ability to induce and maintain safe anesthesia, even in neonates, gives the modern day laryngologist advantages never before available to his predecessors.

II. INSTRUMENTATION

Diagnosis and treatment of lesions of the pediatric airway requires a broad range of equipment and instrumentation. A full complement of Jackson and Holinger laryngoscopes is required, as well as the complete set of microscopic laryngoscopes, such as the Jako and Dedo binocular laryngoscopes and the Healy subglottiscope (Fig. 1). An operating microscope with a 400-mm lens is also required. A complete set of fiberoptic bronchoscopes with appropriate telescopes should be

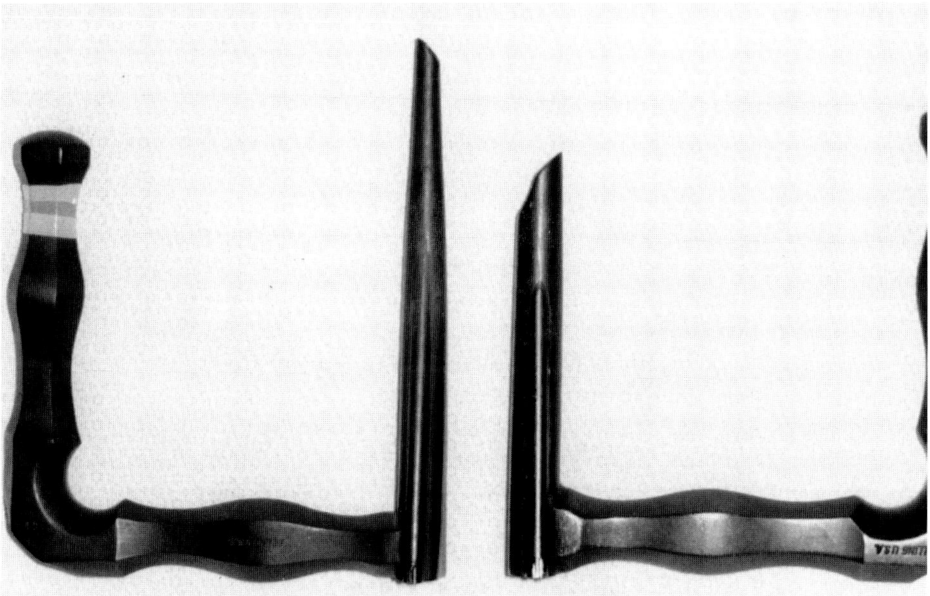

Figure 1 Standard Jako pediatric binocular laryngoscope on right and Healy subglottiscope on left.

available. In addition, appropriate laser bronchoscopes should be included. Flexible laryngoscopes are also important to assess laryngeal function, but as yet, flexible fiberoptic bronchoscopes that allow ventilation and suction capability are still not available for patients under the age of 2.

Any CO_2 laser that has bronchoscope capability should be sufficient for use in the pediatric airway.

A Lewy laryngoscope holder will suffice for almost all cases of microlaryngoscopy in the pediatric age group. Difficult problems of visualization frequently encountered in the adult larynx are usually not seen in children, and therefore other types of holders are not required.

The same microinstrumentation that is used in adult microlaryngeal surgery can also be applied to the pediatric population. Few special instruments are required for pediatric procedures.

III. ANESTHESIA

General anesthesia should be employed in all patients when microlaryngeal techniques are contemplated. Anesthesia in the pediatric patient is usually induced with a mask technique employing halothane, nitrous oxide, and oxygen. After a satisfactory plane of anesthesia is achieved, an intravenous catheter is placed. A muscle relaxant such as succinylcholine should not be employed if there is any question that the airway cannot be safely intubated.

After intubation is accomplished, the larynx is visualized, the laryngoscope suspended, and the larynx is brought into view by use of the operating microscope. At this point it is advantageous to remove the endotracheal tube and substitute Venturi ventilation. A 16- or 18-gauge jet needle, mounted on a clamp, is attached to the laryngoscope and the patient is then ventilated by a Venturi technique. The usual pressure setting in the pediatric population varies between 8 and 12 psi. The end point should be the pressure setting at which chest movements are visually observed and good breath sounds are heard by auscultation. To utilize this technique, the larynx and trachea must be in a straight line and the anterior portion of the larynx must be cleared of any obstruction. If obstructing lesions, such as papilloma, are present in the glottic region, it is advisable to intubate first, remove as much of the lesion anteriorly as possible, and then remove the tube for Venturi ventilation (3).

The cardiac status must be continually monitored by electrocardiography because frequently the first signs of hypoxia will be the sudden onset of an arrhythmia or bradycardia. The surgeon must always be attentive to this and not leave it solely in the hands of the anesthesiologist. It should also be remembered that children have an active vagal reflex, and the use of atropine sulfate must be planned in most cases.

If a Venturi ventilation has been employed, the vocal cords should be lightly sprayed with topical 0.5% lidocaine (Xylocaine), at the termination of the procedure, both to reduce laryngospasm and also to moisten the larynx, which has probably become dry because of repeated jet injection. At this point it is advisable to reintubate the patient so that he may be adequately oxygenated during the waking phase. The larynx and pharynx must be meticulously cleared of all blood and secretions by the surgeon while these areas are under direct vision. It is also wise to pass a nasogastric tube to empty the stomach of all air that may have accumulated during Venturi ventilation.

When employing Venturi ventilation, the surgeon must be constantly aware of all potential complications with this technique (4).

Recovery room attention is critical and all patients should be monitored closely and placed in an area with adequate humidification. This preferably means placing patients in a mist tent which can be conveniently attached to the bed (Fig. 2).

The use of corticosteroids is frequently helpful in preventing or reducing iatrogenic edema, and should be administered about 4 hr before the procedure. Dexamethazone (1 mg/kg up to a maximum dose of 20 mg) is usually used. Inasmuch as the drug has a half-life of 36-54 hr, one large bolus should be adequate to accomplish the anti-inflammatory effect that is desired.

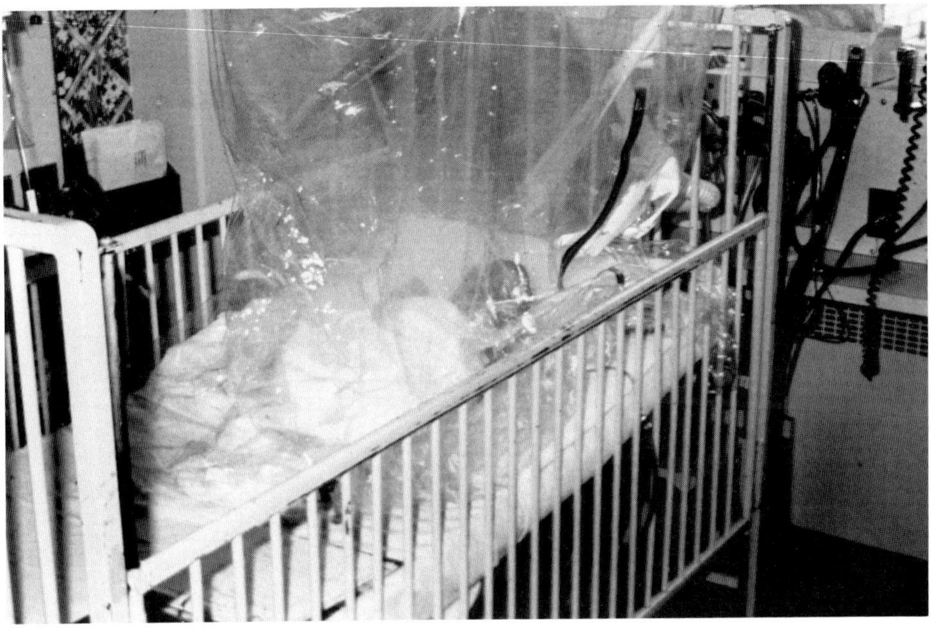

Figure 2 Mist tent used for all postoperative laser cases.

The technique for bronchoscopy may be somewhat different depending upon the patient's situation. In patients with both laryngeal and tracheobronchial lesions, it is probably advisable to first clear the upper airway and then pass the bronchoscope for treatment of lesions below. This is especially true when the CO_2 laser is being used.

In patients in whom only the bronchoscope is to be employed, the anesthesia should be induced by the mask ventilation technique, as described earlier, and then the bronchoscope passed in the same way as one would pass an endotracheal tube. The patient's anesthesia is then maintained through the ventilating arm of the bronchoscope.

The surgeon must be continuously aware of the fire hazard that may be present during laser surgery. The use of flammable endotracheal tubes and tracheostomy tubes is strongly discouraged. Appropriate nonflammable tubes are available and should be employed in every instance (5).

IV. PROCEDURE

After induction of anesthesia, the patient is positioned for laryngoscopy. Positioning is extremely important both for the adequate visualization of

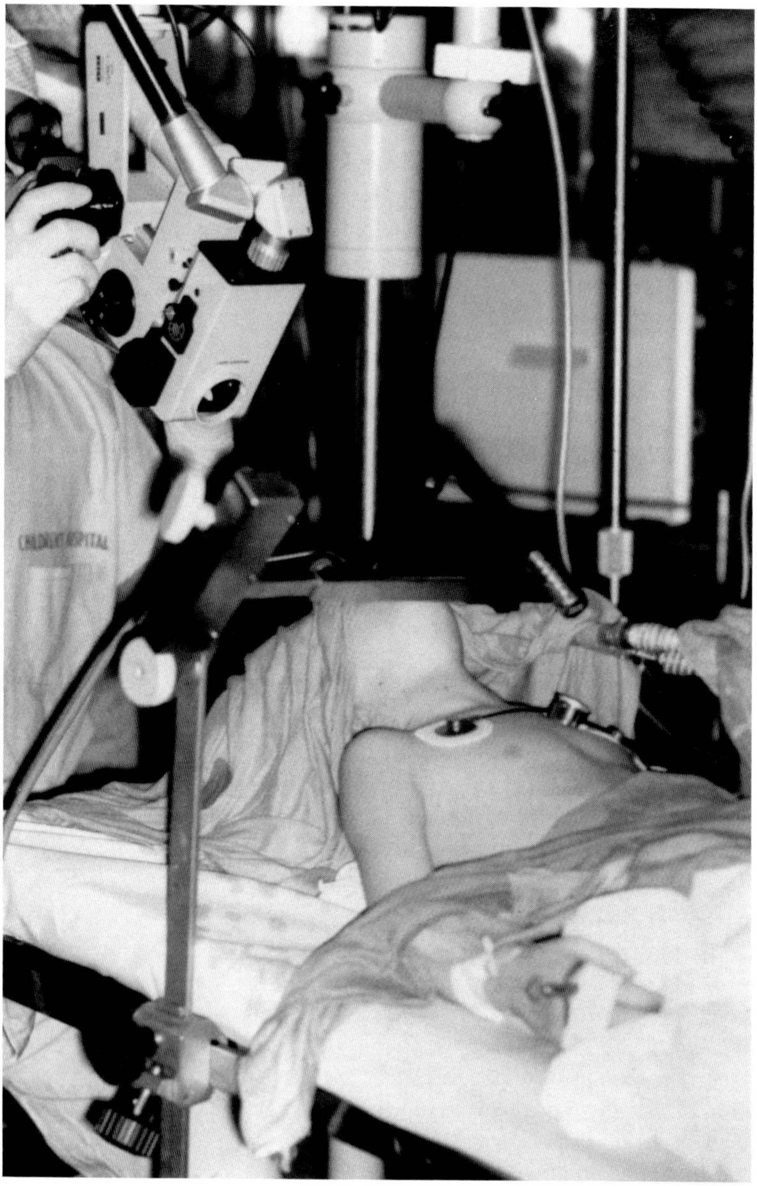

Figure 3 Patient in typical 20° reverse Trendelenberg position for endoscopic laser surgery.

the lesion in question and for the comfort of the patient and the surgeon.

Pediatric patients are best positioned by elevating the shoulders and hyperextending the head. This position may not be as advantageous as others for adults. The patient should be placed on the "foot" of the operating table so that the instrument stand can be easily placed over the operating table for placement of the suspension apparatus if that is employed. The operating table is then placed at approximately a 20-25° reverse Trendelenberg position (Fig. 3). This permits straight-line visualization from the operating microscope to the airway. In bronchoscopic procedures it enables the surgeon to be comfortable and to easily visualize the tracheobronchial tree.

Care must be taken to protect the teeth, and in small children, this is best done using moist gauze pads after checking for any loose teeth that might be dislodged during the procedure. The laryngoscope is then gently inserted until the vocal cords are visualized, and the suspension apparatus is then applied. At this point, the surgeon should always check to be sure ventilation is adequate and that the laryngoscope is not kinking the endotracheal tube. If the CO_2 laser is to be used, the patient's face, eyes, and surrounding tissues must be adequately protected with moistened towels that are continuously remoistened by the surgical assistant.

V. APPLICATIONS

Laser laryngoscopy and bronchoscopy has achieved a wide variety of applications in the pediatric population. This technique also enables delineation of a wide variety of supraglottic, glottic, and subglottic lesions, as well as documentation by photography, television, and videotape recording.

The most common lesion of the pediatric airway to be treated by laser endoscopy has been that of recurrent respiratory papillomatosis of the larynx. Microscopic laser techniques allow a precise removal of this lesion and thus diminish as the risk of damage to underlying normal tissues. A wide variety of other lesions of the pediatric airway, including those listed in Table 1, have been identified or treated, or both, with this technique.

Lesions, such as acquired subglottic stenosis and subglottic hemangioma, that have formally caused difficulty in both diagnosis and management, have responded dramatically to this modality in children (6,7).

Thirty-two patients with capillary subglottic hemangioma have now been successfully treated with the CO_2 laser utilizing a microscopic technique. If patients had not had a previously placed tracheostomy, it has usually not been necessary to perform this procedure either before or after laser therapy. The patient can usually be intubated with a small-sized endotracheal tube and the larynx

Table 1 Lesions of the Pediatric Airway: The Larynx

Lesion	Number of cases
Papilloma	301
Hemangioma	41
Subglottic stenosis	58
Congenital cyst	12
Neurofibroma	3
Lymphangioma	3

visualized microscopically. The endotracheal tube is then removed and Venturi jet ventilation is used while biopsy specimens of the subglottic hemangioma are taken and the hemangioma then vaporized.

In 30 of 39 patients with acquired subglottic stenosis, decannulation has been successfully achieved utilizing the CO_2 laser as a therapeutic modality. The stenotic area is usually vaporized, and then depending upon the degree of stenosis, a small stent is placed and fixed over a through-and-through laryngeal wire. These stents are usually left in position from 4-6 weeks. During this time the patients are maintained with prophylactic antibiotic therapy, but corticosteroids are not employed, to avoid any retardation of reepithelialization. If the stenosis is circumferential, but not complete in terms of total occlusion of the airway, it is then probably appropriate to remove one side of the lesion at a time so that two opposing raw surfaces will not be created. In this situation a stent need not be placed (8).

In the pediatric airway, it is important that tissue be impacted with a large amount of energy over an extremely short time. The laser is best used at approximately 25-30 W in intermittent bursts of between 0.1 and 0.05 sec.

VI. BRONCHOSCOPY

The CO_2 laser has found limited usefulness in the pediatric tracheobronchial tree. Its main advantage has been in treating obstructing benign lesions such as papillomata. The laser has also been useful in dealing with granulation tissue resulting from long-term intubation or tracheostomy.

Mild, weblike stenoses also have been fairly responsive to the CO_2 laser, but thick, circumferential scarring has not responded satisfactorily.

The need for the deep coagulation ability of the Nd:YAG laser is almost never necessary in the pediatric population. Occasionally, a patient with diffuse hemangiomatosis of the upper respiratory tract may be amenable to this type of therapy.

However, these lesions usually originate from the mediastinum, and thus in most patients a major surgical resection is required (6).

VII. SAFETY

Safety considerations must carry a high priority whenever surgical lasers are being employed. Both the patient and operating room personnel must be protected against inadvertent injury. This includes protection of the eyes and all exposed skin areas of the patient, as well as the eyes of the operating room personnel.

Anesthesia-related fires seem to be the greatest hazard posed by laser surgery (5). Meticulous attention must be paid to this possibility, and thus all combustible materials must be adequately protected or removed. Polyvinyl chloride endotracheal tubes should never be used in the airway, even if protected by metallic nonflammable tape. In children, the use of Venturi ventilation allows the removal of the endotracheal tube from the airway, and thus provides a safe access for surgery at the vocal cord level.

Only through diligence and meticulous attention can the surgeon hope to prevent any serious complications related to laser surgery.

VIII. CONCLUSIONS

The development of fiberoptic instrumentation and the ability to utilize this in conjunction with the operating microscope have opened new avenues of diagnosis and treatment in lesions of the pediatric airway. The addition of the CO_2 laser has markedly reduced the morbidity and mortality associated with treating many of these lesions. As technology develops, it is clear that we are standing on a vast frontier in which lasers will play a major role in the treatment of lesions of the pediatric airway.

REFERENCES

1. J. Tucker, et al. A clinical perspective on the development and anatomical aspects of the infant larynx and trachea, in *Laryngotracheal Problems in the Pediatric Patient* (G. B. Healy and T. J. McGill, eds.). Charles C Thomas, Springfield, Illinois, 1979.
2. Strong MS: Microscopic laryngoscopy: A review and appraisal. *Laryngoscope 80*:1540-1552, 1970.
3. Healy GB et al.: The use of the CO_2 laser in the pediatric airway. *J. Pediatr. Surg. 24*:735-740, 1979.
4. O'Sullivan T and Healy GB: Complications of Venturi jet ventilation during microlaryngeal surgery. *Arch. Otolaryngol. 111*:127-131, 1985.

5. Norton M et al.: Endotracheal intubation and Venturi (jet) ventilation for laser microsurgery of the larynx. *Ann. Otol. Rhinol. Laryngol.* *85*:656-663, 1976.
6. Strong MS et al.: Endoscopic management of laryngeal stenosis. *Otol. Clin. N. Am.* *12*:797-805, 1979.
7. Healy GB et al.: Carbon dioxide laser management of subglottic hemangioma—an update. *Ann. Otol. Rhinol. Laryngol.* *93*:370-373, 1984.
8. Healy GB: An experimental model for the endoscopic correction of subglottic stenosis with clinical applications. *Laryngoscope* *92*:1103-1115, 1982.

19

Complications of Laser Surgery

MARVIN P. FRIED and LOUIS BURKE
Harvard Medical School, Boston, Massachusetts

I. INTRODUCTION

As can be appreciated from the previous chapters, lasers have found application in numerous surgical specialties. Some procedures can be carried out equally as well with other techniques without the laser; however, many situations exist in which the laser is not only preferable but may be the only method available. Accurate control, diminished bleeding, and minimal tissue reaction are benefits that enhance the value of the laser in the surgical armamentarium. Advances in instrument design and surgical applications enhance the usefulness of the laser, leading to its broader use by more surgeons. As the frequency of laser surgery increases, potential complications will rise concomitantly. The number of these untoward effects can be limited by awareness of all operating room personnel of the potential risks and hazards and how they may be avoided.

A general survey of laser complications in all disciplines is beyond the scope of this chapter. We will attempt to discuss complications in a general sense and then more specifically in our specialties of otolaryngology and gynecology, stressing endoscopic considerations.

II. LASER LIMITATIONS AND SOFT-TISSUE INTERACTION

Of the current clinical lasers—carbon dioxide (CO_2), argon, neodymium:yttrium-aluminum-garnet (Nd:YAG), the CO_2 laser is the one most often used. Each laser, however, causes specific tissue effects that are unique to the source of the photons and the wavelength of emitted radiation. The CO_2 laser evaporates cellular water. Therefore, moisture on the surface of the area to be evaporated will interfere with the effect of the laser beam. The char that is produced by tissue destruction must be removed to prevent the overheating of local tissue that occurs when the laser energy is absorbed by the carbon debris. Vessels larger than 0.5 mm in diameter cannot be sealed with the CO_2 laser. They must be cauterized individually, ligated, or clipped (1,2). The use of vasopressin or phenylephrine (Neo-Synephrine) in gynecology will shrink these larger vessels on the cervix, in the vagina, and of the vulva and allow evaporation with good hemostasis. The utilization of these vasopressors intraabdominally, especially in fallopian tube and uterine surgery, enhances the application of the CO_2 laser as both a knife and cauterizing instrument.

The shorter wavelength of the argon laser permits a higher power density as a result of a smaller spot size but requires red pigmentation for absorption. This property limits the usefulness of the argon laser in poorly vascularized structures despite the higher energy intensity capabilities. The Nd:YAG laser exerts its effect deeper into tissue than the other two units, producing greater tissue destruction. However, the diminished control and precision is a definite shortcoming. In gynecology its asset is its ability to be transmitted through fiberoptic bundles and through liquid media. Thus, it can be used by way of the laparoscope to eradicate areas of endometriosis or through the hysteroscope to eradicate the endometrium of the uterus as well as small neoplastic lesions (3,4).

Potential complications can therefore be intrinsic to the surgical laser and its particular physical characteristics. The surgeon must be mindful of the effect of the laser on the operative site. Shrinking, curling, and cavity formation are produced with carbonization of the solid components of the surface tissue. The depth of necrosis and size of the ulcer crater are directly proportional to the duration of exposure as well as to the power output (5). The deeper the laser penetration, the greater is the potential for scarring or perforation. This is particularly hazardous when the laser is used on the continuous mode. Rapid pulsation can produce a similar effect.

III. PATIENT CONSIDERATIONS

Before any procedure is undertaken, the surgeon must consider whether or not the laser is the best method to take care of the problem. For example, excessive tissue manipulation for excision of small laryngeal lesions may not be advantageous; the weight of the laser apparatus on a bronchoscope may be too cumbersome for removal of a small amount of tracheal granulation. In gynecology, in the treatment of vaginal intraepithelial neoplasia, 5-fluorouracil (5-FU) cream and surgical excision are alternatives to the use of the CO_2 laser. On the cervix, cryosurgery may be advocated by some as an alternative to the use of the laser for cervical intraepithelial neoplasia (6). Numerous techniques have withstood the test of time and may yield results similar or even better than a laser procedure.

After these and other considerations are weighed and the laser is believed to be the most appropriate surgical instrument, certain patient constraints require attention. For patients undergoing upper aerodigestive procedures, mandible configurations as well as spinal column flexibility may limit the type of endoscope that can be used and therefore the degree of visualization. Patients with ischemic cardiovascular disease may not be able to withstand the prolonged laryngeal suspension that stimulates the vagus nerve receptors and may produce subsequent cardiac arrhythmia as well as silent myocardial infarction. Similarly in gynecology, for example, the overall medical state of the patient is important. If the laser technique is to be used in the vagina or vulva, the type of anesthesia is a prime consideration because these lesions cannot be eradicated without some form of anesthesia. Patients in poor physical health may be treated under local or regional block rather than general anesthesia. In patients who have had previous radiation to the vagina, visualization of both the cervix and the vagina may be difficult through the usual bivalve speculum. In general, whether in otolaryngology or in gynecology, a complete evaluation of the extent of the medical condition of the patient is necessary before the application of the technique.

IV. INSTRUMENTATION

Any endoscopic technique naturally limits the surgeon's capabilities. Access may be gained to various structures; however, binocular vision may be difficult to obtain and maneuvering instruments in a confined space may be difficult to achieve. The use of a stainless steel or gold mirror in the vagina or in the abdomen will permit access to hidden recesses so that the beam can be applied to these areas. Currently, no efficient means are available to transmit the CO_2 laser through a fiberoptic unit. The Nd:YAG laser has been the mainstay of the fiberoptic techniques, although it has been used as well with the rigid bronchoscope with much success (7). Its use in endometriosis and metropathia hemorrhagicum in gynecology has already been alluded to. In otolaryngology, the possible combustion of the fiberoptic endoscope has been recently documented; a bronchoscope and endotracheal tube ignited when a severe tracheal stenosis was being treated with the Nd:YAG laser (8). This problem may be avoided if the surgical site is well visualized and free of debris, moisture, blood, and vapor. Steam and laser tissue must be evacuated constantly so that carbon debris is not deposited, raising tissue temperatures and energy absorption as well as obscuring the field. Hemostasis must be maintained throughout the procedure (9). The endoscope should be prevented from overheating by avoiding the continuous mode and employing frequent passes. This is not as vital in gynecology; the beam is usually used in a continuous or pulsed mode (10). A pulsed mode is safer for intraabdominal removal of adhesions and endometrial implants (11).

V. ANESTHETIC CONSIDERATIONS

In many laser procedures, the type of anesthetic agent administered or the method of delivery of the agent has little bearing on the operation performed. This is not so in laser surgery of the head and neck or of the tracheobronchial tree. Here, it is imperative that the surgeon be aware of the anesthesiologist's concern in each particular case so that a team approach can be established. A full discussion of various possible techniques is advisable before the operation (12). The method used should be compatible with the anesthesiologist's need to ventilate the patient as well as with the surgeon's goals so that both can be accomplished easily. Both individuals should be aware of the suspected abnormality, especially if the patency of the air passage is in question. A dialogue should continue throughout the procedure concerning the status of the ventilation, the amount of bleeding encountered, the timing of both the laser's use and of respiration, as well as the patient's response. Although these considerations for otolaryngology are not as important in gynecology, it is imperative that the anesthesiologist be made aware of the use of various vasopressor substances and the extent of administration of the various agents should be discussed. As a general rule, vasopressin is diluted to a concentration of 20 units in 30 ml of saline solution and for the most part no more than 5-8 units (7-10 ml) of the solution are used. This will minimize the effects on the patient's cardiovascular state and enable vasoconstriction to occur and ensure hemostasis.

The type of anesthetic technique chosen is usually the province of the anesthesiologist; however, the surgeon must be cognizant of potential hazards. A metallic, foil-wrapped endotracheal tube does not guarantee safety. Dehiscences in wrapping or sharp edges may occur. Polyvinyl chloride (PVC) tubes, even if wrapped, are hazardous because of flammability and potential blowtorchlike effects. Reflective tape produced by some manufacturers may be made of Mylar and may not deflect a laser beam. All tapes and tubes should be tested by the surgeon before their use. Similarly, gynecological equipment should be of the nonreflective type. Accessory instruments must be both light and heat absorbing. The best absorbing materials are constructed of titanium or anodized aluminum. An irregular surface promotes dispersion rather than reflection of the laser beam. Spheric design also promotes dispersion of the laser beam.

The use of a Venturi system will remove the risk of combustion of an endotracheal tube. The injection system may be placed either proximally or distally on a laryngoscope or metal endotracheal tube (13-15). This cuffless system may be difficult in patients with restrictive pulmonary disease. If a high pressure is delivered in close proximity to the site of laser application, a small rent in the mucosa may be enlarged with anesthetic agents dissecting outside of the airway causing subcutaneous emphysema, pneumomediastinum, or a pneumothorax. Standard intubation may be needed at the outset of a procedure or at its completion despite the use of the Venturi system, thereby increasing the manipulation of the larynx.

If the patient has a tracheotomy in place, the use of a metal tracheotomy tube during a laser procedure is safe and reliable. Similarly, metal endotracheal tubes have been designed but are not always suitable or available (16). It therefore remains that an endotracheal tube of some sort will be used in many, if not most, laser endoscopic procedures of the aerodigestive tract. The incendiary characteristics of various types of tubes have been compared (17), and it appears that a red rubber tube is the safest. Besides adequate reflective wrapping, inflating the cuff with normal saline solution acts as an indwelling heat sink and fire extinguisher (18). Endotracheal tubes are now being designed specifically for use with the laser, with the potential of decreasing the likelihood of ignition. If a spark should occur, the appropriate environment is needed to support a flame. Nitrous oxide as well as oxygen can support combustion; however, a mixture of oxygen and helium may give adequate ventilation and retard ignition. The helium acts as a heat sink, reducing the combustion point of the endotracheal tube (19). At high power outputs, however, even this added benefit may be insufficient to prevent ignition. In gynecology, the only closed-space are in which a similar situation exists consists of laser in the anal canal. Methane from the lower digestive tract theoretically can be ignited and cause combustion. Therefore, it is advisable when working in this area that some form of moist plug be inserted into the anus preventing the dispersion of gas downward into the operative field. In the abdomen, irrigating water serves to disperse absorbed heat of the various manipulating instruments so that high power densities can be used over a long period.

Although the argon laser has found little utility in the aerodigestive tract, its effect on endotracheal tubes has been demonstrated experimentally. Using a 2:1 nitrous oxide/oxygen mixture, minimal damage occurs to red rubber or PVC tubes, with the transparent PVC offering a theoretical advantage (20).

VI. EXAMPLES OF DISEASE-RELATED LIMITATIONS

As an otolaryngologist and gynecologist, it is impossible for us to cover the entire spectrum of disorders that can be treated by endoscopic and hand-held laser surgery. Rather, specific examples will be discussed briefly to highlight possible limitations of laser techniques imposed by the disease.

Scarring of the larynx and trachea can be caused by many factors, such as external trauma (crush injury) or internal trauma (prolonged intubation). Correction of the stenosis can be managed by open procedures as well as endoscopically. Severe narrowing leading to respiratory distress at rest is difficult to treat by laser surgery alone; however, many cases are amenable to laser endoscopy, avoiding an external approach. The laser may delay collagen formation by allowing rapid epithelization; however, the benefit is not as important when large stenotic areas are involved (21). As the stenosis occurs progressively from supraglottic, to glottic, to subglottic-tracheal, the degree of difficulty in acquiring an adequate repair increases (22). At the glottic level, overzealous resection of scar tissue may lead to an incompetent larynx and subsequent aspiration. Subglottic and tracheal stenosis can be managed endoscopically if adequate external support exists as well as some degree of luminal aperture. An attempt to accomplish more than is reasonable can only lead to failure and restenosis.

Cervical scarring is rarely a problem after evaporation conization of the cervix, except in the patient who has been exposed in utero to diethylstilbestrol (DES) (23). Evaporation ablation to depths of 7 mm or even 1 cm can rarely cause stenosis. This is especially true if the endocervical canal is not impinged on too freely. The use of "button technique" in which a small moat is made around the endocervical canal and everting it will result in the cervix being on the portio and open. This is not the case in women who were exposed in utero to diethylstilbestrol. Apparently the connective tissue and vascular makeup of their cervices are different from the non-DES-exposed female. Any surgical manipulation of these cervices tends to result in excess scar formation and cervical stenosis (24). Although laser application is less likely to result in this complication, these patients should be treated only when absolutely necessary.

Scarring and coaptation of the vagina rarely occur if good postoperative care is undertaken (25). Inasmuch as only the vaginal epithelium is usually ablated, scarring seldom occurs in the vagina. However, if large denuded areas are left unattended, coaptation and interference with coital function may result. The application and insertion of various vaginal creams, such as sulfa creams, estrogen creams, and Aci-Gel, several times daily will keep the walls apart and permit simple healing.

Scarring of the vulva is a more complex situation. The nonhair-bearing areas usually have only the epithelium involved and respond similarly to the vagina as noted earlier. In general if the ablation is kept to the epithelium (1-2 mm), no scarring will result (26). If the stroma is inadvertently destroyed, scarring and subsequent dyspareunia may result. In the hair-bearing areas, if the skin appendages are involved in the disease process, the ablation must go deeper (3-5 mm) and scarring will occur. However, the disfigurement is usually much less than if a cold-knife excision were employed to remove the diseased tissue.

Vascular tumors, whether benign (hemangiomas) or malignant (with sepsis), can bleed excessively, and the surgeon must be prepared to achieve hemostasis by methods other than the laser. Inadequate control of the airway, especially in the presence of bulky exophytic tumors, can be fraught with hazard. Attempts at avoiding a tracheotomy with the hope that laser excision will simply take care of the problem may be risky if adequate tumor resection is not achieved. Similarly, incomplete endoscopic resection of laryngeal carcinoma in the hope of succeeding with one session or method of treatment, albeit inadequate, is unwise. Overtreatment, with excessive resection, can lead to poor phonatory function and quality.

While the coagulation effect of the CO_2 laser will seal capillaries and arterials of 1 mm or less, larger vessels must be shrunk with infiltration of vasoconstrictors or bleeding will ensue. Many methods can be employed to correct the immediate bleeding that can occur. Decreasing the power density, applying pressure and heating the tissue around the vessel, and continuously applying the laser to the bleeding points with increased power density are some of the maneuvers that are used to control immediate bleeding. However, other methods to control bleeding should not be avoided simply because one is using the laser. The use of ferric subsulfate (Monsel's solution), silver nitrate, or even suturing should be considered if bleeding persists. The CO_2 laser has been used experimentally on metastatic implants of the vagina and debulking nonresectable cancers of the cervix (27). These tumors should be frozen with cryosurgery first to cause thrombosis of the vasculature, and then the tumor mass can be evaporated with the CO_2 laser beam. If attempted before the vessels are controlled, massive hemorrhage might ensue.

VII. COMPLICATIONS OF LASER SURGERY

Considering the potential hazards associated with endoscopic laser surgery in otolaryngology, the frequency of complications has been surprisingly low. This may well be due to the meticulous observation to detail by early clinical investigators. In a series of over 4400 endoscopic procedures, a complication rate of 0.2% was reported by Healy and associates (28). Six of these were related to fires, one was a facial burn caused by an overheated bronchoscope, and two resulted from hemorrhage.

Ossoff and colleagues have had a similarly low complication rate (12 of 204 cases) (29). Six laser-related problems involved retained metallic tape in the oropharynx, laryngeal perichondritis with posterior web formation, laser burn of a tooth, and two instances of postoperative airway obstruction (one from edema and one from a retained pharyngeal pack). Two minor complications were sustained by the operating room staff. The remaining complications were related to the laryngeal suspension system and not directly to the laser.

Fried (30) reported a survey of a senior otolaryngology society's members who represented a broad field of experience in laser use; 229 questionnaires were mailed, and a response of 92% was obtained. Of the physicians responding, 49% had used the laser without complications. An additional 27% did not use the laser at the time of the survey. The remaining 49 physicians reported a total of 81 complications. These included endotracheal explosions (28); laryngeal web formation (15); facial burns (9); pneumothorax (5); hemorrhage (3); and laryngeal (4), subglottic (3), and tracheal (1) stenosis. Increased experience with the laser did not necessarily guarantee fewer complications. The general number of complications reflected the number of procedures performed for a month.

Overall complications following gynecological laser surgery of the lower generative tract have been extremely low (31). Currently, complications attendant with the intraabdominal and endoscopic use of the laser in gynecology are unknown. The number of procedures performed are too few to make any judgment about the rates of the complications in these settings. Overall, minor bleeding that results in application of some form of pressure or in application of an astringent will occur in about 10% of the patients. Major bleeding requiring hospitalization, repeat laser application, or suturing will occur in fewer than 1% of patients treated. Pain after cervical and vaginal procedures is nonexistent. However, pain after vulva procedures is almost 100%. This may be minimized by proper postoperative care. Because of the sterilizing nature of the beam and because no slough is left behind as a culture medium, infection is almost nonexistent. However, patients who have a history of pelvic inflammatory disease should be treated with antibiotics. Patients who have an intrauterine device should have it removed several weeks before treatment. Vulval lesions will frequently have a superficial area suggesting some form of mixed bacterial infection. The use of bioocclusive dressings, silver sulfadiazine (Silvadene) cream, and topical antibiotic cream will minimize this type of change. Bleeding in the vulva will be seen in approximately 4% of patients and is usually minor. Cervical and introital stenosis and labial agglutination will occur in less than 0.5% of patients if proper postoperative care is applied.

As already noted, the incidence of complications intraabdominally is unknown. However, proper packing away of viscera, such as small or large bowel, and protecting the bladder adequately with wet drapes can prevent the development of fistulas from these organs when the laser is used intraperitoneally. When ablating intraabdominal lesions, such as endometriosis, careful attention to operative detail can also prevent late complications. The CO_2 laser should be used intraabdominally only by the most experienced gynecological laser surgeons.

In all laser procedures, the eyes of the patient, surgeon, and operating room personnel must be protected. Any plastic or glass eyewear is satisfactory for use with the CO_2 laser; however, special glasses that attenuate light specifically at 1.06 μm are needed for the Nd:YAG laser. All personnel must be warned that the laser is being used. Under no circumstances should anyone look directly into the laser aperture.

All combustible or flammable substances must be moistened or protected with reflective material. The patient's face should be covered with a moist cloth or gauze for laser laryngoscopy. During lower generative tract ablation, the thighs, buttocks, and suprapubic area are protected with wet towels. The laser itself should be on standby whenever not in use and should have been fully tested before beginning the procedure. Preoperative washing of the lower generative tract with various preparatory solutions should be avoided. Obviously, anything containing inflammable substances, such as ether or alcohol, should never be used. The use of chlorhexidine (Hibiclens) as an antiseptic cleanser should also be avoided inasmuch as the application of the laser to this solution when wet produces an extremely irritating plume and may result in skin burns to the operating room personnel. Even when dried, a toxic fume may result from its use. Cleansing with 3-5% acetic acid is usually all that is necessary to sterilize the operative area (32).

Above all, the surgeon must constantly be mindful of the potentials for complications from before the start of the procedure until after it is over and the patient safely recovered. The physician should be aware of the tissue response and observe the characteristics of the tissue being lased. The laser team must be in constant communication during the procedure so that the advances achieved by this device are not hazardous to them or the patient.

Severe complications from laser surgery for gynecological lesions should be infrequently encountered even by the neophyte. Careful attention to safety factors as well as meticulous postoperative care will reduce the chance for error. If complications are to be avoided, careful training in the use of the laser must precede any clinical use of this new modality.

REFERENCES

1. Gillis TM and Strong MS: Surgical lasers and soft tissue interaction. *Otolaryngol Clin N Am 16*:775-784, 1983
2. Vaughan CW: Transoral laryngeal surgery using the CO_2 laser: Laboratory experiments and clinical experience. *Laryngoscope 88*:1399-1420, 1978.
3. Goldrath MH, Fuller TA, and Segal S: Laser photovaporization of endometrium for the treatment of menorrhagia. *Am. J. Obstet. Gynecol. 140*: 14, 1981.
4. Lomano JM: Photocoagulation of early pelvic endometriosis with the Nd: YAG laser through the laparoscope. *J. Reprod. Med. 30*:77, 1985.
5. Mihashi S, Jako GJ, Incze J, Strong MS, and Vaughan CW: Laser surgery in otolaryngology: Interaction of the CO_2 laser in soft tissue. *Ann. N.Y. Acad. Sci. 267*:263-293, 1976.
6. Townsend DE and Richart RM: Cryotherapy and carbon dioxide laser management of cervical intraepithelial neoplasia: A controlled comparison. *Obstet. Gynecol. 61*:75, 1983.

7. Dumon JF, Shapshay S, Bourcereau J, Cavaliere S, Meric B, Garbi N, and Beamis J: Principles for safety in application of neodynium-YAG laser in bronchology. *Chest 86*:163-168, 1984.
8. Casey KR, Fairfax WR, Smith SJ, and Dixon JA: Intratracheal fire ignited by the Nd:YAG laser during treatment of tracheal stenosis. *Chest 84*:295-296, 1983.
9. Fried MP: Limitations of laser laryngoscopy. *Otolaryngol. Clin. N. Am. 17*: 199-207, 1984.
10. Burke L: The use of the carbon dioxide laser in the therapy of cervical intraepithelial neoplasia. *Am. J. Obstet. Gynecol. 144*:337, 1982.
11. Baggish MS and Chong AP: Intraabdominal surgery with the CO_2 laser. *J. Reprod. Med. 28*:269, 1983.
12. Edelist G and Alberti PW: Anesthesia for CO_2 laser surgery of the larynx. *J. Otolaryngol. 11*:107-110, 1982.
13. Goforth AJ, Cooke JE, and Putney FJ: An anesthesia technique for laser surgery of the larynx. *Laryngoscope 93*:822-823, 1983.
14. Ruder CB, Raphael NL, Abramson AL, and Oliveris RM: Anesthesia for carbon dioxide laser microsurgery of the larynx. *Otolaryngol. Head Neck Surg. 89*:732-737, 1981.
15. Woo P and Strong MS. Venturi jet ventilation through the metal endotracheal tube: A nonflammable system. *Ann. Otol. Rhinol. Laryngol. 92*:405-407, 1983.
16. Norton ML and deVos P: New endotracheal tube for laser surgery of the larynx. *Ann. Otol. Rhinol. Laryngol. 87*:554-557, 1978.
17. Duncavage JA, Ossoff RH, Roman WC, Toohill RJ, Almagro VA, Carl GA, and Aronson S: Injuries to the bronchi and lungs caused by laser-ignited endotracheal tube fires. *Otolaryngol. Head Neck Surg. 92*:639-643, 1984.
18. LeJeune FE Jr, Guici C, Letard F, and Marice H: Heat sink protection against lasering endotracheal cuffs. *Ann. Otol. Rhinol. Laryngol 91*:606-607, 1982.
19. Cassissi NJ, Pashagan AG, and Gravenstein JS: Helium-oxygen mixture (Gator gas) prevents laser fires. Presented at the Annual Meeting of the American Academy of Otolaryngology, Head and Neck Surgery, Las Vegas, 1984.
20. McLaren ID, Bellman MH, and Cooley J: Effects of the argon laser in anaesthetic gases and endotracheal tubes. *Br. J. Anaesth. 55*:1001, 1104, 1983.
21. McGee KC, Nagle JW, and Toohill RJ: CO_2 laser repair of subglottic and upper tracheal stenosis. *Otolaryngol. Head Neck Surg. 89*:92-95, 1981.
22. Strong MS, Healy GB, Vaughan CW, Fried MP, and Shapshay S: Management of laryngeal stenosis. *Otolaryngol. Clin. N. Am. 12*:787-805, 1979.
23. Baggish MS: Complications associated with carbon dioxide laser surgery in gynecology. *Am. J. Obstet. Gynecol. 139*:568, 1981.
24. Haney AF and Hammond MG: Infertility in women exposed to diethylstilbestrol in utero. *J. Reprod. Med. 28*:851, 1983.
25. V. Jobson. Prevention of complications in gynecologic laser surgery, in *Basic and Advanced Laser Surgery in Gynecology* (M. Baggish, ed.). Appleton-Century-Crofts, Norwalk, Connecticut, 1985.

26. Dorsey J: Understanding CO_2 laser surgery of the vulva. *Colpo. Gynecol. Laser Surg.* *1*:205, 1984.
27. H. F. Schellhas. Invasive gynecologic cancer treated by laser, in *Basic and Advanced Laser Surgery in Gynecology* (M. Baggish, ed.). Appleton-Century-Crofts, Norwalk, Connecticut, 1985.
28. Healy GB, Strong MS, Shapshay S, Vaughan C, and Jako G: Complications of CO_2 laser surgery of the aerodigestive trace: Experience of 4416 cases. *Otolaryngol. Head Neck Surg.* *92*:13-18, 1984.
29. Ossoff RH, Hotaling AJ, Karlas MS, and Sisson GH: CO_2 laser in otolaryngology—head and neck surgery; a retrospective analysis of complications. *Laryngoscope* *93*:1287-1289, 1983.
30. Fried MP: A survey of the complications of laser laryngoscopy. *Arch. Otolaryngol.* *110*:31-34, 1984.
31. Reid R: Laser safety I. Avoidance of surgical misadventure with the CO_2 laser. *Colpo. Gynecol. Laser Surg.* *1*:117, 1984.
32. J. C. Fisher. Principles in laser surgery and therapy, in *Basic and Advanced Laser Surgery in Gynecology* (M. Baggish, ed.). Appleton-Century-Crofts, Norwalk, Connecticut, 1985.

20

Establishment of a Laser Center

JOHN F. BEAMIS, Jr., and NANCY RIZZO
Lahey Clinic Medical Center, Burlington. Massachusetts

I. INTRODUCTION

The dramatic increase in the application of laser technology in medical therapeutics and diagnostics in the last 15 years has been the result of much labor on the part of clinicians, researchers, physicists, nurses, businessmen, hospital administrators, and many others. Similarly, the development of a laser program in a hospital requires the energy and cooperation of individuals from the institution's medical, surgical, nursing, administrative, and support staffs. Although the programs are usually initiated through the leadership of a highly motivated individual, the secret of a successful laser program is teamwork.

This chapter reviews the steps to be taken to develop a laser program. The recommendations are based on our experience in developing a laser surgery program and the Eleanor Naylor Dana Research Laboratory at the Lahey Clinic Medical Center. In general, the same principle should apply whether a multidisciplinary, multilaser center is being formed or whether the institution only plans to support a single application for one laser instrument. Three phases are necessary to organize an institution's laser program: planning, development, and implementation.

II. PLANNING PHASE

Careful and thorough initial planning sets the stage for a successful laser program. During this phase, the institution and involved physicians develop short-range goals and long-range plans, identify costs to the institution, estimate methods of reimbursement, determine space requirements, identify potential sites for application of the laser, and choose a leader to coordinate laser activity within the institution. The physician leader should have demonstrated a strong interest in laser technology either by past experience or by recent study. This person must be dedicated to developing an institutional laser program rather than fostering or expanding his or her practice. This individual must also be able to stimulate physicians of similar and other specialties and be willing to assist them in obtaining the necessary instruction and training to incorporate laser surgery into their practice safely and effectively.

One member of the institution's administrative staff should be selected to assist in the development of the laser program. This person should be aware of the physical and personnel resources of the institution, the local and national regulatory standards regarding medical devices, and the increasingly complicated field of reimbursement for hospitals and physicians.

The first decision these key members of the laser program should make is whether the institution should establish a multidisciplinary laser center geographically distinct within the institution or whether existing facilities, such as general operating rooms, procedure rooms, bronchoscopy suites, or ophthalmology clinics, should be modified for laser use. Because of physical constraints and uncertainties about commitments of physicians, most institutions initially choose to modify existing space for laser surgery and then expand gradually as need develops. At the Lahey Clinic Medical Center, two rooms within the main operating pavillion were equipped with special electrical and plumbing outlets for carbon dioxide (CO_2) and neodymium:yttrium-aluminum-garnet (Nd:YAG) lasers for use by several specialists. Relatively inexpensive modifications of the pulmonary bronchoscopy suite and the ophthalmology clinic have permitted laser application for outpatient use. Other institutions (1,2) have reported on experiences in setting up laser programs.

As physicians from various specialties voice an interest in laser surgery, a cost analysis of each specific application should be undertaken. This should include estimates of patient demand, capital equipment, renovation, educational programs, and annual operating costs, including salaries of personnel, service contracts, and supplies. Estimates of income are based on the estimated number of patients treated in conjunction with current and projected fee schedules. Difficult to determine, but possibly meaningful, would be savings to the institution by reduction in hospital stay afforded by laser surgery when reimbursement is under the prospective payment system using diagnosis-related groups.

If clinical or animal research with lasers is to be carried out, a separate research laboratory must be planned. Research procedures can be performed only if an Institutional Review Board and an animal research committee are established with specific policies and procedures mandated by the National Institutes of Health, the Federal Drug Administration, and the Department of Health and

Human Services. If a research laboratory is not feasible for the institution, an early affiliation should be obtained with an established laser research laboratory to provide the medical and surgical staff with instruction and methods of developing new techniques of laser application.

III. DEVELOPMENTAL PHASE

During the second or developmental phase of organization, eight objectives should be accomplished: establishment of a laser committee, allocation of space for laser surgery, proper modification of facilities, acquisition of laser equipment, development of policies and procedures for laser surgery and research, appointment of a laser safety officer, establishment of policies to ensure safety, and identification and training of a laser team.

Although universal standards for laser surgery have not been mandated by national organizations, a number of groups are currently developing standards or have made recommendations for laser use. These include the American National Standard for the Safe Use of Lasers in the Health Care Environment, the Association of Operating Room Nurses, and the American Society for Laser Medicine and Surgery. The Joint Commission on Accreditation of Hospitals does not address laser technology specifically but requires that safety and quality assurance mechanisms be followed as with any other therapy.

The establishment of an active laser committee is a most important step in ensuring compliance with future regulations and in setting local standards for laser safety. Because laser surgery crosses many specialties, the membership of the laser committee must include several disciplines (Table 1). The functions of the laser committee include development and enforcement of institutional policies and procedures for the use of lasers in the hospital and outpatient areas; development of laser technology in conjunction with the Institutional Review Board; establishment of processes for accrediting physicians, nurses, and allied health professionals; data collection concerning laser use; development and implementation of quality assurance plans regarding the use of laser technology within the institution and regular reporting of results to the institution's quality assurance department; periodic reviews of complications associated with laser use and filing of a formal report to the Division of Surgery on a regular basis; ongoing determination of the institution's needs regarding laser application; and the planning and developing of future laser programs.

Our laser committee reports to the chairman of the Division of Surgery and subsequently to the Chief Executive Officer and the Board of Governors (Table 2). The medical chairman of the laser committee is responsible for all activities of the committee and reports to the chairman of the surgical division. The chairman of the laser committee coordinates the process of accrediting physicians within the

Table 1 Membership of Laser Committee

Chairman
Administrator
Secretary to the committee
Laser safety officer
Operating room supervisor
Chief laser technician
Medical staff representative(s)
Surgical staff representative(s)
Outpatient nurse representative(s)
Operating room nurse instructor
Staff committee members

Table 2 Organizational Chart for Laser Committee at Lahey Clinic Medical Center

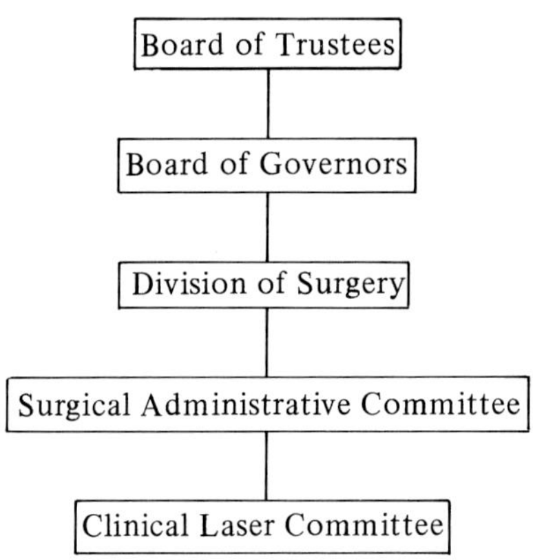

Department of Education, acts as an on site preceptor for the medical and surgical staff at their initial operations using the laser, and establishes medical policies related to laser technology.

Physicians' requests for laser equipment are directed to the laser committee where the request is acted on in relation to the needs and capabilities of the institution. Once a laser wavelength is approved, space requirements and room modifications, such as water and electrical access, are determined and recommendations forwarded to the hospital administration for purchase and implementation.

A policies and procedures manual should be developed to address such topics as laser safety; accreditation of medical, nursing, and allied medical staff; orientation programs for nurses and technical staff; quality assurance procedures; and infection control. If a research laboratory is planned, a separate set of policies and procedures must be developed.

The laser safety officer is an essential member of the laser committee whose duty is to establish and enforce safety procedures that include warning labels and the protection of personnel, patients, and animals. The laser safety officer must ensure that the hospital staff is aware of infection control procedures and disaster and fire plans. Most important, the laser safety officer is responsible for the calibration and proper maintenance of laser equipment. This person need not be a physician but must be knowledgeable in laser technology and appreciative of the potential benefits and harms of lasers in medicine.

Identification and training of members of the laser team should take place during the developmental phase. Members of the laser team need not be on the laser committee. Any person certified to operate lasers or to assist physicians in learning laser technique is a qualified member of the team, including clinical and nursing staff, physicians, or other professional employees. Personnel should become familiar with the lasers that have been or may be purchased and should attend inservice sessions organized by the education departments of companies supplying the laser equipment. At this time, the medical and surgical staffs should be encouraged to attend instructional courses outside of the institution regarding specific applications of lasers, to participate at more than one hands-on course, and to visit another institution with an established laser program in their specialty. The committee should also establish guidelines for ongoing educational programs for members of the laser team.

IV. IMPLEMENTATION PHASE

With the laser committee functioning properly and the laser equipment purchased and installed, the implementation phase begins with final certification of physicians—the most important yet potentially the most controversial function of the laser committee. At our institution, physicians seeking privileges for laser surgery must provide documentation that they have attended at least one hands-on course in their specialty and if possible have visited a center using lasers in that specialty. We also require several visits to our laser research laboratory during which time the physician becomes familiar with laser equipment and interactions between lasers and tissues in animals. Once this educational background has been confirmed, the physician may begin laser surgery with the stipulation that either a previously accredited member of that physician's department or the chairman of the laser committee is in attendance during the physician's first four procedures. The physician receives written approval of privileges in the appropriate laser wavelength, and a recommendation is sent to the surgical administrative committee for confirmation. No "generic" approval for the use of all laser wavelengths exists.

Because some laser committees may not wish to be solely responsible for the certification of physicians and the always controversial task of denying privileges, extramural accrediting bodies are being developed. In 1984, the American Board of Laser Surgery was incorporated to assist local laser committees in ensuring that candidates for privileges in laser surgery have achieved basic knowledge of lasers and application of lasers in their specialty. The first examination is planned for 1986 and will consist of a written examination on broad topics of laser physics, laser safety, and tissue interaction, followed by an oral examination given by an examiner in the candidate's specialty (J. Fisher, personal communication, 1985).

Once the laser program is in operation, some consideration should be given to marketing. Direct advertising in the media or by direct mailing to patients is not an acceptable practice to many patients and referring physicians. The best marketing tool is excellence in patient care and safe, successful results with laser surgery. Speaking engagements at local hospitals or at national and regional meetings by physicians and other members of the laser team help greatly in establishing a referral practice. Publication of results in the appropriate medical literature is the most accepted way of "advertising."

At this point, the laser committee begins to implement previously developed policies and procedures and guidelines for laser safety. The number of procedures and complications should be monitored closely and periodic reports made to the hospital's quality assurance department. The laser safety officer, in cooperation with the hospital's medical equipment repair department and the laser vendor's maintenance personnel, should carry out proper maintenance of equipment and report to the laser committee.

V. CONCLUSION

The application of lasers to medicine is a new and exciting field that should be entered only after careful organized planning by physicians, administrators, and support personnel. A laser program is developed in three occasionally overlapping phases: planning, development, and implementation. The combination of a highly motivated and highly skilled medical director, a strong multirepresentative laser committee, and an interested, supportive administration can best ensure a successful program in laser surgery.

ACKNOWLEDGMENT

The authors wish to acknowledge the support of the Eleanor Naylor Data Charitable Trust which generously agreed to support a laboratory for the investigation of laser technology at the Lahey Clinic Medical Center.

REFERENCES

1. McCaughan JS: Development of a laser center. *Med. Instr. 17*:417-418, 1983.
2. Chavoin JP, Costagliola M, Laffitte F, and Vachaud M: Prototype of a multipurpose trilaser unit in a teaching hospital surgical department. *Lasers Surg. Med. 3*:1-8, 1983.

SUGGESTED READING

1. Dorsey JH: Initiating a hospital CO_2 laser program. Part 1. *Colpo. Gynecol. Laser Surg. 1*:102-108, 1984.
2. Schwartz D (ed.): Laser credentialing: Too much of a good thing? *Clin. Laser Mon. 2*:57-60, 1984.
3. Schwartz D (ed.): Nationwide laser credentialing board in formative stages. *Clin. Laser Mon. 2*:109-110, 112, 1984.
4. Schwartz D (ed.): Laser marketing is key to boosting patient loads, referrals. *Clin. Laser Mon. 2*:153-155, 1984.
5. Schwartz D (ed.): ANSI standards delayed; push for national regs continues. *Clin. Laser Mon. 3*:1-3, 1985.
6. Schwartz D (ed.): Cost and reimbursement: Reports from the laser battlefield. *Clin. Laser Mon. 3*:17-20, 22, 1985.

Index

Numbers in italics indicate illustrations. Tables are designated by a lower-case "t" following the numbers.

A

Abrazement, for tissue fixation, 401
Abscess, after HPD phototherapy, 228
Absorbance, spectroscopic studies of, *372*
Absorption
  with CO_2 laser, 214
  in living tissue, 96-100, *101*
  related to pigmentation, 245
Accessibility
  of bladder, 361
  of bronchi, 203-204
  in orthopedic surgery, 402
  of trachea, 203
Accreditation, of staff, 479
Adapters, endoscopic, 209, 211
Adenocarcinoma
  at cardia, 248
  of esophagogastric junction, 245
  gastric, 248
  gastric, mucinous, 248
[Adenocarcinoma]
  of lung, photodynamic therapy for, 436
  of stomach, in dog, *268-269*
  tracheobronchial, *202-203,* 204
Adenoidectomy, laser vs classic technique of, 169, 170
Adenomas
  bronchial, 214-215, *216*
  colorectal, villous, 264
  gastric, 248
Adhesions
  endometrial, 330
  intraabdominal, 298
  intrauterine, 327, 344
  pelvic and peritubal, 334-335, *345-347*
Aiming beam, helium-neon, 210, 211-212
Airway obstruction, postoperative
  avoidance of, 217
  after HpD phototherapy, 225, 228
  as laser-related complication, 469
Albarran bridge, *359,* 362

Aldrette scale, 196
Anal fistula, 265-266
Anastomosis, microvascular, 414-415, *416*, 417
Anesthesia
  designing a protocol, 194
  in laryngeal laser microsurgery, 175
  in pediatric patients, 455-457
  potential hazards of, 466-467
  in small larynx, 177
  using universal endoscopic coupler, 214
Angiodysplasia, multiple, in cecum, 257, 260
Angiofibromas, nasopharyngeal, 168
Angiography, during colonoscopy, 260
Angiomas, tracheobronchial, 204
Angioplasty, 369-384
  as alternative to bypass surgery, 370-371
  excimer laser for, 141
  simulated intraoperative technique of, *374*
Angiosarcoma, of nasal septum, 168
Angioscopy, for controlling beam emission, 392
Anorexia, postoperative, 247
Argon, laser wavelength from, 139t
Arteries, coronary
  atherosclerotic features in, *382*
  intraoperative laser irradiation of, 383-384
Arteries, peripheral, percutaneous laser irradiation of, 383
Arteriovenous malformations
  colonic, pretreatment and post-treatment, *258-259*, 260
  of external nose, 164-166, 170
Arthroscope, laser-competent, *414*
Arthroscopy, 402-405
Articulating arms
  configuration of, *146*
  difficulties with, 145-147
  features of, 145-147
[Articulating arms]
  in laryngeal surgical laser systems, 174
Arytenoidectomy, with CO_2 laser, 179
Aspiration, in treatment of gastrointestinal tumor, 246
Atelectasis, of lung, 436
Atherolysis, using laser, in vitro, 371-380
Atherosclerotic plaque, vaporization of, in vitro, 371-380
Autoclaving, of glass spit windows, 212
Autolytic tissue changes, after partial pancreatectomy, in rat, 289

B

Backscatter, hazards from, 277
Bacteremia, as laser-related complication, 247
Beam-manipulating mirror, in direct coupling, 147
Bile ducts, laser surgery of, 281-300
Biopsy, excisional
  for carcinoma of lateral tongue, *158*
  for carcinoma of oral cavity, 152-160
  in cervix, 306-310
  for endocervical lesions, 306-309
  in larynx, 176-177
  layer-by-layer, 181-182
Biostimulation, related to energy density, 257t
Biosuppression, related to energy density, 257t
Bleeding (*see also* Hemorrhage)
  control of, 468-469
  laryngeal, 176, 469
  as laser-related complication, 247
  from metastatic lesions, 242
  from neoplasms, 242
  rectal, 260
  tracheobronchial, 206

[Bleeding]
in upper gastrointestinal tract, 233-241, 249
Blood gas analysis, during laser resection, 197
Bone, laser incision of, 410-411
Bowen's disease, of skin, photodynamic therapy for, 433-435
Brachytherapy (*see also* Radiotherapy)
endobronchial, after HpD phototherapy, 225
after laser therapy, 206
Bronchoscope(s)
flexible fiberoptic, advantages and disadvantages of, 217
rigid, *192*
advantages and disadvantages of, 207, 215-217
ventilating, coupled to CO_2 laser, 209, 211
with universal endoscopic coupler, 212-214, *213*
Bronchoscopy
for cleanup, after photodynamic therapy, 436
laser, advantages and disadvantages of, 215
laser, indications and contraindications for, 214
Bronchi, access to, 203-204
Burns
minimizing, 174
of skin, after HpD phototherapy, 228
of skin, as laser-related complication, 470
tracheal, avoidance of, 217
Bursting pressure, colonic, with CO_2 laser, 271

C

Calculi, extrahepatic and intrahepatic, 292-294
Cancers (*see also* Carcinomas; Lesions; Tumors)
advantages of laser surgery for, 181
of bladder, photodynamic therapy for, 439-441
of breast, 267
bronchial, 267
colonic, *263*
cutaneous and subcutaneous, 433-435
endobronchial, fiber probe implantation for, 221
esophageal, photodynamic therapy for, 437-439
esophageal, pretreatment and posttreatment, *246*
of false vocal cord, 180
of head and neck, photodynamic therapy for, 441-442
laryngeal, 179-185
of larynx, verrucous, 179
of lung, photodynamic therapy for, 435-437
pancreatic, 295
of penis, 356, 358, 366
supraglottic, 180
of vocal cord, 179-181
Cannula, laser, for arthroscopic meniscectomy, *410*
Carbon dioxide laser
in colorectal surgery, experimental, 271
construction of, 50-53, *56-57*
development of, 132
flowing-gas vs sealed, 132-137
laser-tissue effects of, 113-119
mechanism of excitation of, 137-138
milliwatt, 356
vs Nd:YAG laser for meniscectomy, 402-405
vs Nd:YAG laser for synovectomy, 413
specifications of, 256t, 282t
for treatment in tracheobronchial tree, 209-217

[Carbon dioxide laser]
for treatment of nasal cavities and nasopharynx, 163-170
Carbonization, with high-power vs low-power density, 175
Carbon residue (*see also* Cleansing)
during excision of cartilage, 405, 409
in histological studies of rat pancreas, 287
after laser laparoscopic vaporization, 331
during laser meniscectomy, 405
overheating from, 464
Carcinomas (*see also* Cancers; Lesions; Tumors)
adenoid cystic
of distal trachea, 198-199, *198*
of lung, photodynamic therapy for, 436
basal cell, of skin, 433-435
of breast, cutaneous, 433-435
colorectal, 261-262
esophageal, *243*
in situ
bronchial, 220
of larynx, 179
of vagina, 319-322, *320*
intraepithelial, of larynx, 179
nasopharyngeal, 169
oat cell, tracheobronchial, 204
of papilla of Vater, 292
renal cell
in solitary kidney, 367
of stomach, metastatic, 242
signet ring cell, gastric, 248
squamous cell
on carina, *200-201*
of esophagus, 245, 247
of lung, 435-436
of penis, 366
of skin, 433-435
tracheobronchial, 204, 220, *223*, 224t, 226-227t
tracheobronchial, 214-215, 219
[Carcinomas]
transitional cell, in solitary kidney, 366
Carcinosarcoma, of larynx, 179
Cardiac arrhythmia, subsequent to laser surgery, 465
Cardiomyopathy, hypertrophic, 393
Cardiovascular disease
ischemic, as contraindication to treatment, 465
laser surgery for, 369-393
Cartilage
articular, debridement and excision of, 405, 409
grafts of, fixation of, 418
Caruncles, urethral, 367
Cavity-dumping, 65, 133 (*see also* Modes, pulsed)
Certification, of physicians, 479-480
Cervical intraepithelial neoplasia
laser treatment of, 301-315
alternatives therapies for, 464
Cervix
cellular patterns of, *303*
excisional procedure in, 306-309
neoplasia of, areas of susceptibility to, *302*
Chemical reactions, in HPD phototherapy, 220
Chemotherapy, adjuvant, for laryngeal tumor, 182
Choanal atresia
nasal, congenital, 167-168
nasopharyngeal stent for, *169*
Choice of laser, related to vascularity of lesion, 160
Choice of therapy
for upper gastrointestinal bleeding, 234-236
for vaginal epithelial neoplasia, 319
Cholangitis, and intrahepatic bile duct stricture, 289
Cholecystectomy, 299
Choledochofiberscope, for treatment of bile duct calculi, 292, *293*

Choledocholithiasis, and intrahepatic bile duct stricture, 289
Choledocholithotomy, 293
Choledochoscope, laser fiber insertion in, experimental, 286
Chondromas, tracheobronchial, 204
Chromatography, gas, *372*
Cleansing (*see also* Carbon residue)
  of carbon particles, 176
  during colonoscopy, 260
  tracheobronchial, during laser resection, 197
Coagulation
  by CO_2 and Nd:YAG laser coupling, 157
  in gastrointestinal tract, 234
Coherence, of laser beams, 71-74, 88
Collimation, of laser beams, 75, 88-89
Colonoscopy, fiberoptic, 260-262
Colostomy, laser therapy as alternative to, 261
Colposcopy
  cervical, 303-304
  clinical, development of, 171
Combination procedure, in cervix, 309-310
Complications, laser-related, 463-471
  abscess formation, 228
  airway obstruction, postoperative, 217, 225, 228, 469
  aspiration, 246
  avoidance of, 202, 228
  bacteremia, 247
  burns, of skin, 228, 470
  burns, tracheal, 217
  cavity formation, 464
  dysphagia, 247
  edema, 228
  expectoration, 228
  exudative response, 228
  fever, 228
  hemorrhage, 197-200, 470 (*see also* Bleeding)
  infection, 202, 266
[Complications, laser-related]
  in oral cavity, 154-155
  perforations (*see* Perforations)
  pneumonia, 228
  pneumopericardial tamponade, 202
  pneumothorax, 470
  pneumothorax, tension, 202
  pseudoaneurysm, de facto, 380, *381*
  sepsis, 247
  sialadenitis, 155
  stenosis, 470
  tissue curling, 464
  tissue shrinking, 464
  tracheobronchial, 217
  after treatment of cervical lesions, 311, 313
  web formation, laryngeal, 470
Computed tomography, of bladder, 365
Condyloma acuminatum
  of genitalia, 356, 362, 366-367
  perianal, 266
  vaginal, 315-319
Contact technique (*see* Endoprobes, contact; Direct-touch technique; Sapphire tip, synthetic; Scalpel, contact laser)
Contracture, of bladder neck, 358, 362, 366
Cordectomy, 182, 183-184
Coring-out, progressive, for upper gastrointestinal tumor, 245
Cost analysis, in establishment of laser center, 476
Cost-effectiveness
  of coronary laser angioplasty, 370-371
  of laser laparoscopy, 299
  of laser treatment of laryngeal cancer, 179
  of laser treatment of rhinological disease, 170
  of Nd:YAG laser, in urological surgery, 359
Countertraction, with CO_2 laser, 154

Coupler, universal endoscopic, for bronchoscopic CO_2 laser surgery, 210-214, *210-213*
Coupling
alterations of, for cardiovascular disease, 388-389
direct, 147-148
optomechanical, types of, 147-148
Crohn's disease, 263
Cross hair siting, with ruby laser, 131
Curative therapy, for neoplasms, 233, 242
Cutting
with Nd:YAG contact laser scalpel, 271, 283
in orthopedic surgery, 402-413
with pulsed vs continuous-wave beams, 175
Cylindroma, tracheal, 225
Cystectomy, after laser irradiation, 365
Cystitis, 362
Cystogastrostomy, laser, 297
Cystoscope, with Nd:YAG laser, 358
Cysts, of vocal cord, 178
Cytoreduction
for cancer of false vocal cord, 180
for cancer of oral cavity, 156-157
Cytotoxic effects, with HPD phototherapy, 220, 267

D

Debridement
of degenerative articular lesions, 402
excimer laser for, 141
of wounds, 418
Debulking
for cancer of oral cavity, 156-157
in larynx, 176
in nasopharyngeal carcinoma, 169
Degenerative tissue changes, after partial pancreatectomy, in rat, 289
Delivery system(s)
flowing-gas vs sealed, 132-137
future developments in, 419
for laser angioplasty, 390-392
with micromanipulator, 173-174
mirror arm vs fiberoptic, 189
one-unit, for laryngeal microsurgery, 174
rigid, advantages and disadvantages of, 215-217
rigid open-tube vs flexible fiber, 191-192, 207
types of, 145-149
universal bronchoscopic, 192
Dermatomes, pelvic, denervation of, 273
Diagnosis
lasers for, 250
of liver tumor, 292
Diathermy, compared with laser and conventional techniques, 271
Dichroic window, of bronchoscopic coupler, 210, *210*
Diffraction, of light, in living tissue, 100, 102-104
Dihematoporphyrin ether
as active component of HPD, 425-426
compared with HPD, 425-427
Direct-touch technique, 286
Diskectomy, anterior cervical, 418
Doppler probe, 392
Doppler velocimetry, 260
Dysphagia
persistence of, postoperatively, 246
radiation-associated, 247
Dysphonia plicae ventricularis, 179
Dysplastic epithelium, of oral cavity, 152

E

Edema
bronchial, after HPD phototherapy, 228
interlobular, in rat pancreas, 289
of vocal cord, 178
Electric fields, 6-8
Electrocautery, monopolar, compared with laser technique, 257
Electrocoagulation, during microlaryngeal surgery, 183
Electromagnetic spectrum, 17, *18*
Electromagnetic waves, 14-17
Electroresection, and laser irradiation for bladder tumor, 361, 364-365
Emission
ionic, discovery of, 138
stimulated (*see* Stimulated emission)
Encystment, of pancreas, in rat, 289
Endarterectomy, 390
Endocardiectomy, laser-assisted, 389, 393
Endometriosis
argon vs Nd:YAG laser for, 326-327
coli, 260
CO_2 laser laparoscopy for, 332-334, *333-344*
vaporization of, 297-298
Endoprobes, contact
through laparoscope, future of, 298
shapes and applications of, *275*
for treatment of calculi, 294-295
in treatment of colorectal tumor, 271
used with Nd:YAG and argon lasers, *273*
Endoscope
considerations in choice of, 465
[Endoscope]
two-channel, with Nd:YAG laser, 242
Endoscopy
laser, intraosseous, 419
in soft-tissue microsurgery, 171
Energy density, for HpD phototherapy, 219, 221, 222
Energy levels, in gases, liquids, solids, 34-35
Epistaxis, 167
Epithelium, atypical cervical, grading of, 303
Equipment (*see also* Instrumentation)
accessory, of recovery room, 193
for CO_2 laser laparoscopy, 328-332
Erythroplasia, 362
Excisional biopsy (*see* Biopsy, excisional)
Excitation
of dye, by optical pumping, 142
of gas
direct current vs radiofrequency, 137-138
longitudinal vs transverse, 137-138, *137*
transverse, advantages of, 138
methods of, in surgical lasers, 135t
Expectoration, after bronchoscopic HpD phototherapy, 228
Experimental studies
of coronary artery, in vitro, 371-380
of hepatobiliary and pancreatic tissues, 282-286
in treatment of atherosclerosis, in vivo, 380-385
Exposure times, in HpD phototherapy, 267
Exudative response, after HpD phototherapy, 228

F

Familial polyposis syndrome, 263-264

Fever, after HpD phototherapy, 228
Fiber(s)
  advantages and disadvantages of, 148-149
  composition of, 191
  geometry of, *149*
  for laser coronary angioplasty, 389
  quartz, for disintegration of bile duct calculi, 292
Fibroids, submucosal, Nd:YAG treatment of, 327
Fibromas, tracheobronchial, 204
Fibrosarcoma, bronchogenic, 225
Filter, near-infrared, for endoscope, 276
Fimbrioplasty, results of, 298
Fire
  as laser-related hazard, 201-202
  prevention of, 174-175
Fistula, anal, 265-266
Fixation, laser for, in orthopedics, 414
Flashing, laser, in treatment of vaginal condyloma, 318
Fluorescein, in dye laser, 142
Fluorescent dyes, use of laser with, 250
5-Fluorouracil (5-FU), in combination with laser treatment, 318-319, 322
Follow-up
  after excisional biopsy of laryngeal cancer, *186*
  after laryngectomy, 184-185
  after laser surgery in tracheobronchial tree, 196
  after treatment of cervical disease, 310-311
  after treatment of vaginal condyloma, 319
Foreign bodies, removal of, 205-206
Future developments, 249-250, 298, 419

G

Gallstone, disintegration of, experimental, 286
Gas flow
  rates of, in surgical lasers, 132-133
  transverse, 133
Gas-in port, in flowing-gas system, *136*
Gastrointestinal tract, laser therapy in
  lower tract, 255-277
  upper tract, 242-250
Gating, in continuous wave mode, 144
Giant pulsing, in Q-switched mode, 144
Gimballed assembly, of bronchoscopic coupler, *210*, 211
Glioblastoma, photodynamic therapy for, 442
Gliomas, HpD phototherapy for, 267
Goggles, protective, *360*
Granulation tissue
  tracheal, 214-215
  tracheobronchial, in children, 460
Granulomas, solitary, of trachea, 205

H

Hand-held adapter, *360*, 366-367
Hard seals, 137
Helium:cadmium, laser wavelength from, 139t
Helium:neon, laser wavelength from, 139t
Hemangiomas
  of external nose, 164-166
  subglottic, in children, 459
  of tongue, capillary, 159
Hemangiomatosis, upper respiratory, in children, 460
Hematoporphyrin derivative (HpD)
  as photosensitizer, 424-426
  preparation and culture of, 221

[Hematoporphyrin derivative]
toxicity studies of, 427-428
uptake and distribution of, 442-444
uptake of, in vitro studies of, 430-431
Hematoporphyrins, in photodynamic therapy, 248-249
Hemilaryngectomy, 182, 184
Hemoptysis, after HpD phototherapy, 225, 228
Hemorrhage (*see also* Bleeding)
acute, in upper gastrointestinal tract, 236-239
control of, during laser resection, 199-200
as laser-related complication, 197-200, 470
lower gastrointestinal, 257-261
Hemorrhoids, 264-265
Hemostasis (*see also* Bleeding; Hemorrhage)
with continuous-wave vs pulsed laser, 144
with CO_2 laser, 173, 215
in laryngeal surgery, 177
in treatment of oral cavity lesions, 157
Hepaticojejunostomy, use of Nd:YAG laser for, 289
Hip, revision surgery of, 419
Histiocytoma, of lung, malignant fibrous, 436
Holography, of bladder, 365
Hydrosalpinx, 335
Hyperkeratosis, 178
Hyperthermia therapy, local, for colorectal tumors, 270-271, *272*
Hypertrophy
of cutaneous epithelium, 163
turbinate, of nose, 168
Hypoxemia, as laser-related hazard, 197
Hysteroscopy, laser, 327, 344, 350

I

Ileal conduit, after laser irradiation, 365
Implantation, of fiber probe, for endobronchial cancer, 221
Implants, endometrial, 330
Infection
as laser-related complication, 266
postoperative, avoidance of, 202
from tuboperitoneal causes, 298
Injury, avoidance of
deep thermal, 166
to eye, 174, 217
from irradiation, 191, 214
Instrumentation (*see also* Equipment)
complications with, 465
for laryngeal laser microsurgery, 172-174
for laser arthroscopy, 405, *414*
for treatment of cervical lymph node metastases, 185
for treatment of pediatric airway, 454-455, *455*
Interstitial technique
for colorectal tumors, 270
in HpD phototherapy, 267
Iridectomy, with ruby laser, 143

J

Jacobsson's classification, of mode of invasion, 152-154
Jet ventilation, with universal endoscopic coupler, 214
Joystick
of bronchoscopic coupler, *210,* 211, *211*
in direct coupling, 147

K

Kaposi's sarcoma, of skin, photodynamic therapy for, 433-435

Keratoplasty, radial, excimer laser for, 141
Krypton
arc lamp, of Nd:YAG laser, 140
laser wavelength from, 139t

L

Lahey Clinic
laser center at, establishment of, 475-481
laser system employed at, 356, *357*, *358*
Laparoscope(s)
second-puncture, 326, 329, *331*
single-puncture, 326, 329
types of, 326, *330*
Laparoscopy
laser, 297-299, 325
using CO_2 laser, 328-350
Laparotomy, microlaser, 336-337
Laryngectomy, 182, 183-185
indications for, 187
Laryngoscope, Jako, *455* (*see also* Microlaryngoscopes, types of; Subglottiscope, Healy)
Laryngoscopy, direct vs indirect, 172-173
Larynx, laser treatment of, 171-187
Laser beam(s) (*see also* Modes)
automatic scanning of, 70
coaxial alignment of, 329
coherence of, 71-74, 88
collimation of, 75, 88-89
monochromaticity of, 74-75, 89-91
pulsed vs continuous-wave, 175
single-mode, 174
Laser head, of Nd:YAG laser, *358*
Laser plume (*see* Smoke evacuation)
Laser(s) (*see also* Delivery systems; Laser systems)
alexandrite, 143
[Laser(s)]
argon
construction of, 53-54, 57, *58*
and coronary arterial bypass surgery, 384
in HpD phototherapy, 270
introduction of, 139
in ophthalmology, 131
specifications of, 256t, 282t
for treatment of nasal cavities and nasopharynx, 163-170
combined, for hemostasis, 401
copper vapor, in HpD phototherapy, 270
CO_2 (*see* Carbon dioxide laser)
diode, medical uses of, 84-85
dye
argon-pumped, in HpD phototherapy, 219-229, *222*
in bronchoscopic HpD phototherapy, 219-229
excimer-pumped, 142
flash lamp-pumped, 142
invention of, 141-142
neodymium-pumped, 142
nitrogen-pumped, 142
pulsed, categories of, 142
rhodamine B, in HpD phototherapy, 270
early therapeutic, 131
excimer, 371
development and surgical applications of, 141
laser-tissue effects of, 119-120
medical uses of, 85-87
free electron, medical uses of, 87
ion
development of, 138-140
transfer of energy in, 139
mercury, contribution of, 138
Nd:YAG (*see* Neodymium:yttrium-aluminum-garnet laser)
neodymium, development of, 140-141

[Laser(s)]
neodymium:glass, discovery of, 140
radiofrequency, unguided and waveguided, 138
rare-gas halide (*see* Lasers, excimer)
ruby
cross hair siting with, 131
early ophthalmic, 131
operation of, 142-143
selectable, in future, 250
solid state, development of, 142-143
tuneable, in future, 250
visible, laser-tissue effects of, 125
waveguided, compared with unguided, 138
Laser systems (*see also* Delivery systems)
gas, types of, *136*
sealed vs flowing-gas, 132-137
surgical, flow rates in, 132-133
types of, *133*
Laser-tissue effects
with CO_2 laser, 113-119
with excimer laser, 119-120
with Nd:YAG laser, 119-125
with visible lasers, 125
Laser-tissue interaction(s)
absorption, 96-100, *101*
diffraction, 100, 102-104
in larynx, experimental studies of, 171
photonic processes, 108-112
reflection, 93-96
related to energy density, 257t
related to wavelength, 190
scattering, 100, 102-104
thermal conduction, 104-108
with CO_2 vs Nd:YAG laser, 175
Lavage (*see* Cleansing)
Leakage
anastomotic, related to surgical technique, 271
[Leakage]
biliary, in hepaticojejunostomy, 289, 292
Leiomyofibromas, tracheobronchial, 204
Leiomyomas, gastric, 248
Lens
continuous variable defocus, 210, *210*
focusing, in direct coupling, 147
focusing, zinc arcenide, 326
Lesions (*see also* Cancers; Carcinomas; Tumors)
benign
of external nose, vascular, 164-166
of larynx, 178-179
of lower gastrointestinal tract, mucosal, 263-266
of oral cavity and oropharynx, 157-160
of oral cavity, bony, 159
of stomach, angiodysplastic, 239-241, *240*
malignant
of anterior commissure, 180
of larynx, 179-185
of oral cavity and oropharynx, 151-157
of palatine arch, 155-156
of retromolar trigone, 155-156
Leukoplakia, 178
Light dosimetry, 431-432
Lipomas, tracheobronchial, 204
Lithotomy
cholangioscopic, experimental, 283, 286
percutaneous transhepatic cholangioscopic, 293
Liver
laser surgery of, 281-300
Nd:YAG laser surgery of, 289-295
tissue from, experimental studies of, 282-283

Liver resection
  adjuvant use of laser in, 289
  contact vs noncontact technique for, *285*
  experimental studies of, 282-283
Lymphangiomas, of tongue, 159

M

Magnetic fields, 8-14, 139-140
Malignant cells, heat-sensitivity of, 271
Mandible, configuration of, and type of endoscope, 465
Marketing, of a laser program, 480
Mayo Clinic, experience with HpD phototherapy at, 220-229
Melanomas, malignant
  of choroid, photodynamic therapy for, 442
  of esophagus, 245
  HpD phototherapy for, 267
  of penis, 366
  of skin, photodynamic therapy for, 433-435
Meniscectomy
  using arthroscopy, 402
  laser, 402-405, *406-413*
Menisci, of knee, laser repair of, 418
Menometrorrhagia, Nd:YAG laser hysteroscopy for, 327, 350
Mercury, laser wavelength from, 139t
Metal protectors, development of, 173
Metastases
  in colorectal carcinoma, 261
  lymph node, 179-180, 182
  malignant, intramedullary, 419
Microlaryngoscopes, types of 173 (*see also* Laryngoscope, Jako; Subglottiscope, Healy)
Micromanipulator(s)
  closed-cube, coupled to laparoscope, 325, 326, 329
  development of laser system with, 173-174
  in direct coupling, 147-148
  electromechanically driven, 148
  mechanically driven, 147-148
Microneurorrhaphy, future of, 417
Microsurgery
  CO_2 laser, applications of, 172
  soft-tissue, development of techniques of, 171
Mirror arm, of CO_2 laser, 189 (*see also* Coupling, optomechanical)
Mist tent, 456, *457*
Mode-locking, 65-69 (*see also* Modes, locked; Pulse width compression)
Mode(s)
  continuous wave
    description of, 143-144
    physics of, 69-71
  gaussian, 77-84
  locked, 145 (*see also* Mode-locking)
  longitudinal, 43-45
  pulsed
    for cardiovascular disease, 385-386
    description of, 144
    for endometrial implants, 334
    physics of, 57-69
  Q-switched (*see also* Q-switching)
    description of, 144-145
    giant pulsing in, 144
  temporal, 57-71
  transverse electromagnetic (TEM), 75-84
Monochromaticity, of laser beams, 74-75, 89-91
Mucosa, vaginal, normal histological features of, *316*
Mycosis fungoides, photodynamic therapy for, 433-435
Myelotomy
  commissural, with CO_2 laser, 273
  by microsurgical laser technique, 273

Myocardial infarction, subsequent to laser surgery, 465
Myomas
  intraabdominal, 330, 344, 350
  submucous, laser hysteroscopy for, 344, 350, *350-351*
  vaporization of, by laser laparoscopy, 344, 350, *350-351*
Myoplasty, using laser, 393

N

Nasal cavities, laser treatment of, 166-169
Nasal septum, reconstruction of, 170
Nasopharynx, laser treatment of, 169-170
Necrosis
  by continuous-wave pulsed laser, 144
  with electrocautery vs laser, 257
  after partial pancreatectomy, in rat, 289
Neodymium:yttrium-aluminum-garnet (Nd:YAG) laser
  advantages of, in hepatobiliary surgery, 289-295
  applications of, in bronchology, 189-207
  vs argon laser in gastrointestinal tract, 233-234
  compared with CO_2 and argon lasers, 190-191
  continuous wave, specific effects of, 120-125
  construction of, 48-50, *54-55*
  vs CO_2 laser for meniscectomy, 402-405
  vs CO_2 laser for synovectomy, 413
  for diagnosis, histological, 292
  early use of, 140
  frequency-quadrupled, specific effects of, 119-120
[Neodymium:yttrium-aluminum-garnet (Nd:YAG) laser]
  operating console of, *357*
  specifications of, 256t, 282t
  for treatment of calculi, 292-294
  for treatment of nasal cavities and nasopharynx, 163-170
  in urologic surgery, 355-367
  wavelengths of, 140
Neoplasms
  of esophagus and esophagogastric junction, 242
  of oral cavity and oropharynx, vascular, 160
  in upper gastrointestinal tract, 233, 241-249
Neoplastic cells, disruption of, with low energy, 249
Neosalpingostomy, laser laparoscopic, 335-337, 348-349
Nephrectomy, partial, 356, 366, 367
Nephroscope, with Nd:YAG laser, 358
Neuroablation, with surgical lasers, 273
Noble gases, emissions from, 138
Nodules, of vocal cord, removal of, 176, 178
Nontouch technique
  advantages of, in bladder, 361
  for pancreatectomy, experimental, *287*
Nose, laser therapy for
  external nose, 163-166
  internal nose, 166-169

O

Obstruction, malignant
  biliary, 292
  tracheal, 214
Operating microscope
  with CO_2 laser, 154, 155, 273

[Operating microscope]
Zeiss, photo and video attachments of, 177
Operative technique
for combination procedure in cervix, 310
for ectocervical intraepithelial neoplasms, 305-306
for endocervical lesions, 307-309
of laser laparoscopy, 326
in tracheobronchial tree, 194-196
for vaginal condyloma acuminatum, 317-318
for vaginal intraepithelial neoplasia, 319, 322
Oral cavity, laser surgery of, 151-160
Oropharynx, laser surgery of, 151-160
Orthopedics, laser surgery in, 397-419
Osler-Weber-Rendu syndrome, 167
Oxygenation, during laser resection, 197

P

Pain, after procedures on vulva, 470
Painting, laser, of vagina, 322
Palliation, of gastrointestinal cancer, 242
Pancreas
histological appearance of, *296*
laser surgery of, 281-300
tissue from, experimental studies of, 286-289
Pancreatectomy
experimental, 286-289, *287*
using Nd:YAG laser, 295, *297*
Pancreatitis
acute, 289
chronic, 295
Papillomas
of larynx, 177
[Papillomas]
tracheobronchial, in children, 460
Papillomatosis, recurrent respiratory
laser endoscopy for, 459
laser excision of, 214-215
Papillomavirus, human
in normal-appearing epithelium, *317*
vaginal condyloma from, 315, 317-318
Paralysis, of bilateral abductor, laryngeal, 179
Patient considerations, in use of laser technique, 464-465
Pediatric population, laser laryngoscopy and bronchoscopy in, 453-461
Penetrability, of light, in tissue, 431-432
Penetration, depth of
with fiberoptic bronchoscope, 228
judgment of, 190
and power density, 289
Perforations
of bladder and bowel, 361
in colorectal surgery, 263
during laser atherolysis, in vivo, 380-384
during laser coronary angioplasty, in vitro, 373-380
as laser-related hazard, 205-206
lower gastrointestinal, avoidance of, 260
tracheal, avoidance of, 217
tracheal, danger of, 214
during treatment of gastrointestinal tumor, 245, 247
during treatment of rectosigmoid tumors, 262
Perichondritis, as laser-related complication, 469
Persistent disease, after treatment of cervical neoplasms, 306, 309
pH level, related to thermal sensitivity, 271
Photoacoustic waves, *109,* 110

Photochemical reaction, in HpD
phototherapy, 220
Photochemolysis, 108, *109*
Photochemosynthesis, 108, *109*
Photocoagulation
for colonic bleeding, 260
in orthopedic surgery, 418
laser, during colonoscopy, 260-261
related to energy density, 257t
of retina, 131, 138
of superficial hemangiomas, 160
Photodisruption, 386-388
Photodynamic therapy, 423-446
(*see also* Phototherapy, HpD)
as adjunct to surgery, 436-437
mechanism of effect in, 423
tumoricidal effects of, 429-430
Photoelectrolysis, *109,* 110
Photoexplosion, *109,* 110
Photography
of larynx, 176, 177
of pelvic pathology, 328, *329*
Photon absorption, by HpD, 220
Photonic processes, in living tissue,
108-112
Photoplasmolysis, *109,* 110-111
Photopyrolysis, 108, *109*
Phototherapy, HpD
for adenocarcinoma, in dog, *268-269*
with argon dye laser, 219-229
in bronchology, 191
cardiovascular applications of,
392
for colorectal tumors, 267-270
cytotoxic mechanism of, 267
efficacy studies of, experimental,
428-429
energy density for, 257t
history of, 423-424
mechanism of action of, 219-220
Photovaporolysis, *109,* 110
Pneumonia, after HpD phototherapy,
228
Pneumothorax, as laser-related complication, 470
Pneumothorax, tension, as laser-related complication, 202
Polypectomy, nasal, 170
Polyps
colorectal, adenomatous, 264, *265*
gastric, 248
laryngeal, 176
nasal, 163, 168
submucous and endometrial, 327
of vocal cord, 178
Popcorn effect, 190
Population inversion, physics of, 38
Port-wine stains, of external nose
and face, 166
Postoperative care
after laser surgery in tracheobronchial tree, 196
after microlaryngeal surgery in child,
456-457
Power
density of, for bronchoscopic HpD
phototherapy, 221, 222
ranges of, in surgical lasers, 135t
Pregnancy, outcome of, after treatment of cervical neoplasms, 313t
Preoperative testing, of laser system,
175
Preparation, preoperative
of bowel, 277
for laser surgery, 193-194
for surgery of external nose, 164
for treatment of endocervical
lesions, 307
Probes
Doppler, 392
titanium suction irrigation, 331,
332
two-dimensional echocardiographic,
390-391
Proctectomy, 264
Protein coagulum, 287
Pseudoaneurysm, de facto, 380, *383*
Pseudocyst, pancreatic, drainage of, 297

Pulse width compression, 145 (*see also* Mode-locking)
Pumping, physics of, 38-42

Q

Q-switching, 63-64, 141 (*see also* Modes, Q-switched)
Quality assurance, procedures for, 479
Quantum theory, of atomic structure, 21-31
Quartz fiber, for disintegration of bile duct calculi, 292
Quartz halogen lamp, in HpD phototherapy, 270
Quartz waveguide, through conventional endoscope, *244*

R

Radiation
  nature of, 2-42
  particle theory of, 17-42
  from real lasers, 71-84
  wave theory of, 3-17
Radiography (*see* Angiography, during colonoscopy; Computed tomography, of bladder; Holography, of bladder; Ultrasonography)
Radiotherapy (*see also* Brachytherapy)
  after laser therapy, 206
  for lymph node metastases, 182
  for nasopharyngeal carcinoma, 169
Rebleeding, 249
Rectal pain, intractable, 271, 273
Recurrence
  of carcinoma of bladder, 362, 364
  of colorectal cancer, 262
[Recurrence]
  of laryngeal cancer, 181
  of perianal condyloma acuminatum, 266
Reflection, of light, in living tissue, 93-96
Resectoscope, 361
Resonators, function of, 42-45
Results of therapy
  for angiodysplasia, 241
  for bladder cancer, 440t
  for bladder tumor, 361-365
  after bronchoscopic HpD phototherapy, 223-228, 224t, 226-227t
  for cervical lesions, 311-313, 312t, 313t
  for colorectal disease, 261-262, 264
  for esophageal cancer, 437, 439
  for hereditary hemorrhagic telangiectasia, 168t
  after intraabdominal laser surgery, 298
  after laser meniscectomy, 403, 405
  for lung cancer, 438t
  for oral cavity disease, 152-155
  after photodynamic therapy, 444-445
  for skin cancers, 434t
  after terminal salpingostomy, 326
  for upper gastrointestinal tumor, 245-248
  for vaginal intraepithelial neoplasia, 321t, 322
Retinacular release, using arthroscopy, 402
Retractors, vocal cord, development of, 173
Retreatment procedures, for vaginal condyloma, 318-319
Rhinological disease, laser treatment of, 163-170
Rhinophyma
  CO_2 laser for, 163-165, 170
  pretreatment and posttreatment, *164, 165*

Rhodamines, in dye laser, 142

S

Safety
with CO_2 laser surgery in tracheobronchial tree, 217
environmental, 419
in firing laser, 276
standards and procedures for, 479-480
in use of laser, 174-175, 206-207, 461, 470-471
Salpingostomy, results of, 298
Salvator method, 190
Sanders injector, for jet ventilation, 214
Sapphire tip, synthetic
with Nd:YAG laser, 271
vs noncontact technique, 275
Scalpel, contact laser
with Nd:YAG laser, 283, *284*
for open anal and colorectal surgery, 271, *274*
Scarring
facial, avoidance of, 166
and limitation of laser technique, 467-468
Scattering, 214
of light, in living tissue, 100, 102-104
with Nd:YAG vs CO_2 laser, 190
Sciatic nerve, repair of, in rat, 417
Sealing effect, of CO_2 laser beam, 214
Sepsis, as laser-related complication, 247
Septae, uterine
CO_2 laser hysteroscopy for, 344
Nd:YAG treatment of, 327
Sialadenitis, as laser-related complication, 155
Singlet oxygen, as cytotoxic agent, 219
Skin, CO_2 laser treatment of, 409-410
Smoke evacuation
in cervix, 305, 307, 317-318, 322, 329, 331
in larynx, 174
Specificity, of wavelengths, 249-250
Spectroscopy, argon laser for, 138
Sphincteroplasty, anal, 266
Spinal column, flexibility of, and type of endoscope, 465
Spit window, of bronchoscopic coupler, *210*, 212, *213*
Stability, physical principles of, 43
Staging, of bladder tumors, 361, *362*
Stenosis
acquired subglottic, in children, 459
as laser-related complication, 470
postadenotonsillectomy, 169
tracheal, 205, 206, 214-215
weblike, in children, 460
Sterilization
laser hysteroscopic, 327
of surgical fields, 418
of wounds, 402
Stimulated emission
first demonstration of, in ruby laser, 142
physics of, 36-38
Stricture
anastomotic, after hepaticojejunostomy, 294
benign, of lower gastrointestinal tract, 262-263
of intrahepatic bile duct, 289
urethral, 356, 358, 362, 366
Subglottiscope, Healy, 173, 177, *455* (*see also* Laryngoscope, Jako; Microlaryngoscopes, types of)
Suction cautery, bipolar bronchoscopic, for hemostasis, 215
Superpulsing, 69-70 (*see also* Modes, continuous-wave)

Surveillance, after treatment of angiodysplastic lesions, 241
Survival, after treatment of esophageal cancer, 247
Synovectomy, using arthroscopy, 402, 405, 411-413
Synovitis, villonodular, 413

T

Tachyarrhythmia, refractory ventricular, 389
Tamponade, pneumopericardial, as laser-related complication, 202
Telangiectasia, hereditary hemorrhagic, 167, 168t
Telangiectasias
  argon vs Nd:YAG laser for, 166
  of external nose, 164-166
  facial, *166*
Thermal conduction, in living tissue, 104-108
Thermal damage, control of, in laser treatment of cervix, 306, 307
Thermal regulation, in laser surgery, 400
Thermal sensitivity, of malignant cells, 271
Thermocouples, used in local hyperthermia therapy, 271
Thrombosis, related to excessive energy, 286
Tissue differentiation, in laryngeal surgery, 185
Tissue-sensitizing agents, future use of, 250
Tonsillectomy, lingual, 159
Trachea, access to, in laser resection, 203
Tracheobronchial diseases, CO_2 laser treatment of, 209-217
Tracheomalacia, 214-215
Tubes
  built-in suction, development of, 173
  endotracheal, complications with, 466-467
  flowing gas-type, development of, 174
  suction, uses of, 193
Tumors (*see also* Cancers; Carcinomas; Lesions)
  adenoid cystic, 204
  of ampulla of Vater, 248
  assessment of, for risk of complications, 245
  of bladder, 355-356, 361-365, *363-365*
  of brain, photodynamic therapy for, 442
  colorectal, 270-271
  of duodenum, 248
  glottic, 182
  gynecological, photodynamic therapy for, 441
  histological type of, influence of, 248
  intraocular, photodynamic therapy for, 442
  of liver, 292, 299
  neovasculature of, and risk of hemorrhage, 199
  of renal pelvis, 356, 358, 366
  of skin and subcutaneous tissue, 219
  of soft palate, 248
  tracheobronchial, 203-205, 220
  of ureter, 356, 358, 366

U

Ulcer, at rectosigmoid junction, 263
Ultrasonography
  of bladder, 365
  in intraoperative laser coronary angioplasty, 390-391
Universal bronchoscope system, 192

Universal holder, for contact laser endoprobes, *273*
Ureterorenoscope, 356
Ureteroscope, with Nd:YAG laser, 358, 366
Urethroscope, and flexible quartz fiber, *359*
Urethrotomy, laser vs cold-knife, 366
Urinary tract, upper, transitional cell tumors of, 366
Uveopalatopharyngoplasty, using CO_2 laser, 159

V

Vacuolation, cytoplasmic, in rat pancreas, 289
Vaginal disease, laser surgery for, 315-322
Vaginal intraepithelial neoplasia, alternatives in treatment of, 464
Vaporization
- of atherosclerotic plaque, in vitro, 371-380
- in cervix, in combination procedure, 309-310
- of ectocervical intraepithelial neoplasia, 304-306, 309-310
- of endometriosis, 332-334, *333-344*
- in gastrointestinal tract, 234, 242-248
- of myoma, 344, 350, *350-351*
- related to energy density, 257t

Vascularity, related to choice of laser, 160
Vasopressin infusion, during colonoscopy, 260
Vasovasostomy, 356
Velocimetry, Doppler, 260
Ventilation, with universal endoscopic coupler, 214
Ventilation ports, of bronchoscopes, 214
Videoendoscopes, future development of, 250
Video recording, in larynx, 177

W

Wavelengths
- future developments in, 249
- in HpD phototherapy, 267, 270
- of surgical lasers, 135t

Web
- formation of, avoidance of, 179
- laryngeal, as laser-related complication, 470
- tracheal, 214

Wedge resection
- of false vocal cords, 179
- ovarian, results of, 298

X

Xenon arc lamp, in HpD phototherapy, 270
Xenon, laser wavelength from, 139t

About the Editor

STANLEY M. SHAPSHAY is Chairman of the Department of Otolaryngology-Head and Neck Surgery, and Director of the Eleanor Naylor Dana Laser Research Laboratory at the Lahey Clinic Medical Center, Burlington, Massachusetts. He is also Clinical Associate Professor of Otolaryngology at Boston University School of Medicine and Clinical Instructor in Otolaryngology at the Tufts University School of Medicine, Boston. The author of more than 90 articles, text chapters, and books, he is a Fellow of the American College of Surgeons, American Academy of Otolaryngology-Head and Neck Surgery, American Bronchoesophagological Association, American College of Chest Physicians, and American Society for Laser Medicine and Surgery. Dr. Shapshay received his B.S. (1964) degree from Brooklyn College and his M.D. (1968) from the Medical College of Virginia.